Fundamentals
of
Ecotoxicology
SECOND EDITION

Fundamentals
of
Ecotoxicology
SECOND EDITION

Michael C. Newman
Michael A. Unger

LEWIS PUBLISHERS

A CRC Press Company
Boca Raton   London   New York   Washington, D.C.

Concerns arising in the 1990s about global amphibian decline and instances of amphibian malformations have led to an increase in ecotoxicological studies of amphibians. Both terrestrial and aquatic life stages are subject to contaminant effects. Most studies have focused on potential contaminant effects to larvae, such as the larvae of the spotted salamander (*Ambystoma maculatum*) shown on the cover. (Photograph by David Scott.)

Library of Congress Cataloging-in-Publication Data

Newman, Michael C.
    Fundamentals of ecotoxicology / Michael C. Newman and Michael A. Unger.—2nd ed.
        p. cm.
    Includes bibliographical references (p.).
    ISBN 1-56670-598-3 (alk. paper)
        1. Pollution—Environmental aspects. 2. Pollution—Health aspects. I. Unger, Michael A. II. Title.

QH545.A1 N49 2002
571.9′5—dc21   2002034105

To Our Families

I do not think myself any further concerned for the success of what I have written, than as it is agreeable to *truth*.

Berkeley (1734)

Preface

TENOR OF THIS BOOK

Nothing is new under the sun. Even the thing of which we say, "See, this is new!" has already existed in the ages that preceded us.

Ecclesiastes 1: 10–11

In contrast to the conceptual doldrums described above by Coheleth, we live in exciting times, rich in new discoveries and novel applications of familiar concepts: Tremendous opportunity exists for intellectual growth. This is so obvious in fields such as molecular genetics and computer sciences that no elaboration is required beyond this simple statement. Several immediately recountable examples affecting our understanding of the world range widely from discoveries in planetary biogeochemistry, mathematics, and geology. Lovelock's **Gaia hypothesis** (the earth's albedo, temperature, and surface chemistry are homostatically regulated by the biota) (Lovelock, 1972, 1988; Margulis and Lovelock, 1989) has stretched our perspective beyond the ecosystem to consider contaminants in a global context. Recently, we learned that a **global distillation** has moved volatile and persistent pesticides from warmer areas of use to cooler areas of the globe, where they had been banned for decades or seldom used (Simonich and Hites, 1995).

Only a few decades ago, nonlinear dynamics and chaos theory revealed the limitations of determinism. We now know that it is impossible to predict precisely the behavior of all but the simplest of systems. Such new concepts allowed us to understand that the bioaccumulation models presented in Chapter 3 do not necessarily predict pollutant accumulation until a single equilibrium concentration is reached. In sharp contrast to current thinking, oscillations in concentrations are expected under certain conditions (Newman and Jagoe, 1996).

Just a few decades ago, French scientists discovered that life's activities had modified the geochemistry in Oklo (Gabon, Africa) 1.8 billion years ago and brought together enough fissile uranium (^{235}U) to reach critical mass (Lovelock, 1988; 1991). Proterozoic life set the stage by influencing the hydrological cycle, creating an oxidizing atmosphere and enhancing uranium accumulation through **biomineralization** (biologically mediated deposition of minerals) (Lovelock, 1988; Milodowski et al., 1990). These **Oklo natural reactors**, producing power for nearly a million years (Choppin and Rydberg 1980), are clear evidence that the nuclear consequences of life's activities began long before humankind appeared on the earth! The present distribution of residue from these natural reactors also provides environmental scientists with clues about the long-term fate and migration of modern waste fission products.

Our reasons for writing this book stem from the realization that new concepts and novel applications of existing ideas appear every day. This book is our effort to expose you to new and useful concepts of ecotoxicology. Not only are these concepts and facts fascinating in themselves, they also provide us with the tools to avoid or solve environmental problems.

… Alert and healthy natures remember that the sun rose clear. It is never too late to give up our prejudices. No way of thinking or doing, however ancient, can be trusted without proof. What everyone echoes or in silence passes by as true to-day may turn out to be a falsehood to-morrow, mere smoke of opinion. …

Thoreau (1854)

CONTENT AND FORMAT OF THIS BOOK

This book is intended for use in a graduate or upper-level undergraduate course and as a general reference. It is a treatment of ecotoxicology ranging from molecular to global perspectives. Reflecting our present imbalance of knowledge and effort, the book retains some bias toward lower levels of ecological organization, e.g., biochemical and organismal topics. Yet, it does extend discussion beyond the ecosystem to include landscape, regional, and biospheric topics. The intent of this extension is to impart a perspective as encompassing as the problems facing us today: Many problems transcend the conventional ecosystem context.

Human and ecological health issues are occasionally interwoven. The reason for taking this unconventional and potentially distracting approach is simple: To do otherwise would be inconsistent with our present approach to assessing hazards and risks from pollutants. Also, many mechanisms of action, internal dynamics, and effects are common to all animal species, including humans. Finally, although we prefer the delusion of transcendency in environmental issues,[1] our decisions regarding ecological effects of contaminants are made with motives as profoundly anthropocentric as those involving human health. Why not discuss them together?

Topics are divided among fifteen chapters. Vignettes written by experts are placed throughout the text to enrich key points or to describe particularly good examples. The first chapter provides general perspective on the field. Chapter 2 details the key contaminants of concern today and explores their fate and cycling in the biosphere. Bioaccumulation and effects of contaminants are detailed at increasing levels of ecological organization in Chapters 3 to 12. The framework of these ten chapters is scientific, not regulatory. Regulatory aspects of the field are covered partially in Chapters 13 and 14, which address the technical issues of risk assessment. Key U.S. and European legislation is discussed in Appendices 3 and 4. The final chapter summarizes the volume.

There is a slight bias toward North American examples in all chapters. This reflects the authors' backgrounds, not a quiet snub of non–North American interests. We recognize, but have no completely satisfactory solution for, this shortcoming of the book relative to British, European, and other non–North American students. Non–North American issues are presented where possible, e.g., European environmental law is described in Appendix 4.

To aid the reader, important terms are highlighted in the text and summarized in the glossary. Some important material has been included in the footnotes for purposes of continuity; consequently, the reader is cautioned not to unintentionally overlook these materials. Opinionative insights of the authors, not generally shared by ecotoxicologists, are clearly identified as "incongruities" throughout the chapters, e.g., the Lorax incongruity discussed below. The reader may decide to ignore most of them with no danger of becoming an uninformed ecotoxicologist. Suggested readings are provided at the end of each chapter, and study questions are collected at the end of the book.

---

[1] Decisions about ecological effects are often perceived as arising from selfless protection of the earth. In fact, our motives are based on the value we give to the heretofore-free services provided by intact ecological systems. These services include generation of clean water and air, production of food and game species, provision of biological materials for medical and genetic uses, and provision of pleasing settings for living and recreation. Decisions are based on the perceived value of these services relative to those of technological services and goods. The delusion of selfless motivation in environmental stewardship and advocacy is sufficiently widespread as to be named the **Lorax incongruity**. (The Lorax is a character in the popular children's book by Dr. Seuss (Geisel and Geisel, 1971) who "speaks for the trees, for the trees have no tongues.") This well-intended but intransigence-inducing delusion is pervasive in society today.

Acknowledgments

We are grateful to Drs. M. Crane, T. Hinton, P. Landrum, and all vignette authors for their intelligent and thorough contributions to this book. These contributions greatly enrich this book and are appreciated. Finally, the wonderful cover photographs for both editions of this book were taken and provided by David Scott.

About the Authors

Michael C. Newman

Dr. Newman is a Full Professor at the College of William and Mary's Virginia Institute of Marine Science. From 1999 to 2002, he served as the Dean of Graduate Studies for the School of Marine Sciences. Before joining the faculty at the Virginia Institute of Marine Science, he was a Senior Research Scientist at the University of Georgia's Savannah River Ecology Laboratory and head of its Environmental Toxicology, Remediation, and Risk Assessment (ETRRA) Group. After receiving B.A. (Biological Sciences) and M.S. (Zoology) degrees from the University of Connecticut, he earned M.S. and Ph.D. degrees in Environmental Sciences from Rutgers University. After postdoctoral fellowships at the University of Georgia and the University of California at San Diego, he joined the research faculty at the University of Georgia (1983). His research interests include toxicity and bioaccumulation models, QSAR-like models for predicting metal bioactivities, toxicant effects on populations, factors modifying toxicity and bioaccumulation, quantitative methods for ecological risk assessment, statistical toxicology, and inorganic water chemistry. He has published nearly 100 scientific articles on these topics. He has also authored three books, *Quantitative Methods in Aquatic Ecotoxicology* (1995), *Fundamentals of Ecotoxicology,* 1st ed. (1998), and *Population Ecotoxicology* (2001), and edited *Metal Ecotoxicology: Concepts and Applications* (1991, with A.W. McIntosh), *Ecotoxicology: A Hierarchical Treatment* (1996, with C.H. Jagoe), *Risk Assessment: Logic and Measurement* (1998, with C.L. Strojan), *Coastal and Estuarine Risk Assessment* (2002, with M. Roberts and R. Hale), and *Risk Assessment with Time to Event Models* (2002, with M. Crane, P. Chapman, and J. Fenlon). He also directed the development of UNCENSOR, a program that produces univariate statistics for data sets containing "below detection limit" observations.

Active in professional societies and teaching, he founded and was first president of the Carolinas Chapter of the Society of Environmental Toxicology and Chemistry (SETAC). He served on (1988–1996) and chaired (1992–1994) the SETAC awards committee. He has served as an editor for the journal *Environmental Toxicology and Chemistry*, and he was on the editorial boards of the journals *Archives of Environmental Contamination and Toxicology* and *Risk Analysis* and two books series, *Advances in Trace Substances Research* and *Current Topics in Ecotoxicology and Environmental Chemistry*. He is currently series editor for the *Environmental and Ecological Risk Assessment Series*. He has taught at the University of Connecticut, University of California at San Diego, University of South Carolina, College of William and Mary, University of Georgia, Royal Holloway University of London (U.K.), University of Antwerp (Belgium), and University of Joensuu (Finland).

Michael A. Unger

Dr. Unger is a Research Associate Professor at the College of William and Mary's Virginia Institute of Marine Science. He received his B.S. (Zoology) from Michigan State University and his M.S. (Environmental Chemistry) and Ph.D. (Marine Science) degrees from the College of William and Mary. After two years working as a Research Environmental Chemist with the Johns Hopkins University (APL Aquatic Sciences group at Shady Side, MD), he came to the Virginia Institute of Marine Science to head the Analytical Chemistry Section of the Division of Chemistry and Toxicology in 1990. Currently a member of the Department of Environmental Sciences, he is active in research, teaching, and advisory services related to environmental contaminant issues in Chesapeake Bay.

His research has focused on contaminant behavior and how this affects transport and toxicity. Research interests include analytical method development, contaminant partitioning behavior, long-term fate studies to assess regulatory actions, and contaminant degradation. He has developed a regularly taught graduate-level course in Environmental Chemistry for the School of Marine Science and contributes to educational outreach programs for teachers and the public. A member of the Virginia Department of Environmental Quality's Science Advisory Committee, he regularly advises the state on environmental issues.

Contents

Section Two

Chapter 3   Uptake, Biotransformation, Detoxification, Elimination, and Accumulation

Chapter 4   Factors Influencing Bioaccumulation

SECTION ONE

General

Introduction

> On the day of the patient's victory at court, someone wrote a headline: "The Day That Tomoko Smiled." She couldn't possibly have known. Tomoko Uemura, born in 1956, was attacked by mercury in the womb of her outwardly healthy mother. No one knows if she is aware of her surroundings or not.
>
> **Smith and Smith (1975)**

I HISTORIC NEED FOR ECOTOXICOLOGY

As pressures mount for fiscal restraint, it is natural and responsible to reconsider the wisdom of our complex system of costly environmental regulations. At this time, it may be difficult to understand why significant amounts of these monies should not be redirected to the national deficit, critical social problems, medical research, technological innovation, education, space exploration, or other worthwhile endeavors (Lomborg, 2001). But, just a few decades ago, it was not hard to understand the need for such expenditures: Tomoko Uemure's mother understood. Tomoko Uemure was born with severe and permanent neurological damage after her mother had unknowingly consumed mercury-laden fish.

As World War II ended, the **dilution paradigm** ("the solution to pollution is dilution") was slowly replaced by the **boomerang paradigm** ("what you throw away can come back and hurt you"). Two horrible epidemics of heavy metal poisoning from contaminated food had occurred in Japan. In the 1950s, organic mercury was transferred through the marine food web to poison hundreds of people. Nearly 1000 people, including Tomoko Uemure, fell victim to **Minamata disease** before Chisso Corp. halted discharge of mercury into Minamata Bay. From 1940 to 1960, Japanese in the Toyama Prefecture were poisoned by cadmium in their rice. This outbreak of what became known as **Itai-Itai disease** was linked to irrigation water contaminated from metal mine wastes. The name *Itai-Itai* reflects the extreme joint pain associated with the disease and literally means "ouch-ouch."

In 1945, open-air testing of nuclear weapons began at Alamogordo, NM, and nuclear bombs exploded over Hiroshima and Nagasaki later that year. Nine years later, the Project Bravo bomb exploded at Bikini Atoll, dropping fallout on thousands of square kilometers of ocean, including several islands and the ironically named fishing vessel, *Lucky Dragon* (Woodwell, 1967). The islands of Ailingae, Rongelap, and Rongerik received radiation levels of 300 to 3000 rem[1] within 4 days of detonation (Choppin and Rydberg, 1980). The rapid hemispheric dispersal and unexpected accumulation of fission products in foodstuffs from these and subsequent detonations (see Chapter 14) created much concern about possible long-term effects to humans. From 1960 to 1965, human **body burden** of [137]Cesium increased rapidly worldwide and then slowly decreased as the United States, the former Soviet Union, France, and China ceased open air-testing (Shukla et al., 1973).

Unreported discharge of radionuclides occurred prior to these overt releases. All were kept from the general public for reasons of national security. On the northwest coast of England, a fire in the Windscale plutonium-processing unit released 20,000 Ci of radioactive iodine (^{131}I) to the surrounding area (Dickson, 1988). Radioactive iodine is of particular concern because it concentrates in the thyroid, causing cancer. After atmospheric release, ^{131}I contaminates local vegetation, can be taken up by dairy cattle, and accumulates in thyroids of humans after consumption of dairy products. At a secret Soviet military plant (Chelyabinsk 40) in the Urals, plutonium processing had secretly discharged 120 million Ci to a nearby lake and enough down the Techa River to induce radiation poisoning in citizens living downriver (Medvedev, 1995). In September 1957, a storage tank explosion at Chelyabinsk 40 released 18 million Ci of radioactive material and forced the evacuation of approximately 11,000 people from a 1000 km[2] area (Trabalka et al., 1980; Medvedev, 1995). From 1944 to 1966, knowledge of releases from the U.S. Atomic Energy Commission's Hanford Site in Washington State was kept from the general public. Between 1944 and 1947, the complex released 440,000 Ci of radioactive iodine (^{131}I) into the atmosphere (Stenehjem, 1990). On May 12, 1963, at the Hanford K-East reactor, 20,000 Ci were released to the Columbia River (Stenehjem, 1990).

Concern about pollutant effects on nonhuman species was also growing. Pesticides such as DDT[2] (dichlorodiphenyltrichloroethane or 2,2-bis-[p-chlorophenyl]-1,1,1-trichloroethane) accumulated in wildlife to alarming concentrations, resulting in direct toxicity and sublethal effects. From 1957 to 1960, Hunt and Bischoff (1960) and Dolphin (1959) documented deaths of Western grebes (*Aechmophorus occidentalis*) resulting from bioaccumulation of the pesticide DDD (1,1-dichloro-2,2-bis[p-chlorophenyl] ethane) from a freshwater food chain (Clear Lake, CA). These pesticides accumulated in the brain until enough was present to cause axonic dysfunction and death. Excessive amounts of pesticide were available to be transferred to the grebes. Dolphin (1959) described the 1949 administration of DDD to control a nonbiting gnat of Clear Lake as "involving introducing approximately 40,000 gallons of a 30% DDD formulation … from drum-laden barges"!

Silent Spring, the extraordinary book by Rachel Carson (1962), drew the attention of the public to these and less obvious consequences of pesticide accumulation in wildlife. Although relatively nontoxic to humans, DDT and DDE (dichlorodichloroethylene or 1,1-dichloro-2,2-bis-[p-chlorophenyl]-ethene) inhibit Ca-dependent ATPases (ATP = adenosine triphosphate) in the shell gland of birds, resulting in shell thinning and increased risk of damage to eggs after being laid (Cooke, 1973; 1979). Birds at higher trophic levels were extremely vulnerable because DDT and its degradation product DDE are relatively resistant to degradation and accumulate in lipids. These qualities result in an increase in concentration with each trophic exchange in a food web. Reproductive failure of raptors and fish-eating birds became a widespread phenomenon. For example,

---

[1] A rem—**roentgen equivalent man**—is a measure of radiation that takes into account the differences in potential biological effects of various types of radiation. It relates the radiation dose received to potential damage. As such, it is a convenient unit for defining allowable radiation exposures. For example, the average person receives approximately 0.360 rem (360 mrem) of radiation annually, and the average radiation worker must not exceed exposures of 600 mrem/month. The rem has been replaced as the official unit by the sievert (Sv). (1 rem = 0.01 Sv.) In contrast, the **curie** (Ci), as used later in this book, is simply a measure of radioactivity. One curie is 2.2×10^6 dpm (disintegrations per minute). Although still used widely as in this book, the curie has been replaced by the **becquerel** (Bq) as the official unit of radioactivity. One curie is 3.7×10^{10} Bq.

[2] DDT was an extremely important tool for disease and agricultural pest control throughout the world. Indeed, Paul Müeller was awarded the 1948 Nobel Prize in medicine for discovery of its value as an insecticide. Its importance in this context is often overshadowed by our present understanding of its adverse effects on nontarget species if used indiscriminately.

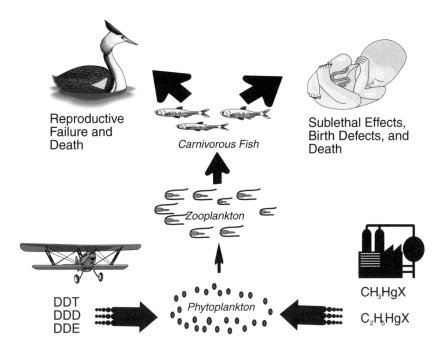

Figure 1.1 Two of the first pollutants to draw attention to the inadequacies of the dilution paradigm were DDT and methylmercury. They became watershed examples of the boomerang paradigm. Both chemicals were returned to humans or to valued wildlife species by transfer through food webs.

the average number of offspring per pair of osprey (*Pandion haliaetus*) nesting on Long Island Sound dropped from 1.71 young per nest (1938–1942) to only 0.07 to 0.40 young per nest by the mid-1960s (Spitzer et al., 1978).[3] Reproductive success of raptor populations decreased in Alaska (Cade et al., 1971) and other regions of the United States (Hickey and Anderson, 1968). Ratcliffe (1967, 1970) reported the same downward trends for falcons (*Falco peregrinus*) and other raptors in the United Kingdom. Reproduction of brown pelicans (*Pelecanus occidentalis*) on the South Carolina coast from 1969 to 1972 fell below that needed to maintain the population (Hall, 1987).

Among these many incidents, the two watershed events that most captured the public's attention and resulted in a paradigm shift (dilution paradigm to boomerang paradigm) were Minamata disease and DDT accumulation in raptors and fish-eating birds (Figure 1.1). Together, they drew some attention away from giddy industrialization and the Green Revolution to the consequences of ignoring pollutants in ecological systems. They were among the first issues to give impetus to the science of ecotoxicology.

II CURRENT NEED FOR ECOTOXICOLOGY EXPERTISE

Everyone would like to feel that the problems described above reflect mistakes made earlier in the techno-industrial revolution that will not be repeated. This is not the case. Environmental problems continue to emerge despite our increased awareness and complex regulations. Indeed, problems seem to extend more and more frequently to transnational and global scales.

Nuclear materials still require our attention and expenditure of monies. The core of Three Mile Island Reactor Unit 2 (Harrisburg, PA) melted on March 28, 1979, releasing approximately 3 Ci of radiation and incurring an estimated $965 million in cleanup costs (Booth, 1987). Nearly 30

---

[3] Osprey populations have rebounded. Ambrose (2001) reports that fewer than 8000 breeding pairs existed in the United States in 1981, but that number increased to 14,246 pairs by 1994.

years after the Chelyabinsk 40 explosion in the Urals, the Chernobyl Reactor 4 core melted down in the Ukraine on April 26, 1986, producing the largest radioactive release in history (301 million curies as estimated by Medvedev [1995]). Fallout from Chernobyl spread rapidly across the Northern Hemisphere. High-level radioactive waste storage tanks like (but much larger than) that which exploded at Chelyabinsk 40 remain unresolved post–Cold War problems for the U.S. Department of Energy nuclear complex. Despite worldwide protest, French underground testing of nuclear devices resumed briefly in Micronesia in 1995. In late 2001, Afghan members of the al-Qaeda were making vague threats about detonating a dirty nuclear weapon. At the moment of revising this chapter, Pakistan and India are rattling their nuclear sabers at each other, prompting pundits to discuss the consequences of a nuclear exchange between these two countries.

Chemical wastes continue to require attention and funds. A myriad of Soviet environmental issues remain as part of the Cold War legacy (Tolmazin, 1983; Edwards, 1994).

Tributyltin (TBT), a widely applied antifouling agent in marine paints, has harmed estuarine mollusks throughout the world (Bryan and Gibbs, 1991). Mercury in fish and game remains a concern, with new sources appearing, e.g., the mercury used in South American gold mining (de Lacerda et al., 1989; Branches et al., 1993; Reuther, 1994). Recently, subsurface agricultural drainage in the San Joaquin Valley of California brought selenium in the Kesterson Reservoir and Volta Wildlife Area to concentrations that were sufficiently high enough to cause avian reproductive failure (Ohlendorf et al., 1986). Efforts to reduce lead in products such as gasoline (human poisoning, e.g., Ember, 1980; Millar and Cooney, 1982; Settle and Patterson, 1980) and lead shot (poisoning of dabbling ducks, e.g., Hawkes, 1977) have only been effective since the late 1970s. Even into the 1980s, debate continued about effects of lead and the need for federal regulation (Anderson, 1978; Anon., 1984a; Ember, 1984; Marshall, 1982; Putka, 1992).[4] At the same time, the controversy about the Hooker Chemicals and Plastics Corp.'s dump sites at Hyde Park and Love Canal became hysterical as the public watched (Anon., 1981; 1982; Culliton, 1980; Smith, 1982). On December 2, 1984, a storage tank at a Union Carbide pesticide plant in Bhopal, India, exploded and released a cloud of methyl isocyanate, killing 2,000 people and harming an estimated 200,000 more (Anon., 1984b; Heylin, 1985; Lepkowski, 1985). On March 16, 1978, the *Amoco Cadiz* supertanker ran aground at Portsall (France) and released roughly 209,000 m^3 of crude oil (Ellis, 1989). On March 24, 1989, the *Exxon Valdez* spilled 41,340 m^3 of crude oil into Prince William Sound. The oil covered an estimated 30,000 km^2 of Alaskan shoreline and offshore waters (Piatt et al., 1990). Marine bird populations are still recovering from this spill (Lance et al., 2001). From August 2, 1990, until February 26, 1991, the largest oil release to have ever occurred was deliberately spilled by Iraqi troops occupying Kuwait. Half a million tons (roughly equivalent to 522,000 m^3) of crude oil from the Mina Al-Ahmadi oil terminal were pumped into the Arabian Gulf (Sorkhoh et al., 1992). Plumes of contaminating smoke from the intentional ignition of Kuwaiti oil wells by the Iraqi troops were visible from space (Figure 1.2).

Other smaller or more diffuse but incrementally more damaging events also require expertise in ecotoxicology. Beyond the intentional release described above, the Arabian Gulf receives 67,000 m^3 of oil annually from smaller leaks and spills (Sorkhoh et al., 1992). The average number of oil spills and the volume per spill in or around U.S. waters from 1970 to 1989 were 9,246 and 47,000 m^3, respectively, with no obvious downward trend in either statistic (Table 8 in Gorman, 1993). Prior to the *Exxon Valdez* spill, a 1978 act of sabotage to the trans-Alaska pipeline had released 2540 m^3 of oil onto the land near Fairbanks. In October 2001, 1081 m^3 of oil gushed from a hole shot in the trans-Alaska pipeline by an intoxicated man. At the time, Rachel Carson was writing *Silent Spring* (ca. 1960) and annual production of synthetic organic chemicals was 43.9 billion kg.

---

[4] Relative to our slow acceptance of lead's adverse effect, Tackett (1987) provides a revealing quote by Benjamin Franklin (July 31, 1786). "This my dear Friend is all I can at present recollect on the Subject. You will see by it, that the Opinion of this mischievous Effort from lead is at least above Sixty Years; and you will observe with Concern how long a useful Truth may be known and exist, before it is generally receiv'd and practis'd on." As recently as one decade ago, 200 years after this quote was made, the value of reducing lead in gasoline was being actively questioned.

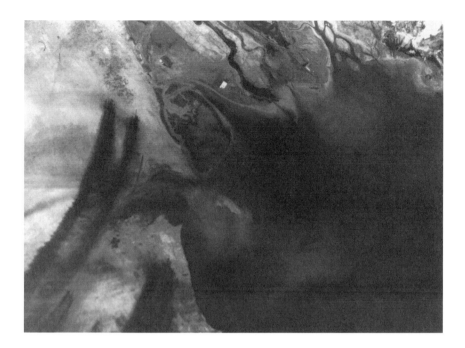

Figure 1.2 Kuwaiti oil wells set afire by Iraqi troops as seen from a U.S. space shuttle flight. Oil wells are seen burning north of the Bay of Kuwait and immediately south of Kuwait City. (Courtesy of NASA.)

By 1970, worldwide production had reached 145.1 billion kg (Corn, 1982). By 1985, the use of pesticides in the United States had roughly doubled from the 227 million kg used in 1964 (Figure 7 in Gorman, 1993). Many persistent pesticides restricted in developed countries are still used in the Third World (Simonich and Hites, 1995).

Contaminants amenable to atmospheric transport have become especially disconcerting. Acid rain is now a transnational problem (Likens and Bormann, 1974; Likens, 1976; Cowling, 1982), damaging both aquatic (Baker et al., 1991; Glass et al., 1982) and terrestrial (Cowling and Linthurst, 1981; Ellis, 1989) ecological systems. Chlorofluorocarbons (CFC), used as propellants and coolants, have been linked to ozone depletion in the stratosphere (Kerr, 1992; Zurer, 1987; 1988), and efforts are being made to greatly reduce their use (Crawford, 1987). But despite the 1987 Montreal Protocol calling for complete elimination of CFC use by 2000, efforts by lawmakers were still under way in the mid-1990s to delay, and perhaps avoid, any U.S. reduction of CFC emissions (Lee, 1995).

New chemicals of concern are emerging. Hale et al. (2001 and Vignette 2.2) identified brominated fire retardants, synthetic estrogens, alkylphenol ethoxylates and their degradation products, manufactured antimicrobial products, and constituents of personal-care products as heretofore ignored contaminants that are currently discharged in large quantities. Only during the last few years have ecotoxicologists begun to take notice of the potentially widespread impacts of these chemicals in natural systems.

This litany of problems is not intended to convince the reader that techno-industrial advancement is an Icarian endeavor incompatible with environmental integrity and human well-being. Rather, it is intended to demonstrate two simple points. First, approximately 50 years ago, the dilution paradigm failed, with clearly unacceptable consequences to human health and ecological integrity. Second, expertise in ecotoxicology is now critical to our well-being. Major environmental problems remain, and new challenges arise daily that are as significant as the historical problems just described. Expertise in ecotoxicology is essential for determining the costs and benefits of the innumerable technological and industrial decisions affecting our lives. Consideration of nonmarket goods and services and the value of natural capital (Odum, 1996; Prugh, 1999) must be incorporated into these decisions. Such services provided *pro bono* by nature are estimated to be in the range of $33 trillion annually, twice the annual gross domestic product of the earth's 194 countries (Rousch, 1997). Investment of time, thought, and resources to avoid damage to service-providing natural systems is economically, as well as ecologically, wise behavior. Complex and costly environmental regulations save human lives, foster sustained economic prosperity, and allow responsible environmental stewardship.

… growth and the environment are not opposites—they complement each other. Without adequate protection of the environment, growth is undermined; but without growth it is not possible to support environmental protection.

Lomborg (2001)

Vignette 1.1   The Emergence and Future of Ecotoxicology

John Cairns Jr.
Virginia Polytechnic Institute and State University, Blacksburg, VA

Quite naturally, toxicology began with concerns about adverse effects of chemical substances on humans. The focus of toxicology gradually extended to organisms domesticated by humans and then to wild organisms of commercial, recreational, or aesthetic value. Unfortunately, these toxicity tests are all homocentric, commendable but inadequate when human society is dependent on an ecological life support system. The close linkage between human health and the health of natural systems makes an ecocentric toxicity component essential. Figure 1.3 illustrates the dimensions of this challenge and illustrates that, despite remarkable progress in the field of ecotoxicology over the last three decades, there is still a long way to go.

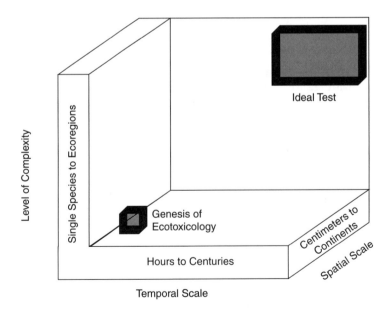

Figure 1.3   Schematic depiction of the genesis of ecotoxicology and the ideal goal about which one can, at present, only speculate. All three major components should be matched to the problem of interest in order to achieve the goal.

In its earliest stages, environmental toxicology depended on short-term laboratory tests with single species that were low in environmental realism but with satisfactory replicability. However, such tests used neither the endpoints nor levels of complexity that are characteristic of ecosystems. Neither did they include cyclic phenomena and many types of variability that are the norm for the complex, multivariate systems known as ecosystems. In short, the "eco" was seriously deficient in the field of ecotoxicology. While simple tests were often used as surrogates for more important properties, toxicity tests at lower levels of biological organization, such as single species, were not readily validated at higher levels of biological organization, such as communities, ecosystems, or landscapes.

Reshaping the Planet

Arguably, the impetus for the development of ecotoxicology resulted from unmistakable environmental transformations that occurred in the 20th century on a scale unique in human history. The world has become increasingly humanized, and ecosystems have become more fragmented and diminished in aggregate size. Consequently, ecosystems have begun to lose resilience and thus need greater protection from threats to their integrity. This situation is an uncontrolled "experiment" on a planetary scale for which the outcome is uncertain. Although environmental change has been the norm for 4 billion years, the planet has been altered by the human species for only 4 million years; however, the rate and intensity of change that occurred in the last century are a cause for deep concern. One might reasonably assert that ecotoxicology is an attempt to provide some rules for the planetary game human society is playing. Almost every sport has elaborate rules that are discussed in great detail by fans. The human, therefore, needs to recognize the natural laws that determine the outcome of the game of life in which all are participants.

Functioning Despite Uncertainty

Both the human condition and the "tools of the trade" (ecotoxicology) are constantly changing. Theories and practices once thought to be sufficient have been shown to be inadequate, often with stunning rapidity. Ecotoxicology can make major contributions in reducing the frequency and intensity of environmental surprises by (1) determining critical ecological thresholds and breakpoints, (2) developing ecological monitoring systems to verify that previously established quality control conditions are being met, (3) establishing protocols for the protection and accumulation of natural capital, (4) providing guidelines for implementing the precautionary principle, (5) developing guidelines for anthropogenic wastes that contribute to ecosystem health, and (6) responding to environmental changes with prompt remedial ecological restoration measures when evidence indicates that an important threshold has been crossed.

Natural Capital and Industrial Ecology

Natural capital consists of resources, living systems, and ecosystem services. Natural capitalism envisions the use of natural systems without abusing them, which is essential to sustainable use of the planet. Sustainable use of the planet requires a mutualistic relationship between human society and natural systems and affirms that a close relationship exists between ecosystem health and human health. Natural capitalism deals with the critical relationship between natural capital—natural resources, living systems, and the ecosystem services they provide—and human-made capital (Hawken et al., 1999).

Industrial ecology is the study of the flows of materials and energy in the industrial environment and the effects of these flows on natural systems (White, 1994; Graedel and Allenby, 1995). The essential idea of industrial ecology is the coexistence of industrial and natural ecosystems. Properly managed, industrial ecology would enhance the protection and accumulation of natural capital in areas now ecologically degraded or at greater risk than necessary. However, the most attractive feature of industrial ecology may be that it would involve temporal and spatial scales greater than those possible with even the most elaborate microcosms or mesocosms. In order to optimize the quality and quantity of information generated by these hybrid systems, some carefully planned risks must be taken, and regulatory agencies must be sufficiently flexible to permit them. Because some ecological damage is inevitable under these circumstances, ecotoxicologists must be knowledgeable of the practices commonly used in ecological restoration. Even industrial accidents can be a valuable source of ecotoxicological information if they are immediately studied by qualified personnel and the information widely shared. Regulatory and industrial flexibility in assessing experimental remedial measures would also enhance the quality of the information base. The obstacles to achieving this new relationship between industry and regulatory agencies boggles the mind, yet there seems to be no comparable, cost-effective means of acquiring the needed ecotoxicological information over such large temporal and spatial scales. Convincing the general public and its representatives of the values inherent in this approach will be a monumental task, but the consequences of making ecotoxicological decisions with inadequate information are appalling.

Speculations

Failure to react constructively to unsustainable practices does not always lie in not knowing what is wrong or even in not knowing what to do about it but, rather, in the failure to take this knowledge seriously enough to act on it. Thus, ecotoxicologists have a responsibility to raise public literacy about their field so that the information they generate is taken seriously and utilized effectively. Ecotoxicologists and other environmental professionals must be aware that their data, predictions, estimates, and knowledge will be used in a societal context that is embedded in an environmental ethos (or set of guiding beliefs). Because science is often idealized as independent of an ethos, there is some level of tension between the essentially scientific task of estimating risk and uncertainty, and the value-laden task of deciding what level and type of risks are acceptable. If sustainable use of the planet becomes a major goal for human society, it is difficult to visualize how the mixture of science and value judgment can be avoided. One hopes that the process of science, with its priceless quality control component, can remain intact for data generation and analysis while producing better and more relevant information to be used with intelligence and reason in making value judgments. However, sustainable use of the planet will require a major shift in present human values and practices. At the same time, ecotoxicology will be evolving, possibly rapidly, so that keeping system and order in the mixture will be difficult.

Although some of the trends in ecotoxicology briefly described here seem inevitable, the direction of the field will almost certainly be determined by environmental surprises, which are ubiquitous and unlikely to occur in convenient places at convenient times. Worse yet, the temptation will be strong to study them entirely with existing methodology, which will probably be inadequate.

A multidimensional research strategy is needed that emphasizes ecosystem complexity, dynamics, resilience, and interconnectedness, to mention a few important attributes. However, major obstacles exist to the development of such a program in the educational system, governmental agencies, industry, and with a citizenry increasingly suspicious of science and academe in general. Society depends primarily on its major universities for the generation of new knowledge; however, this function has become a commodity produced for sale, which means the research direction is all too often a function of marketability. An unfortunate consequence is that writing grant proposals (now often termed contracts) consumes an ever-increasing proportion of the time of ecotoxicologists and other environmental professionals. Ecotoxicologists may often postpone visionary, long-term projects whose outcomes are highly uncertain for short-term projects of severely limited scope determined by the perceived needs of the funding organization rather than being truly exploratory undertakings.

Some counter trends exist to these discouraging developments, often occurring outside of "mainstream" science. A number of new journals are challenging the fragmentation of knowledge, and publications are espousing the consilience (literally "leaping together") of knowledge. Increasingly within academe the focus is on issues that transcend the capabilities of a single discipline or even a few disciplines. Also, environmental professionals, such as ecotoxicologists, are finding ways to minimize the effects of budgetary constraints. However, it still seems likely that one or more major environmental catastrophes will be needed to persuade decision makers that a major shift in approach is needed to cope more effectively with the ecotoxicological and other uncertainties that human society now faces and are likely to increase substantially in the future.

Acknowledgments

I am indebted to Eva Call for transcribing the handwritten draft and to Darla Donald for editorial assistance.

III ECOTOXICOLOGY

The subject of this book could have been titled either environmental toxicology or ecotoxicology, because definitions of both terms are rapidly converging. Often, use of the term *environmental toxicology* implies that only the effect of environmental contaminants on humans will be discussed. Such an implication would be inappropriate for this book. Some definitions of ecotoxicology seem to exclude discussion of humans except as the source of contaminants (Table 1.1), but the original definition given to ecotoxicology by Truhaut (1977) includes effects to humans. The term *ecotoxicology* was selected here with the intention of including effects to humans as well as to ecological entities. *Ecotoxicology* was chosen with reservation because a strong and confining ecosystem emphasis is apparent in many of its definitions. For example, the recent textbook by Connell et al. (1999) specifies that the scope of ecotoxicology includes "organisms, populations, communities and ecosystems" (Table 1.1). More and more, this conventional individual → population → community → ecosystem context is necessarily being stretched to include contexts of landscape → ecoregion → continent → hemisphere → biosphere. Because the conventional framework is gradually becoming insufficient to contain all germane subjects, ecotoxicological discussion in this book extends beyond the ecosystem to the biosphere. As applied here, **ecotoxicology** is the science of contaminants in the biosphere and their effects on constituents of the biosphere, including humans.

Table 1.1   Definitions of Ecotoxicology and Environmental Toxicology

Definition	Reference
Environmental Toxicology	
1. The study of the effects of toxic substances occurring in both natural and man-made environments	Duffus (1980)
2. The study of the impacts of pollutants upon the structure and function of ecological systems (from molecular to ecosystem)	Landis and Yu (1995)
Ecotoxicology	
1. The branch of toxicology concerned with the study of toxic effects, caused by natural and synthetic pollutants, to the constituents of ecosystems—animals (including human), vegetable, and microbial—in an integrated context	Truhaut (1977)
2. The natural extension from toxicology, the science of poisons on individual organisms, to the ecological effects of pollutants	Moriarty (1983)
3. The science that seeks to predict the impacts of chemicals upon ecosystems	Levin et al. (1989)
4. The study of the fate and effect of toxic agents in ecosystems	Cairns and Mount (1990)
5. The science of toxic substances in the environment and their impact on living organisms	Jørgensen (1990)
6. The study of toxic effects on nonhuman organisms, populations, and communities	Suter (1993)
7. The study of the fate and effect of a toxic compound on an ecosystem	Shane (1994)
8. The field of study which integrates the ecological and toxicological effects of chemical pollutants on populations, communities, and ecosystems with the fate (transport, transformation, and breakdown) of such pollutants in the environment	Forbes and Forbes (1994)
9. The science of predicting effects of potentially toxic agents on natural ecosystems and nontarget species	Hoffman et al. (1995)
10. The study of the pathways of exposure, uptake, and effects of chemical agents on organisms, populations, communities and ecosystems	Connell et al. (1999)
11. The study of harmful effects of chemicals on ecosystems	Walker et al. (2001)

IV ECOTOXICOLOGY: A SYNTHETIC SCIENCE

IV.A   Introduction

Ecotoxicology is a synthetic science drawing from many disciplines (Figure 1.4). Questions about effect are posed at the molecular (e.g., enzyme inactivation by a contaminant) to the population (e.g., local extinction) to the biosphere (e.g., global warming) levels of biological organization. Questions of fate and transport are addressed from the chemical (e.g., dissolved metal speciation) to the habitat (e.g., contaminant accumulation in depositional habitats) to the biosphere (e.g., global distillation of volatile pesticides) levels of physical scale. Sometimes, this can produce a confusing complex of scales and associated specialties. The key to maintaining conceptual coherency in this complex of interwoven and hierarchical topics was articulated by Caswell (1996), who wrote that "processes at one level take their mechanisms from the level below and find their consequences at the level above. ... Recognizing this principle makes it clear that there are no truly 'fundamental' explanations, and make it possible to move smoothly up and down the levels of the hierarchical system without falling into the traps of naive reductionism or pseudo-scientific holism." Understanding fates and effects at all levels is essential for effective environmental stewardship (Newman, 2001).

Although all levels are equally important, they contribute differently to our efforts and understanding (Figure 1.5). Questions dealing with lower levels of the conceptual hierarchy, e.g., biochemical effects of toxicants, are more tractable and have more potential for easy linkage to a specific cause than effects at higher levels such as the biosphere. Changes in δ-aminolevulinic

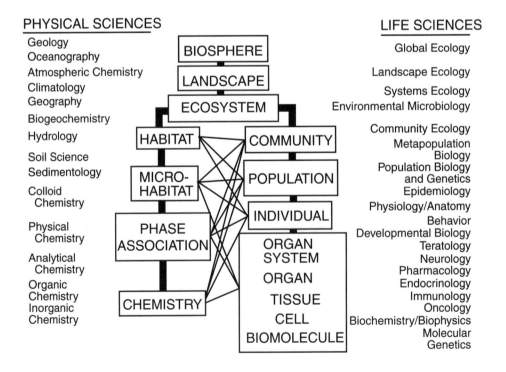

Figure 1.4   Hierarchical organization of topics addressed by ecotoxicology. Disciplines contributing to understanding abiotic interactions are listed on the left side of the diagram and those contributing to understanding biotic interactions are listed on the right. Important interactions, denoted by lines connecting components, occur between biotic and abiotic components.

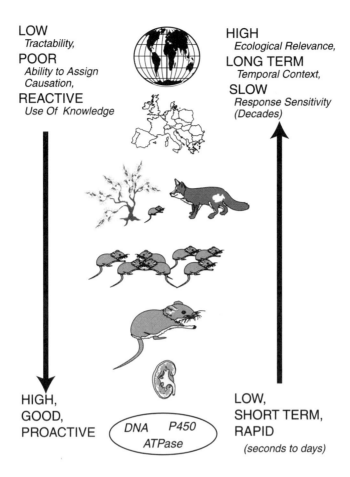

Figure 1.5 Hierarchical organization of topics in ecotoxicology relative to ecological relevance, general tractability, ability to assign causation, general use of knowledge, temporal context of consequence, and temporal sensitivity of response.

acid dehydratase (ALAD) activity in red blood cells can be assayed inexpensively and quickly linked to lead exposure. The general loss of fish species from Canadian lakes was much more challenging to document and to link to acid-precipitation-generating sulfur and nitrogen oxide emissions. As a consequence, effects at lower levels of the ecological hierarchy are used more readily in a proactive manner than are effects at higher levels. They can indicate the potential for emergence of an adverse ecotoxicological effect, whereas effects at higher levels are useful in documenting or prompting a regulatory reaction to an existing problem. Although highly tractable and sensitive, the ecological relevance of effects at lower levels is much more ambiguous than effects at higher levels of organization. A 50% reduction in species richness is a clear indication of diminished health of an ecological community, but a 50% increase in metallothionein in adults of an indicator species provides an equivocal indication of the health of populations in the associated community. Relative to those at higher levels of biological organization, effects at lower levels tend to occur more rapidly after the stressor appears and to disappear more quickly after it is removed. Considering all of these points, it is clear that information from all relevant levels of biological organization should be used together, as described by Caswell (1996) and Newman (2001).

IV.B   Science, Technology, and Practice

Now the world's turned foul; my happiness rots.
… [but] there's a text on repentance, I seem to recall.
But what? What is it? I've forgotten the words,
Don't have a book; and there's no one to guide
My footsteps here in this trackless wood.

Ibsen (1875)

There is a general lack of unanimity in ecotoxicology that leads to much confusion. Much like the quote above, there appears to be no single "text" or clear voice indicating the direction one should go. The main objective of this section is to describe how the outwardly inconsistent activities of ecotoxicologists come together to address three intermeshing and equally important goals.

Ecotoxicology has diverse goals (scientific, technological, and practical) in addition to diverse contributing disciplines (Slobodkin and Dykhuizen, 1991; Newman, 1995). The diversity of subjects prompts most ecotoxicologists to specialize in particular areas and look only peripherally at other information. Subsets of ecotoxicologists must block out of consideration major portions of our collective knowledge structure in order to move effectively toward their respective and more focused goals. For example, distinct but overlapping subsets of information and methodologies are used by the scientist, analyst, and regulator. Whereas a scientist may rely heavily on the hypothetico-deductive method, this approach would be an impediment to a regulator who might employ a weight-of-evidence approach instead to expeditiously assess the need for remediation at a particular site. The use of a standard method with a suboptimal detection limit might hinder achievement of a scientific goal, such as accurately quantifying rates at which a micropollutant moves between ecosystem components, yet be necessary for achieving another goal, e.g., generating a consistent water quality database for regulatory purposes.

Slobodkin and Dykhuizen (1991) observed that much confusion is generated if the distinct intentions and approaches are not understood and respected by the professionals in any applied science. Difficulties arise if methods of one ecotoxicologist are judged unacceptable by another without recognition of their differing goals. Because the goals of ecotoxicologists are tightly intermeshed and boundaries are not easily drawn, there is sometimes the appearance of inconsistency or confusion in the field. Again, the purpose of this chapter is to dispel some of this apparent inconsistency by delineating these three goals or sets of intentions (scientific, technological, and practical) and by generally describing the means by which they are pursued. Although one extremely important aspect of practical ecotoxicology—environmental law—will not be discussed, key U.S. and European laws are summarized in Appendices 3 and 4.

IV.B.1   Scientific Goal

The goal of any science is to organize knowledge based on explanatory principles (Nagel, 1961). It follows that the scientific goal of ecotoxicology is to organize knowledge, based on explanatory principles, about contaminants[5] in the biosphere and their effects. The approaches used to reach this scientific goal are well established. Still, they are worth reviewing because they are taught informally and, consequently, many unkept opinions exist regarding the conduct of science.

---

[5] As discussed in Chapter 2, terms such as *pollutant*, *contaminant*, *xenobiotic*, and *stressor* have specific and distinct connotations. A **pollutant** is "a substance that occurs in the environment at least in part as a result of man's activities, and which has a deleterious effect on living organisms" (Moriarty, 1983). A **contaminant** is "a substance released by man's activities" (Moriarty, 1983). There is no implied adverse effect for a contaminant, although one may exist. A **xenobiotic** is

The discussion below is condensed from Newman (1995, 1996, 2001) who synthesized the works of Sir Karl Popper, and extraordinary articles by Platt (1964) and Chamberlin (1897) in the context of ecotoxicology.

In the early history of science, untested or weakly tested theories were used to explain specific phenomena (Chamberlin, 1897). A question was presented to an acknowledged expert, and explanation was given based on some prevailing or "ruling" theory. This was all that was required to fit the phenomenon or observation into the existing knowledge structure. Facts gradually accumulate around the ruling theory, fostering a sense of consistency. This sense of consistency enhances belief. The cumulative effect of such uncritical acceptance of an explanation based on a ruling theory (**precipitate explanation**) is considered inappropriate in modern science. In fact, Descartes's method of proper scientific reasoning included as the first of his four crucial rules:

... never to accept anything as true that I did not know to be evidently so: that is to say, carefully to avoid precipitancy and prejudice.

Descartes (1637) (Translated by Sutcliffe [1968])

Unfortunately, precipitate explanation reappears periodically in most scientific disciplines. Consequently, it is important to recognize precipitate explanation in any field and avoid it in your own behavior.

Modern sciences have replaced the ruling theory with the working hypothesis. The **working hypothesis** is never accepted as true and only serves to enhance the development of facts and their relations by functioning as the focus of the falsification process (Chamberlin, 1897). Experiments and less-structured experiences are used to test the working hypothesis. The falsification process is often conducted using the null-hypothesis-based statistics developed by Fisher in the 1920s. The working-hypothesis approach still has a proclivity toward precipitate explanation because a central theory or hypothesis tends to be given favored status during testing. Chamberlin (1897) suggested application of the **method of multiple working hypotheses** to reduce this tendency. The method of multiple working hypotheses reduces precipitate explanation and subjectivity by considering all plausible hypotheses simultaneously so that equal amounts of effort and attention are provided to each. In fields where multiple causes or interactions are common, it also reduces the tendency to stop after "the cause" has been discovered.

In any modern science, a hypothesis is never assumed to be true regardless of the approach used, but it can gain enhanced status after repeated survival of rigorous testing. Status is not legitimately enhanced unless tests also have high powers to falsify. Unfortunately, consistent application of weak testing can lead to the progressive dominance of an idea by repetition alone. Weak testing is occasionally used to promote an idea or approach; consequently, members of any science such as ecotoxicology must be able to recognize false paradigms that emerge from weak testing and to avoid weak testing in their own work. Further, tests involving imprecise or biased measurement should be avoided because they frequently generate false conclusions and foster confusion.

Gradually, observational and experimental methods produce a framework of explanatory principles or paradigms about which facts are organized. These **paradigms** (generally accepted concepts in a healthy science that withstood rigorous testing and, as a result, hold enhanced status as causal explanations) are learned by members of a discipline and define the major directions of inquiry in

---

"a foreign chemical or material not produced in nature and not normally considered a constitutive component of a specified biological system. [It is] usually applied to manufactured chemicals" (Rand and Petrocelli, 1985). A **stressor** is that which produces a stress. Stress "at any level of ecological organization is a response to or effect of a recent, disorganizing or detrimental factor" (Newman, 1995). As will be discussed, the terms have slightly different legal definitions too.

the field. They act as nuclei around which ancillary concepts are formulated and as a framework for further testing and enrichment of fact. Unlike ruling theories, these paradigms remain subject to future scrutiny, revision, rejection, or replacement. They are explanations that are currently believed to be the most accurate and useful reflections of truth, but they are not absolute truths. This is an important distinction to keep in mind. For example, the paradigm of matter conservation (matter cannot be created nor destroyed) was an adequate explanation of phenomena until Einstein demonstrated its conditional nature. It was then incorporated into a more inclusive paradigm (relativity theory) with the qualification that relativistic mass (mass + mass equivalent of energy) is constant in the universe, but that mass can be converted to energy and vice versa $[\Delta E = \Delta m(c^2)]$. In a field with mixed goals such as ecotoxicology, the conditional status of scientific explanations or paradigms is sometimes forgotten.

Two general and interdependent types of behavior occur in any science: normal and innovative science (Kuhn, 1970). **Normal science** works within the framework of established paradigms and increases the amount and accuracy of our knowledge within that framework. The contribution of normal science is the incremental enhancement of facts and articulation of ideas with which paradigms can be reaffirmed, revised, or replaced by new paradigms (Kuhn, 1970). Most scientific effort is normal science, and the collective work of ecotoxicologists is no exception. In contrast, **innovative science** questions existing paradigms and formulates new paradigms. Innovative science is completely dependent on normal science and can only occur after the incremental enrichment of knowledge brought about by normal science has uncovered inconsistencies between facts and an established paradigm.

Although normal science tends to be more important in a young field, an excessive preoccupation with details ("tyranny of the particular" [Medawar, 1967]) or measurement (*idola quantitatus* [Medawar, 1982]) can slow the maturation and progress of a science. Conversely, insistence on rigorous hypothesis testing prior to the accumulation of enough facts and establishment of accurate measurement techniques can lead to premature rejection of a hypothesis that might otherwise be accepted. Both normal and innovative science must be balanced in a healthy science. In ecotoxicology, many areas still require more normal science before innovative science can be applied effectively. In many other areas of this maturing science, the tyranny of the particular and *idola quantitatus* exist at the expense of much-needed innovative science (Newman, 1996). A balance between normal and innovative science is essential to effectively achieve the scientific goal of ecotoxicology. The long-term benefit of such a healthy balance will be optimal efficiency and effectiveness in environmental stewardship.

IV.B.2 Technological Goal

The technical goal of ecotoxicology is to develop and apply tools and methods to acquire a better understanding of contaminant fate and effects in the biosphere. Often, some activities in technology are indistinguishable from normal science. Nevertheless, their goals are distinct. Relative to plainly scientific endeavors, the benefits to society of technology are more immediate but slightly less global. Although analytical instrumentation is an obvious component of ecotoxicological technology, other components include standard procedures and approaches as well as computational methods. Many of these technologies can also become pertinent to the practical goals of ecotoxicology *when used to address specific problems*. Consequently, the distinction between technology and practice is also based on context.

The development of analytical instrumentation able to detect and quantify low concentrations of contaminants in complex environmental matrices has been essential to the growth of knowledge. For example, the number of commercial atomic absorption spectrophotometers (AAS) increased exponentially in the 1950s and 1960s, making possible the rapid measurement of trace-element contamination in diverse environmental materials (Price, 1972). Flameless AAS methods lowered detection limits for most elements and enhanced analytical capabilities even further. Now, a wide range of atomic emission, atomic absorption, atomic fluorescence, and mass spectrometric techniques are available for the study of elemental contaminants at levels ranging from mg g^{-1} to pg kg^{-1} concentrations. Gas chromatography (GC) techniques allowed study of the more-volatile organic contaminants. Techniques including GC coupled with a mass spectrometer (GC-MS), more-effective columns for separation, and improved detectors have all enhanced our understanding of fate and effects of organic contaminants. For organic compounds less amenable to GC-related techniques, innovations such as advanced separation columns and high-pressure pumps have quickly improved high-pressure liquid-chromatographic (HPLC) methods. Overarching all of these advances have been computer-enhanced sample processing, analytical control, and signal processing. These and a myriad of instrumental techniques have appeared in the last few decades and allowed rapid advancement of scientific ecotoxicology.

Again, procedures and protocols are also important components of environmental technology. Pertinent procedures vary widely. As an example, they can include such activities as the mapping of **ecoregions**—relatively homogeneous regions in ecosystems or associations between biota and their environment—as a means of defining sensitivity of U.S. waters and lands to contaminants (Omernik, 1987; Hughes and Larsen, 1988). These naturally similar regions of the country are grouped for development of a common study or management strategy. Another important example is a crucial technology created through seminal papers such as the series defining the generation and analysis of aquatic toxicity data (e.g., Sprague, 1969; 1970; Buikema et al., 1982; Cherry and Cairns, 1982; Herricks and Cairns, 1982). The recent establishment of a procedural paradigm for ecological risk assessment (e.g., EPA, 1991a) constitutes a technological advance as well as a contribution to ecotoxicology's practical goals. General methods to **biomonitor** (use of organisms to monitor contamination and to imply possible effects to biota or sources of toxicants to humans) (e.g., Phillips, 1977; Goldberg, 1986) and apply **biomarkers** (cellular, tissue, body fluid, physiological, or biochemical changes in extant individuals that are used quantitatively during biomonitoring to imply presence of significant pollutants or as early warning systems for imminent effects) (e.g., McCarthy and Shugart, 1990) are also important technologies developed in the last several decades. Most biomonitoring programs are only possible now because of the advances in analytical instrumentation described above.

Experimental design schemes, statistical methods, and computer technologies are also important here. Valuable descriptions of experimental designs and statistical methods are provided by professional organizations (e.g., APHA, 1981) and government agencies (e.g., EPA, 1985a; 1988a; 1989a; 1989b), facilitating effective data acquisition to enhance our understanding of contaminant fate and effects. These often have easily implemented computer programs associated with them (e.g., EPA, 1985e; 1988; 1989b). Other computer programs have been developed by EPA to enhance scientific progress. One example is the MINTEQA2 program (EPA, 1991b), which predicts speciation and phase association of inorganic toxicants such as transition metals. Numerous programs for statistical analysis of toxic effects data are available from EPA (e.g., EPA, 1985; 1988a; 1989b) and commercial sources. As discussed in Chapter 12, geographic information system (GIS) technologies have been developed to study nonpoint source contamination over large areas such as watersheds (e.g., Adamus and Bergman, 1995).

Some technology-related approaches are difficult to understand if an inappropriate context is forced upon them. Some are focused primarily on supporting scientific goals, while others are designed to support the practical goals of ecotoxicology. Unfortunately, the complex blending of goals in ecotoxicology makes this a common area of confusion. The designed use of any technology must be kept clearly in mind to avoid confusion and generation of misinformation. For example, standard or operational definitions (e.g., acute versus chronic effect, sublethal versus lethal exposure) may have dubious scientific value relative to predicting the actual impact of toxicants in an ecological context. The 96-h duration of the conventional acute toxicity test was selected because it fits conveniently into the workweek, not because it has any particular scientific underpinnings. The operational distinction made between acute and chronic exposures is also partially arbitrary.

A sublethal exposure could produce a lethal cancer. Regardless, when appropriately and thoughtfully applied, such standard definitions and associated tests are invaluable in applying our technology to various scientific subjects, e.g., using standard acute toxicity endpoints to determine if the free metal ion is the most toxic species of a metal.

The qualities valued in technologies are effectiveness (including cost effectiveness), precision, accuracy, an appropriate level of sensitivity, consistency, clarity of results, and ease of application. As discussed below, several of these qualities are also important in practical ecotoxicology.

IV.B.3   Practical Goal

The practice of ecotoxicology has as its goal the application of available knowledge and technologies to documenting or solving specific problems. Scientific information is integrated into the practice of ecotoxicology with the intent of addressing a specific problem. During the process, some scientific knowledge will be marginalized to expeditiously solve the problem. What might seem to be from a rigid scientific vantage an incomplete definition of the truth is, in fact, the most expeditious means of defining the major problem and resolving it. Many technologies are relevant to practical ecotoxicology; however, the goal of their application is also to solve or document a particular environmental situation. Techniques appropriate for the practical ecotoxicologist can include general methods, such as for determining contaminant leaching from wastes (e.g., Anon., 1990). Predictive software such as the QUAL2E program (EPA, 1987a), which estimates stream water quality under specific discharge scenarios, can also be an important tool in achieving practical goals. Other tools include specific steps to take during implementation of a method, e.g., biomarker-based biomonitoring on U.S. Department of Energy sites (McCarthy et al., 1991). They might involve guidelines for the practice of risk assessment on hazardous waste sites (EPA, 1989c) or for waste basin closure. In each of these instances, the goal is not to understand the ecotoxicological phenomena more completely or to develop a technology to better study a system. The goal is to address and resolve a specific problem. Indeed, attempts to conduct scientific work in such efforts or to expend resources in developing a novel technology could delay progress toward the practical goal—solving the immediate problem and removing potentially harmful pollutants.

Practical tools may also include criteria and standards for regulation of specific discharges or water bodies. For example, water quality **criteria** are estimated concentrations of toxicants based on current scientific knowledge that, if not exceeded, are considered protective for organisms or a defined use of a water body (or some other environmental medium). Criteria are developed for individual contaminants, e.g., aluminum (EPA, 1988b), cadmium (EPA, 1985b), copper (EPA, 1985c), lead (EPA, 1985d), and zinc (EPA, 1987b) using a standard approach (i.e., EPA, 1985e). Based on scientific knowledge, they are used to recommend toxicant concentrations not to be exceeded as a result of discharges into waters. Although the example here is that for water, criteria are also defined for air (see Clean Air Act in Appendix 3) and solid media such as sediments (Shea, 1988; Di Toro et al., 1991).

Based on criteria and the specified use of a water body, water quality standards can be set for contaminants. **Standards** are legal limits permitted by each state for a specific water body and thought to be sufficient to protect that water body. Both criteria and standards are designed with the intent to avoid specific problems, but they are based partially on existing scientific knowledge. Consequently, a healthy growth of scientific knowledge in ecotoxicology is essential to improving our progress toward this practical goal of ecotoxicology. Indeed, criteria and standards (EPA, 1983) are revised periodically to accommodate new knowledge.

Effectiveness, precision, accuracy, sensitivity, consistency, clarity, and ease of application are valued in practical ecotoxicology as well as in technical ecotoxicology, as discussed above. Also important to practical ecotoxicology are the following: unambiguous results, safety, and clear documentation of progress during application.

V SUMMARY

At the close of the Second World War, the dilution paradigm failed because of its clearly unacceptable consequences to human health and ecological integrity. Expertise in ecotoxicology is now critical to our well-being. Ecotoxicology—the science of contaminants in the biosphere and their effects on constituents of the biosphere—has emerged to provide such expertise.

Ecotoxicologists have overlapping yet distinct scientific, technological, and practical goals that must be understood and respected. Our current knowledge available for achieving these goals (Figure 1.6, upper panel) requires further expansion and more integration. Although the knowledge applied to these goals overlaps, there remain many instances of inappropriate or inadequate integration. For example, present regulations are biased toward single-species tests done in the laboratory, yet our scientific knowledge indicates that results from multiple-species tests are at least as valuable to understanding risk. Recognizing the continual need for reintegration of knowledge, lawmakers have incorporated periodic review and revision into major legislation and associated regulations. Further, new or improved technologies are continually drawn into our scientific efforts, e.g., new molecular technologies applied to assay genetic damage. The scientific foundations of the field should also expand and come into balance with technology and practice. This is done most effectively using the methods described in this chapter. Passé and flawed behaviors such as precipitate explanation, overdominance of normal science, the tyranny of the particular, and *idola quantitatus* should be avoided as impediments to scientific progress. An understanding of and respect for the different goals of ecotoxicologists must also prevail in order to appropriately apply science and technology to practical problems of environmental stewardship.

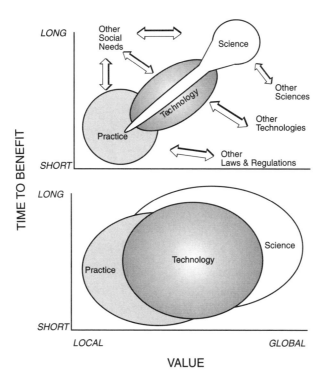

Figure 1.6   Present (top panel) and ideal (bottom panel) balance among scientific, technological, and practical components of ecotoxicology. The relative amount of effort in each is reflected by area on the plots of scale of value (local to global) and time to realize (short to long term) benefit from these components of the field.

Ecotoxicology is rapidly becoming a mature discipline and, hopefully, will soon achieve an effective balance of knowledge to best address its scientific, technical, and practical goals (Figure 1.6, lower panel). Scientific understanding must expand in all directions, especially toward more global and long-term phenomena. Technology and practice must do the same. The sound laws presently implemented in the United States should and do accommodate this evolution of technology and science.

SELECTED READINGS

Carson, R., *Silent Spring,* Houghton-Mifflin Co., Boston, 1962, p. 368.

Chamberlin, T.C., The method of multiple working hypotheses, *J. Geol.,* 5, 837–848, 1897.

Mackenthun, K.M. and J.I. Bregman, *Environmental Regulations Handbook,* Lewis Publishers, Boca Raton, FL, 1992, p. 297.

Marco, G.J., R.M. Hollingworth, and W. Durham, *Silent Spring Revisited,* American Chemical Society, Washington, D.C., 1987, p. 214.

McGregor, G.I., *Environmental Law and Enforcement,* Lewis Publishers, Boca Raton, FL, 1994, p. 239.

Newman, M.C., *Quantitative Methods in Aquatic Ecotoxicology,* Lewis Publishers, Boca Raton, FL, 1995, p. 426.

Newman, M.C., Ecotoxicology as a science, in *Ecotoxicology: A Hierarchical Treatment,* Newman, M.C. and Jagoe, C.H., Eds., CRC/Lewis Publishers, Boca Raton, FL, 1996.

Platt, J.R., Strong inference, *Science,* 146, 347–353, 1964.

Smith, W.E. and A.M. Smith, *Minamata,* Holt, Rinehart and Winston, New York, 1975, p. 192.

Woodwell, G.M., Toxic substances and ecological cycles, *Sci. Am.,* 216, 24–31, 196.

CHAPTER 2

Environmental Contaminants

> We have subjected enormous numbers of people to contact with these poisons, without their consent and often without their knowledge. It is the public that is being asked to assume the risks that the insect controllers calculate. The public must decide whether it wishes to continue on the present road, and it can only do so when in full possession of the facts.
>
> **Rachel Carson (1962)**

I INTRODUCTION

The contaminants of interest to ecotoxicologists can be divided grossly into two categories: organic and inorganic. Both categories were initially based on whether the chemical was obtained from living organisms (organic) or mineral sources (inorganic). However, this distinction is imprecise. For example, carbon dioxide is considered an inorganic gas, yet it is produced by organisms. Further distinctions can be made: Organic compounds can generally be considered those containing carbon and involving at least one C-H covalent bond, i.e., methane (CH_4) but not carbon dioxide (CO_2). But graphite is an exception to this rule. One might then begin to make the distinction by including reduced carbon as part of the definition, but carbon tetrachloride (CCl_4) is considered an organic compound. (The interested reader is referred to Larson and Weber [1994, p. 2] for further discussion of this ambiguity.) Fortunately, the distinction becomes much clearer with compounds composed of carbon chains or rings.

Some important inorganic contaminants are carbon monoxide, carbon dioxide, nitrogen oxides, and sulfur dioxide. Most of the anthropogenic nitrogen oxides entering the atmosphere are generated by the internal-combustion engine (Manahan, 2000). Sulfur and nitrogen oxides are converted to sulfuric and nitric acids in the atmosphere and contribute to acid precipitation. Important inorganic water contaminants are nitrogenous wastes and phosphates. These are essential plant nutrients that, in excess, can lead to eutrophication.

Metallic elements are a major class of inorganic contaminants for which various related terms are used to differentiate them in the literature. **Metals** are elements known for their lustrous appearance, malleability, ductility, and conductivity. With the exception of hydrogen, they make up the left two-thirds of the periodic table. **Metalloids** are intermediate in properties between the metallic and nonmetallic elements and line up between them in the periodic table. They have a less lustrous appearance than metals, are semiconductors, and include elements such as silicon, arsenic, antimony, and selenium. They are often grouped with metals when discussed in the environmental literature. The term **heavy metal** appeared in early studies of the harmful effects of metallic elements such as mercury, lead, and cadmium, which all had densities greater than iron. This classification has since been applied to other metallic elements of environmental concern regardless of their density. The term **trace metal** is sometimes seen in the literature and simply implies that the concentration of the metallic element measured is very low (≤ppm).

Organic compounds can originate from natural or anthropogenic sources. They are divided into a variety of chemical classes based on molecular structure. Organic compounds have a wide range of chemical and physical properties that affects their fate and toxicity. Some synthetic compounds are based on naturally occurring analogs and degrade rapidly in the environment. Others (i.e., kepone) are unlike anything found in nature and can persist for years.

When does a compound or element become an environmental contaminant? A **contaminant** has been defined as "a substance released by man's activities" (Moriarty, 1983) and more specifically "a substance present in greater than natural concentration as a result of human activity" (Manahan, 2000). Although this second definition seems very specific, it can be difficult to define the natural concentration for a substance and to determine a significant deviation from it. Many of the organic compounds discussed in this chapter are synthetic or man-made (i.e., **anthropogenic**). If the only source is human activity, the compound can be called a **xenobiotic**, which is defined as "a foreign chemical or material not produced in nature and not normally considered a constitutive component of a specific biological system" (Rand and Petrocelli, 1985). Regardless of the concentration detected in the environment, a xenobiotic automatically fits our specific definition of contaminant, as there is no natural concentration for it.

Some of the organic compounds that become important environmental contaminants can be produced by both natural processes and human activities. The polycyclic aromatic hydrocarbons (PAH, discussed below) are a good example. They are produced by the incomplete combustion of organic matter and can be found in high concentrations in the smoke from a forest fire or in the oil deposits deep within the earth. As these sources are natural and have existed for millions of years, there is a low-level background concentration of PAH in the environment. When these same oil deposits or forests are collected, refined to fossil fuels, or burnt, they are converted to energy and by-products (i.e., exhaust or emissions). Combustion is never a completely efficient process, and the emissions can contain high concentrations of incomplete combustion products such as PAH. The distribution of PAH concentrations in the environment often follows gradients, with the highest concentrations near human population centers, giving evidence that they fit the definition of environmental contaminants (Figure 2.1).

Metals and metalloids are naturally occurring elements and, therefore, are not synthesized by industrial processes. They become contaminants if concentrations in the environment are altered from natural distributions through human activities. This can occur during mining and refining, or through release to the environment in industrial effluents and vehicular emissions. Careless disposal of metallic waste has also contaminated terrestrial, groundwater, and aquatic environments. The elemental distribution in the Earth's crust is not uniform, so some knowledge of local geological conditions is necessary when assessing metal contamination in the environment.

Both natural and anthropogenic sources can be important contributors to metallic element emissions to the atmosphere. When comparing atmospheric loading on a global scale, elements such as selenium, mercury, and manganese have greater emissions through natural sources than anthropogenic sources. However, on a regional basis, anthropogenic sources can be many times greater than natural inputs (Pacyna, 1986), and these metallic elements can become contaminants on a local scale. This illustrates the importance of identifying the sources, sinks, and processes that contribute to contaminant transport on the scale of the system under study.

When does a contaminant become a pollutant? A **pollutant** has been defined as "a substance that occurs in the environment at least in part as a result of man's activities, and which has a deleterious effect on living organisms" (Moriarty, 1983). Therefore, for a substance to become a pollutant it must first fit our specific definition of a contaminant (exceeding natural concentration) and must further be shown to cause or be able to cause an adverse biological effect. In cases like the unfortunate poisoning of individuals with mercury in Minamata Bay (see Chapter 1) or wildlife

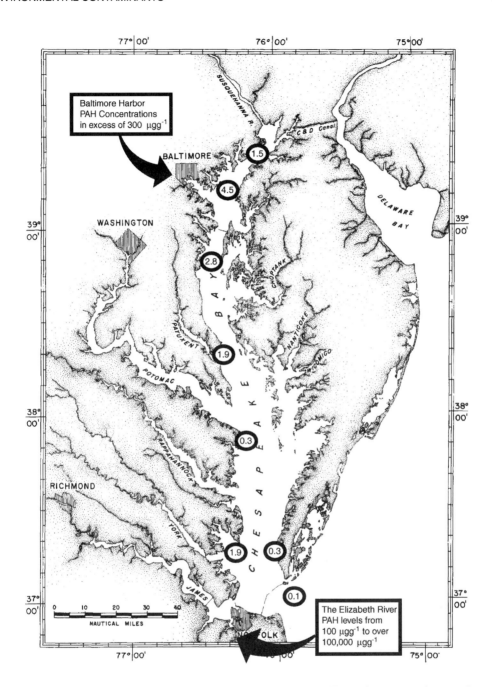

Figure 2.1 PAH can be derived from natural or anthropogenic sources. Trends in concentrations can be used to identify sources and determine if PAH are environmental contaminants. The PAH trend in surface sediments of Chesapeake Bay shows increasing concentrations (μg g^{-1}) near population centers, giving evidence of PAH contamination. Extremely high concentrations at "hot spots" result from historical spills and can be several orders of magnitude higher than ambient levels. These have been implicated for toxic effects to biota in places like the Elizabeth River in Virginia.

mortality resulting from an oil spill, establishing the cause-and-effect relationship is not difficult. But when biota are exposed to a wide variety of contaminants and other stressors in a complex and dynamic environment, identifying the causative agent for a biological effect can be difficult. Section Three of this book, "Toxicant Effects" (Chapters 6 to 12), describes methods for establishing the relationship between measured contaminant concentrations and biological effects on the sub-organismal, organismal, population, community, and global scales.

Under some circumstances, the complexity of the ecosystem and the interactions of the chemical species within it make it difficult to distinguish between contaminants and pollutants. In the case of acid precipitation, toxic effects to biota in acidified streams are often the result of aluminum toxicity, not the acidity alone. In this case the toxic agent, aluminum, is naturally occurring in the sediments and bedrock of the watershed. So, pH does not strictly meet our definition of a pollutant. The true pollutants in this case are atmospheric releases of nitrogenous and sulfur compounds, which are converted to acids in the atmosphere and then are transported to streams as acidic precipitation. The reduced pH in the stream changes the aluminum speciation, increases the concentration of dissolved aluminum, and produces the toxic effect.

Some contaminants become pollutants only after exceeding a certain level or concentration in the environment. Also, the context as well as the concentration can define a pollutant. Nitrite in drinking waters becomes a problem only if our activities bring concentrations to abnormal levels. Indeed, several of the metals listed in this chapter are essential to life, but above certain concentrations they produce adverse consequences.

A mammalian toxicologist would not consider some of the chemicals listed to be toxicants, as they do not directly poison individuals. However, contaminants such as excessive amounts of phosphorus and nitrogen nutrients in a lake can have pronounced adverse consequences to the ecological community of that lake. Similarly, global changes in atmospheric gases can have a pronounced influence on ecosystems of the Earth but no direct toxicity to humans. Putrescible compounds also might not directly kill aquatic biota, but they can indirectly kill large numbers of organisms by removing dissolved oxygen from receiving waters. Consequently, these must be considered as potential contaminants and pollutants along with the more conventional toxicants in the context of ecotoxicology.

II ENVIRONMENTAL FATE OF CONTAMINANTS

Two factors determine contaminant fate in the environment: the physical/chemical properties of the specific element or compound, and the conditions of the surrounding ecosystem. The diverse group of contaminants described in this chapter spans a wide range of physical and chemical properties, and we will discuss how that affects their distribution and persistence under various environmental conditions. Accurate information on the specific conditions of the system under study is as important but more difficult to obtain. The first step for the investigator is to define the scale and boundaries of the system of interest. As the reader can see from the various topics in this text, this could span from an organism to a lake to the atmosphere to the entire biosphere. If considering contaminant fate on an organismal scale, we would describe the processes involved in bioaccumulation, which are covered in detail in Section Two, "Bioaccumulation" (Chapters 3 to 5). Factors affecting contaminant transport on a global level are covered briefly in Chapter 12.

Defining the system under study is a critical step to understanding contaminant fate because specific processes can dominate in a particular ecosystem while being relatively unimportant in others. For example, strong ultraviolet radiation can be an important contributor to photolysis reactions in the atmosphere, but it is not a significant factor affecting chemical degradation in aquatic sediments. Although this example is obvious, other factors like aerobic versus anaerobic conditions in sediment can be difficult to measure and are affected greatly by scale. Figure 2.2 illustrates some of the important processes affecting contaminant fate in various environmental compartments.

Characterization of environmental conditions is especially important to understanding the fate of metals and metalloids. An element's chemistry and the conditions within the surrounding environment affect its speciation. Most metals and metalloids can exist in solution as free cations or as inorganic and organic complexes. Some, such as chromium or arsenic, occur as oxyanions such as

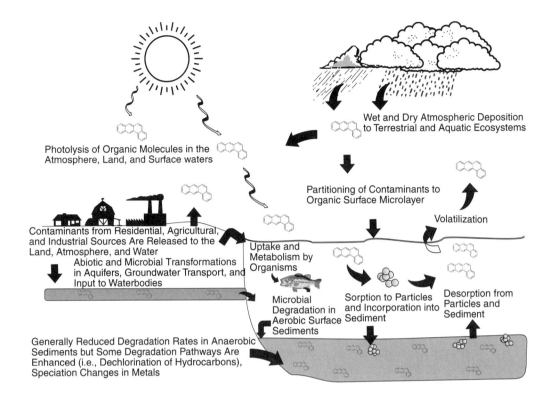

Figure 2.2   Some of the important physical and chemical processes that govern the fate and transport of environmental contaminants.

chromate and arsenate. Speciation controls the solubility, sorption, and volatility of metallic elements. Solution parameters that affect speciation are ionic strength, temperature, pH, Eh, concentrations of competing cations, and ligand concentrations. Natural dissolved organic compounds (DOCs, i.e., humic and fulvic acids) are good chelating agents and will form complexes with metals in solution. This can alter metal sorption to particulate material and sediments. Partitioning will either increase or decrease, depending on the chemistry of the specific element, the nature of the particles, and the solution chemistry. Consequently, speciation also affects the bioavailability and toxicity of metals and metalloids (see Chapter 4). Therefore, total concentrations alone are not always useful for predicting transport and toxicity of metals and metalloids. For those interested in speciation, a useful approach using Eh-pH diagrams has been developed to illustrate some of the thermodynamically stable metallic species present under various environmental conditions (e.g., Brookins, 1988).

II.A Contaminant Partitioning

Contaminant transport is greatly influenced by partitioning of the compound or element between the various phases or compartments. Some examples of important partitioning processes are the distribution between gas and aqueous phases, aqueous and sediment/particulate phases, and dissolved and liquid/solid phases. Even though thermodynamic equilibrium seldom exists in the natural environment, partitioning can be rapid enough for us to assume near steady-state local conditions, allowing description and prediction of these processes with equilibrium expressions. The distribution relationship for a given contaminant (X) at equilibrium is described by Equation 2.1:

$$X_{(phase\ a)} \leftrightarrow X_{(phase\ b)} \qquad (2.1)$$

If the system is at equilibrium, the **distribution or partition coefficient** (K_d or K_p) describes the relative amount of contaminant in each phase. The partition coefficient is simply the concentration or activity of the contaminant in one phase divided by the concentration or activity in the other, as seen in Equation 2.2:

$$[X_{(phase\ b)}/X_{(phase\ a)}] = K_d\ or\ K_p \qquad (2.2)$$

If this relationship describes the partitioning of a contaminant between the aqueous phase (a) and the particulate phase (b), the partition coefficient is specifically called the **sorption coefficient**. The term **adsorption** is used if the contaminant is associated with the surface of the particle. Because most of our contaminant measurement methods rely on exhaustive extraction techniques and macroscopic evaluations, it is difficult to determine if the partitioning mechanism is entirely a surface phenomenon. Sorption is a very complex process that will depend on the structure or speciation of the contaminant, the microscopic composition of the particulate material, and the water chemistry. For this reason, the more general term *sorption* is the most appropriate and does not imply mechanistic knowledge.

Sorption coefficients have been estimated empirically under laboratory conditions for a variety of contaminants and particulate phases. These calculated K_p values are useful tools for comparing the relative tendency for contaminants to accumulate in sediment or on solids, and they give insight into the potential fate of contaminants. It is important to remember, however, that sorption coefficients derived from measurements made in the laboratory might not always be accurate predictors of concentrations of contaminants in sediment and water under environmental circumstances. Natural systems might not be in equilibrium. Also, the structure and quality of natural particulate material varies greatly, as does natural water chemistry.

Two of the major factors governing the partitioning of a contaminant between the aqueous and the vapor phases include its vapor pressure (P) and its water solubility (C). These properties can be used to calculate the **Henry's law coefficient** (H), as seen in Equation 2.3:

$$H = P/C \qquad (2.3)$$

Henry's law coefficient increases as the tendency of a compound to partition into the vapor phase increases. Either low water solubility or a high vapor pressure can increase the Henry's law coefficient for a contaminant. A high Henry's law coefficient tells us that as a contaminant comes to steady state, it will most likely accumulate in the atmosphere. Once in the atmosphere, long-distance transport can become a factor contributing to its distribution (see Chapter 12). Conversely, if this same contaminant was discharged as a component of an effluent into a water body, its aqueous concentration is likely to decrease with time or distance from the source. Its partitioning could keep it from becoming a significant water pollutant. Henry's law coefficient is particularly useful when comparing the relative air/water partitioning behavior of contaminants regardless of their structure. When a group of related compounds (e.g., polychlorinated biphenyls [PCB] and PAH) with diverse water solubilites and/or vapor pressures are introduced into the environment as a mixture, Henry's law coefficients can be used to predict their relative partitioning. This can help explain the changes in relative concentrations of individual compounds in the mixture over time, a process known as **weathering**.

II.B Degradation

Organic compounds undergo a variety of degradation processes in the natural environment. These usually involve the direct reaction with oxygen (oxidation) or water (hydrolysis). Degradation processes either occur as abiotic or biologically mediated reactions. Biologically mediated reactions that occur within higher organisms are often catalyzed by enzymes, which are key to the detoxification, activation, and elimination of contaminants (see Chapter 3).

Biologically mediated reactions controlled by microorganisms are the most important processes controlling the degradation of environmental contaminants. This is because of the increased degradation rates provided by microbial enzymes. Reactions that would occur very slowly abiotically can be accelerated many orders of magnitude by enzymatic action. This is facilitated through steric alignment of reactants or by the production of reactive intermediates (Schwarzenbach et al., 1993). Contaminants utilized by the bacteria as sole carbon sources can be degraded completely to carbon dioxide, water, and inorganic salts in a process is known as **mineralization**. This rarely occurs abiotically. The rate of microbial degradation of a contaminant will be governed by adaptation of the microbial community either through species selection or enzyme induction. Although microbial enzymatic systems are diverse and adaptive, some synthetic organic compounds have structures unlike any naturally occurring analogs and, as a consequence, are resistant to microbial attack. Polychlorinated hydrocarbons are good examples of microbially resistant compounds, yet some of these molecules can undergo a degree of microbial dechlorination under anaerobic conditions.

Abiotic reactions include those that occur chemically in the dark and reactions that are initiated by light (**photolysis**). Hydrolysis reactions occur through the reaction of a water molecule with the organic compound, and they usually result in the addition of a hydroxyl group and the cleavage of other substituents. Addition of the hydroxyl group and fragmentation of the molecule can alter the biological activity of the parent molecule and will increase its water solubility. Hydrolysis reactions usually proceed very slowly and are rarely significant contributors to contaminant degradation unless the reaction is catalyzed by microorganisms or light.

Photolysis reactions are photochemical processes in which light energy is absorbed by the molecule, with consequent breaking of chemical bonds and degradation of the contaminant. Light with enough energy to cause these reactions will be in the ultraviolet to visible range. For molecules to absorb this light energy, they must have internal energy states that correspond to the incoming light energy. Therefore, certain chemical groupings and classes of compounds are more likely to undergo photolysis. Compounds with conjugated double bonds and aromatic ring structures such as PAHs are good UV absorbers and are photolabile. Photodegradation can occur either through **direct photolysis** or **indirect photolysis**. During direct photolysis, the contaminant itself absorbs the light and undergoes fragmentation or oxidation. During indirect photolysis, other molecules are the light absorbers, and they form reactive species that, in turn, facilitate the degradation of the contaminant. Naturally occurring dissolved organic matter such as humic acids are good UV absorbers, can form reactive hydroxyl radicals, and can increase degradation rates of dissolved contaminants. For this reason, characterization of naturally occurring compounds (DOC) as well as basic water chemistry and contaminant concentrations can be important when determining the fate of contaminants under environmental conditions.

III MAJOR CLASSES OF CONTAMINANTS

The large range of possible environmental contaminants makes it impossible to provide much detail in one chapter. Many of the organic classes covered here include a group of related compounds that are formed as a complex mixture in commercial products and are introduced into the environment as mixtures. We have selected examples that are significant because they are prominent in the literature, are a current or perceived future source of concern, or are historically important as case studies. Information on potential sources is provided to inform the reader about the possible significance of each contaminant for future ecotoxicological study.

III.A Metals and Metalloids

Metal and metalloids are naturally occurring elements that become contaminants when human activity raises their concentrations in the environment above natural levels. Environmental concentrations can be compared with normal or average background levels to provide evidence for contamination.

In a review of trace-element distribution in soils of Western Europe, Angelone and Bini (1992) found that metal concentrations in soils were greatest in the neighborhood of industrial factories and in the vicinity of busy roads, giving evidence for metal contamination. Metals are released into the water, air, and terrestrial environments as wastes from industrial manufacturing, mining, combustion products, and agricultural pesticides. The inorganic chemical manufacturing industries (i.e., chlor-alkali, inorganic acids, pigments, copper sulfate) are potential sources for metallic wastes to aquatic ecosystems, so they are regulated to minimize inputs (Manahan, 2000). Important anthropogenic sources to the atmosphere are coal combustion, oil combustion, mining and nonferrous metal production, steel and iron manufacturing, waste incineration, phosphate fertilizers, cement production, and wood combustion (Haygarth and Jones, 1992). Once released to the atmosphere, metals can be transported in a gaseous phase or sorbed to particles and will enter aquatic or terrestrial environments through wet or dry atmospheric deposition (Haygarth and Jones, 1992).

III.A.1 Aluminum

Aluminum is the second most abundant metallic element in the Earth's crust (8%). Under normal environmental pH ranges of 6 to 9, aluminum is found primarily as a component of mineral phases (e.g., gibbsite, $Al_2O_3 \cdot 3H_2O$ and kaolinite, $Al_2Si_2O_5(OH)_4$) (Stumm and Morgan, 1981). Low pH conditions, such as those resulting from acid precipitation or mine drainage, can increase free aluminum (Al^{3+}) to unusually high dissolved concentrations that can kill aquatic species (see Chapter 12).

III.A.2 Arsenic

This metalloid and its compounds are used in numerous products, including metal alloys, pesticides (e.g., $Pb_3(AsO_4)_2$), wood preservatives, plant desiccants, and herbicides (e.g., $Na_3As_3O_3$). It is associated with coal fly ash and is also released during gold and lead mining. It is often present as an oxyanion, e.g., $HAsO_4^{2-}$, AsO_4^{3-}, and $H_2AsO_4^-$. Arsenic can be methylated by some fungi, producing methyl, dimethyl, and trimethyl arsines. These derivatives are volatile and highly toxic compounds (Atlas and Barta, 1981). Arsenic is also a carcinogen.

III.A.3 Cadmium

Unlike many other elements, cadmium is not essential to organisms; indeed, it is toxic (see Itai-Itai disease, Chapter 1) and carcinogenic. It is used in alloy production, plastic stabilizers, electroplating, galvanizing, pigments, batteries, and numerous other products. Cadmium is found at low concentrations in crustal rocks and soils, but because it is chemically very similar to zinc, it is generated as a by-product during zinc ore processing. Therefore, cadmium production is controlled by the zinc industry (Laws, 1993). Sewage sludge is often contaminated with cadmium, and land application of this material can make a significant contribution to agricultural soils (Jackson and Alloway, 1992). Smokers are exposed to high levels of cadmium in cigarettes. It is transported in the atmosphere, primarily in the particulate phase.

III.A.4 Chromium

Chromium is used in alloys, catalysts, pigments, and wood preservatives. It also is used in product tanning. It may be present as Cr(VI) or Cr(III). These are referred to as hexavalent and trivalent chromium, respectively. Hexavalent chromium is carcinogenic and is the more toxic of the two forms. Chromium is often present as an oxyanion, e.g., CrO_4^{2-} and CrO_7^{2-}.

III.A.5 Copper

Copper is used extensively for wiring and electronics and for plumbing. It is also used to control growth of algae, bacteria, and fungi. It is a biocide in marine antifoulant paints and in wood preservatives. It is toxic at high concentrations, but it is easily complexed by dissolved organic matter in solution, which reduces the biologically available fraction. Electron spin resonance studies have shown that copper cations in the +2 valence state are bound to carboxyl, phenolic hydroxyl, and carbonyl functional groups on dissolved humic materials (Senesi, 1992).

III.A.6 Lead

This poisonous metal is ubiquitous due to its long-term and widespread use in gasoline, batteries, solders, pigments, piping, ammunition, paints, ceramics, caulking, and numerous other applications. Greater than 70% of the total lead consumption in the United States is used for the manufacture of batteries (Laws, 1993). Lead is transported in the atmosphere primarily in the particulate phase (Haygarth and Jones, 1992). A concern for adverse effects to humans and wildlife from lead exposure has prompted regulations limiting its use in several products. Lead causes anemia and neurological dysfunction with chronic exposure. Regulations eliminating lead additives to gasoline have reduced inputs from that source over the past 30 years. Reduced human exposure by eliminating lead in residential plumbing and other products may have resulted in lower concentrations in human tissues in recent decades (Manahan, 2000). Decreases in exposure to airborne lead from reduced combustion of lead-amended gasoline might have also contributed to this trend. Exposure to lead in air derived from the combustion of leaded gasoline has been shown to raise body burdens (Davies, 1992). Poisoning of birds can occur by ingestion of lead shot, which has led to regulations restricting its use over water. Lead is present in natural waters in the +2 oxidation state.

III.A.7 Mercury

Mercury is used in electronics, dental amalgams, chlorine-alkali production, gold mining, and paints. Phenylmercury compounds and mercury salts are used as fungicides for seed treatments and growth inhibition in numerous industries such as pulp mills (Atlas and Barta, 1981). It is an excellent industrial catalyst and, because it is liquid at ambient temperatures, it is used as a component in electrolysis. It can be released to the environment as waste from laboratory chemicals, batteries, fungicides, and pharmaceutical products and as a component in sewage effluent (Manahan, 2000). Speciation greatly influences the transport and bioavailability of mercury. The same microbial system that is involved with the anaerobic generation of methane is capable of methylating mercury under the anaerobic conditions found in sediments (Atlas and Barta, 1981). This results in the generation of monomethyl- and dimethyl-mercury, with a corresponding increase in water solubility, volatility, and bioavailability.

III.A.8 Nickel

Nickel is used in alloys such as stainless steel and for nickel plating. It also has innumerable other uses, including battery production (Ni-Cd batteries). At sufficiently high concentrations, nickel is both toxic and carcinogenic.

III.A.9 Selenium

The metalloid, selenium, is used in the production of electronics, glass, pigments, alloys, and other materials. It is a by-product of mining for gold, copper, and nickel, and is also found in high concentrations in coal fly ash. It can enter the environment through the disposal of wastes from these industries. Atmospheric inputs come from fossil fuel burning, smelting, burning vegetation, and volcanism (Frankenberger and Karlson, 1992). Selenium can undergo microbial transformation to produce alkylselenides, which greatly affects the transport and toxicity of this element. The alkylated forms are more volatile, but unlike organoderivatives of other elements, they are much less toxic than the pure elemental form. Biomethylation has even been proposed as a detoxifying technique to remediate sediments and soils contaminated with high levels of selenium (Frankenberger and Karlson, 1992).

III.A.10 Zinc

This metal is used extensively in protective coatings and galvanizing to prevent corrosion. It is also used in alloys and as a catalyst for some polymer synthesis reactions. It is less toxic than most metals listed here.

III.B Inorganic Gases

Carbon dioxide (CO_2) from combustion is a concern because concentrations in the atmosphere are slowly increasing through time. This increase has been linked to global warming. Nitrogen oxides (NO_x) and sulfur dioxide (SO_2) are also produced by combustion at stationary (e.g., coal power plants) and mobile (e.g., automobiles) sources. Nitrogen oxides and sulfur dioxide react in the atmosphere to produce low-pH precipitation or acid rain. There is epidemiological evidence of adverse health effects of these gases, such as linkage to various pulmonary diseases. High levels of sulfur dioxide also can cause plant damage, e.g., leaf necrosis.

III.C Nutrients

An excess of one or more of nitrogen and phosphorus nutrients in aquatic systems, and some terrestrial systems, can change the structure and functioning of associated ecological communities. Cultural eutrophication is the classic example of this process. Nitrogen species can contribute to the dysfunction of ecosystems receiving an excess of nutrients. They have other adverse effects if present at sufficiently high concentrations. Nitrate can enter water bodies from runoff or sewage discharges. High concentrations in drinking water can cause **methemoglobinemia** ("blue-baby syndrome," resulting from the reaction of nitrite to hemoglobin to convert it to methemoglobin, which is incapable of normal transport of oxygen) in newborn infants. (The nitrite is produced from nitrate in the baby's stomach.) Nitrosamines, potent carcinogens, can form from nitrogen compounds in drinking waters. Nitrite is very toxic to aquatic biota and can also cause methemoglobinemia in babies. Ammonia can cause toxicity to aquatic biota near sources such as sewage discharges. Ammonia toxicity is very pH dependent.

Vignette 2.1 Nutrient Intoxication—Ecological Consequences of Overenrichment in Aquatic Ecosystems

Hans W. Paerl

Institute of Marine Sciences, University of North Carolina at Chapel Hill, Morehead City, NC

The nutrients nitrogen (N) and phosphorus (P) fuel the fertility and biodiversity of freshwater, estuarine, and marine systems, enhancing the commercial and recreational value of these systems. However, excessive amounts of nutrients can lead to impairment or "nutrient intoxication." Nutrient overfertilization can greatly accelerate plant production, or primary production, at the base of the food web (Likens, 1972; Ryther and Dunstan, 1971). One can view this as too much of a good thing, where excess production is not effectively utilized by the food web, leading to the accumulation of excess organic matter. Nutrient-enhanced productivity, leading to an increase in organic-matter content is termed eutrophication (Nixon, 1995). The symptoms of eutrophication include nuisance algal blooms, producing severe oxygen depletion (hypoxia) and resultant fish kills (Paerl, 1988; Elmgren, 1989; Smetacek et al., 1991; Nixon, 1995).

Increased nutrient loading and eutrophication have been linked to human population growth and resultant urban, agricultural, and industrial expansion. Most aquatic ecosystems are sensitive to nutrient enrichment, which exacerbates problems associated with eutrophication (Vollenweider and Kerekes, 1982; Nixon, 1995). In nutrient-enriched water bodies having restricted circulation and exchange, and resultant long water residence times, nutrients tend to accumulate, enhancing eutrophication potentials. One example is North Carolina's Albermarle-Pamlico Sound, North America's largest lagoonal ecosystem (4350 km^2), which drains approximately half the state's freshwater runoff. This semi-enclosed system has a water residence time of approximately one year and is sensitive to even small amounts of nutrient enrichment (Copeland and Gray, 1991). In most aquatic ecosystems, nutrients are effectively cycled between the water column and sediments, enhancing their availability for plant growth. Collectively, nutrient input dynamics along with hydrological and nutrient cycling characteristics determine a water body's biological response and sensitivity to nutrient enrichment. From a water quality management perspective, anthropogenic nutrient inputs are most easily addressed, while geomorphology and hydrology are largely controlled by natural events (i.e., earthquakes, volcanoes, wet or dry periods, storms, and hurricanes) and are far less manageable.

Where Do Nutrients Come from and How Are They Linked to Eutrophication?

Nitrogen and phosphorus are the nutrients of greatest concern because they control eutrophication and their inputs reflect human activity (Ryther and Dunstan, 1971; Likens, 1972; Vollenweider and Kerekes, 1982; Nixon, 1995). These nutrients are delivered by (1) surface water discharge, i.e., surface runoff or wastewater discharge delivered via creeks and rivers, (2) subsurface discharge of groundwater, and (3) atmospheric deposition (rainfall or particle-associated dryfall).

In agricultural and rural watersheds, such as those in the U.S. Midwest farm belt, Mid-Atlantic, Southeastern Atlantic, and Gulf of Mexico regions, nutrient inputs are dominated by diffuse, nonpoint sources (Castro et al., 2000). For example, in Coastal North Carolina, at least 70% of N and P input to the rivers and estuaries is from nonpoint sources, including surface runoff, rainfall, and groundwater (Dodd et al., 1993). Accounting for the rest are point sources, including wastewater treatment effluent and industrial and municipal discharges. In the Neuse River Basin, a key tributary of the Pamlico Sound, approximately 50% of the N input originates from nonpoint source surface runoff and groundwater. Nearly 40% can be attributed to atmospheric deposition, with the remaining 10% coming from point sources (Figure 2.3)

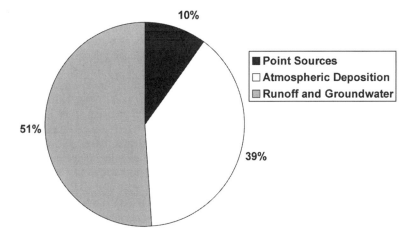

Figure 2.3 Nitrogen inputs to the Neuse River, NC. (Data are based on Dodd, R.C. et al., Watershed Planning in the Albemarle-Pamlico Estuarine System, Report No. 93-01, NC Dept. of Environment, Health and Natural Resources, Research Triangle Institute, Research Triangle Park, NC, 1993; and on Whitall, D.R. and H.W. Paerl, *J. Environ. Qual.,* 30, 508–1515, 2001.)

Figure 2.4   Examples of nuisance algal blooms on systems receiving excessive nutrient loading. Left photograph: A bloom of the toxic blue-green algae (cyanobacteria) *Nodularia* on the Baltic Sea (Gulf of Finland). Right photograph: A bloom of the freshwater cyanobacteria *Microcystis* on the Neuse River, NC.

(Dodd et al., 1993; Paerl and Fogel, 1994; Whitall and Paerl, 2001). In contrast, N and P loadings in urban watersheds (e.g., Narragansett Bay and Long Island Sound) are dominated (>50%) by point sources. Watersheds encompassing both urban centers and intensive agriculture (e.g., Chesapeake Bay and San Fransisco Bay regions) exhibit a more even distribution of these source types (Boynton et al., 1995; Nixon, 1995). Seasonal changes in delivery mechanisms and relative contributions from sources arise from variability in discharge rates and rainfall.

Phosphorus is the major nutrient of concern in freshwater (Vollenweider and Kerekes, 1982), while N is paramount in estuarine and coastal waters (Nixon, 1995). The rates of either P or N inputs control plant growth in respective systems because these nutrients are in shortest supply. Therefore, high N or P loading rates usually promote high rates of plant production. If plant production exceeds consumption by animals, plant matter accumulates in the receiving water body. This imbalance can lead to massive blooms of free-floating microscopic algal communities that result in green, yellow, or red water discoloration (Figure 2.4). Blooms are considered undesirable because they are malodorous, release toxins that can adversely affect animal and human health, foul waters and shorelines, and deplete oxygen from the ecosystem (Paerl, 1988; 1997). This can lead to large-scale decreases in biodiversity, including loss of bottom fauna and flora (Diaz and Rosenberg, 1995; Pihl et al., 1991; Lenihan and Peterson, 1998).

Sudden changes in temperature, nutrient depletion, and light-availability constraints can cause blooms to die (Paerl, 1988; Paerl et al., 2002). In poorly flushed systems, dying blooms sink to the bottom, where bacterial decomposition consumes large amounts of oxygen. If decaying blooms are large enough and bottom water oxygen cannot be replenished by downward mixing of oxygenated surface waters, bottom waters and sediments can exhibit dangerously (to both animal and plant life) low oxygen levels (hypoxia) or a complete loss of oxygen (anoxia) (Officer et al., 1984; Rabalais et al., 1994; 2001; Paerl et al., 1998). Hypoxia and anoxia events have been linked to fish kills in a variety of aquatic habitats (Pihl et al., 1991; Winn and Knott, 1992; Paerl et al., 1998). This is particularly problematic in brackish estuaries, where low-density freshwater riverine discharges can be layered over a denser, deeper saltwater layer (i.e., salt wedge), impeding effective vertical mixing and thus sufficient oxygenation of bottom waters. If persistent for weeks to months, these conditions greatly enhance the likelihood of hypoxia or anoxia (Paerl et al., 1998).

In highly productive reservoirs, lakes, and estuaries, hypoxia and anoxia are natural occurrences. Historical accounts of "bad water," or anoxic water with a rotten-egg sulfide smell, are common for many estuaries. This situation can, however, be aggravated by excessive N loading that leads to enhanced algal growth, bloom formation, and increased potential for hypoxia/anoxia development and persistence. Accelerated nutrient loading has exacerbated this undesirable condition. A worst-case scenario for systems like the seasonally eutrophic Neuse River Estuary in North Carolina is unusually large N-enriched freshwater discharge into the broad, slow-flowing lower estuary that coincides with optimal algal growth and bloom periods (spring-summer). These conditions promote the maximum potentials for hypoxia/anoxia and fish kills (Paerl et al., 1998).

Recognizing the vital role that excessive N loading plays in causing these undesirable conditions, North Carolina has mandated a 30% reduction in N loading to the Neuse Estuary. Similar N reductions are under consideration for other North Carolina estuaries impacted by accelerating anthropogenic N loads. A 40% N reduction has been targeted for Chesapeake Bay for similar reasons (D'Elia et al., 1986; Boesch, 2001).

What Are the Nutrient Management Options and Actions?

Clearly, anthropogenic point and nonpoint sources are large targets for nutrient reduction. The relative proportions among these nutrient sources, and hence priorities for cutbacks, vary according to human activities, locations and distributions of population centers, and the route of nutrient discharge (i.e., surface, subsurface, and atmospheric). In point-source-dominated watersheds, the emphasis is placed on improved treatment of wastewater from sewage and industrial effluent and the removal of both N and P. In nonpoint-source-dominated coastal watersheds, the current focus of N reduction strategies is surface runoff, especially that originating from agricultural operations. These include (1) a range of best-management practices, including prudent and timely applications of fertilizers and soil conservation; (2) establishment of riparian vegetative buffer zones; and (3) use of wetlands to enhance "stripping" of runoff N by incorporation into terrestrial vegetation and "venting" of N as harmless nitrogen gas by microbial denitrification (Lowrance et al., 1984; Gilliam et al., 1997). Recent data for the Neuse River Basin indicates that we are beginning to reap the benefits of these strategies in terms of reduced N loading rates.

Until recently, atmospheric deposition has been left out of most budgets and models as major N sources. However, this has changed substantially over the past decade, as local and regional studies have shown this important source to be large, increasing, and widespread (Valigura et al., 2000). Atmospheric emissions originate from two dominant human sources, fossil fuel combustion (e.g., nitrogen oxide emissions from power plants and automobiles) and agriculture (e.g., ammonia gas released from animal wastes and fertilizers). Controls on these emissions have lagged behind treatments for land-based runoff because of expense (e.g., catalytic converters on automobiles and scrubbers on smokestacks), a lack of effective technology (i.e., controlling N emissions from animal operations and stored wastes), and regional economic and political complications (i.e., emission sources frequently come from outside of river basins, and cross state boundaries). Management of atmospheric N emissions is in a state of infancy, but it is under increasing state and federal scrutiny and consideration for legislation.

Nutrient input from point sources is an issue that is under intense local, state, and federal management scrutiny. Recent legislative action has led to the implementation of strict N and P discharge limits from wastewater treatment plants, and those plants failing to meet these standards are under considerable pressure to upgrade. In addition, a phosphate detergent ban was enacted in the mid-1980s in North America, Europe, and parts of Australia. This has led to marked decreases in P loading in many watersheds. Current strategies aimed at reducing nonpoint N discharge—including riparian buffers, constructed wetland, and soil conservation—also retard the movement of P to nutrient-sensitive waters (Gilliam et al., 1997). Therefore, the primary aim of pursuing N input constraints will yield significant parallel reductions in P.

Much work is needed to further identify, characterize, and manage nutrient inputs that control freshwater, estuarine, and coastal water quality and fisheries resources. Research and monitoring are providing the key information essential for formulating and implementing long-term nutrient management strategies aimed at protecting and preserving the high standards of water quality and resourcefulness that we value. As we enter the next millennium, fostering a process-based understanding of nutrient–water quality interactions and their implications for effective management are among our primary social, economic, and political responsibilities, challenges, and investments for a functional and sustainable global ecosystem.

III.D Organic Compounds

The organic contaminants include those used intentionally as poisons (e.g., insecticides and herbicides) and those that are products, by-products, or wastes from industrial processes. Some are naturally occurring compounds, and others are synthetic molecules unlike anything found in nature. They span a wide range of physicochemical properties and environmental persistences.

III.D.1 Polycyclic Aromatic Hydrocarbons

The polycyclic aromatic hydrocarbons (PAH) are compounds composed of two or more fused aromatic rings. The PAHs are generated by the incomplete combustion of organic matter and can span a wide range of structures with two to more than six aromatic rings. They can vary widely in their physical properties due to the great range in their sizes and molecular weights.

Naphthalene	Anthracene	Phenanthrene	Pyrene

Chrysene   Benzo[j]fluoranthene   Benzo[a]pyrene

Perylene   Coronene

Higher-molecular-weight PAHs are considered hydrophobic compounds that can accumulate in sediments and organisms. Although PAHs can be metabolized by vertebrate organisms with active cytochrome P-450 systems (see Chapter 6), they can accumulate in invertebrate organisms that have reduced metabolic activity. The lower-molecular-weight compounds have acutely toxic properties, and some of the higher-weight PAHs (i.e., benzo[a]pyrene) are known carcinogens or important precursors of carcinogenic metabolites. Production of PAH can occur as part of a natural process such as forest burning after a lightning strike or the slow breakdown of natural organic matter under the elevated temperature and pressure of geologic processes involved in the formation of petroleum (**catagenesis**). Generation of PAH can also take place as a direct result of human activities, and they become important contaminants when this occurs. Regardless of the mechanism of formation, the complex nature of the source material and the varying conditions under which PAH are formed lead to the production of many related molecules during PAH genesis. When PAHs are detected in the environment, a large suite of related compounds are usually present.

Emissions of PAH as combustion by-products can be an important source for PAH to the atmosphere. These can then be transferred to terrestrial or aquatic environments via wet or dry deposition (Dickhut and Gustafson, 1995). Mixtures of PAH derived from the direct combustion process are characterized by unsubstituted ring structures and are termed **pyrogenic PAH**. The PAH mixtures derived from the slow degradation of natural organic matter tend to contain higher precentages of molecules with alkyl substitutions to the basic ring structure. This characterization can be used to distinguish PAH derived from oil spills (highly alkylated) from PAH derived from pyrogenic sources in environmental samples (Figure 2.5).

1,3-Dimethyl-naphthalene   1,3,5-Trimethyl-naphthalene

2,6-Dimethyl-phenanthrene   2,3,7-Trimethyl-phenanthrene

Another potential source for PAH contamination to the environment is through the misuse of creosote. Creosote is a dense liquid formed by the distillation of coal tar and has been used for more than a century to impregnate wood products to make them resistant to rot and attack by boring pests. Creosote can contain up to 90% PAH by weight (Nestler, 1974), and the toxic nature of the pyrogenic PAH and related compounds within creosote contribute to its preservative properties. Long-term operation of wood-treatment facilities and the associated spillage of large amounts of creosote have produced extremely elevated levels of PAH in the environment. For example, PAH concentrations exceeding thousands of ppm have been measured in the Elizabeth River in Virginia near wood-treatment facilities (Huggett et al., 1997).

III.D.2 Polychlorinated Biphenyls

The polychlorinated biphenyls (PCB) are a group of related chlorinated hydrocarbons formed by the chlorination of a biphenyl base molecule with molecular chlorine. The PCB can be formed with one to ten chlorine atoms substituted around the biphenyl ring structure, making possible 209 different related compounds or **congeners**.

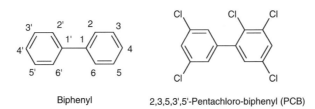

Biphenyl   2,3,5,3',5'-Pentachloro-biphenyl (PCB)

The formation process is not highly selective, so complex mixtures of congeners are produced during synthesis. Commercial-grade mixtures of PCB are identified based on the approximate percentage of chlorine content on a mass basis. Monsanto was a principal manufacturer of PCB in the United States and sold their PCB mixtures under the trade name of Aroclor, with number designations identifying the various mixtures (e.g., 1221, 1232, 1242, 1248, and 1260). The designation Aroclor 1260, for example, would indicate a PCB mixture (1260) with 60% chlorine content (1260) on a mass basis. Similar PCB mixtures were marketed in Europe and Japan under the trade names Clophen and Kanechlors. Like PAH, PCBs are found in the environment as complex mixtures, and the distribution pattern of individual congeners in these mixtures can be used to help identify possible source materials (Schultz et al., 1989). Unlike the PAHs, which are produced by natural processes, PCBs are synthetic compounds that are not readily metabolized or degraded.

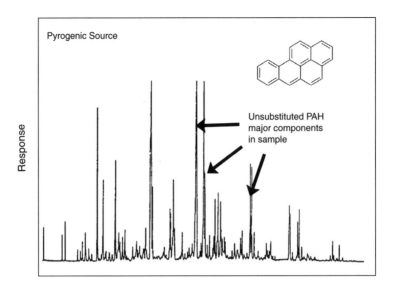

Retention Time

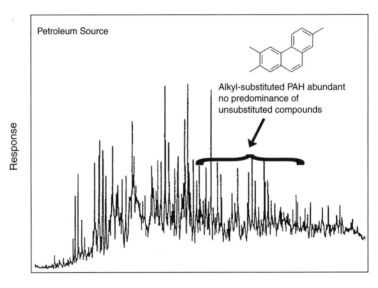

Retention Time

Figure 2.5   Chromatograms of PAH in sediment samples can be used to differentiate the sources of the PAH contamination. The PAH from pyrogenic sources (i.e., creosote or combustion products) show a predominance of unsubstituted molecules, while those from petroleum-based sources contain a more complex mixture of compounds, including a high percentage of alkyl-substituted PAH.

The high molecular weight, low vapor pressure, and high stability of PCBs led to their use as lubricants, heat conductors in electrical transformers, and plasticizers. Their toxic effects to humans and wildlife are a major concern, with some biota such as mink being very sensitive. PCBs degrade very slowly and are soluble in fats and oils. Consequently, they accumulate in biota to high concentrations.

The closely related compounds, the **polychlorinated terphenyls** (PCT), have also been used as heat stabilizers in cooling oils. With one more phenyl ring, the possible number of congeners increases to over 2000.

2,5,2',2'',5''-Pentamethyl-[1,1';4',1'']terphenyl

Like PCBs, PCTs have found their way into the environment and accumulate in sediments and biota due to their high lipid solubility (Gallagher et al., 1993). The congeners with high degrees of chlorination appear less likely to accumulate in biota, probably the result of limited transport across biological membranes (Hale et al., 1990; see Chapter 4 for more details).

III.D.3 Polybrominated Biphenyls

Polybrominated biphenyls (PBB) are a group of brominated hydrocarbons that are directly analogous to the structure of PCB but have bromine atoms taking the place of chlorine around the biphenyl backbone.

2,3,5,3',5'-Pentabromo-biphenyl (PBB)

They have similar physical properties to PCB and, therefore, have a similar environmental fate. Mixtures of PBB were manufactured and sold as fire retardants that were added to hard plastics and incorporated into televisions and other electronic appliances. Although PBBs are not typically found in significant concentrations in environmental samples today, they are infamous for one of the greatest environmental contamination accidents in history. In the early 1970s, an accidental switching of bags containing PBB mixture with those containing a livestock feed supplement at a chemical distribution center introduced PBB into the agricultural food chain in Michigan. The resultant contamination produced toxic effects in livestock and humans and led to widespread environmental distribution of PBB in the agricultural community. It has been estimated that the losses to livestock and poultry alone were $75 million to over $100 million (Carter, 1976). An interesting account of the tragedy and the detective work used to identify the chain of events leading to the mix-up can be found in Carter (1976).

III.D.4 Chlorofluorocarbons

Chloroflourocarbons (CFC) are used in refrigeration and in the production of foams such as Styrofoam. These gaseous contaminants contribute to ozone depletion in the stratosphere.

III.D.5 Organochlorine Alkenes

Organochlorine alkenes are compounds that are used in large amounts as solvents and degreasers. Examples include tetrachloroethene, a dry-cleaning solvent that sometimes contaminates drinking water, and trichloroethene, a degreaser that is denser than water, very insoluble in water, and frequently associated with groundwater contamination.

III.D.6   Chlorinated Phenols

Chlorinated phenols include compounds that are purposely synthesized for commercial uses and compounds that are formed as by-products during other industrial processes. Commercial chlorinated phenols are used as wood preservatives (e.g., trichlorophenol [TCP] and pentachlorophenol [PCP]) and fungicides (e.g., PCP) and as precursors for the synthesis of other chlorinated hydrocarbons such as 2,4,5-TCP.

2,4,5-Trichlorophenol (TCP)   2,3,4,5,6-Pentachlorophenol (PCP)

Technical-grade chlorophenols are rarely pure and usually contain a variety of related compounds formed during synthesis. Mixtures of PCP can have parts per million concentrations of chlorodibenzo-p-dioxins, chlorodibenzofurans, or chlorodiphenylethers (Verschueren, 1983). They have been released into the environment as industrial wastes and have contaminated atmospheric, aquatic, and terrestrial systems. Chlorophenols are moderately persistent in the environment, with decomposition rates for PCP in soils and by pseudomonas species in wastewater treatment plants on the order of days to months (Verschueren, 1983). Various chlorinated phenols have been identified in the effluents of paper mills processing and bleaching kraft wood pulp. Formed from degrading natural organic matter and the chlorine used in the bleaching process, they can accumulate in sediments downstream of discharges. They include chlorinated catechols (o-dihydroxybenzenes), chlorinated guaiacols (o-methoxyphenols), and chlorinated syringols (dimethoxyphenols) (Gundersen and MacIntyre, 1996). The hydroxyl group of chlorophenols will dissociate under specific pH conditions, greatly influencing their sorption and bioaccumulation (Lee et al., 1990). The acid dissociation constants for some of the chloroguaiacols are in the 7 to 9 range, so pH changes typically encountered in aquatic environments should be considered in predicting the environmental fate of these compounds (Gundersen and MacIntyre, 1996).

Chlorinated   Chlorinated   Chlorinated
catechols   guaiacols   syringols

III.D.7   Chlorination Products

Chlorine compounds or gas used to disinfect drinking waters can produce chlorinated organic compounds with potential toxic or carcinogenic effects. For example, chlorination produces trihalomethanes such as the carcinogen, chloroform ($CHCl_3$).

III.D.8   Organochlorine Pesticides

Organochlorine pesticides are synthetic compounds that have some amount of chlorine substituted for hydrogen on a hydrocarbon backbone. They are also known as chlorinated hydrocarbons or chlorohydrocarbons, but also included in this group are a few compounds that have oxygen incorporated into their structure (i.e., methoxychlor, dieldrin, kepone). Some organochlorine pesticides that might be found in environmental samples today are aldrin, chlordane, dieldrin, lindane, methoxychlor, and toxaphene.

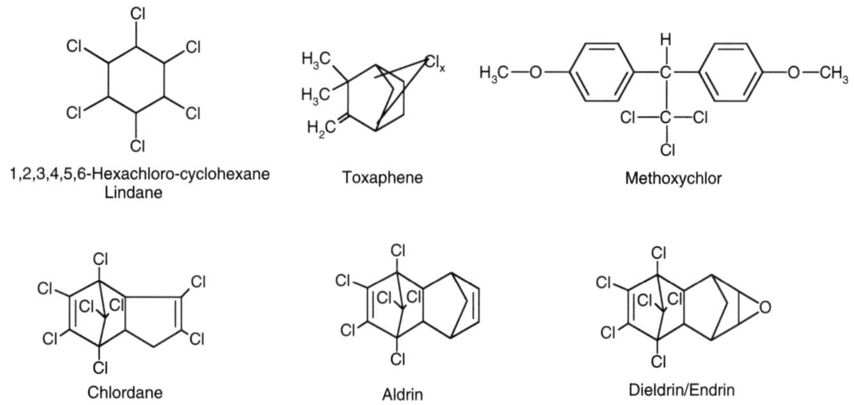

1,2,3,4,5,6-Hexachloro-cyclohexane
Lindane

Toxaphene

Methoxychlor

Chlordane

Aldrin

Dieldrin/Endrin

These synthetic pesticides do not resemble most naturally occurring organic compounds in their structure, so they tend to degrade slowly in the environment. Most have high molecular weights and nonpolar structures. Consequently, they have low water solubilities and are very soluble in lipids, such as those in organisms. This results in bioaccumulation and possible biomagnification. One of the first widely used organochlorine pesticides was dichloro-diphenyl-trichloroethane (DDT). DDT was heralded for its toxicity to a wide range of insect pests, low mammalian toxicity, and persistence in the environment.

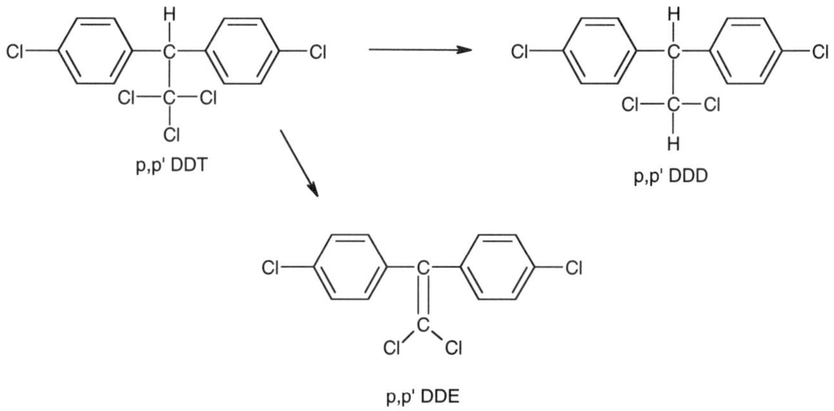

p,p' DDT

p,p' DDD

p,p' DDE

It was used extensively during World War II to control lice, flies, fleas, and mosquitos to prevent malaria, typhus, and other serious diseases among servicemen and civilians. This enhanced its reputation, so DDT became the pesticide of choice for agricultural and commercial use after the war. Long-term use of DDT led to insect resistance that necessitated higher application rates. Adverse environmental side effects were eventually identified. The EPA banned the application of DDT to crops at the end of 1972 based on these concerns and the availability of more-effective and less-harmful alternatives. DDT undergoes degradation in the environment to DDD and DDE, with a half-life on the order of years. Environmental samples today usually contain predominately the DDD and DDE isomers, but occasionally DDT is found as the primary contaminant in some samples, suggesting that isolated DDT use continues 30 years after the ban.

Another organochlorine pesticide, kepone (deachlorooctahydro-1,3,4,-methano-2H-cyclobuta [cd]-pentalen-2-one), was spilled into a tributary of the Chesapeake Bay in the 1970s and caused one of the most significant economical impacts ever attributed to pesticide manufacturing.

Kepone

Unmindful practices at a small production facility in Hopewell, VA, led to contamination of employees and to kepone release into the air, soil, and wastewater near the facility. Storm drains carried the kepone to the nearby James River, where it contaminated sediments and accumulated in the resident biota. Concern over possible health effects from eating contaminated seafood led to the FDA setting action levels for kepone in seafood and the closing of the James River to commercial fishing. Monitoring of the environment showed that kepone readily sorbed to bottom sediments, degraded slowly, and had accumulated in shellfish, fish, birds, and small mammals (Huggett and Bender, 1980). Even though kepone was shown to degrade very slowly in sediments, the natural depositional processes in the James River estuary have buried the most severely contaminated sediments, thus reducing the bioavailability of the contaminant. This reduced availability is evident in the decreasing trend in kepone concentrations measured in striped bass fillets collected from the river. Monitoring has continued for over 25 years since kepone was discovered in the James River, and while concentrations have decreased well below action levels, kepone is still detectable in almost all striped bass tissue samples.

III.D.9 Polychlorinated Dibenzodioxins (PCDD) and Dibenzofurans (PCDF)

Polychlorinated dibenzodioxins (PCDD) and dibenzofurans (PCDF) are compounds that are quite different from most of the organic contaminants on our list, as they are not intentionally manufactured and have no commercial value. Like the PCBs, they are a group of related compounds with varying degrees of chlorination around a multiple ring system. There are 75 possible PCDD congeners and 135 possible PCDF congeners.

Dibenzo-*p*-dioxin   2,3,7,8-Tetrachloro-dibenzo-*p*-dioxin   2,3,7,8-Tetrachloro-dibenzofuran

Toxicity of PCDD and PCDF vary greatly with target organism, raising some controversy about their significance as environmental contaminants. The 2,3,7,8,-tetrachlorinated congeners are the most toxic form and are often the compounds that are considered when the more general terms *dioxin* or *dibenzofuran* are used. They are formed as by-products of chemical synthesis or during the combustion of certain organic materials. Dioxins can be formed as by-products during the production of chlorophenols and the herbicide 2,4,5,-T.

2,4,5-Trichlorophenol (TCP)   2,3,7,8-Tetrachloro-dibenzo-p-dioxin

Dioxins are also formed as combustion products and during the bleaching process at kraft pulp mills. Dioxins partition to organic matter and can be transported with particles in aquatic environments and accumulate in sediments. They are persistent, thermally stable compounds that do not readily biodegrade.

III.D.10   Organophosphate Insecticides

Organophosphate insecticides are compounds that evolved from early research on the development of nerve gases for warfare. Derivatives of orthophosphoric acid, they all contain phosphorus and ester bonds or can include sulfur in the place of oxygen. Examples include methyl parathion, diazinon, and chlorpyrifos.

Orthophosphoric acid   Methyl parathion

Diazinon   Chlorpyrifos

The mode of action for these compounds is the inhibition of acetylcholine esterase activity, which causes death by hindering nerve function and ultimately muscle response. This enzymatic system is common to all organisms and is the reason early organophosphate compounds (i.e., parathion) were nonspecific in their toxicity. They were powerful insecticides that also posed risks to humans, especially those handling the pesticide during application. The danger of using compounds with high mammalian toxicity led to the development of alternate organophosphate structures that are preferentially metabolized by vertebrate organisms. Invertebrates are not capable of detoxifying the compounds metabolically, so the compounds retain insecticidal activity. Organophosphates generally have much higher water solubility (>1 mg l^{-1}) than organochlorine compounds and are less likely to accumulate in sediments. Degradation occurs through base-catalyzed hydrolysis, so they are also less persistent in soils or sediment. Half-lives range from days to weeks.

III.D.11   Carbamate Insecticides

Carbamate insecticides are synthetic derivatives of carbamic acid. Because of their chemical structure, carbamates are more polar and much more water soluble (100 to more than 1000 mg l^{-1}) than chlorinated hydrocarbons.

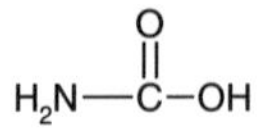

Carbamic acid

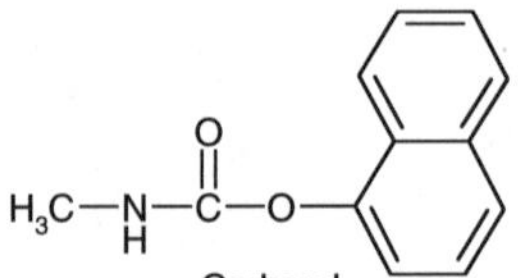

Carbaryl

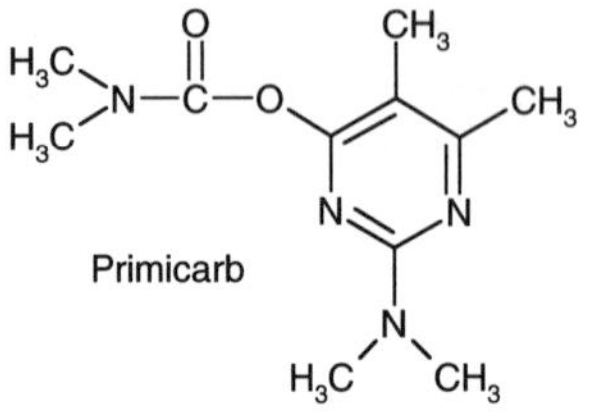

Primicarb

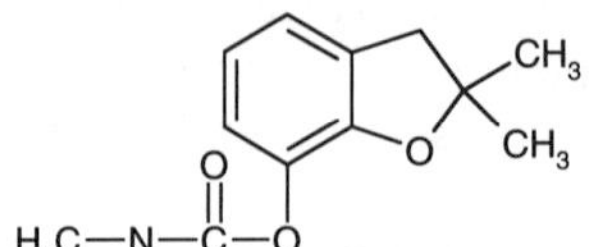

Carbofuran

Like organophosphate insecticides, carbamate compounds degrade rapidly in the environment and cause neural dysfunction by inhibiting acetylcholine esterase, an enzyme essential to neuron function. The carbamate functional group can easily hydrolyze in water, so half-lives are typically on the order of weeks in terrestrial and aquatic environments. This hydrolysis reaction also readily occurs metabolically, which will reverse the acetylcholine esterase inhibition activity of the molecule and is the reason that carbamates are generally less toxic to mammals than are organophosphates.

III.D.12 Pyrethroid Insecticides

Pyrethroid insecticides are synthetic analogs of the naturally produced pyrethrins. Pyrethrins are produced in the flowers of chrysanthemum or "pyrethrum" plants and have long been recognized for their insecticidal properties. Because they are natural products, they usually undergo rapid biodegradation in the environment and are metabolized by mammals. The pyrethroid insecticides were synthesized to mimic the basic structure and function of the pyrethrins.

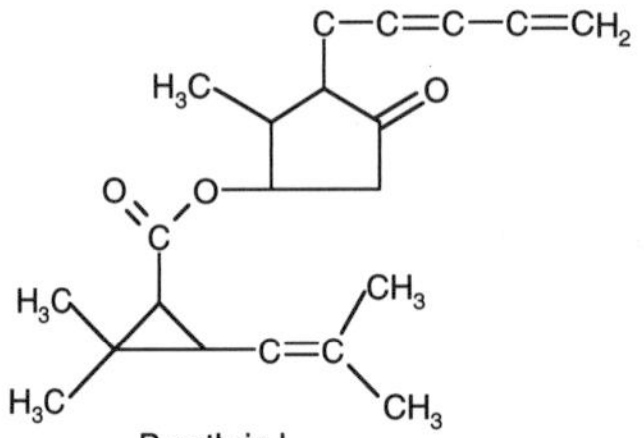

Pyrethrin I

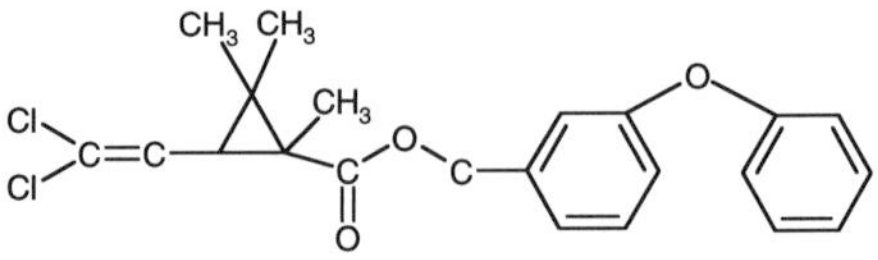

Permethrin

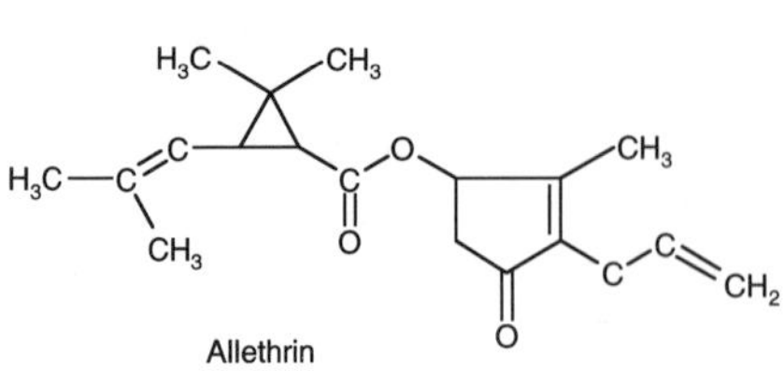

Allethrin

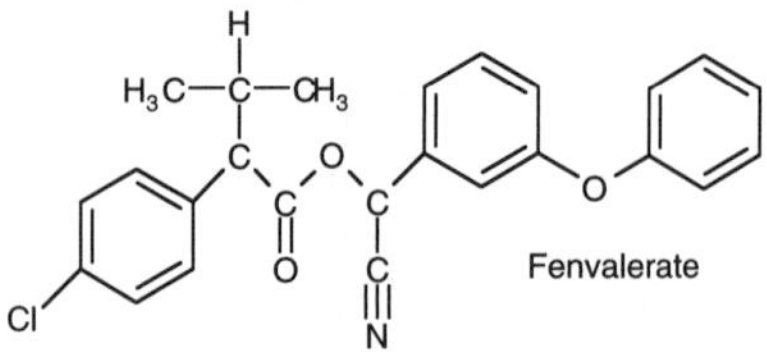

Fenvalerate

Pyrethroids are more stable than their natural predecessors but are still considered relatively short-lived insecticides and, consequently, most problems are associated with acute exposure. They have low water solubility (fenvalerate = 0.085 mg l^{-1}) and will sorb to particulate material, which could influence transport and long-term biodegradation in aquatic environments. Pyrethroids have low mammalian toxicity but can cause toxic effects to fish and invertebrates in the low μg l^{-1} range.

In 1999, spraying of pyrethroid pesticides near Long Island Sound to control mosquitoes and the spread of West Nile Virus raised the concern for potential effects in the local lobster population. Other factors such as disease outbreaks have also been implicated for the higher-than-normal mortalities in the lobster population during this period. This has spurred new research to try to establish if there was a causal relationship between pyrethroid exposure and effects in these commercially important invertebrates.

III.D.13 Aromatic Herbicides

Aromatic herbicides are a diverse grouping of compounds that are applied to farmland and residential communities to control unwanted plant growth. They can enter the aquatic environment through runoff or atmospheric deposition. They are also of concern as potential contaminants to groundwater supplies. Some of these herbicides are made up of nitrogen heterocyclic-based molecules, which give them a polar structure. This structure contributes to a high water solubility and the potential to contaminate aquifers. Paraquat and diquat are examples of herbicides made up from two pyridine rings (bipyridine herbicides).

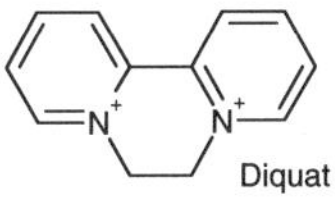

Diquat

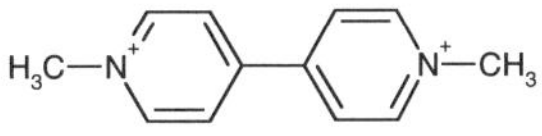

Paraquat

They form positively charged species in solution and can readily partition to negatively charged particles such as clays in soils or sediments. This can reduce their effectiveness and will influence their transport. They are fast-acting herbicides and must be applied directly to the plant because of their tendency to partition to soils.

The triazines are another class of nitrogen heterocyclic compounds that have widespread herbicidal use. They have aromatic ring structures with three nitrogen atoms.

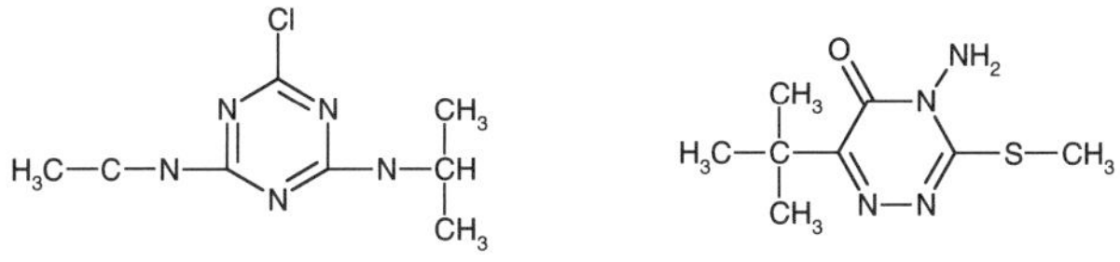

Atrazine

Metrabuzin

The most commonly used triazine compound is Atrazine. Atrazine is metabolized by monocot plants more effectively than by dicots, so it is effective as a preemergence herbicide for corn, sugarcane, pineapple, and macadamia nut. It has a water solubility around 70 mg l^{-1} and can become an important aquatic contaminant if runoff from agricultural fields is not controlled. Atrazine degrades via hydrolysis or microbial dealkylation and has a half-life on the order of weeks to months in soils, depending on conditions (Verschueren, 1983). Degradation in groundwater can be much slower. Another triazine compound, Irgarol 1051, has recently been used as an additive to marine antifoulant paints.

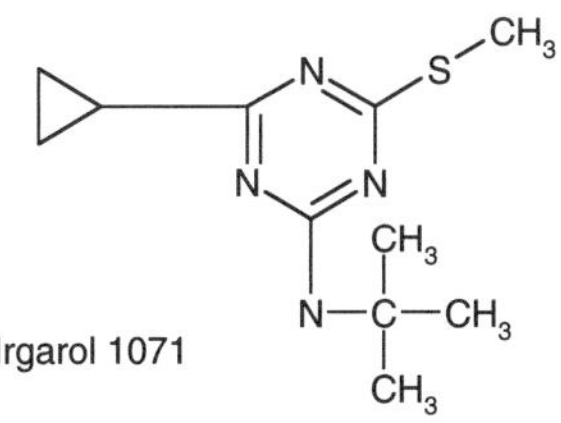

Irgarol 1071

Antifoulant biocides such as tributyltin (TBT) (see organometallics, discussed later in this chapter) are very effective at preventing marine invertebrates from growing on ship hulls but do not prevent the growth of marine algae. Irgarol 1051 has been added to paints as an algaecide, and it has been found in coastal waters and biota (Sargent et al., 2000).

Phenoxy herbicides are also important in the control of dicots and function by disrupting plant growth regulation. They are generally very water soluble (2,4-D, 890 mg l^{-1}), have low aquatic toxicity (>100 mg l^{-1}), and degrade on the order of days to weeks in soils. They include compounds like 2,4-D and 2,4,5-T, which are used for weed control and are infamous as major components of the military defoliant, Agent Orange.

(2,4-Dichloro-phenoxy)-acetic acid
"2,4-D"

(2,4,5-Trichloro-phenoxy)-acetic acid
"2,4,5-T"

Use of Agent Orange defoliant during the Vietnam War has been linked to human health effects in military personnel exposed to the herbicide. It was discovered that Agent Orange could have high concentrations of the highly toxic compound 2,3,7,8-TCDD (dioxin). The phenoxy acetic acid herbicides 2,4-D and 2,4,5-T are synthesized from their chlorophenol analogs and chloracetic acid. If reaction conditions are not monitored carefully, dioxins can be produced instead of the targeted product. Phenoxy herbicides are now monitored to ensure dioxin levels are below allowed thresholds.

Some representative nitrogen-substituted herbicides are propanil, alachlor, and trifluralin. These chemicals are widely used in agricultural applications to control weeds and are sold under trade names such as Lasso® (45.1% alachlor) and Treflan® (43% trifluralin).

Propanil

Alachlor

Trifluralin

It is not uncommon for farmers today to use a long list of commercial agricultural products to protect their crops from infestations and to increase yield. These are rarely pure compounds, typically containing mixtures and added surfactants to enhance product performance. A good reference for identifying the composition of commercial pesticide and herbicide chemical formulations is the *Crop Protection Reference*. It is published annually by Chemical and Pharmaceutical Press, Inc., and lists the specific chemical make-up and application instructions for each product. This information is also available on the Internet (www.greenbook.net). It can provide good starting information when assessing the possible chemicals of concern in agricultural areas, and application information can give clues to possible modes of transport.

III.E   Organometallic Compounds

These molecules are interesting because they are metal atoms covalently bonded to organic moieties that greatly influence their fate and toxicity. Often, the addition of organic groups to the metal increases its ability to partition to lipids and can therefore increase its bioavailability and toxicity. Some organometallics are purposely synthesized for industrial use (e.g., TBT), and others are the products of microbial transformations that occur in the environment (e.g., methyl mercury). An element like tin, which is relatively nontoxic, can be transformed in the laboratory to tributyltin, one of the most toxic compounds released into the environment.

III.E.1   Organolead

Organic compounds of lead such as tetraalkyllead have been synthesized and used extensively as antiknock additives to gasoline.

Tetraethyl Lead

As a result, they became important air pollutants in areas with concentrated vehicular traffic. At high concentrations, these compounds (a second compound, trialkyllead, is produced via liver metabolism from tetraalkyllead) can cause neurological dysfunction and other problems. Regulations restricting the addition of organolead to gasoline have greatly reduced the flux of tetraalkyl lead into the environment.

III.E.2   Organomercury

Organomercury compounds are produced as a consequence of natural processes as well as by direct synthesis. Synthesized organomecury compounds are manufactured as biocides in products such as seed coatings. Elemental mercury is used in batteries and in industrial processes. Mercury is usually released to the environment in the elemental form and is methylated in sediments by bacterial metabolism. This is a process described as **biomethylation**. Once methylated, the organomercury readily sorbs to suspended particulate material and sediments. It also can bioaccumulate in muscles and tissues and has been shown to increase in concentration with age of the organism or to increase with trophic level. Methyl mercury is found in many species consumed by humans, leading to concern because these compounds cause neurological damage (see Chapter 1).

III.E.3   Organotins

Organotins are organometallic compounds that have been used as catalysts, polymer stabilizers, insecticides, fungicides, bactericides, wood preservatives, and antifouling agents. Most are synthesized for industrial use, but methyltin compounds have been produced in the environment through biomethylation. Organotins such as trimethyltin (TMT) and triethyltin (TET) are neurotoxicants. Dibutyltin is added to polyvinyl chloride (PVC) plastics as a stabilizer. Tributyltin (TBT) is a general-purpose biocide that has been used as a mildewicide, a fungicide, and as an additive to antifoulant paints to prevent the growth of barnacles and other fouling organisms.

Tri(n)butyltin chloride [TBT]

Unfortunately, TBT is toxic to nontarget invertebrate organisms in very low concentrations (ng l^{-1}). Vertebrate organisms are capable of metabolizing TBT and do not show the extreme toxicity evident in some invertebrates, especially crustaceans and mollusks. TBT can cause extensive damage to molluscan populations at very low concentrations, resulting in shell abnormalities in oysters and modification of sexual characteristics of snails.

Tributyltin degrades to dibutyltin (DBT), monobutyltin (MBT), and ultimately inorganic tin, detoxifying the molecule more with each subsequent step. It has a half-life in the water column on the order of days to weeks, but it is much more stable in sediment, with a half-life on the order of several months. Tributyltin will sorb to sediments and has partition coefficients in the range of 10^2 to 10^4. In anaerobic sediment, TBT is very stable, and degradation rates are estimated to be on the order of years. Concern for adverse effects to aquatic ecosystems from the use of TBT antifoulant paints has resulted in regulations restricting, but not totally banning, their use in several countries. This has reduced inputs in marina areas, but environmental concentrations can still exceed those shown to cause adverse effects in some organisms. The International Maritime Organization (IMO) has recently (October 8, 2001) passed a convention calling for the ban of TBT antifouling paints and their complete removal from, or encapsulation on, ships by 2008. A good reference for the reader interested in learning more about organotin fate and effects is Champ and Seligman (1996).

III.F Emerging Contaminants of Concern

Each year, new contaminants are listed in the literature or make the news as potential suspects for adverse environmental effects. These are often compounds designed to take the place of banned or outdated chemicals, or they can be chemicals that have been in use for years but have recently been discovered as environmental contaminants accumulating in portions of the ecosystem, where the potential for harm is increased (e.g., polybrominated diphenyl ethers, PBDE). Often they become apparent merely due to changes in analytical methods that had previously targeted other specific compounds. The list of emerging contaminants will obviously change as years pass and analytical methods evolve. Another possible source of such contaminants is chemicals that are altered in the environment through physical or biologically mediated degradation to more toxic or bioavailable forms (e.g., alkylphenols, perfluorooctanyl sulfanates [PFO]). This last group is the most difficult to predict and identify, as the permutations for degradation products of the myriad contaminants that are released to the environment are astronomical. Regulatory actions rarely target degradation products, and analytical methods often follow this logic, Thus little data are available on the fate and effects of degradation products in the environment.

III.F.1 Polybrominated Diphenyl Ether (PBDE or BDE)

Polybrominated diphenyl ethers (PBDE or BDE) are similar in basic structure to PCB and PBB but with an oxygen atom separating the two phenyl rings to produce an ether configuration.

2,2',4,4'-Tetrabrominated-diphenylether (BDE-47)

Like PCB and PBB, PBDE are formed as complex mixtures, with different formulations identified by the degree of halogenation. Used as flame-retardants, they are added to products like urethane foams in high concentrations and, curiously, have been studied very little as environmental contaminants until recently. They have been shown to accumulate in fish and sediments, but little is known about their long-term fate or effects in the environment.

III.F.2   Alkylphenols

Alkylphenols are good examples of compounds that are derived from the degradation of common starting products, surfactants. They are largely the by-products of the alkylphenol ethoxylate (APE) biodegradation that occurs in wastewater treatment plants. Nonylphenol ethoxylates are the most widely used surfactants, and they degrade both aerobically and anaerobically to nonylphenols (Bjorn et al., 1997).

Nonylphenol

Alkylphenols have low water solubility (low mg l^{-1}) and readily sorb to suspended solids or sediment. Thus they can contaminate sediments downstream from sewage treatment plant outfalls. Land-based disposal of biosolids from wastewater treatment plants provides a means for terrestrial contamination by alkylphenols, which are of environmental concern because their chemical structure resembles some estrogenic compounds. Alkylphenols can interfere with normal development in aquatic species at concentrations as low as $\mu g\ l^{-1}$.

III.F.3   Perfluorooctane Sulfonates (PFOS)

Perfluorooctane sulfonates (PFOS) are fluorinated compounds that have only recently been studied as potential environmental contaminants. Fluorinated organic compounds have been used in a diverse range of products, including refrigerants, surfactants, adhesives, paper coatings, fire retardants, and lubricants. Sulfonyl-based fluorochemicals are synthetic polymers that have been used primarily as surface protectors for papers and fabrics.

Perfluorooctane sulfanate [PFOS]

These sulfonyl-based fluorochemicals degrade in the environment, with PFOS being one of the primary degradation products that have been detected. PFOS are difficult to detect by common analytical procedures, and they were originally not considered as an environmental problem because the surface-active properties of the molecules were thought to prevent biological uptake. However, recent studies have shown that PFOS are widespread in wildlife and can accumulate in higher trophic levels (Giesy and Kannan, 2001).

Vignette 2.2   Emerging Environmental Pollutants

Robert C. Hale

Virginia Institute of Marine Science, College of William and Mary, Gloucester Point, VA

To preempt future catastrophic scenarios, environmental agencies have assembled chemical monitoring lists, e.g., the U.S. EPA priority pollutants. Chemicals on these lists have generally been chosen based on their known presence in the environment, production statistics, and toxicological potential. However, such lists typically encompass only the most obvious contaminants of the total number of chemicals of potential concern. The monitoring lists are generally restricted to compounds that are "contractually and analytically manageable," i.e., those that can be measured relatively easily and for which reference standards are commercially available (Keith and Telliard, 1979). In addition, we are hindered in creating such lists by our lack of knowledge concerning potential toxic effects. As new modes of toxicity have been uncovered, it has become apparent that some chemicals are less benign than originally suspected, and these merit further scrutiny. Established lists are also prone to become outdated if they are not periodically reviewed. The EPA priority pollutant list contains 129 compounds and has changed little since its initial development in the mid-1970s. In contrast, approximately 100,000 chemicals are in commercial use, and up to 1000 additional chemicals are introduced annually (Hale and La Guardia, 2002).

There are numerous examples of "nonpriority" pollutants emerging as compounds of concern. Some have particular regional relevance, such as kepone in the James River, VA (Huggett and Bender, 1980). Others have global significance, such as tributyltin oxide (TBT), which is a potent endocrine disruptor, a mode of action receiving renewed appreciation. TBT is known to cause reproductive and developmental problems in some aquatic invertebrates at the parts-per-trillion (ng l^{-1}) level. Although TBT is widely used as a marine antifoulant, it has also been used in a variety of other products, including textiles, plastics, and wood preservatives (Fent, 1996). Thus, its entry point into the environment can extend well beyond boats and shipyards. Although TBT appears to be the most toxic, it is only one in a family of organotin compounds. Less is known regarding the effects and fates of the other members, some of which are used in large amounts and whose cumulative impact might be significant. An unexpectedly wide range of chemicals could impact endocrine function, including phthalates (Sonnenschein and Soto, 1998). Ironically, phthalates were responsible for a major portion of the Chisso Corporation's (see Minamata disease, Chapter 1) postwar profits, and dioctylphthalate is also one of the rare additions to EPA's originally proposed priority pollutant list (Keith and Telliard, 1979).

Degradation products can also be a problem. For example, octyl- and nonylphenol polyethoxylates are used extensively as surfactants (Sonnenschein and Soto, 1998). Although the parent polyethoxylates are not persistent, bioaccumulative, toxic chemicals (i.e., PBT chemicals), they can degrade in the environment or during wastewater treatment to octylphenol (OP) and nonylphenol (NP). Both alkylphenols exhibit PBT properties and are endocrine disruptors. Natural, conjugated estrogens released to wastewater can also become more potent if deconjugated during treatment.

Polybrominated diphenyl ethers (PBDE) are highly persistent and bioaccumulative flame-retardant additives. They are structurally similar to PCB and polybrominated biphenyls (PBB). PBB were accidentally introduced into livestock feed in the 1970s in Michigan, resulting in human exposure and necessitating the destruction of large numbers of contaminated animals. The PBDE are also structurally similar to thyroxine, a critical thyroid hormone. Consequently, toxicological research is concentrating on their effects on development and thyroid function. A ban on the most bioaccumulative PBDE formulation—known as "Penta-"—is expected in the European Union in 2003. A driving force for this action is the observation that concentrations of its constituents have been increasing exponentially in human breast milk in Sweden. Recently, these same components have been detected at high concentrations in U.S. fish and in land-applied sewage sludges in the United States (Hale et al., 2001). Levels in sludge were

10 to 40 times those in European samples, in agreement with the greater North American commercial demand for Penta-. Reports now suggest proportionally elevated PBDE concentrations in breast milk of U.S. and Canadian citizens (Betts, 2001).

While our focus traditionally has been on chemicals released from industry, the fate of chemicals used and released by domestic consumers are an emerging concern. Pharmaceuticals are specifically engineered to be biologically active. Natural and synthetic estrogens, e.g., ethinylestradiol (the active ingredient in birth-control pills), have been associated with a variety of reproductive and developmental impacts on wild fish populations near sewage treatment plants (Hale and La Guardia, 2002). Although widely prescribed for humans, large amounts of drugs are also used to prevent disease in livestock. Little is known regarding the effects of pharmaceuticals on nontarget organisms or their fate in the environment (Jørgensen and Sorensen, 2000). Drugs and antibacterial agents (e.g., triclosan)—present in increasing amounts as components of household cleaning products—might impact normal bacterial assemblages or select for resistant strains. Antibiotic-resistant bacteria have been reported in fish from a number of areas near sewage outfalls (Miranda and Zemelman, 2001). Some personal-care product ingredients, such as musk compounds, have also been detected in the environment and are highly bioaccumulative (Hale and La Guardia, 2002). Consequences of this uptake are unknown.

Chlorinated persistent organic pollutants (POP) have been the focus of international treaties due to their potential for transboundary transport. However, some chemicals containing bromine and fluorine (e.g., PBDE) and fluorinated surface protectants (e.g., perfluorooctane sulfonate) behave similarly (Giesy and Kannan, 2001).

As noted, chemicals accumulate in sewage sludge, particularly chemicals that are persistent and hydrophobic. Millions of tons of these biosolids are applied each year on agricultural and other lands (Hale et al., 2001). Driving this practice is the cost of alternative disposal methods. On-site "land-farming" of contaminated industrial soils, such as soils from oil refineries, has been practiced for decades as a means of reducing pollutant loads. However, in the case of biosolids, the organic waste is being sold as a beneficial recycled product containing nutrients and soil-conditioning properties. The stated goal for these organic products is to improve plant growth at low cost, not to eliminate sludge-associated contaminants. Given the variability and diversity of inputs and subsequent treatment, biosolids are less well characterized than refinery waste. Burdens of some metals and pathogens are regulated, and steps have been taken to reduce these further by improved wastewater pretreatment and sludge stabilization (e.g., composting, anaerobic digestion, and alkaline or heat treatment). However, organic chemicals in these sludges have received far less attention. Sludges can contain several problematic classes at part-per-million concentrations or above, including PBDE, TBT, and NP (Hale et al., 2001; Fent, 1996; Hale and La Guardia, 2002). Although our increasingly specific analytical methodologies (e.g., selective-ion-monitoring mass spectrometry) have facilitated lower detection limits and higher precision for target compounds, their great selectivity may actually mask the presence of other chemicals. For example, numerous studies on the concentrations of PCB in sludges have been conducted. In contrast, the comparatively high concentrations of the structurally related PBDE have, until recently, gone unreported in U.S. biosolids, despite their similar behavior in subsequent extraction and purification protocols (Figure 2.6).

The discussion of emerging chemicals of concern in a textbook can only highlight today's issues. Thus, the chemicals discussed above should be viewed as examples showing where established lists have failed to be sufficiently inclusive. Researchers and regulators must incorporate new data, generated locally and in other countries, in decision-making and be cognizant of chemical consumption patterns, as demonstrated by the PBDE and NP. We should also recognize that most of our analytical approaches, toxicological studies, and regulations focus on a limited number of all contaminants and that interactive effects between these contaminants are likely. We should give chemicals that exhibit potentially hazardous structures more attention. For example, the presence of halogens imparts heightened persistence and bioaccumulative potential to many molecules. Thus, it should not be surprising that some brominated and fluorinated compounds exhibit the same problematic environmental properties as the relatively well-studied and regulated chlorinated ones. We also need to consider chemicals in the context of their use. Dispersive applications and routes of disposal lead to increased environmental release. Examples include the historical use of PCB in hydraulic fluids, ongoing inclusion of TBT in marine antifouling paints, and land application of biosolids. Finally, degradation of commercially produced chemicals generates an additional level of contaminants. Although breakdown is often beneficial, in some cases the result is intermediates with even more undesirable PBT properties (e.g., NP).

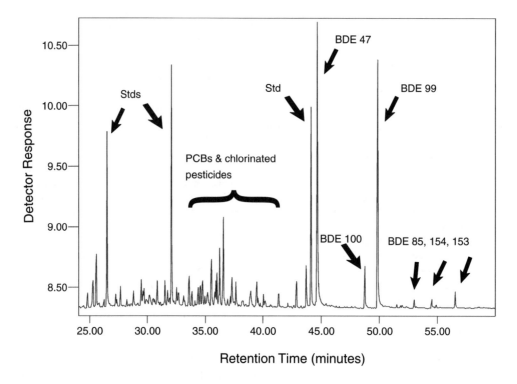

Figure 2.6 A chromatogram from the analysis of a biosolids sample showing the relative concentrations of BDE to PCB and chlorinated pesticides found in the sample. Despite the high concentrations of BDE and other emerging contaminants often present in environmental samples, they are often not reported as part of standard analytical procedures, which preferentially target priority pollutants.

IV SUMMARY

This chapter introduced the various classes of environmental contaminants while defining the basic types of contaminants and some of the mechanisms controlling contaminant fate and transport. Specific classes and structures of contaminants were presented along with their potential sources, relative persistence, and reasons for environmental concern. Specific examples were selected based on their prominence in the environmental literature, current or perceived future concern for adverse effects, or importance as case studies from a historical perspective.

SELECTED READINGS

Champ, M.A. and P.F. Seligman, Eds., *Organotin: Environmental Fate and Effects,* Chapman & Hall, London, 1996.

Connel, D.W., *Basic Concepts of Environmental Chemistry,* CRC Press, Boca Raton, FL, 1997.

Crop Protection Reference, Chemical and Pharmaceutical Press, New York (published annually).

Manahan, S.E., *Environmental Chemistry,* CRC Press, Boca Raton, FL, 2000.

Schwarzenbach, R.P., P.M. Gschwend, and D.M. Imboden, *Environmental Organic Chemistry,* John Wiley & Sons, New York, 1993.

SECTION TWO

Bioaccumulation

CHAPTER 3

Uptake, Biotransformation, Detoxification, Elimination, and Accumulation

Models are, for the most part, caricatures of reality, but if they are good, then, like good caricatures, they portray, though perhaps in a distorted manner, some of the features of the real world.

Kac (1969)

I INTRODUCTION

It is important to understand and to be able to predict the accumulation of contaminants in biota because effects are a consequence of concentrations in target organs or tissues. Moreover, because human exposure often occurs through the consumption of tainted food, prediction of accumulation in these potential vectors of exposure is key to avoiding human poisonings. The basic processes resulting in bioaccumulation (uptake, biotransformation, and elimination) are discussed in this chapter. Chapters 4 and 5 contribute detail and depth to this framework.

Bioaccumulation is the net accumulation of a contaminant in and, in some cases, on an organism from all sources including water, air, and solid phases in the environment. Solid phases include food, soil, sediment, and fine particles suspended in the air or water. Bioaccumulation is not the same as the more restricted term, **bioconcentration**, that has come to mean the net accumulation of a contaminant in (or, in some cases, on) an organism from water only. These two terms and their distinctions arise from an earlier, somewhat-dated debate in aquatic toxicology regarding the relative importance of water and food sources of contaminants to aquatic organisms.

The treatment of bioaccumulation has always relied heavily on mathematical models. This has greatly enhanced progress, and understanding of these tools has become essential to prediction of bioaccumulation and consequent effects. However, as expressed in the quote above, it is important to understand also that these models are only useful caricatures, and understanding of the models does not constitute a sufficient and accurate understanding of the associated phenomena.

Figure 3.1 depicts the rendering of bioaccumulation to a simple mathematical model. As with any mathematical model, only the details essential to understanding and accurately predicting the behavior of the system are included in the model. More detail might be needed if the model fails to do so. At the top of this figure, a fish is exposed to a contaminant through the water passing over its gills and the food passing through its gut. Dermal exposure may also be important, depending on the contaminant qualities (Landrum et al., 1996) and the surface-to-volume ratio of the fish (Barron, 1995; Landrum et al., 1996). In this example of a largemouth bass, only one major source of constant concentration is assumed for the sake of numerical expediency. All other sources

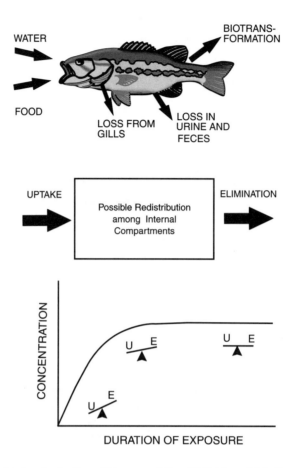

Figure 3.1   A simplified conceptualization of bioaccumulation. At the top of this figure, the fish (largemouth bass) is thought to potentially take in contaminants from its food and water and to lose contaminants via the gills, urine, and feces. There may be internal redistribution or biotransformation of the contaminant. This process is rendered to a simple box-and-arrow diagram (center graph). Here, only uptake from water is assumed to be significant, and all elimination processes are described by one elimination process. The most common mathematical description of this model predicts a gradual increase in contaminant in the fish until a steady-state concentration is obtained, as depicted in the graph at the bottom of this figure.

are considered insignificant or adequately reflected in the one uptake coefficient. Some contaminant may be lost from the gills or pass over the gills without being taken up. It could also be excreted in the urine or eliminated in the feces after passage through the gut or entering the gut through biliary excretion. The contaminant may be transformed and redistributed to various compartments within the fish, e.g., from the gill to the plasma to the kidney. In this example, only one elimination coefficient is included. Multiple processes are mathematically represented in this one coefficient or assumed trivial relative to a single, dominating elimination process. Also, no biotransformation or effects to the organism are occurring here. Kinetics and rate coefficients are assumed to remain constant over the time of accumulation. These details can be rendered to a box-and-arrow model (center of figure) with associated simplifying assumptions regarding the mathematics of uptake, elimination, and internal transformations and exchanges. In this example, uptake and elimination take place for one compartment in which the contaminant is mixed homogeneously and instantly. (These assumptions and simplifications are subject to testing during model development.) Mathematical expression of this simplest of models predicts a time course for bioaccumulation such as that shown at the bottom of Figure 3.1. This curve of a gradual and monotonic increase to a maximum concentration is a result of the change in relative influences of uptake and elimination processes on the change in internal concentration during the course of exposure. At the beginning of the exposure, little contaminant is contained in the fish and available for elimination: uptake (U) dominates relative to elimination (E) in the initial dynamics, and internal concentrations increase. But, as more and more contaminant accumulates in the fish, more contaminant becomes available for elimination. Elimination becomes increasingly important, and the rate of increase in internal concentration begins to decline. Eventually, a balance between uptake and elimination results in a steady-state[1] concentration in the fish that will be maintained as long as conditions remain constant.

Although described here in terms of concentration (amount of contaminant per amount of organism, such as 25 µg of Pb per gram of tissue), such models are also developed in terms of **body burden**, the mass or amount of contaminant in (or sometimes also on) the individual (e.g., 2500 µg of Pb per individual). In this chapter, the details and mathematical expression of this general model and models derived from it will be developed primarily in terms of concentration.

II UPTAKE

II.A Introduction

Uptake (the movement of a contaminant into, and sometimes onto, an organism) can occur by several mechanisms and can involve the dermis, gills, pulmonary surfaces, or the gut. In all cases, the process begins with interactions with the cells of tissues. Simkiss (1996) categorized uptake by a cell into three general routes: the lipid, aqueous, and endocytotic routes. The lipid route encompasses the passage of lipophilic contaminants through the bilayer of membrane lipids. Small, uncharged polar molecules such as CO_2, glycerol, and H_2O also diffuse readily through the lipid bilayer (Alberts et al., 1983). The aqueous route employs two general types of **membrane transport proteins** that (1) form channels (**channel proteins**) or act as **carrier proteins** in the membrane and (2) transfer hydrophilic contaminants into cells. Some channels (**porins**) are nonspecific; others are specific relative to the substances passing through them. The functioning of some channels can be influenced by the presence of other chemical substances, including other contaminants (Simkiss, 1996).

Possible mechanisms for uptake include adsorption, passive diffusion, active transport, facilitated diffusion or transport, exchange diffusion, and endocytosis (Newman, 1995). Several of these processes may be important for any particular combination of contaminant, exposure scenario, and species.

But, before a contaminant can be taken up, it must first interact with some surface of the organism. Movement onto the organism can involve and be modeled as adsorption. **Adsorption** is the accumulation of a substance at the common boundary of two phases (e.g., solution onto a solid surface). An adsorption example might be metal ion exchange with a hydrogen ion associated with a ligand on the surface of an insect's integument. The more general term **sorption** will be used instead of adsorption if the specific mechanism by which a compound in solution becomes associated with a solid surface is unknown or poorly defined.

Two equations are used to define adsorption: the **Freundlich and Langmuir isotherm equations**. The Freundlich equation (Equation 3.1) is an empirical relationship, and the Langmuir equation (Equation 3.2) is a theoretically derived relationship.

(Empirical relationship)

$$\frac{X}{M} = KC^{\frac{1}{n}}$$
(3.1)

Freundlich

---

[1] Landrum and Lydy (personal communication, 1991; Landrum et al., 1992) comment that the terms *steady state* and *equilibrium* are often used incorrectly. *Steady state* refers to a constant concentration in an organism resulting from processes (e.g., uptake, elimination, and internal exchange among compartments), including those requiring energy. However, equilibrium concentrations resulting from chemical equilibrium processes do not require energy to be maintained. Steady-state concentrations resulting from bioaccumulation can be considerably higher than those predicted for chemical equilibrium.

where X = amount adsorbed, M = the mass of adsorbent, K = a derived constant, C = the concentration of solute in the solution after adsorption is complete, and n = a derived constant.

$$\frac{X}{M} = \frac{abC}{1+bC}$$ (Theoretic relationship) Langmiur Isotherm (3.2)

where a = the adsorption maximum (amount) and b = affinity parameter reflecting bond strength.

A plot of C (abscissa or *x*-axis) vs. X/M (ordinate or *y*-axis) will result in a curve shaped like the one at the bottom of Figure 3.1. Such relationships can be linearized with equations such as Equation 3.3 (Freundlich isotherm) or Equation 3.4 (Langmuir isotherm) to facilitate data fitting by linear regression.

$$\log \frac{X}{M} = \log K + \frac{\log C}{n}$$ (3.3)

$$\frac{C}{X/M} = \frac{1}{ab} + \frac{1}{a} C$$ (3.4)

Adsorption theory and these equations have been used successfully to define toxicant movement onto diverse biological surfaces such as those of unicellular algae (Crist et al., 1988), fish gills (Pagenkopf, 1983; Janes and Playle, 1995), periphyton (Newman and McIntosh, 1989), and zooplankton (Ellgehausen et al., 1980). Crist et al. (1988) found that uptake of H^+ by algae involved rapid adsorption followed by a slow diffusion into the cell. Langmuir equation constants (Equation 3.2) were used by Crist and coworkers to define relative metal interactions with algal cell surfaces.

Diffusion is the movement of a contaminant down an electrochemical gradient.[2] It may be simple diffusion of a charged ion via a channel protein, or it may involve the passage of a lipophilic molecule through the lipid route (Figure 3.2). Simple diffusion does not require expenditure of energy. If diffusion involves a protein channel, passage through a channel is influenced by ion charge and size, including the size of the hydration sphere about the ion. Channels can be gated and respond to various chemical or electrical conditions. Diffusion could also be facilitated by a carrier protein. **Facilitated diffusion** occurs down an electrochemical gradient, requires a carrier protein, does not require energy, and is faster than predicted for simple diffusion. Because a carrier protein is involved, facilitated diffusion may be subject to saturation kinetics and competitive inhibition. Some facilitated diffusion involves the exchange of ions across a membrane (**exchange diffusion**). Diffusion accurately describes uptake of many nonpolar organic compounds or uncharged inorganic molecules such as ammonia (NH_3) (Fromm and Gillette, 1968; Thurston et al., 1981) or $HgCl_2^0$ (Simkiss, 1983) by the lipid route (Spacie and Hamelink, 1985; Barber et al., 1988; Erickson and McKim, 1990) and many charged moieties such as dissolved metals and protons (H^+) via the aqueous route (Spacie and Hamelink, 1985). Facilitated diffusion is expected for some metals and organic contaminants (Landrum and Lydy, 1991). Diffusion back and forth across a membrane can be complicated if the chemical is changed after crossing that membrane. For example, pentachlorophenol may pass though a membrane into the bloodstream but then be converted to the charged pentachlorophenate. This compound is much less capable of passing back across the membrane and out of the organism.

Diffusion is described most readily by Fick's Law (Equation 3.5).

$$\frac{dS}{dt} = -DA\frac{dC}{dX}$$ Fick's Law (Diffusion) (3.5)

[2] The term *electrochemical gradient* means a concentration, activity, or electrical gradient.

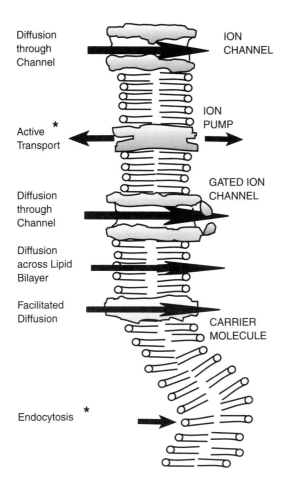

Figure 3.2   Mechanisms of uptake of contaminants into cells. Simple diffusion can occur across the lipid bilayer or through an ion channel formed by a channel protein. Channels may be gated and their functioning influenced by chemical and electrical conditions. Facilitated diffusion occurs via a carrier protein. Active transport passes the solute up an electrochemical gradient. Here the Na$^+$, K$^+$, ATPase pump is illustrated. Potassium is pumped in as sodium is pumped out of the cell. The last mechanism for cellular uptake is endocytosis. As indicated by an asterisk, endocytosis and active transport require energy.

See Math notes from Hydrology.

where dS/dt = rate of contaminant movement across the surface, D = a diffusion coefficient, A = the area of surface through which diffusion is occurring, and dC/dX = the concentration (or some other type of) gradient across the boundary of interest, e.g., the difference in concentration between the two sides of a cell membrane. This equation is incorporated into many models of bioaccumulation.

Active transport requires energy to move the contaminant up an electrochemical gradient. Because it involves a carrier molecule, active transport can be subject to saturation kinetics and competitive inhibition. The best example of active transport is cation transport by membrane-bound ATPase (ATP = adenosine triphosphate). ATPase, using energy from the hydrolysis of ATP, acts as a coupled ion pump to simultaneously remove some ions from the cell (e.g., sodium) while moving other ions (e.g., potassium) into the cell. Radiocesium, a chemical analog of potassium, can be taken up by this mechanism (Newman, 1995). Some metals (e.g., cadmium taken up as an analog of calcium) and some large, hydrophilic compounds can also be subject to active transport (Cockerham and Shane, 1994).

ATPase moves Na from cells K into cell

Radiocesum mistaken for K
Cadmium mistaken for Ca

Endocytosis (pinocytosis and phagocytosis) can be important in uptake, especially for contaminants entering an organism in food. Simkiss (1996) details an excellent example of contaminant metals being taken up by the transferrin route for iron assimilation. Iron, and perhaps other metals, is bound to the membrane-associated transferrin protein. The metal-transferrin complex moves to a specific region of the cell surface, where it becomes engulfed and incorporated into a vesicle. In the cell, the vesicle fuses with a lysosome, and the associated metal is released.

II.B   Reaction Order

The specific kinetics of uptake is often defined in terms of reaction order; therefore, a short review of reaction order is required here. In the context of a reaction involving only one reactant, reaction order refers to the exponent (n) to which the reactant concentration (C) is raised in the equation describing the reaction rate, $dC/dt = kC^n$. In the case of a zero-order reaction, $n = 0$ and $dC/dt = kC^0$. This reduces to $dC/dt = k$, with units of C/h. The concentration of the product will increase independent of reactant concentration with zero-order kinetics. For first-order kinetics ($dC/dt = kC$), the rate will change with reactant concentration, and units for k are h^{-1}. First-order reaction kinetics are the most commonly observed and applied kinetics for bioaccumulation modeling. Higher-order reaction kinetics are occasionally warranted. However, also common are saturation kinetics, such as Michaelis-Menten kinetics. Saturation kinetics is relevant with enzyme-mediated processes such as those associated with detoxification. Saturation kinetics is also applied for active or facilitated transport. Above a certain concentration of reactant, a system is saturated and cannot proceed any faster than a maximum velocity (V_{max}), i.e., zero-order kinetics. However, the kinetics shifts to first order as reactant concentrations drop below saturation conditions. For example, Mayer (1976, as reported in Spacie and Hamelink, 1985) saw evidence of saturation for the elimination of di-2-ethylhexylphthalate from fathead minnows (*Pimephales promelas*). Differential Equation 3.6 describes the change in reactant concentration by Michaelis-Menten kinetics:

$$-\frac{dC}{dt} = \frac{V_{max}\, C}{k_m + C} \tag{3.6}$$

where V_{max} = the maximum rate of reactant change (C/h), and k_m = half-saturation constant. A plot of the velocity of concentration change (V, *y*-axis) against concentration (C, *x*-axis) would look very much like the plot at the bottom of Figure 3.1. The velocity of concentration change increases with reactant concentration only to a certain point. There would be a maximum velocity (V_{max}) that could not be exceeded regardless of any further increase in concentration. The concentration at which V was equal to $V_{max}/2$ would be the k_m in Equation 3.6. Again, although zero-order, first-order, or saturation kinetics are applied to uptake, first-order uptake is the most common.

III   BIOTRANSFORMATION AND DETOXIFICATION

III.A   General

Once a contaminant enters an organism, it becomes available for **biotransformation**, the biologically mediated transformation of a chemical compound to another. Biotransformation involves enzymatic catalysis and, as a consequence, can be subject to saturation kinetics and competitive inhibition. Biotransformation can lead to enhanced elimination, detoxification, sequestration, redistribution, or activation. Biotransformation can enhance the rate of loss from the organism, as is often the case if a lipophilic xenobiotic is converted to a more hydrophilic compound, e.g., naphthalene oxidation to naphthalene diol. The contaminant may be rendered to a nontoxic form.

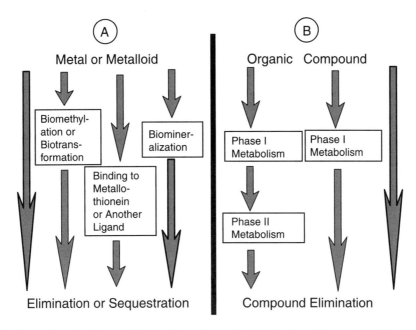

Figure 3.3   General mechanisms of biotransformation and detoxification of inorganic (A) and organic (B) contaminants. See text for explanation.

Some contaminants may be transformed to a form that is retained within the organisms but is sequestered away from any site of possible adverse effect. With **activation**, the adverse effect of a contaminant is made worse by biotransformation, or an inactive compound is converted to one with high, adverse bioactivity. For example, the organophosphorus pesticide, parathion, undergoes oxidative desulfuration to form the very potent paraoxon (Hoffman et al., 1995). General processes associated with biotransformation (Figure 3.3) are described below.

III.B   Metals and Metalloids

Although the term *biotransformation* is used almost exclusively relative to transformation of organic compounds, metals and metalloids are subject to "biotransformation" of a kind, resulting in elimination from or sequestration within the individual (Figure 3.3A). However, an ion can quickly bind to a plasma-associated ligand and become available for removal from the organism without any transformation. Lyon et al. (1984) demonstrated the role of ligand binding to metal ions in the blood of crayfish in determining the relative rates of such elimination for a series of metal ions.

Microbes genetically adapted to metal-contaminated environments can possess an enhanced ability to add methyl or ethyl groups to the metal ion, e.g., ionic mercury conversion to methyl mercury (Wood and Wang, 1983). Trivalent (As^{3+}) and pentavalent (As^{5+}) arsenic entering plants or animals might be rendered less toxic by methylation to monomethylarsonic and dimethylarsinic acids (Nissen and Benson, 1982; Peoples, 1983). Arsenic can also be converted to an arseno-sugar (i.e., trimethylarsonium lactate), arsenobetaine, or phospholipid (O-phosphatidyl trimethylarsoni-umlactate) (Cooney and Benson, 1980; Edmonds and Francesconi, 1981). Selenium entering a plant as selenite (SeO_3^{2-}) can be reduced and converted to selenocysteine. If incorporated into proteins, such selenoamino acids result in protein dysfunction, providing one mechanism for selenium toxicity (Brown and Shrift, 1982). In selenium-tolerant plants, large amounts of the nonprotein amino acids (Se-methylselenocysteine and selenocystathionine) are produced, suggesting that high rates of synthesis of these amino acids serve as a means of selenium detoxification (Brown and Shrift, 1982).

Metals can be bound and sequestered from sites of toxic action by metallothioneins and similarly functioning molecules. **Metallothioneins** are a class of relatively small (circa 7000 Da) proteins with approximately 25 to 30% of their amino acids being sulfur-rich cysteine and possessing the capacity to bind six to seven metal atoms per molecule (Hamilton and Mehrle, 1986; Shugart, 1996; Vignette 6.1 in Chapter 6). They are commonly induced by metals, including cadmium, copper, mercury, and zinc.[3] Silver, platinum, and lead also have been reported to induce metallothioneins (Garvey, 1990). In addition to their role in essential metal homeostasis, they also can be induced by toxic metals, bind these metals, and reduce the amount available to cause a toxic effect (Fraüsto da Silva and Williams, 1991). Metallothioneins and metallothionein-like proteins are found in many vertebrates and invertebrates, and they also have been reported in higher plants (Grill et al., 1985). A class of metal-binding polypeptides, **phytochelatins**, is similarly induced in plants by metals (Grill et al., 1985). Grill et al. (1985) found that phytochelatins were induced by the metal ions Cd^{2+}, Cu^{2+}, Hg^{2+}, Pb^{2+}, and Zn^{2+}. They can also function in regulation and detoxification of metals.

Finally, metals and other cations can be sequestered or eliminated through biomineralization. Metals such as lead (e.g., Bercovitz and Laufer, 1992; Newman et al., 1994) and radionuclides such as radiostrontium or radium (Grosch, 1965) can be incorporated into relatively inert shell, calcareous exoskeleton, or bone. Some metals are incorporated into exoskeleton (including the gut lining) of soil invertebrates and lost upon molting (Beeby, 1991). Other metals such as lead can be incorporated into molluscan shell (Beeby, 1991; Newman et al., 1994) and rendered unavailable for interaction at target sites. Radiostrontium from atmospheric fallout accumulates in bone along with calcium. Children whose bodies are actively building and reworking bone during development are particularly vulnerable. The consequence of such sequestration in bone can be extensive damage, such as in the case of another bone-seeking radionuclide, radium. In 1925, an outbreak of radium poisoning was reported among young women employed at brushing radium-laced paint onto watch faces.[4] They ingested radium while licking the paintbrush bristles to a fine point as they worked. The gamma-emitting radium accumulated in bone, causing extensive bone lesions or fatal anemia by damaging the bone marrow (Grosch, 1965).

Metals can be sequestered by incorporation into a variety of granules or concretions in addition to sequestration in structural tissues (Mason and Nott, 1981; Simkiss and Taylor, 1981; Pynnönen et al., 1987). Such granules are usually associated with the midgut, digestive gland, hepatopancreas, Malpighian tubules, and kidneys of invertebrates (Roesijadi and Robinson, 1994). They are also found in other specialized cells of invertebrates and in the connective tissues of vertebrates and invertebrates (Roesijadi and Robinson, 1994).

Hopkin (1986) described four general categories of granules in invertebrates. Type A granules are intracellular granules 0.2 to 3 μm in diameter built as concentric layers of calcium (and magnesium) pyrophosphate and an organic matrix of lipofuscin. Metal precipitation with phosphate is mediated by pyrophosphatase (Howard et al., 1981). These Type A granules are found in most invertebrate phyla (Roesijadi and Robinson, 1994). Type A granules accumulate Class A and intermediate metals such as manganese and zinc. Type B granules are also intracellular granules of roughly the same size as Type A granules and have high concentrations of sulfur. Consequently, they have high concentrations of Class B and intermediate metals (copper, cadmium, mercury,

---

[3] As pointed out by Simkiss and Taylor (1981), metallothioneins are involved primarily with the detoxification of Class B metals. **Class B metal cations** (Nieboer and Richardson, 1980) have filled d orbitals of 10 to 12 electrons and low electronegativity. They form "soft" spheres readily deformed by adjacent ions, and they easily form covalent bonds with donor atoms such as sulfur. Class B metals include Cu^+, Ag^+, Au^+, Au^{3+}, Cd^{2+}, and Hg^{2+}. **Class A metal cations** (e.g., Li^+, Na^+, K^+, Mg^{2+}, Ca^{2+}, and Sr^{2+}) have inert-gas electron configurations, high electronegativity, and hard spheres. Intermediate between Class A and B metals are **borderline metal cations** such as Fe^{2+}, Co^{2+}, Ni^{2+}, Cu^{2+}, Mn^{2+}, and Fe^{3+} (Stumm and Morgan, 1981; Brezonik et al., 1991).

[4] Our collective ignorance of the effects of radionuclides such as radium was staggering at that time. An even more shocking example is the intentional ingestion of radium as a patented medicine, Radithor. Consumption of Radithor was responsible for numerous painful ailments and unnecessary deaths, and its use was banned only after the fatal poisoning of a prominent New York socialite (Macklis, 1993).

silver, and zinc) that avidly bind to sulfur-containing groups. Type C granules are intracellular granules rich in the iron-containing products of ferritin. The last granule (Type D) is extracellular, can be 20 μm in diameter, is composed of calcium carbonate, and functions to buffer the hemolymph (blood) of molluscs.

The various types of granules differ in their locations in the organism. In the marine snail, *Littorina littorea*, Type A granules are found in basophil cells of the digestive diverticulum, and Type D granules are located in calcium cells of connective tissues of the foot and other tissues.

Type A, B, and C granules can serve as storage sites of metals and can be important in the elimination of metals by their discharge from cells into the gut (Hopkin, 1986; Beeby, 1991). Elimination can involve exocytosis (fusion of intracellular vesicles with the cell membrane and emptying of vacuole content to the cell exterior) or cell lysis with release of associated granules.

III.C   Organic Compounds

Depending on their qualities, organic contaminants can be eliminated rapidly or be subjected to metabolism,[5] with subsequent excretion of metabolites. During biotransformation, lipophilic compounds are often, but not always, made more amenable to excretion by conversion to more hydrophilic products. Biotransformations of organic contaminants can be separated into Phase I and Phase II reactions (Figure 3.3B). Generally, reactive groups such as –COOH, –OH, –NH$_2$, or –SH are added or made available by **Phase I reactions**, thus increasing hydrophilicity. Although Phase I comprises predominantly oxidation reactions, hydrolysis and reduction reactions are also important (George, 1994). One of the most common Phase I reactions involves addition of an oxygen to the xenobiotic by a **monooxygenase** (mixed-function oxidases or MFOs) (Hansen and Shane, 1994). After formation, the products of Phase I reactions can be eliminated or enter into **Phase II reactions** (Figure 3.3B). Conjugates are formed by Phase II reactions, which inactivate and foster elimination of the compound. Compounds conjugated with xenobiotics include acetate, cysteine, glucuronic acid, sulfate, glycine, glutamine, and glutathione (Hansen and Shane, 1994; Landis and Yu, 1995).

These Phase I and II reactions can be illustrated with the metabolism of naphthalene (Figure 3.4). Phase I oxidation (naphthalene $\rightarrow$ naphthalene epoxide) and hydrolysis (naphthalene epoxide $\rightarrow$ naphthalene 1,2-diol) are shown in this example to produce more water-soluble metabolites of naphthalene. Subsequent Phase II conjugation with glucuronic acid is shown for the naphthalene 1,2-diol. Many more reactions than shown here are involved in the metabolism of xenobiotics, as discussed in greater detail in Chapter 6. Many involve inducible enzymatic systems subject to the complex regulation typical of biochemical pathways. Subsequently, the modeling of bioaccumulation and elimination of the original xenobiotic may be complicated by the dynamics of associated metabolites, especially if a radiotracer is used in the experimental design. But, in the case of an activated compound, the dynamics of a metabolite may be more important than that of the original compound.

IV ELIMINATION

IV.A   Elimination Mechanisms

Elimination is the excretion or biotransformation of a contaminant, resulting in a decrease in the amount of contaminant within an organism. Although often used synonymously, depuration and clearance do not have precisely the same meaning as elimination (Barron et al., 1990). **Depuration**

---

[5] Lech and Vodicnik (1985) object to using the term *metabolism* in this context and suggest *biotransformation* as the preferred word for biochemically-mediated conversion of xenobiotics. They feel that the term *metabolism* should be restricted to biochemical reactions of "carbohydrates, proteins, fats and other normal body constituents."

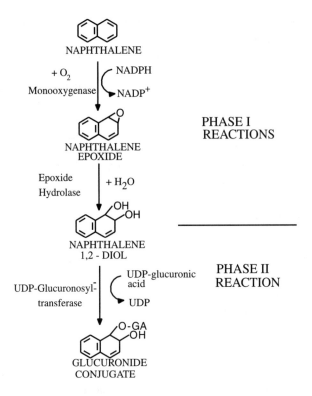

Figure 3.4   Metabolism of naphthalene including Phase I and Phase II reactions. The first Phase I reaction (naphthalene → naphthalene epoxide) is an oxidation reaction, and the second Phase I reaction (naphthalene epoxide → naphthalene 1,2-diol) is a hydrolysis reaction. UDP-glucuronic acid is formed by condensation of UTP (high-energy nucleotide, uridine triphosphate) with glucose-6-phosphate. UDP is uridine diphosphate and GA is glucuronic acid. (Composite figure of Figures 1, 5, and 14 in Lech, J.J. and M.J. Vodicnik, Biotransformation, in *Fundamentals of Aquatic Toxicology,* Rand, G.M. and Petrocelli, S.R., Eds., Hemisphere Publishing Corp., Washington, D.C., 1985.)

is a term associated with a particular experimental design in which the organism is placed into a clean environment and allowed to lose contaminant through time. **Clearance**, as will be discussed shortly, is a term used when modeling bioaccumulation kinetics and reflects the rate of substance movement between compartments normalized to concentration. Clearance has units of volume time^{-1}, e.g., mlh^{-1}.

If one monitors concentration in individuals over time, another phenomenon resulting in a decrease in concentration must be defined, i.e., **growth dilution**. The concentration of contaminant can decrease in a growing organism because the amount of tissue in which the contaminant is distributed has increased. Growth dilution is not a component of elimination because the total amount of contaminant (body burden) has not changed as a result of growth.

The relative importance of specific elimination mechanisms varies among plants, vertebrates, and invertebrates as well as among contaminants. Plants can lose contaminant by leaching, evaporation from surfaces, leaf fall, exudation from roots, or herbivore grazing (Duffus, 1980; Newman, 1995). Animals can eliminate contaminants by transport across the gills, exhalation, secretion of bile from the gall bladder, secretion from the hepatopancreas, secretion from the intestinal mucosa, shedding of granules, molting, excretion via the kidney or an analogous structure, egg deposition, or loss in hair, feathers, and skin. Aluminum bound to gill mucus is rapidly lost by mucus sloughing (Wilkinson and Campbell, 1993). In higher animals, elimination may involve loss in sweat, saliva, and genital secretions (Duffus, 1980). The liver bile, gills, and kidney are often the primary elimination routes for animals.

Elimination from the gills can be rapid for xenobiotics with low lipophilicity, i.e., those with log K_{OW}[6] values of 1 to 3 (Barron, 1995). Nonpolar organic compounds resistant to biotransformation tend to diffuse slowly across the gill (Spacie and Hamelink, 1985). Spacie and Hamelink (1985) note that DDT, di-2-ethylexylphthalate (DEHP), phenol, and pentachlorophenol can be eliminated in significant amounts this way.

Large, nonpolar molecules and associated metabolites can be eliminated from the liver, into the bile, and lost in the feces (Duffus, 1980; Spacie and Hamelink, 1985; Barron, 1995). In humans, compounds with molecular weights exceeding approximately 300 Da are eliminated in significant amounts in the bile (Gibaldi, 1991). Many metals (e.g., aluminum, cadmium, cobalt, mercury, and lead) and metalloids (e.g., arsenic and tellurium) complexed with proteins or other biochemical compounds in the plasma are incorporated into bile (Camner et al., 1979).

After passage into the liver, a contaminant enters the hepatic sinusoids, where it can be absorbed by parenchymal cells and biotransformed. The compound or its metabolites either return to the sinusoids or become incorporated into bile. Those incorporated in bile can be eliminated in the feces; however, some compounds entering the small intestine in bile may be subject to reabsorption and repeated passage through the liver. This **enterohepatic circulation** increases persistence of some compounds in the body and can increase damage to the liver (Duffus, 1980; Gibaldi, 1991). Although a minor complication of most elimination processes, compounds incorporated into saliva can also establish a similar cycle (Wagner, 1975). A Phase II reaction metabolite can be reabsorbed after being deconjugated, e.g., hydrolyzed, in the intestine (Wagner, 1975) and also lead to cycling. Metals can exhibit enterohepatic circulation, depending on the metal complex size. Trivalent arsenic and methyl mercury exhibit significant enterohepatic circulation (Camner et al., 1979). For arsenic, this involves active transport. Competition between compounds during incorporation into bile, and factors affecting bile formation and enterohepatic circulation, dictate the effectiveness of bile elimination.

Kidney excretion tends to be important for compounds with molecular weights less than approximately 300 Da. It is also the primary route of metal (e.g., cadmium, cobalt, chromium, magnesium, nickel, tin, and zinc) excretion by mammals (Roesijadi and Robinson, 1994); however, there are exceptions in which **gastrointestinal excretion** dominates. Metals such as cadmium and mercury can be excreted directly through the intestinal mucosa by active or passive processes (Camner et al., 1979). Gastrointestinal excretion can also involve loss by normal cell sloughing of the intestine wall.

Renal elimination involves three processes: glomerular filtration, active tubular secretion, and passive reabsorption. Filtration through capillary pores allows passage of most xenobiotics except those bound to plasma proteins. Elimination of some metals and lipophilic compounds by this passive filtration mechanism is inhibited by such binding. Weak organic acids and bases can be actively secreted to the urine or reabsorbed, depending on urine pH (Spacie and Hamelink, 1985). Beryllium elimination involves tubular secretion (Camner et al., 1979). In contrast to filtration, renal secretion is an energy-requiring movement up a concentration gradient that depends on the contaminant concentration, concentrations of competing compounds, pH of the urine, delivery rate to carrier proteins in the proximal tubule, and the relative affinity of the compound for the carrier proteins of the tubule and the plasma proteins (Gibaldi, 1991). Because secretion involves carriers, it can be subject to competitive inhibition and saturation kinetics (Gibaldi, 1991).

After a high concentration gradient is established, compounds can be passively reabsorbed into the blood, a process that favors lipid-soluble compounds and is less favorable to ionized or water-soluble compounds (Gibaldi, 1991). Cadmium bound to metallothionein is reabsorbed in the renal tubules (Camner et al., 1979). Some compounds can be actively reabsorbed (Gibaldi, 1991). Obviously, pH of the urine in the distal tubule will strongly influence reabsorption of weak acids and bases (Spacie and Hamelink, 1985) and metals such as lead and uranium (Camner et al., 1979).

---

[6] K_{OW} is the partition coefficient for a compound between n-octanol and water. It is used to reflect the lipophilicity of a compound and to imply relative partitioning of a xenobiotic between aqueous phases of the environment and lipids in an organism. It is sometimes designated by the capital letter, P.

Other routes of elimination may be important, depending on the organism and contaminant. Volatile organic compounds can be lost in expired breath, and lipophilic compounds can be lost from species such as fish by deposition in lipid-rich eggs. Arthropods have the capacity to eliminate contaminants such as metals by molting (Lindqvist and Block, 1994), peritrophic membrane sloughing from the midgut and hindgut cuticle loss during molting (Hopkin, 1989), or by discharge of metal-rich granules. Birds can incorporate metals into feathers, e.g., lead (Amiard-Triquet et al., 1992; Burger et al., 1994) and mercury (Becker et al., 1994). Mammals can eliminate lipophilic compounds (e.g., DDT) or calcium analogs (e.g., strontium) in milk. Fin whales transfer some of their body burden of PCB and DDT to their young by normal placental processes prior to giving birth and, after birth, transfer more to the calf in milk (Aguilar and Borrell, 1994). Some metals and metalloids are incorporated into skin and hair (e.g., Roberts et al., 1974).[7]

IV.B Modeling Elimination

Given the variety of elimination mechanisms, it is surprising that first-order kinetics are used in most models of elimination. Zero-order and saturation kinetics are used less frequently and tend to be used more in complex model formulations.

Compartment models of elimination similar to that shown in Figure 3.1 can be formulated as rate-constant, clearance-volume, or fugacity models. All three approaches are equivalent for the simple, single-source models shown here, and the associated constants can be interconverted (Newman, 1995). Statistical-moments methods for quantifying elimination without formulation of a specific compartment model are also common in pharmacology (Yamaoka et al., 1978) and have application to ecotoxicology (Barron et al., 1990; Newman, 1995).

Rate-constant-based models employ constants such as the first-order rate constants described earlier to quantify the rate of change of concentration (or amount) of toxicant in one or more compartments. Frequently, compartments are not physical (e.g., excretion from the liver) but, rather, are mathematical compartments (e.g., the "fast" elimination component). However, some interpretations of mathematical compartments imply associated physical compartments. Consequently, it is important to determine exactly the type of compartment model being described when assessing this type of research.

A simple, first-order rate-constant model can be used to describe elimination of contaminant from an organism after it has been moved to a clean environment and allowed to depurate (top of Figure 3.5). The model can be expressed in terms of concentration (C) or body burden (amount in the individual or X).

$$\frac{dC}{dt} = -kC \tag{3.7}$$

where C = concentration in the compartment, k = rate constant for the concentration-based formulation (h^{-1}), and t = time or duration of elimination.

$$\frac{dX}{dt} = -kX \tag{3.8}$$

where X = amount in the compartment, k = rate constant for the amount-based formulation (h^{-1}), and t = time or duration of elimination. Equations 3.7 and 3.8 are integrated to Equations 3.9 and

---

[7] Incorporation of elements into hair is used to monitor exposure, as typified by two studies of historical significance. Analysis of hair from Napoleon Bonaparte (Maugh, 1974) indicates that, sometime during his exile on Elba or in Saint Helena, he was slowly being poisoned with arsenic, a common political tool of the time. In the second case, hair from Sir Isaac Newton was found to have extremely high levels of mercury, a common element used in his experiments. Newton habitually tasted his chemicals during experimentation. Poisoning is now identified as the most probable reason for his "year of lunacy" (1693), a year filled with abnormal behaviors characteristic of mercury's neurological effects (Broad, 1981).

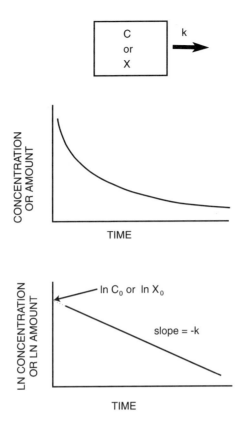

Figure 3.5   Elimination of a contaminant under a depuration scenario, including linearization to extract k and C_0 or X_0.

3.10 to allow prediction of elimination from an initial concentration (C_0) or amount (X_0) to any time (t) during elimination.

$$C_t = C_0 e^{-kt} \tag{3.9}$$

$$X_t = X_0 e^{-kt} \tag{3.10}$$

Figure 3.5 (center) shows the exponential elimination of a contaminant described by these equations. To fit depuration data to these equations, the ln of concentration or body burden is plotted against time (bottom of Figure 3.5). The resulting line has a y-intercept equal to the ln of the initial concentration or body burden. The slope is an estimate of $-k$. If linear regression of the logarithm of concentration or logarithm of body burden vs. time is used to estimate C_0 or X_0, a commonly ignored bias exists in these estimates that can and often should be corrected (Newman, 1995).

The time required for the amount or concentration to decrease by 50% ($t_{1/2}$, **biological half-life**) is (ln 2)/k. The **mean residence time** of a particle of compound (τ) in the compartment is $1.44t_{1/2}$ or, more directly, $k = 1/\tau$.

If two or more elimination mechanisms are responsible for elimination from the compartment, they are easily included in the models:

$$C_t = C_0 e^{\Sigma -k_i t} \tag{3.11}$$

$$X_t = X_0 e^{-\Sigma k_i t} \tag{3.12}$$

where k_i = the i individual elimination rate constants.

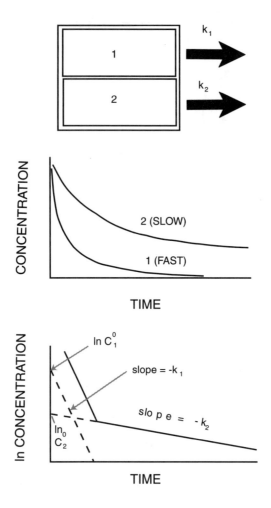

Figure 3.6 Elimination involving two compartments and two elimination constants, including backstripping, to calculate k_1, k_2, C_1, and C_2.

For models described by Equations 3.11 and 3.12, the **effective half-life** (k_{eff}) is $(\ln 2)/\Sigma k_i$. Equations 3.11 and 3.12 are particularly useful if a radiotracer is used to quantify elimination because the (first-order) radioactive decay rate constant (λ) can be included in the formulation, i.e., $\Sigma k_i = k + \lambda$ for a one-component elimination model with radioactive decay of the tracer.

Rate-constant-based models can be developed for biexponential or multiexponential elimination. One such conceptual model (top and middle of Figure 3.6) has elimination from two different compartments within an organism displaying distinct ("fast" and "slow") elimination kinetics. The integrated forms of this model are Equations 3.13 and 3.14:

$$C_t = C_1^0 e^{-k_1 t} + C_2^0 e^{-k_2 t} \tag{3.13}$$

where C_t = total concentration assuming equal volumes or masses for Compartments 1 and 2, C_1^0 = initial concentration in Compartment 1, C_2^0 = initial concentration in Compartment 2, k_1 = elimination rate constant for Compartment 1, and k_2 = elimination rate constant for Compartment 2. Obviously, the assumption of equal sizes for the two compartments is an inconvenient assumption of this model. This constraint will be resolved later using clearance-volume models. However, the assumption of equal compartment sizes is not a difficulty in the case of the model based on amount in the various compartments:

$$X_t = X_1^0 e^{-k_1 t} + X_2^0 e^{-k_2 t} \qquad (3.14)$$

where X_t = total amount of contaminant in the organism, X_1^0 = initial amount in Compartment 1, X_2^0 = initial amount in Compartment 2, k_1 = elimination rate constant of Compartment 1, and k_2 = elimination rate constant for Compartment 2.

With the linearizing method just described for monoexponential elimination, biexponential elimination produces a line with a distinct break in slope (bottom of Figure 3.6, solid line). A **backstripping or backprojection procedure** can be used to extract the model parameters from a multiexponential elimination curve. First, the region of the ln-concentration-vs.-time curve—where predominantly slow elimination occurs (the straight part of the line after the break in slope in Figure 3.6, bottom)—is used to fit a line for Component 2. The slope of this line is $-k_2$, and the y-intercept estimates ln C_2^0. The linear equation just derived for this second component (ln C_2 = ln $C_2^0 - k_2 t$) is then used to predict concentrations of contaminant in Compartment 2 for all sampling times during depuration. These predicted concentrations in Compartment 2 are subtracted from the original data observed for all times prior to the break in the line to estimate the concentrations present in Compartment 1 during elimination; these "stripped away" data for Component 1 produce a straight line (Figure 3.6, bottom panel, dashed line) with a slope of $-k_1$ and y-intercept of ln C_1^0. Wagner (1975) and Newman (1995) detail this and more accurate means of backstripping exponential curves of elimination. Computer programs are available to implement these methods and assess the results statistically.

Other multiple-compartment models can be fit with this rate-constant-based approach. One of the more common two-compartment elimination models (Figure 3.7, top) can be fit in a similar manner. In this model, a contaminant is introduced as a bolus or single mass into a central compartment (e.g., a single injection of the dose into the blood), and the compound is then subject to passage into and out of another peripheral or storage compartment. The compound can only be eliminated from the central compartment. The resulting elimination curves for these two compartments are illustrated at the bottom of Figure 3.7. Initially, the concentration in the central compartment (1) is high, and there is no compound in the peripheral compartment (2). The concentration in the central compartment decreases, and the concentration in the peripheral compartment begins to increase. The concentration in Compartment 1 eventually describes a biexponential curve, and the curve for Compartment 2 eventually runs parallel to the end component of Compartment 1 because elimination is occurring from Compartment 1 only. If the two dashed lines shown in the bottom panel of Figure 3.7 are used to obtain intercepts and slopes for the apparent biexponential elimination dynamics from Compartment 1, the microconstants (k_{10}, k_{21}, and k_{12}) can be estimated with the following equations. For example, one could monitor blood concentrations through time and fit the results to Equation 3.13 with backstripping methods. The resulting model is the same as Equation 3.13, except that A, B, α, and β are substituted into the equation. Obviously, these constants do not retain the same physical interpretation as C_1, C_2, k_1, and k_2:

$$k_{21} = \frac{A\beta + B\alpha}{A + B} \qquad (3.15)$$

$$k_{10} = \frac{\alpha\beta}{k_{21}} \qquad (3.16)$$

$$k_{12} = \alpha + \beta - k_{21} - k_{10} \qquad (3.17)$$

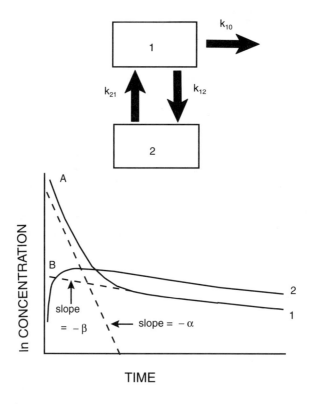

Figure 3.7   Elimination from a two-compartment model after bolus introduction of the contaminant into compartment 1. A, B, α, and β are used to estimate k_{12}, k_{10}, and k_{21}, as described in the text.

where A and B = antilogs of the y-intercepts for Components 1 and 2, respectively, and α and β =
slopes of Components 1 and 2, respectively. The biological half-life for a contaminant using this
model is ln $2/\beta$ (Barron et al., 1990).

Many similar multiple compartment models can be developed under the rate-constant-based
modeling scheme. However, modeling can be made easier as more compartments are added if one
uses the **clearance-volume-based model**. In this approach, the substance is distributed in and
cleared from compartments of different volumes. Clearance (Cl) is expressed as flows, the volume
of compartment completely cleared per unit time (volume time^{-1}).

The apparent-volume-of-distribution concept is used in clearance-volume-based models. It
can be envisioned with the example of the introduction of a dose of compound into a compartment. After introduction, the dose distributes itself within the compartment, and a compartment
concentration is realized. The compartment volume is V = dose/concentration. For example, a
dose of a chemical is injected into the blood and allowed to distribute in the circulatory system.
The **apparent volume of distribution** (V_d) for blood would be dose/concentration. Estimation
becomes more involved as the compound becomes distributed in many compartments, but
methods developed in pharmacology for estimation of apparent volumes and clearance rates
are available (Newman, 1995). If other compartments are involved, their apparent volumes of
distribution are derived mathematically and expressed in units of volume of the reference
compartment. Often the reference compartment is the blood or plasma compartment. Consequently, these volumes are not physical volumes *per se*, but mathematical volumes. The total
dose (D_T) in the organism is distributed among compartments as defined by the concentrations

in the compartments (C) and the compartment V_d's (V). For example, if blood is the reference compartment and there are n additional compartments, then:

$$D_T = C_b V_b + \sum_{i=1}^{n} C_i V_i \qquad (3.18)$$

where subscripts of b and i denote the b (blood) and the i (nonblood) compartments, respectively. The k_i values of the rate-constant formulation are equal to Cl_i/V_i for the clearance-volume-based formulation.

The clearance-volume-based formulation allows development of more-complex models and parameterization of these models. They have been used in pharmacology to describe the internal kinetics of drugs (**pharmacokinetics**) and poisons (**toxicokinetics**) for many years. A wealth of data, techniques, software, and expertise has been developed around this approach. Also, clearance-volume-based models allow direct incorporation of physiological parameters into models, and thus enhance predictive capabilities of models under different conditions and for different species. (Pharmacokinetic models that include physiological and anatomical features in describing internal kinetics are called **physiologically based pharmacokinetics [PBPK] models**.)

A final compartment model that uses different units but is equivalent to the rate-constant-based and clearance-volume-based formulations is the fugacity model. It is based on the escaping tendency of a compound in a compartment or its fugacity. Fugacity (f) is expressed as a pressure (Pa) and is related to concentration in a phase, $C = fZ$, where Z is the fugacity capacity (mol, $m^{-3} Pa^{-1}$) of the phase. The rate of transport between two compartments $(1 \rightarrow 2)$ is N (mol h^{-1}) = $D(f_1 - f_2)$ where D is a transport constant (mol, $h^{-1} Pa^{-1}$). The k_i values for the rate-constant-based model are equivalent to the D_i/V_iZ_i of fugacity models, where V is the compartment volume. The major advantage of the fugacity model is that units are the same regardless of the phases being considered. Consequently, wide differences in concentrations are easily accommodated in complex models such as those including water, sediment, food, and biological compartments (Mackay and Paterson, 1982; Gobas and Mackay, 1987; Mackay, 1991).

Finally, there exists in the pharmacology literature a body of methods for calculating elimination qualities such as mean residence time for substances in individuals without assuming a specific model. Yamaoka et al. (1978) introduced a statistical-moments approach to pharmacokinetics that uses only the area under the curve (AUC) of a plot of concentration (y-axis) vs. time (x-axis). The mean residence time, its associated variance, and other parameters can be estimated with the AUC. Although underutilized in ecotoxicology (Barron et al., 1990; Newman, 1995), these methods can be very useful if an exact model is judged to be unnecessary, impractical, or impossible to define. Also, because of their general utility, these methods can be used if the exact model is known.

V ACCUMULATION

Bioaccumulation is the net consequence of uptake, biotransformation, and elimination processes within an individual. The simplest, rate-constant-based model includes first-order uptake from one source into one compartment and first-order elimination from that compartment:

$$\frac{dC}{dt} = k_u C_1 - k_e C \qquad (3.19)$$

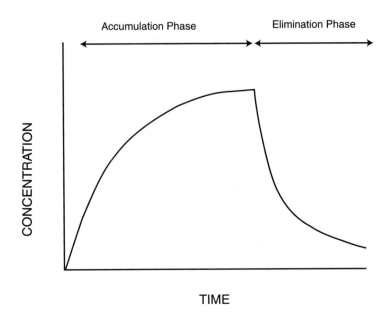

Figure 3.8 Simple bioaccumulation through time as described by Equation 3.20, followed by elimination as described by Equation 3.9. Data of this type are generated with an accumulation-elimination experiment. In the accumulation phase, kinetics is dictated by both uptake and elimination. The organism is then allowed to depurate, and kinetics in this elimination phase is solely a consequence of elimination.

where C_1 = concentration in the source (e.g., 1 = water), C = concentration in the compartment (e.g., fish), k_u = uptake clearance (ml [$g^{-1} h^{-1}$]),[8] and k_e = elimination rate constant (h^{-1}). This equation integrates to Equation 3.20:

$$C_t = C_1 \left(\frac{k_u}{k_e} \right)(1 - e^{-k_e\,t})$$ (3.20)

The concentration in organisms can be predicted at any time (t) based on Equation 3.20. The resulting bioaccumulation curve is shown in Figure 3.1 (bottom panel) and in the accumulation phase of Figure 3.8. The clearance-volume-based and fugacity equivalents of Equation 3.20 are given in Equations 3.21 and 3.22, respectively:

$$C_t = V_d C_1 \left(1 - e^{-\frac{Cl}{V_d}t} \right)$$ (3.21)

where V_d and Cl = V_d and C_1 for the organism, respectively, and

$$C_t = C_1 \left(\frac{Z}{Z_1} \right)\left(1 - e^{-\frac{Dt}{VZ}} \right)$$ (3.22)

[8] The units of k_u are often expressed as h^{-1} under the assumption of equal density of the source (e.g., water) and compartment (e.g., tissue). A milliliter of source volume and a gram of tissue then cancel each other out, and the units become h^{-1}. This makes k_u appear as a rate constant. However, k_u reflects a clearance from the source (Barron et al., 1990; Landrum et al., 1992) and, consequently, should retain units of flow normalized to mass. Details supplied by Peter Landrum are provided in Appendix 5 for students interested in the derivation of these units for k_u.

where Z and Z_1 = fugacity capacities for the organism and source, respectively; D = transport constant for the organism; and V = volume of the organism.

As the concentration in the organism approaches equilibrium (the rightmost bracketed term $[1 - e^x]$ in Equations 3.20 to 3.22 approaches 1), the relationship between the concentration in the organism (C at steady state or C_{ss}) and the source (C_1) is described by Equation 3.23:

$$\frac{C_{ss}}{C_1} = \frac{k_u}{k_e} = V_d = \frac{Z}{Z_1} \tag{3.23}$$

These terms define the steady-state concentration in the organism relative to the source. If the environmental source is water, the k_u/k_e, V_d, or Z/Z_1 estimate the **bioconcentration factor** (BCF). This ratio is defined by many workers as the **bioaccumulation factor** (BAF or BSAF)[9] if the source of contaminant is sediment. Normalization can be done for some nonpolar contaminants in these models. If the concentrations are normalized to amount of lipid in the organism and to the amount of organic carbon in sediments, the ratio is often referred to as the **accumulation factor** (AF).

Bioaccumulation models are often fit to the curve of concentration in the organism vs. time of exposure using nonlinear regression methods. For the rate-constant-based formulation, this method estimates k_u and k_e simultaneously. Although often adequate, this can lead to some difficulty in computations (i.e., poor convergence on a solution, lowered estimate precision, and a degree of covariance between estimates of k_u and k_e). A combined accumulation-elimination design (Figure 3.8) is used to avoid some of these difficulties, as in the studies of Yamada et al. (1994) and Murphy and Gooch (1995). In this design, organisms are exposed to a contaminant and allowed to accumulate until internal concentrations approach practical steady-state concentration (i.e., 95% of C_{ss}). Then, the organisms are removed from the contaminated arena and allowed to depurate in a clean setting. The drop in internal concentration is measured over time during this elimination phase. With these data, the k_e can then be estimated independent of k_u from the elimination-phase data. This estimate of k_e can then be used in the nonlinear fitting for the accumulation-phase data. Now, the nonlinear regression need only estimate k_u from the accumulation data.

Other formulations of these bioaccumulation models allow inclusion of more detail. If there had been an initial concentration of contaminant in the organism prior to the trial exposure, the concentration of contaminant can be predicted through time by combining Equations 3.20 (bioaccumulation from source 1) and 3.9 (elimination of initial concentration, C_0):

$$C_t = \frac{k_u}{k_e} C_1 (1 - e^{-k_e t}) + C_0\ e^{-k_e t} \tag{3.24}$$

Multiple elimination components can be included, as seen in Equation 3.25:

$$C_t = \frac{k_u}{k_{e1} + k_{e2}}\ C_1\ (1 - e^{-(k_{e1} + k_{e2})t}) \tag{3.25}$$

Under the expedient, but often overly optimistic, assumption that growth dilution can be described as e^{-gt} (g = a rate constant akin to the elimination rate constant), growth dilution can be included as a component in a model such as Equation 3.25.

Uptake from food and water can be incorporated into these models using an estimated assimilation efficiency (α, amount absorbed per amount ingested in food) and specific ration (R, amount

---

[9] As discussed in Chapter 5, the term *bioaccumulation factor* is also used in a more general context to mean accumulation from other sources as well. It is often used if the exact source remains undefined or poorly defined, as in a field survey. Some workers will use the more specific term **biota-sediment accumulation factor** (BSAF) to emphasize that the factor relates accumulated contaminant to that in sediments (Spacie et al., 1995).

of food consumed per amount of organism). Assuming concentrations in water (C_1) and food (C_2) are constant, Equation 3.26 applies:

$$C_t = \frac{k_u C_1 + \alpha R C_2}{k_e}(1 - e^{-k_e t}) \qquad (3.26)$$

A general model incorporating multiple sources (k_{uj}), many elimination components (k_{ei}), and an initial concentration (C_0) can be defined as shown in Equation 3.27:

$$C_t = \frac{\sum_{j=1}^{m} C_j k_{uj}}{\sum_{i=1}^{n} k_{ei}}\left(1 - e^{-\left(\sum_{i=1}^{n} k_{ei}\right)t}\right) + C_0 e^{-\left(\sum_{i=1}^{n} k_{ei}\right)t} \qquad (3.27)$$

As models become more complex, the data requirements become increasingly difficult to meet. Consequently, most pharmacokinetic models are useful compromises between reality and expediency. This is important to understand in order to extract the fullest understanding from modeled systems without making the unintentional transition to naive overinterpretation.

VI SUMMARY

This chapter described the methods and mathematics associated with the uptake, biotransformation, and elimination of contaminants. The general mechanisms of uptake — adsorption, passive diffusion, active transport, facilitated diffusion, exchange diffusion, and endocytosis — were described. Biotransformation for metals, metalloids, and organic compounds was described briefly. Binding proteins and peptides as well as biotransformation and sequestration of metals and metalloids were described in detail. Phase I and Phase II reactions in the transformation of organic compounds were discussed. Elimination by a variety of mechanisms was discussed for plants and animals. Details of elimination via liver, kidney, and gill were highlighted.

Models were developed based on three different formulations: rate-constant-based, clearance-volume-based, and fugacity-based formulations. All are equivalent in their basic forms, but each formulation has its own advantages and disadvantages. There are also useful statistical-moments methods that do not require a specified model. Although rate-constant-based models have the longest history in ecotoxicology, clearance-volume-based models have much promise in allowing linkage to the pharmacokinetics literature and techniques. They also are most amenable to generation of physiologically based pharmacokinetics (PBPK) models. Fugacity models have an advantage for extremely different phases/compartments because they express contaminant levels in identical units for all compartments. Statistical-moments methods require the least amount of information, and no model is required.

SELECTED READINGS

Barron, M.G., G.R. Stehly, and W.L. Hayton, Pharmacokinetic modeling in aquatic animals, I: models and concepts, *Aquatic Toxicol.,* 18, 61–86, 1990.

Gibaldi, M., *Biopharmaceutics and Clinical Pharmacokinetics,* Lea and Febiger, Philadelphia, 1991, p. 406.

Hansen, L.G. and B.S. Shane, Xenobiotic metabolism, in *Basic Environmental Toxicology,* Cockerham, L.G. and Shane, B.S., Eds., CRC Press, Boca Raton, FL, 1994.

Himmelstein, K.J. and R.J. Lutz, A review of the applications of physiologically based pharmacokinetics modeling, *J. Pharmacokinet. Biopharm.,* 7, 127–145, 1979.

Landrum, P.F., H. Lee II, and M.J. Lydy, Toxicokinetics in aquatic systems: model comparisons and use in hazard assessment, *Environ. Toxicol. Chem.,* 11, 1709–1725, 1992.

Lech, J.J. and M.J. Vodicnik, Biotransformation, in *Fundamentals of Aquatic Toxicology,* Rand, G.M. and Petrocelli, S.R., Eds., Hemisphere Publishing Corp., Washington, D.C., 1985.

Mackay, D., *Multimedia Environmental Models: The Fugacity Approach,* Lewis Publishers, Chelsea, MI, 1991, p. 257.

Newman, M.C., *Quantitative Methods in Aquatic Ecotoxicology,* Lewis Publishers, Boca Raton, FL, 1995, p. 426.

Spacie, A. and J.L. Hamelink, Bioaccumulation, in *Fundamentals of Aquatic Toxicology,* Rand, G.M. and Petrocelli, S.R., Eds., Hemisphere Publishing Corp., Washington, D.C., 1985.

CHAPTER 4

Factors Influencing Bioaccumulation

Without question! The chemical company called Chisso poisoned the fishing waters of Minamata, poisoned the aquatic food chain, and eventually poisoned a great number of inhabitants. Chisso poured industrial poisons through waste pipes until Minamata Bay was a sludge dump, the heritage of centuries destroyed.

Smith and Smith (1975)

I INTRODUCTION

I.A General

The degree to which a contaminant accumulates in biota is a function of the qualities of the contaminant, the organism, and the environmental conditions under which the organism and contaminant are interacting. The qualities of the contaminant determine the chemical forms in which it is present in the environment and the degree to which it is available to be taken up, biotransformed, and eliminated. Physiological, biochemical, and genetic qualities determine an organism's ability to minimize uptake, biotransform, and eliminate the contaminant. Developmental or sex-related changes can influence bioaccumulation, e.g., age- and sex-correlated lipid content will influence accumulation of a lipophilic contaminant. Ecological and behavioral characteristics of an organism determine routes of exposure and efficiency of uptake from each potential source; e.g., a predator is exposed through its prey, but a pelagic species is not exposed directly to sediment-associated contaminant. The milieu in which interaction between a contaminant and organism takes place can affect speciation and phase association of the contaminant and, consequently, availability for accumulation. Environmental conditions may also directly modify the functioning of an organism. For example, temperature has clear effects on rates of pertinent physiological and biochemical processes taking place within the organism. Other factors such as salinity and pH strongly modify ion regulation and osmoregulation and, in so doing, influence the uptake of many contaminants. Qualities of the microenvironment at the site of interaction may be as important, or more important, than those of the general environment. The water chemistry at the surface microlayer of the gill (e.g., Janes and Playle, 1995) or that of interstitial waters immediately around an infaunal species (e.g., Campbell et al., 1988) strongly influence uptake.

Basic qualities of chemicals, organisms, and the environment that have the strongest influence on bioaccumulation are outlined in this chapter. They are discussed first for inorganic contaminants and then for organic contaminants. Finally, general biological processes that transcend this dichotomy are discussed.

I.B   Bioavailability

Bioavailability is the extent to which a contaminant in a source is free for uptake (Newman and Jagoe, 1994). In many definitions, especially those associated with pharmacology or mammalian toxicology, bioavailability of a contaminant implies the degree to which the contaminant is free to be taken up *and to cause an effect at the site of action.*[1] This is a reasonable qualification in the context of these sciences, but the more general definition given here seems warranted in ecotoxicology. In ecotoxicology, availability for bioaccumulation in a food or prey species in which no effect is seen could be as much a concern as availability to an organism that is directly affected. Even with this broader definition, the term can still be used in the context of the degree to which a toxicant is available to have an effect, e.g., availability of metal in sediments to kill benthic species (Carlson et al., 1991).

Bioavailability is measured or implied in many ways. Relative bioavailability of a contaminant in two types of food can be implied from measured differences in bioavailability for individuals exposed to similar concentrations of contaminant in the different foods. This qualitative approach has been used for contrasting different foods (Newman and McIntosh, 1983; Reinfelder and Fisher, 1991), sediments (Luoma and Bryan, 1978; Langston and Burt, 1991), and water chemistries (Wright and Zamuda, 1987; Driscoll et al., 1995). Landrum and Robbins (1990) suggested that bioavailability in different sediments can be compared by measuring the amount of a contaminant in sediments and the amount in organisms inhabiting the sediments, or by measuring uptake clearance rates of organisms in the various sediment types. Using this approach, Hickey et al. (1995) compared the impact of feeding modes on bioaccumulation of polychlorinated biphenyls (PCB) and polynuclear aromatic hydrocarbons (PAH) in the deposit feeder, *Macomona liliana,* and filter feeder, *Austrovenus stutchburyi.*

In Chapter 3 (Equation 3.26), assimilation efficiency was used to quantify bioavailability from food. Assimilation efficiency or percentage retention (efficiency expressed as a percentage) can be determined by feeding known amounts of contaminant, most often using a radiotracer, and measuring the increase in contaminant within the fed individual. For example, bioavailability of organotin antifouling agents (tributyltin and triphenyltin) in Red Sea bream (Yamada et al., 1994) and *trans* and *cis* isomers of chlordane in channel catfish (Murphy and Gooch, 1995) from food sources were quantified in this manner. The mass incorporated into tissue divided by the mass fed to the individual is calculated assuming that all of the contaminant measured in the organism has become interspersed or incorporated in its tissues. Normally, much effort is made in designing this type of experiment to minimize the amount of unassimilated contaminant in the gut, hepatopancreas, or other similar sites. Without this precaution, the estimated assimilation efficiency would be biased upward from the true assimilation efficiency.

Bioavailability is measured another way in studies of drug pharmacokinetics. Such studies are concerned with determining the **effective dose** of a drug, i.e., the amount entering the blood and available to have an effect. To do this, a dose can be administered orally, and the amount appearing in the blood is compared with that in the blood after the same dose is injected intravenously. Bioavailability measured in this context involves both the amount of the drug or toxicant and the rate at which it enters the organism (Gibaldi, 1991): Quantification of bioavailability takes both the amount and rate of drug delivery into account. This can be done by comparing the areas under the curves (AUC) of concentration-vs.-time plots obtained for the different routes of administration (Figure 4.1). The **absolute bioavailability** is estimated from the AUC for any route or form of the compound divided by the AUC for the compound after direct injection into the bloodstream.

---

[1] Together, the contaminant concentration and bioavailability in a source dictate the exposure that an organism will experience. This exposure then dictates the probability and intensity of effects to the organism, i.e., the risk associated with exposure.

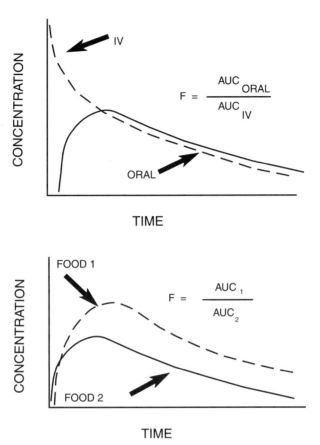

Figure 4.1 Estimation of absolute bioavailability of an ingested dose by comparison of AUCs for ingestion and intravenous injection (top panel) and relative bioavailability by AUC comparison for the same dose administered in two different foods (bottom panel).

For example, Equation 4.1 might be used to estimate the bioavailability of a dose of ingested compound:

$$F = \frac{AUC_{oral}}{AUC_{iv}} \tag{4.1}$$

where AUC_{oral} = AUC for dose D administered orally and AUC_{iv} = AUC for dose D administered intravenously. Figure 4.1 (top panel) shows two curves from which AUCs would be generated with methods such as trapezoidal estimation (Gibaldi, 1991; Newman, 1995). Similarly, **relative bioavailability** can be estimated for two different sources, with one source serving as a reference. Figure 4.1 (bottom panel) shows the blood concentration-time curves used to estimate relative bioavailability of a dose (D) administered in Food 1 and then in Food 2.

With some modification, bioavailability can also be estimated if different doses were used for the various routes of introduction or administered forms (Newman, 1995). Assuming that AUC is a linear function of dose within the range of doses used:

$$F = \frac{AUC_{oral} D_{iv}}{AUC_{iv} D_{oral}} \tag{4.2}$$

where D_{iv} and D_{oral} = doses administered by intravenous injection and orally, respectively (Gibaldi, 1991). However, if half-lives were different for the two routes of administration, the associated difference in clearance is usually corrected by using the following equation instead of Equation 4.2 (Gibaldi and Perrier, 1982):

$$F = \frac{AUC_{oral} D_{iv} (t_{1/2})_{iv}}{AUC_{iv} D_{oral} (t_{1/2})_{oral}}$$ (4.3)

Other AUCs can be applied to estimate bioavailability under various conditions. For example, AUCs for concentration-time curves of urine or bile could be used effectively for some compounds.

With statistical-moments analysis of the AUCs or parameter estimation for the models described in Chapter 3, mean residence times (MRT, the mean time of drug residence in a compartment, similar to τ described in Chapter 3) can be calculated (Yamaoka et al., 1978). The rate of drug or contaminant absorption is estimated with MRTs for various routes of introduction as the difference between MRTs. The **mean absorption time** (MAT) for a drug or contaminant in food is $MRT_{food} - MRT_{iv}$ (Gibaldi, 1991). Assuming first-order kinetics, an **absorption rate constant** (k_a) can be calculated to be $MAT = k_a^{-1}$ (Gibaldi, 1991). Such parameters are meaningful for estimating how quickly an ingested, inhaled, imbibed, or otherwise-introduced toxicant becomes available at a target site within the organism.

II CHEMICAL QUALITIES INFLUENCING BIOAVAILABILITY

II.A Inorganic Contaminants

II.A.1 Bioavailability from Water

Water chemistry affects bioavailability by modifying the chemical species present and the functioning of uptake sites. For example, pH has an obvious effect on the equilibrium, $NH_3 + H^+ \leftrightarrow NH_4^+$. The resulting distribution of ammonia species is important to understand because the neutral NH_3 passes much more readily through the cell membrane than the charged NH_4^+ (Lloyd and Herbert, 1960) and, consequently, is the more bioavailable form of ammonia. Similarly, the bioavailability of cyanide (HCN), a weak acid but an extremely potent inhibitor of cytochrome oxidase, is influenced by the effect of pH on the equilibrium, $H^+ + CN^- \leftrightarrow HCN$ (Broderius et al., 1977). (Bioavailablity of cyanide can also be modified by iron (Fe(II)), which combines with cyanide to form a less-toxic ferrocyanide, $Fe(CH)_6^{-4}$ [Manahan, 1993].) Prediction of the amount of bioavailable toxicant is complicated for both cyanide and dissolved sulfide (H_2S, HS^-, and S^{2-}) because more than one species contributes simultaneously to toxicity (Broderius et al., 1977).

The bioavailabilities of dissolved metals and metalloids are also affected by chemical speciation. Metal cations compete with other cations for dissolved **ligands**,[2] anions, or molecules that form coordination compounds and complexes with metals (Newman and Jagoe, 1994). Ligands forming complexes with metals include dissolved organic compounds and inorganic species. Natural organic ligands such as the humic and fulvic acids have a wide range of functional groups. Among the most important in complexation are carboxylic and phenolic groups. The major inorganic species involved in metal complexation in freshwater and saltwater include $B(OH)$, $B(OH)_4^-$, Cl^-, CO_3^{2-}, HCO_3^-, F^-, $H_2PO_4^-$, HPO_4^{2-}, NH_3, OH^-, $Si(OH)_4$, and SO_4^{2-} (Öhman and Sjöberg, 1988; Newman and Jagoe, 1994). The ligands, NH_3, HS^-, and S^{2-}, are important in anoxic waters. Of course, H_2O is an important ligand that forms a hydration sphere around cations and, in so doing, influences bioavailability. As discussed in Chapter 3, size and charge of a hydrated cation influences its passage

---

[2] Ligands share electron pairs with metals. The ligand is a **monodentate ligand** if only one pair is shared, and it is multidentate if more than one pair is shared. **Multidentate ligands** are also called **chelates**. Multidentate ligands are prefixed, e.g., bi- or tridentate, to indicate the number of electron pairs involved.

through membrane protein channels. At thermodynamic equilibrium, the distribution of a particular dissolved cation among its various species can be estimated as a function of competing cation concentrations, pH, ligand concentrations, temperature, and ionic strength. These predictions, or directly measured concentrations of free (aquated) ion, can be used to normalize metal concentrations under a variety of conditions to better estimate bioavailable metal.

As a general rule, bioavailability or toxicity is correlated with the free-metal concentration (Allen et al., 1980; Andrew et al., 1977; Dodge and Theis, 1979; Borgmann, 1983). This common observation has led to the reasonable suggestion that the free ion is often the most bioavailable form of a dissolved metal. In fact, this concept is sufficiently prevalent as to be given a name, the **free-ion activity model**[3] or FIAM. Campbell and Tessier (1996) define the FIAM as "the universal importance of free metal ion activities in determining the uptake, nutrition and toxicity of all cationic trace metals." It must be kept in mind that the concentrations of other species are also correlated with that of the free-ion concentration (or activity), and their bioavailability is often difficult to define independently as a consequence. Further, it is not always the case that the free ion is the most, or only, available form of a dissolved metal. Simkiss (1983) noted that neutral complexes of some Class B metals (e.g., $Hg(Cl)_2^0$) can be very lipophilic relative to charged species (e.g., Hg^{2+}), and this extreme lipophility combined with the dominance of neutral chloro complexes in marine systems can greatly enhance their bioavailability.

Competition at uptake sites is also modified by water chemistry. Crist et al. (1988) described the competition of H^+ on the initial adsorption of metals to sites (carboxylic groups of pectin) on algal cells. They also showed the interaction of dissolved metals with Ca^{2+}, Mg^{2+}, and Na^+ at algal cell binding sites. The influences of pH and major cations on silver (Janes and Playle, 1995) and aluminum (Wilkinson and Campbell, 1993) uptake on gills have been quantified based on competition for binding to gill surface sites (surface-associated ligand groups). Campbell and Tessier (1996) discussed several studies of competition of metals with hardness cations (Ca^{2+} and Mg^{2+}) at various biological surfaces. Although not the complete explanation, competition between hardness cations and metals is suggested as one reason for the decrease in bioavailability or toxicity of metals often measured following an increase in freshwater hardness.

The water chemistry of the microlayer at a biological surface can be extremely important. Excretion of NH_4^+, NH_3, HCO_3^-, and CO_2 from the gills can rapidly modify the chemistry of water as it passes over gill surfaces (Newman and Jagoe, 1994). Depending on the bulk water chemistry, the shift in water chemistry at surfaces can be sufficient to modify bioavailability of metals such as aluminum (Exeley et al., 1991; Neville and Campbell, 1988; Playle and Wood, 1989; 1991).

II.A.2 Bioavailability from Solid Phases

The bioavailability of inorganic contaminants in aerosols, food, sediments, and other solid phases of the environment is difficult to predict precisely. However, some general themes do emerge from the literature. The direct availability from solid phases is only one part of the story. For example, bioavailability from a particular solid phase such as sediments can be determined by a metal's capacity to partition into the interstitial waters. These general phenomena will be discussed with examples here.

The bioavailability of metals or metalloids in aerosols and larger particulates suspended in air is determined not only by their chemical forms in the solid but also by the size of the particulates and the distribution of the element within the particulates. As an example, arsenic tends to condense onto outer layers of smaller coal fly ash particles as they move up the smokestacks of coal-burning power plants, and because of this surface deposition, the arsenic is more available than if it were

---

[3] The activity is used here instead of concentration to encompass situations where there is significant nonideal behavior of ion concentrations due to interionic interactions. In very dilute solutions, the distinction is not as important as in more concentrated solutions such as seawater where the activity coefficients are necessary to relate concentration and activity. By convention, the activity coefficient (activity/concentration) is 1 with infinite dilution.

uniformly distributed throughout small to large ash particles (Hulett et al., 1980; Wangen, 1981). Lead halides in automobile exhaust are more readily available for dissolution in the lung after inhalation than lead in road dust that has weathered to compounds such as lead sulfate (Laxen and Harrison, 1977). Also, lead is present at highest concentrations in small particles of road dust that gain deeper access to the lungs than large particles (Biggins and Harrison, 1980). In humans, larger particles are removed by nasal hair, and associated contaminants are unavailable as a consequence. Particles with diameters of 5 to 10 µm gain entry only to the region of the pharynx. However, those with diameters of 1 µm or less go much deeper into the terminal bronchioles and aveoli (Cordasco et al., 1995). Clearly, the depth of passage and consequent bioavailability of contaminants in inhaled particles are related to particle size.

Bioavailability of contaminants in food is a function of many factors. Just as with inhaled particulates, the size of a food particle can determine bioavailability as well as the chemical form of the affiliated contaminants. Particle size of materials passing though the human gut can modify bioavailability of some contaminants and drugs (Gibaldi, 1991). Bivalve molluscs have complex sorting mechanisms on the gills and palps, in the gut, and in the digestive diverticula that are strongly affected by particle size. The size of the food particle will determine the degree to which it participates in these different digestive processes. Some small particles, such as iron oxide and iron saccharate, can even be taken into phagocytes while still on the gill of oysters (Galtsoff, 1964). This process of sorting and digestion is further complicated by environmental factors such as temperature and tidal rhythm, which modify feeding behavior and digestive processes of bivalves (Morton, 1970).

Literature describing the bioavailability of elements ingested by humans provides several telling examples of additional factors modifying assimilation. Diet can have a strong impact on bioavailability; we have all been instructed about the do's and don't's of eating while taking various medications. Chronic zinc deficiency and consequent dwarfism found in regions of the Middle East are linked to the poor zinc availability in the predominantly cereal diet of these peoples and the custom of eating clay (Sandstead, 1988). Zinc is more available in meats relative to cereals, and clay in the diet sequesters zinc during its passage through the intestine. Protein in the diet increases zinc availability, likely due to enhanced absorption of zinc that is chelated by histidine and cysteine (Sandstead, 1988). In contrast, a protein-rich diet reduces calcium bioavailability.

Bioavailability of ingested lead to humans and nonhuman species is a topic of much-deserved attention. The lead in paint chips is all too bioavailable to small children ingesting them. Many avian poisonings also involve lead. The bird species of concern are those feeding in wetlands or fields spattered with shotgun pellets from sporting activities. Lead sinkers from the bottom of fishing ponds may also be a source of lead to large waterfowl such as swans or geese (Pain, 1995). Raptors feeding on birds containing lead shot are exposed too (Pain, 1995). Indeed, 338 of 4300 dead bald eagles that were examined in a survey by the U.S. Department of Interior were found to have been poisoned by lead in shot or bullet fragments from their prey (Franson et al., 1995). Shot is retained in the gizzard of birds as gizzard stones, and lead is slowly released under the associated grinding and acidic conditions (Amiard-Triquet et al., 1992).

Bioavailability of metals and metalloids in sediments is not easily estimated (Luoma, 1989); however, it is thought to be determined by concentrations in interstitial water and concentrations in various solid phases. The solid-phase concentrations influence bioavailability by dictating the concentrations of metal in the interstitial waters surrounding the biota and by direct ingestion of solids by benthic species.[4] Because total sediment-associated metal concentration can be a poor indicator of available metal (Tessier et al., 1984), most estimates of bioavailable metals depend on partial

---

[4] Based on this premise, Campbell et al. (1988) classified sediment-associated organisms as Type A (those in contact with sediments but unable to ingest particles) and Type B (those capable also of ingesting particles). Examples of **Type A organisms** include benthic algae or rooted macrophytes. Examples of **Type B organisms** are detritivores or many suspension feeders. With this distinction, bioavailability is discussed for the two classes relative to interstitial water concentrations only (Type A) or solid phase plus interstitial water (Type B).

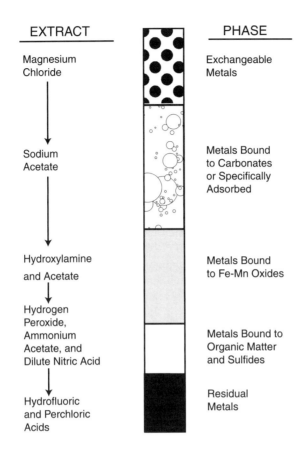

| EXTRACT | PHASE |

Magnesium Chloride → Exchangeable Metals

Sodium Acetate → Metals Bound to Carbonates or Specifically Adsorbed

Hydroxylamine and Acetate → Metals Bound to Fe-Mn Oxides

Hydrogen Peroxide, Ammonium Acetate, and Dilute Nitric Acid → Metals Bound to Organic Matter and Sulfides

Hydrofluoric and Perchloric Acids → Residual Metals

Figure 4.2   Fractionation of sediment-associated metals by sequential extraction. In the scheme shown here (Tessier et al., 1984), an aliquot of sediment is sequentially extracted with magnesium chloride; sodium acetate; hydroxylamine and acetate; hydrogen peroxide, ammonium, acetate and nitric acid; and, finally, strong acid to produce five different extracts. These extracts are thought to grossly reflect the amount of metal in the forms noted at the right side of the figure. These are operational definitions, and the phase descriptions often do not accurately reflect the true phase association of the extracted metals.

extractions, such as a $1 \ N$ HCl extract (Krantzberg, 1994) or a series of sequential extractions of sediments thought to grossly separate particular metal-binding fractions of the solid sediments (e.g., Tessier et al., 1979; Babukutty and Chacko, 1995). As an example of a single extractant, sediment-bound metals extracted into an EDTA solution were correlated with bioavailability to several marine invertebrates (Ray et al., 1981). Relative to sequential extractions, Tessier et al. (1984) found that bioaccumulation of sediment-associated metals in a freshwater mussel was best correlated with metal concentrations in the more easily extracted fractions in the extraction series shown in Figure 4.2.

For oxic sediments, several general trends can be identified regarding metal bioavailability. First, easily extracted ($1 \ N$ HCl) iron, notionally reflecting iron hydrous oxides, tends to inhibit metal bioavailability (Newman, 1995). Presumably this reflects the avid binding of metals to oxides in oxic sediments. Consequently, metal concentrations in a $1 \ N$ HCl sediment extract can be normalized to the simultaneously extracted ("easily extracted") iron to account for this effect (e.g., Luoma and Bryan, 1978). Although less consistent than this effect of iron, an increase in sediment organic carbon can diminish bioavailability for some metals (Crecelus et al., 1982), and metal concentrations can be normalized to sediment organic carbon content. Finally, more easily extracted fractions in sequential extractions tend to be more bioavailable than more tightly bound metals (Tessier et al., 1984; Rule and Alden, 1990; Young and Harvey, 1991; Newman, 1995).

For anoxic sediments, bioavailabilities of some metals (e.g., cadmium, chromium, lead, mercury, and nickel) are correlated with sulfide concentrations, as reflected by the **acid volatile sulfides** (AVS, sulfides extracted with cold HCl, believed to be predominately iron and manganese sulfides). The presence of sufficient amounts of AVS sequesters the sediment-associated metals as highly insoluble metal sulfides. An equilibrium between the extremely insoluble metal sulfides and the large amounts of iron and manganese sulfides is established, favoring metal precipitation (right side of the equation): $Cd^{2+} + FeS_{(S)} \leftrightarrow CdS_{(S)} + Fe^{2+}$. This maintains very low interstitial water concentrations of toxic metals and, according to Di Toro et al. (1990), renders them unavailable, especially those metals that bind avidly to sulfur. For anoxic sediments, a cold HCL extract can be analyzed for AVS and simultaneously extracted metals (SEM). The metal concentrations (SEM) can be normalized to AVS (e.g., Ankley et al., 1991; Carlson et al., 1991). Di Toro and coworkers argue that, for values of SEM/AVS less than 1, the metal is precipitated as sulfide and relatively unavailable to have a toxic effect on associated benthic species. However, the metals to which this method has been applied have been Class B metals, and generalizations are not yet warranted about all heavy metals. Recently, Long et al. (1998) indicated that SEM/AVS normalization was not more accurate than simply normalizing amounts of metal to dry weight of sediment. Lee et al. (2000) published an article in *Science*, concluding that "[their] evidence refutes the prevailing view that the bioavailability of toxic metals is regulated by metal interactions with porewater and reactive sulfides." The conclusion can only be that the increasingly common application of the SEM/AVS approach is unwarranted without further study of its limitations and appropriate application. It would be risky to apply this promising technology to produce many more scientific statements about bioavailability or to regulate sediment-associated toxic metals without further scrutiny.

II.B   Organic Contaminants

II.B.1   Bioavailability from Water

Bioavailability of organic compounds from water and other sources has been described with **structure-activity relationships** (SARs) that use molecular qualities of the organic compound to predict activity, i.e., bioavailability. Such qualitative relationships often predict changes in activity of a drug or toxicant with changes to a parent molecular structure, such as the addition of a chloride atom or the removal of a methyl group. If expressed quantitatively, SARs become **quantitative structure-activity relationships** or QSARs. A QSAR is a quantitative, often statistical, relationship between molecular qualities and bioactivity, i.e., bioavailability or toxicity. Molecular qualities include measures of lipophilicity, steric conformation, molecular volume, reactivity, etc. But, in ecotoxicology, the most commonly used are those based on measures of lipophility, such as K_{OW}, water solubility, or K_{TW} of organic compounds (Figure 4.3). Historically, partitioning between n-octanol and water has been the most common phase-partitioning procedure thought to reflect partitioning between water and lipids of organisms (e.g., Mackay, 1982). Chiou (1985) used triolein (glyceryl trioleate)-water partitioning (i.e., K_{TW}) to better reflect partitioning between water and triglycerides in organisms (Figure 4.3C). N-octanol–water partitioning was found to accurately reflect partitioning between triolein and water to a log K_{OW} of 6. Above 6, the K_{OW} dropped slightly below expectations for perfect linear concordance between K_{TW} and K_{OW}. This suggested that application of log K_{OW} to prediction partitioning with lipids might be slightly biased above a log K_{OW} of 6.

In the simple K_{OW} approach, the organism is envisioned as a membrane-enveloped pool of emulsified lipids. In this conceptual model, uptake and elimination are controlled by permeation of the membrane and/or permeation through aqueous phases (Connell, 1990), with the predominant

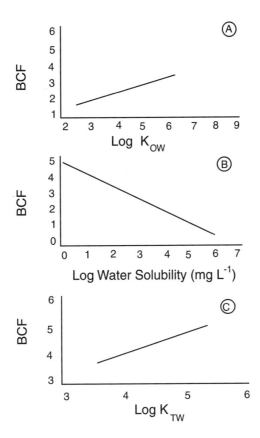

Figure 4.3   Relationships between bioconcentration factors (BCF) for a variety of organic compounds and log K_{OW} (trout muscle), log water solubility (marine mussel), and K_{TW} or log of triolein-water partition coefficient (log K_{TW}, pooled data for rainbow trout and guppies). (Panels A, B, and C are modified, respectively, from Figure 1 in Neely, W.B. et al., *Environ. Sci. Technol.*, 8, 1113–1115, 1974; Figure 1 in Geyer, H. et al., *Chemosphere*, 11, 1121–1134, 1982; and Figure 3 in Chiou, C.T., *Environ. Sci. Technol.*, 19, 57–62, 1985.)

process being dictated by the qualities of the specific compound in question. Small hydrophilic molecules are controlled by membrane permeation, but large hydrophobic compounds are controlled by permeation through aqueous phases. Very large molecules (e.g., molecular size ≥9.5 Å) (Landrum et al., 1996) will not pass through the lipid membrane. Uptake across the membrane is dictated by Fick's Law (Equation 3.5).

The consequence of these behaviors of organic compounds varying in size and lipophilicity can be illustrated by the work of Connell and Hawker (1988) (Figure 4.4). In the top panel of this figure, log of the uptake constant is plotted against log K_{OW} for a series of chlorinated hydrocarbons accumulating in three species of fish. Among log K_{OW} values of approximately 3 to 6, uptake increases linearly with K_{OW} and is thought to be controlled by membrane permeation. Above a log K_{OW} value of 6, the rate of increase slows, and eventually uptake begins to decrease with increasing K_{OW} as the large molecular size of the most-lipophilic compounds begins to impede diffusion in aqueous phases of the fish. Elimination (log k_e^{-1}) is controlled by membrane diffusion at low K_{OW} values and then becomes linear (diffusion controlled, see Equation 3.5) between log K_{OW} values of approximately 2.5 to 6.5. Above this point, the lipid solubility begins to deviate from a perfect correlation with n-octanol-based predictions: K_{OW} becomes an increasingly poorer surrogate of

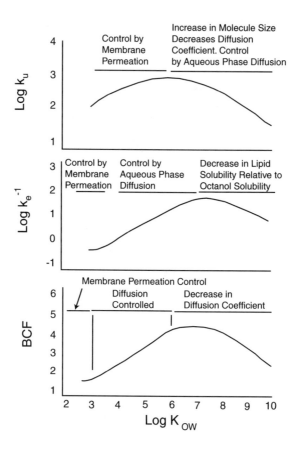

Figure 4.4 A summary of the changes in uptake, elimination, and bioaccumulation (bioconcentration factor, or BCF) as related to K_{OW}. (Modified from Figures 3, 4, and 5 in Connell, D.W. and D.W. Hawker, *Ecotoxicol. Environ. Saf.,* 16, 242–257, 1988.)

lipid-water partitioning as compounds become increasingly lipophilic. Chiou (1985) explained this deviation quantitatively with the **Flory-Huggins theory**, relating solubility to solvent molecular size. The discrepancy between molecular size of n-octanol and lipids of organisms becomes increasingly important for larger, more lipophilic compounds. The net result of these factors on uptake and elimination is shown in the bottom panel of Figure 4.4. Below approximately log K_{OW} of 3, bioaccumulation of the most water soluble compounds is controlled by permeation of the membrane. Bioaccumulation is determined primarily by diffusional processes between log K_{OW} values of 3 to 6. Above this range, the relationship is strongly influenced by effects of molecular size on diffusion, and BCF drops with increasing log K_{OW}.

Other factors contribute to bioavailability of organic compounds from water. The above model would not apply for compounds undergoing extensive biotransformation. Bioavailability of ionizable organic compounds would be influenced by pH, as already described, and for ingested, ionizable contaminants (soon to be described in greater detail). Models (QSARs) can be developed to include lipophilicity and ionization (e.g., Lipnick, 1985). Other factors can also be included in QSARs. Classic models based on the Hansch equation equate bioactivity to hydrophobicity, electronic, and steric qualities of molecules (Lipnick, 1995). **Linear solvation energy relationships** (LSER) are based on molecular volume, ability to form hydrogen bonds, and polarity or ability to become polarized (Blum and Speece, 1990).

II.B.2 Bioavailability from Solid Phases

For ionizable contaminants, gastric pH can influence availability, with the direction of effect being determined by the pK$_a$ of the contaminant.[5] (The pK$_a$ is $-\log_{10}$ of the ionization constant (K$_a$) for a weak Brønsted acid, where K$_a$ = ([H$^+$][X$^-$])/[HX].) Weakly acidic organic compounds with pK$_a$ values greater than 8 are un-ionized in the human gastrointestinal tract, but bioavailability of those with pK$_a$ values between 2.5 and 7.5 is pH-sensitive (Gibaldi, 1991). According to the **pH-partition hypothesis** (Shore et al., 1957; Wagner, 1975), bioavailability is determined by diffusion of the un-ionized form from the gastrointestinal lumen across the "lipid barrier" of the gut lining and into the tissues, as determined by pK$_a$ and pH. The proportion of the compound remaining un-ionized can be estimated with the **Henderson-Hasselbalch relationship** for monobasic acids (Equation 4.4) and monacidic bases (Equation 4.5) (Wagner, 1975):

$$f_u = \frac{1}{1 + 10^{\,pH - pK_a}} \tag{4.4}$$

$$f_u = \frac{1}{1 + 10^{\,pK_a - pH}} \tag{4.5}$$

However, the ionized form can also contribute to bioavailability (Wagner, 1975; Gibaldi, 1991). In such cases, the diffusion rates and amounts of both un-ionized and ionized forms of the contaminant contribute to estimation of bioavailability.

The K$_{OW}$ influences bioavailability of lipophilic compounds in food. Spacie et al. (1995) noted a maximum availability of contaminant in food at log K$_{OW}$ values of 6, with uptake being lower for very hydrophobic and large compounds. They discuss several studies indicating low bioavailability to fish of many lipophilic organic contaminants in food. Donnelly et al. (1994) indicated that organic compounds in soils with log K$_{OW}$ values of 4 to 7, such as PCBs, are quickly absorbed to soils and, consequently, are not readily available to terrestrial plants. In contrast, compounds such as many pesticides with log K$_{OW}$ values of approximately 1 to 2 are taken up more easily by plants.

In sediments, bioavailability to benthic species usually decreases with increasing log K$_{OW}$ (Landrum and Robbins, 1990). This is likely due to the enhanced partitioning of nonpolar organic compounds to the sediment solid phases, with consequent low concentrations in the interstitial waters. Any increase in sediment organic carbon content can diminish the bioavailability of nonpolar organic compounds, much as AVS decreases bioavailability of metals in anoxic sediments.[6] A maximum bioavailability has been noted at a log K$_{OW}$ of approximately 6 for some series of chlorinated hydrocarbons (Landrum and Robbins, 1990).

---

[5] Of course, the bioavailability as influenced by gut pH is more complicated than this because the pH conditions of various regions of the gut are quite different. Weakly basic drugs will be more rapidly absorbed in the small intestine than in the acidic stomach. However, many acidic compounds can also be absorbed more effectively in the small intestine than the stomach because of the large amount of surface area of the small intestine relative to the stomach. Consequently, **gastric emptying rates** (the rate at which the contents of the stomach are emptied into the small intestine) will also influence bioavailability for many substances, e.g., paracetamol (Heading et al., 1973), by modifying the time that a compound remains under different pH conditions.

[6] This general partitioning of contaminants between solid and dissolved phases of sediments with consequent effects on bioavailability and toxicity (bioactivity) is the foundation for the equilibrium partitioning approach to sediment criteria development. See Shea (1988) and Di Toro et al. (1991) for more details.

III BIOLOGICAL QUALITIES INFLUENCING BIOACCUMULATION

III.A Temperature-Influenced Processes

Temperature is perhaps the most widely studied and important factor affecting the general physiology of individual organisms. This being the case, it should be no surprise that temperature has also been found to influence biochemical and physiological processes associated with bioaccumulation. Indeed, the strong positive relationship noted in the 1960s between metal (zinc or radiocesium) excretion rate and temperature-dictated metabolic rates of poikilotherms (Mishima and Odum, 1963; Williamson, 1975) and homeotherms (Pulliam et al., 1967; Baker and Dunaway, 1969) led researchers to explore ^{65}Zn elimination as a means of measuring metabolic rates of free-ranging individuals. Unfortunately, such use was compromised because free-ranging animals and laboratory-maintained animals often differed in other ways that significantly influenced elimination rates (e.g., Pulliam et al., 1967). These early studies identified much inexplicable variability and bias in results. It should be remembered that temperature also determines important rates, such as those for feeding, growth, and egestion in addition to important cellular qualities such as membrane fluidity and lipid composition.

Generally, increases in temperature within normal physiological ranges have been shown to increase bioaccumulation, e.g., mercury in mayfly nymphs (Odin et al., 1994), cadmium and mercury in molluscs (Tessier et al., 1994), cadmium in Asiatic clams (Graney et al., 1984), and DDT in rainbow trout (Reinert et al., 1974). Cesium (^{134}Cs) uptake was highest at temperatures optimal for food consumption and growth of rainbow trout (Gallegos and Whicker, 1971). The biological half life ($t_{1/2}$) for elimination of ^{134}Cs from rainbow trout increased with increased water temperatures according to an exponential relationship,[7] $t_{1/2} = (\text{Constant})e^{-0.106t}$, where t = temperature in °C (Ugedal et al., 1992). For the rainbow trout, retention of methyl mercury was approximately 1.5 times longer at 0.5 to 4.0°C than at water temperatures of 16 to 19°C (Ruohtula and Miettinen, 1975).

Some studies report no effect of temperature changes on bioaccumulation kinetics. For example, methyl mercury uptake and elimination by freshwater clams were not significantly affected by temperature (Smith and Green, 1975). Effects of temperature are sometimes complex and inconsistent with the general trends noted here.

Watkins and Simkiss (1988) gave a fascinating example of such a complication. They, like others, found enhanced accumulation of a contaminant (zinc in the marine mussel, *Mytilus edulis*) with an increase in water temperature (10°C increase to 25°C). But they also examined the effect of fluctuating temperatures between 15 and 25°C, and they found that bioaccumulation of zinc was even higher than the constant 25°C treatment! They hypothesized that this result was linked to the shifts in zinc among various pools within the mussels. Both free and ligand-associated zinc are present in the mussels: $Zn^{2+} + L^{2-} \leftrightarrow ZnL$. When water temperatures increase, the equilibrium for this complexation shifts to favor more zinc existing free of the biochemical ligands. This zinc then becomes more available for incorporation into granules. Upon cooling, zinc entering from outside the animal establishes equilibrium with ligands again, thus replenishing the zinc lost to granules. As temperatures increase and decrease, the process is repeated with enhanced accumulation of zinc in granules. The cyclic association with ligands, shift toward dissociation from ligands, and sequestering in granules ratchets zinc concentrations higher than if the mussel were left at a higher temperature of 25°C.

---

[7] Both power and exponential relationships will be discussed in this chapter. A **power relationship** is a mathematical relationship in which the Y variable is related to the X variable raised to some power. For example, $Y = aX^b$. A power relationship can be linearized and conveniently fit by linear regression by taking the log X and log Y in the regression (log $Y = b \log X + \log a$). In contrast, an **exponential relationship** is a mathematical relationship in which the Y variable is related to some constant raised to the X variable, i.e., $Y = a10^{bX}$. An exponential relationship is transformed to the form log $Y = bX + \log a$ as done earlier to fit exponential (first order) elimination kinetics. Newman (1993) provides details on fitting these models using linear regression.

III.B Allometry

Allometry, the study of size and its consequences (Huxley, 1950), is also important in estimating bioaccumulation. Metabolic rate and a myriad of other anatomical, physiological, and biochemical qualities of organisms change with size (see Adolph [1949] and Heusner [1987] for more detail), and in so doing, uptake, transformation, and elimination rates are modified. The commonly observed consequence is size-dependent bioaccumulation. Unfortunately, because age and size are correlated in most species, allometric effects are often confused with age or exposure duration effects in surveys of bioaccumulation. Regardless, many studies detail allometric effects on contaminant uptake (Newman and Mitz, 1988; Schultz and Hayton, 1994), biotransformation (Walker, 1978) (Figure 4.5), elimination (Mishima and Odum, 1963; Reichle, 1968; Gallegos and Whicker, 1971; Ugedal et al., 1992), and general bioaccumulation (Boyden, 1974; 1977; Landrum and Lydy, 1991; Warnau et al., 1995). These have been reviewed for metals bioaccumulation by Newman and Heagler (1991), for linkage to metabolic rate by Fagerström (1977), and for scaling of pharmacokinetic parameters by Hayton (1989). Most resort to the classic power model for **scaling**, the manipulation of allometric data (size versus some physiological, morphological, or biochemical quality) to produce a quantitative relationship.

The general power function used in allometric scaling is depicted in Equation 4.6:

$$Y = aX^b \tag{4.6}$$

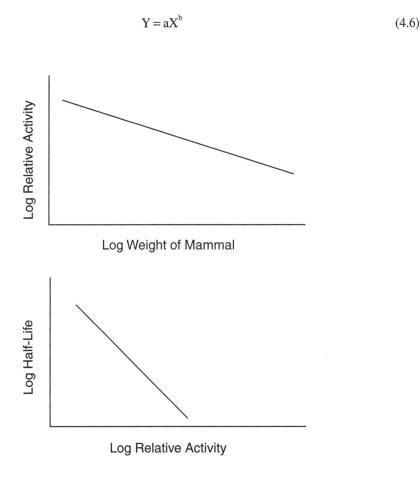

Figure 4.5 The scaling of monooxygenase activity (top panel) and the relationship between relative monooxygenase activity and biological half-life of xenobiotic metabolism (bottom panel). (Composite and modification of Figures 2 and 3 in Walker, C.H., *Drug Metab. Rev.*, 7, 295–323, 1978.)

where Y = some quality being scaled to size, X = some measure of individual (or species) size, and a and b are constants normally derived by regression analysis. Equation 4.6 can be used to link size to morphological (e.g., gill surface area), physiological (e.g., blood flow rate or gill ventilation rate), or biochemical (e.g., monooxygenase activity) qualities. It has also been used to model contaminant body burden to animal size. To express the relationship for bioaccumulation in concentration units, Equation 4.6 can be easily converted by dividing both sides by mass to generate Equation 4.7:

$$Y = aX^{1-b} \qquad (4.7)$$

where X, a, and b are defined as in Equation 4.6, but Y = concentration, not body burden. Much has also been made of the values estimated for b for scaling bioaccumulation and associated parameters, as this parameter is a major theme in the classic physiological literature.

In the mid-1970s, Boyden (1974; 1977) compiled metal body-burden data for mollusks and established a power model for scaling contaminant accumulation. He identified three classes of models based on their associated b-values (Figure 4.6). One class had b-values of 1 and reflected a simple proportionality. According to Boyden, there was a constant number of binding sites in the tissues regardless of size: Body burden increased linearly with weight of the organism, and concentration was independent of weight. Another relationship with b-values in the general range of 0.77 had body burdens and concentrations that changed in a nonlinear fashion with weight, with higher concentrations

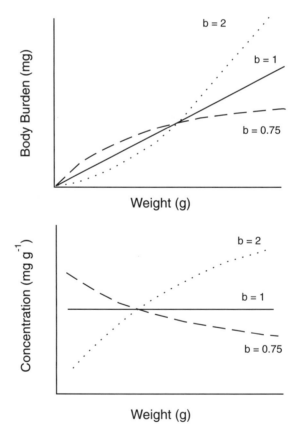

Figure 4.6   The relationships between organism size (weight) and body burden (amount per individual) (top panel) and concentration (amount per g of tissue) of metal (bottom panel). Modified from Figure 13 of Newman (1995) with slight exaggeration of curvature of the concentration-vs.-size plots to make clear the nonlinear nature of two of these curves (b = 2 and b = 0.75). The point of intersection for both plots occurs where X = 1. (Reprinted with permission from *Quantitative Methods in Aquatic Ecotoxicology.* Copyright 1995, Lewis Publishers, an imprint of CRC Press, Boca Raton, Florida.)

for small individuals than large individuals. The b-value of 0.77 suggested a linkage to metabolic rate that, in general, has a b-value of approximately 0.75 during scaling. However, Fagerström (1977) quickly pointed out that one cannot compare b-values for states (body burden) and fluxes (metabolic rate) in this manner. He demonstrated mathematically that one would expect a b-value of 1 for contaminant burdens driven by metabolic rate. Thus the elegant hypothesis of Boyden was shown to be an inadequate explanation. The final class noted by Boyden had a b-value of 2 or greater (burden related to the square of body weight) and reflected a gradual increase in concentration with an increase in size. He suggested that elements (such as cadmium) that are subject to rapid removal from circulation and very avid binding in some tissues conform to this class of relationships. He suggested incorrectly that these relationships (b-values) would be constant for species-element pairs. Soon after Boyden proposed this constancy of b-values, Cossa et al. (1980) and Strong and Luoma (1981) demonstrated considerable spatial and temporal variation in b-values. Further, Newman and Heagler (1991) added to and reanalyzed Boyden's original data set, and they found no clear evidence for distinct classes of relationships based on b-values. Indeed, they found a generally skewed distribution of b-values with medians in the range of 0.80 to 0.83. They also identified several sources of bias in Boyden's approach. Regardless, Boyden made a major contribution by clearly establishing the power model for scaling body burdens and by hypothesizing plausible and testable underlying mechanisms. His approach is used often to normalize body-burden/concentration data taken during surveys of organisms of differing sizes. In such cases, the empirically derived constants are adequate for normalization of data. Regardless of its value, it should not be forgotten that the foundation for such scaling is empirical.

Another equally important use of scaling is computer modeling to predict general behaviors of bioaccumulation dynamics and to isolate important factors controlling bioaccumulation kinetics. For example, a power model was used to model ^{137}Cs half-life in humans of different sizes (Eberhardt, 1967). In some such cases, estimation of the exact values for parameters may not be critical, and the range of expected values may be sufficient. Today, more-complicated allometric explanations and relationships than Boyden's, usually embedded in physiologically based pharmacokinetics (PBPK) models, are used to predict size effects on bioaccumulation (Barber et al., 1988; Hayton, 1989; Schultz and Hayton, 1994). Examples of such complicated models are given in Figure 4.7. Temperature, scaling, and other important factors can be incorporated into these PBPK models. However, as models become more complicated, the number of estimated parameters for the system being modeled increases.

III.C   Other Factors

Many other factors can complicate prediction of bioaccumulation. For example, bioaccumulation can differ between sexes. The female fin whales studied by Aguilar and Borrell (1994) eliminate organochlorine compounds via transfer to young before birth and during nursing. This is obviously not an avenue of elimination for males. Consequently, older female whales have lower concentrations than older male whales. Diet can also influence bioavailability. Shifts in diet associated with ecological or developmental qualities of an organism have consequences relative to bioaccumulation. Acclimation of euryhaline species can change their ability to cope with a contaminant; e.g., killifish uptake of pentachlorophenol and excretion of a pentachlorophenol glucuronide conjugate decreased with the lowering of salinity (Tachikawa and Sawamura, 1994). Lipid content of an individual can also influence bioaccumulation. Rainbow trout with high total body lipid content (21%) accumulated more pentachlorophenol and had lower elimination rate constants than those with lower (13%) lipid content (van den Heuvel et al., 1991).

Inorganic toxicants can be influenced by normal regulatory processes associated with essential chemicals. For example, potassium body burden of humans has a strong influence on the accumulation of its analog, cesium (Leggett, 1986). Other essential elements[8] such as zinc are carefully regulated within the body. For example, a **cysteine-rich intestinal protein** (CRIP) exists to enhance

---

[8] Mertz (1981) lists the **essential elements** as H, Na, K, Mg, Ca, V, Cr, Mo, Mn, Fe, Co, Ni, Cu, Zn, Cd(?), C, Si, Sn(?), N, P, As, O, S, Se, F, Cl, and I.

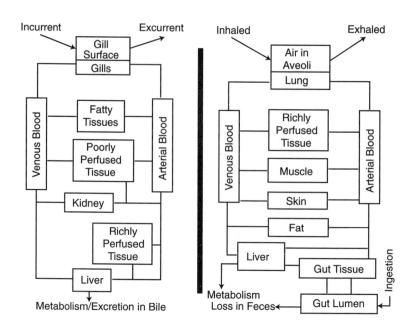

Figure 4.7   Two PBPK models. The model to the left involves the fish uptake of a xenobiotic across the gills, distribution among five compartments and blood, and loss via the liver (Nichols et al., 1990). The model on the right involves styrene inhalation and ingestion by a mammal, distribution into a series of compartments, and loss from liver metabolism and defecation (Paterson and Mackay, 1987). In the original publications, the fish model was formulated as a clearance volume-based model and the mammalian model was formulated as a fugacity-based model.

the uptake of zinc by cells in the intestinal wall of mammals (Roesijadi and Robinson, 1994). This regulation of essential elements influences the availability or effects of nonessential analogs such as cadmium. An excess of an essential element analog such as cadmium can cause symptoms of an apparent (zinc) deficiency (Neathery and Miller, 1975). On the other hand, an excess of zinc can lessen the toxic effects of cadmium (Leland and Kuwabara, 1985). Similarly, copper and zinc interact relative to effects on algal cell function (Rueter and Morel, 1981). Many essential elements and some of their analogs are defined as **biologically determinant**; i.e., their concentrations in the organisms remain relatively constant over a wide range of environmental concentrations (Reichle and Van Hook, 1970). Other elements are **biologically indeterminant**, and their concentrations in organisms are directly proportional to environmental concentrations.

Perhaps less direct, but as telling, is the role of phosphate biochemistry on the accumulation of arsenate in marine species. In the phosphate-deficient waters of the Great Barrier Reef, arsenate enters into the normal biochemical pathways involved in phosphate assimilation and is incorporated into zooxanthellae that live in the mantel of the giant clams, *Hippopus hippopus*, *Tridacna maima*, and *T. derasa* (Benson and Summons, 1981). Arsenic then accumulates to extraordinary levels, gaining entrance into the clam via the biochemistry of the symbiotic algae.

Vignette 4.1   Biological Factors Influencing Bioaccumulation in Bivalves

Michael H. Salazar

Applied Biomonitoring, Kirkland, WA

Introduction

Bioaccumulation is the link between the environment and the organism, and factors affecting the physiological condition of organisms can affect bioaccumulation. Therefore, understanding physiological processes is crucial to understanding the factors affecting bioaccumulation and the interaction between chemical and biological processes. Understanding the significance of dietary exposure pathways in aquatic organisms has been one of the most significant recent advancements in the field. Dietary exposure, seasonal cycles, physiological condition, species, behavior, and biological regulation are all important factors controlling bioaccumulation.

Readers may have noticed that approximately one-third of the references cited in the chapter were related to bivalves, the highest representation of any taxonomic group. Bivalves are useful organisms to help identify biological factors controlling bioaccumulation because their biological processes are reasonably well understood, and many of these biological processes facilitate the accumulation of chemicals in bivalve tissues. Because they filter large volumes of water for food and respiration, bivalves are exposed to dissolved and particulate-sorbed chemicals. Filter- and deposit-feeding bivalves selectively capture, sort, and retain small particles in a particular size range that tends to have the highest concentrations of contaminants. Bivalves can also identify which particles are potentially valuable as food and should be passed to the gut for further processing. They can even determine the length of time specific particles remain in the gut for future processing. Bivalves concentrate and integrate these chemicals in their tissues, and their ability to metabolize most chemicals is more limited than that of other animal groups.

Dietary Exposure

The importance of food as a factor affecting bioaccumulation in aquatic invertebrates has only recently been characterized and understood. As an example, algae often constitute the principal food source for filter-feeding and deposit-feeding organisms, and these algae can enhance the bioavailability of chemicals to aquatic invertebrates. Several studies have shown that many species of benthic invertebrates, including bivalves, are facultative feeders with respect to water and sediment exposures, and they can alter their feeding mode depending on short- or long-term changes in environmental conditions. Filter-feeders have the ability to concentrate chemicals from the water column in their tissues (i.e., the dissolved pathway) and from suspended particulate matter (i.e., the particulate pathway). However, assimilation efficiency is often greater from algae than from suspended sediments. Some chemicals present in the water column originate from the sediments, particularly at the sediment/water interface, where particles are easily suspended by natural processes. Because the smallest particles generally have the highest concentrations of chemicals—and because bivalves have the ability to utilize these particles as food—the particulate phase becomes a significant pathway of exposure. Many bivalve species integrate water, sediment, and dietary pathways of exposure, although not all species utilize all pathways under all conditions.

Digestive Physiology

Digestive physiology can be the dominant process controlling bioaccumulation, even though the biological availability of chemicals is controlled by complex interactions between chemistry, food, and organisms. Important factors include digestive enzymes, gut amino acids, gut retention time, and species-specific responses. Whereas much of the literature for other species suggests that organically bound chemicals (metals or organic compounds) are generally not biologically available, organic binding can increase contaminant availability for organisms like bivalves by facilitating transport across biological membranes. Many marine organisms have a digestive system that maximizes extraction of all possible organic carbon from ingested material, particularly when available food sources are low. For example, filter-feeding bivalves can utilize up to 20% of their diet from the organic coatings on inorganic particles under certain conditions. Chemicals associated with those organic coatings are potentially available for bioaccumulation. Although the mechanisms of absorption across the gut are not fully known, ingested chemicals often become biologically available as they are desorbed from sediment particles during digestive processes. This occurs if the pH decreases from ambient external conditions to internal conditions of approximately 5 or 6. However, bivalves can alter relative uptake based on the relative proportion of dissolved chemical and available food. Biomimetic approaches have been developed to estimate bioavailability by testing extracted gut fluids, but this innovative approach can only mimic static rather than dynamic conditions within the invertebrate gut.

Seasonal Cycles

By contrast, it is now well established that concentrations of both metals and organic chemicals in bivalve tissues change seasonally, even when the external exposure concentrations do not change substantially. For metals, these changes have been primarily attributed to seasonal changes in soft-tissue weight. For the deposit-feeding clam *Macoma balthica*, metal concentrations are generally highest in the fall and winter, when tissue weights are low, and lowest during spring and summer, when tissue weights are high. The magnitude of these seasonal differences in metal concentration can vary geographically, but they have been shown to reach a maximum of about five at one test location. For organics, seasonal changes have been attributed to lipid content, which fluctuates with the annual spawning period. For example, organochlorine concentrations in the filter-feeding mussel *Mytilus edulis* are generally highest in spring and summer, when lipids are high, and lowest in fall and winter, when lipids are low. As with metals, the magnitude of these seasonal differences in organochlorine concentration varied geographically, but they also were shown to reach a maximum of five at the test location.

Physiological Condition

As obvious as this may seem in terms of factors affecting bioaccumulation, it is important that an organism survive chemical exposure for bioaccumulation to occur. Unhealthy animals tend to accumulate lower concentrations than expected. Although most of the examples in this vignette have discussed biological factors affecting bioaccumulation, it should be mentioned that accumulated chemicals can act in a toxic fashion to alter filtration, physiology, and growth and to reduce the proportional uptake of additional chemicals. Physiological condition can substantially alter the uptake of contaminants that might otherwise be predictable from their chemical characteristics. Bivalves are extremely flexible in altering their physiological state, adapting to current conditions, controlling the processing of particles, and managing chemical stress. Accurately interpreting these complex interactions requires a good understanding of associated biological processes under a variety of conditions. Understanding the biological processes influencing bioaccumulation has been difficult because the organism influences its internal chemical environment through biological processes.

Species

There are significant differences in the ability of species within a taxonomic group to accumulate chemicals. For example, Mussel Watch monitoring programs have shown that mussels and oysters differ in their ability to accumulate metals. These differences are attributable to the metabolic requirements and need for selected metals. Oysters accumulate significantly higher concentrations of zinc, copper, and silver, whereas mussels accumulate higher concentrations of chromium and lead. These generalities appear to be consistent across genera of the two families. Interestingly, scallops can accumulate high concentrations of cadmium in their tissues, regardless of the external exposure concentration. It is not clear why scallops accumulate such high cadmium concentrations, but as with most other examples provided here, it is probably a combination of chemical and biological factors. Zebra mussels generally accumulate higher concentrations of chemicals than other bivalves due to their efficient retention of smaller particles, higher filtration rates, and higher growth rates. They also accumulate higher concentrations of organic chemicals because of higher concentrations of lipids in their tissues.

Behavior

The behavior of organisms significantly affects their ability to bioaccumulate chemicals. The most direct behavior is avoidance of chemical exposure. For example, a number of marine and freshwater bivalves, including both filter- and deposit-feeders, have been reported to close their valves for short periods to avoid exposure. This phenomenon probably occurs in most, if not all, bivalves regardless of their feeding strategy. Without active filtration and feeding, they cannot be exposed to either aqueous or particulate-bound chemicals. There is also evidence that bivalves can detect threshold concentrations and will actively filter and feed as long as the external concentration is below this threshold. However, even feeding (filter and deposit) at sub-threshold concentrations of chemicals in water or sediment can lead to adverse responses as tissue burdens increase to thresholds where effects begin to occur. By closing their valves and not feeding, growth is also reduced. If representative results are to be obtained in the field, the animals must have been in the study area for a significant period of time. In the laboratory, the animals must have been exposed to the chemicals without their closing, moving out of sediment to avoid exposure to sediments, or cessation of eating to avoid exposure to either chemical-associated food or sediment.

Biological Regulation of Metal Uptake and Elimination

It is important to note that (1) bioaccumulation is the net result of uptake and elimination and (2) there is a continuum in this biological regulation, tempered by the chemical characteristics of each metal. Even in bivalves there are differences associated with metals, species, seasons, and geography. Bivalve responses to metal exposure can vary from accumulation to regulation. Each case must be considered independently. Some metals may accumulate, unregulated throughout the life of the bivalve, while others may be regulated to some degree. Many of these generalizations need more research to verify the effects of other unmeasured factors. It might be useful to assume that most bivalves are at least partial regulators of most metals and are able to compensate for metal uptake through a variety of elimination processes. By measuring accumulation in bivalve tissues over a range of exposure concentrations, it may be possible to predict where the regulatory system becomes overwhelmed and regulation is no longer possible. Interestingly, this also appears to be the water or sediment concentration where adverse biological effects begin to occur.

Prospectus

Characterizing and understanding the processes controlling bioaccumulation is important to interpreting the environmental significance of bioaccumulation data. Due to the interaction between chemical and biological factors and inherent differences among species, environmental conditions, and exposure regimes in laboratory and field testing, there are many pitfalls in attempting to make generalizations about specific biological factors controlling bioaccumulation. However, we are closer to defining those factors and making useful generalizations. The intent of this vignette is to aid in the refinement of data interpretation, the use of bioaccumulation data, and the design of future experiments to collect bioaccumulation data to enhance their utility in monitoring, assessment, and subsequent regulatory and decision-making processes.

Suggested Supplemental Readings

Carey, J., P. Cook, J. Giesy, P. Hodson, D. Muir, J.W. Owens, and K. Solomon, *Ecotoxicological Risk Assessment of the Chlorinated Organic Chemicals,* SETAC Press, Pensacola, FL, 1998.

Meador, J.P., J.E. Stein, W.L. Reichert, and U. Varanasi, A review of bioaccumulation of polycyclic aromatic hydrocarbons by marine organisms, *Rev. Environ. Contam. Toxicol.,* 143, 79–165, 1995.

Phillips, D.J.H., *Quantitative Aquatic Biological Indicators: Their Use to Monitor Trace Metal and Organochlorine Pollution,* Applied Science Publishers Ltd., London, 1980.

Phillips, D.J.H. and P.S. Rainbow, *Biomonitoring of Trace Aquatic Contaminants,* Elsevier Applied Science, London, 1993.

U.S. EPA, *Proceedings National Sediment Bioaccumulation Conference,* February 1998, U.S. EPA, Office of Water, EPA 823-R-98–002, Bethesda, Maryland, 1998.

IV SUMMARY

This chapter briefly discusses factors influencing bioaccumulation and bioavailability of contaminants along with several methods of measuring bioavailability. Solid and dissolved sources of organic and inorganic contaminants were detailed relative to bioavailability. Finally, the importance of temperature- and size-dictated changes in biological functions and structures were discussed.

Other important factors influencing bioaccumulation were identified, including sex, diet, lipid content, and elemental essentiality. Many details associated with accumulation from trophic exchange were omitted because they are discussed thoroughly in Chapter 5.

SELECTED READINGS

Campbell, P.G.C., A.G. Lewis, P.M. Chapman, A.A. Crowder, W.K. Fletcher, B. Imber, S.N. Luoma, P.M. Stokes, and M. Winfrey, *Biologically Available Metals in Sediments,* NRCC No. 27694, NRCC/CNRC Publications, Ottawa, Canada, 1988, p. 298.

Hamelink, J.L., P.F. Landrum, H.L. Bergman, and W.H. Benson, Eds., *Bioavailability: Physical, Chemical and Biological Interactions,* CRC Press, Boca Raton, FL, 1994, p. 256.

Hansen, L.G. and B.S. Shane, Xenobiotic metabolism, in *Basic Environmental Toxicology,* Cockerham, L.G. and Shane, B.S., Eds., CRC Press, Boca Raton, FL, 1994.

Luoma, S.N., Can we determine the biological availability of sediment-bound trace elements? *Hydrobiolia,* 176/177, 379–396, 1989.

Newman, M.C. and M.G. Heagler, Allometry of metal bioaccumulation and toxicity, in *Metal Ecotoxicology, Concepts and Applications,* Newman, M.C. and McIntosh, A.W., Eds., Lewis Publishers, Chelsea, MI, 1991.

Spacie, A., L.S. McCarty, and G.M. Rand, Bioaccumulation and bioavailability in multiphase systems, in *Fundamentals of Aquatic Toxicology: Effects, Environmental Fate, and Risk Assessment,* 2nd ed., Rand, G.M., Ed., Taylor & Francis, Washington, D.C., 1995.

CHAPTER 5

Bioaccumulation from Food and Trophic Transfer

Far too often food chains have been envisioned as mechanisms operating solely to concentrate pollutants as they move from prey to predator. Less often are they objectively recognized as ecological processes, with the net effect of concentration or dilution of materials during their transport along the food chain being dependent upon a complex of biological variables.

Reichle and Van Hook (1970)

I INTRODUCTION

As evidenced by Minamata disease and DDT poisoning of birds, the transfer of contaminants through trophic webs[1] can have undesirable consequences to top predators. Some contaminants such as mercury and DDT display **biomagnification**,[2] an increase in contaminant concentration from one trophic level (e.g., prey) to the next (e.g., predator) due to accumulation from food. The possibility of biomagnification must be considered in any thorough assessment of ecological or human risk. However, because biomagnification played a pivotal role in our awakening to environmental issues, it is sometimes invoked as a ruling theory when equally plausible, alternative explanations are present (Moriarty, 1983; Beyer, 1986; Laskowski, 1991). For example, predators tend to live longer than prey species and thus have more time to accumulate contaminants than do prey. The result may be higher concentrations in predators than prey (Moriarty, 1983). Predators are often larger than prey, and allometric effects on bioaccumulation can result in higher concentrations of some contaminants in predators relative to prey (Moriarty, 1983). Also, for lipophilic contaminants, higher lipid content in predators than prey can also result in increases in contaminant concentrations with trophic level. Lower food-web organisms tend to grow faster than those higher in the food web; thus growth dilution may be more pronounced at lower levels than at higher levels

---

[1] The term *food web* is preferable to *food chain* in discussions of trophic transfer. The concept of a web of interactions is often more accurate than that of an orderly transfer from one level to the next highest level only. Many species feed on an array of prey that, in turn, have equally complex feeding strategies.

[2] The terms **bioamplification** and **trophic enrichment** are infrequently used instead of *biomagnification*. The general term, *bioaccumulation,* is occasionally used incorrectly as a synonym for *biomagnification*. Sometimes *biomagnification* is used to describe field observations of increasing concentrations with increasing trophic level, regardless of the ambiguity about the magnitude of uptake from food relative to water. The term *biomagnification* is not used in that context here, as the historical literature clearly associates trophic transfer of contaminant with biomagnification. Results of biomagnification studies using this term otherwise can be needlessly confusing and should be scrutinized carefully to determine the inferential strength of associated conclusions.

(Huckabee et al., 1979). Difficulties in defining trophic status, especially if species change their feeding habits with age (Huckabee et al., 1979), can confound and render subjective conclusions of biomagnification. Also, most field studies of bioaccumulation do not distinguish between water and food sources, pressing many researchers toward unjustified speculation about biomagnification. Finally, some communities show such wide variation in concentrations among species of a particular trophic level that the trends noted among trophic levels are often questionable (Beyer, 1986; Laskowski, 1991). Such calculations and their interpretations are frequently biased toward bio-magnification. Reports of biomagnification must be read carefully in order to peel away uninten-tional biases of investigators. Yet, because of the important consequences of biomagnification, the concept is well worth investigating.

Biomagnification is not required in order to have adverse effects as a consequence of trophic transfer. If contaminant concentrations are very high in a food item, species farther up the trophic web might still be exposed to concentrations sufficient to produce an adverse effect. As discussed in the first chapter, this was the case with humans afflicted with Itai-Itai disease. Another example involves metals and metalloids that accumulate to extremely high concentrations in materials covering submerged surfaces in lentic and lotic systems (Newman et al., 1983; 1985). These materials, including the associated *aufwuchs* (periphyton), are ingested by an important grazer/scraper guild of freshwater organisms and have the potential to cause adverse effects even in the absence of biomagnification (Newman and Jagoe, 1994). Also, the flux into an organism, not only the net concentration within that organism, may contribute to determining the adverse outcome of toxicant exposure.

Biomagnification is only one of three possible outcomes for trophic transfer of contaminants. Concentrations may be similar in both predator and prey, with no statistically significant upward or downward trend in concentrations. Alternatively, as is frequently the case, the contaminant concentration may decrease as trophic level increases. During each transfer, the required balance among ingestion rate, uptake from food, internal transformations, and elimination does not exist for the conservative transfer of a contaminant. In such situations, concentrations decrease with each trophic exchange. Diminution with increasing trophic level is called **trophic dilution** (Reichle and Van Hook, 1970) or **biominification** (Campbell et al., 1988). **Bioreduction** (Nott and Nicolaidou, 1993) has also been used in describing trophic dilution, but the meaning seemed to shift slightly to focus on rendering the contaminant less bioavailable in the prey biomass, with a consequent ineffective assimilation in the predator.

This chapter was developed to specifically address bioaccumulation from food because of the importance of this phenomenon, especially in terrestrial systems, where food can be the predominant source of contaminants to biota. The theme brought out by the above quote—that bioaccumulation from food is a complex process that has been viewed too simplistically in the past—will be enriched in this chapter. In contrast to Chapter 4, the focus is on the sequential transfer of contaminants among species rather than bioaccumulation within an individual. Details about differential transfer of radioisotopes and organic isomers are provided. Means of measuring trophic status are also discussed because of its crucial role in bioaccumulation studies.

II QUANTIFYING BIOACCUMULATION FROM FOOD

II.A   Assimilation from Food

Assimilation of a contaminant in food by individuals is one estimate of contaminant transfer between members of different trophic levels. It can be quantified as already described for bioavail-ability from food and used to estimate the magnitude of trophic transfer between the levels represented by the food item (e.g., prey or primary producer) and consumer (e.g., predator or grazer). The amount of contaminant in the organism after a defined time, usually the time to reach a practical steady-state concentration, is divided by the total amount of contaminant fed to the organism to estimate assimilation efficiency (see Chapter 3 for details). Similarly, the amount ingested and amount egested over a period of time can be used to estimate assimilation efficiencies. Less frequently, the area under the curve (AUC) method is used.

Assimilation can be measured by a **twin-tracer technique**, wherein a radiotracer of the substance to be assimilated is introduced simultaneously with an inert radiotracer to which assimilation is compared (Weeks and Rainbow, 1990). The inert tracer is not assimilated to any appreciable amount, and the retention of the assimilated tracer is quantified relative to it. For example, the assimilation of a ^{14}C-labelled organic compound can be compared with the amount of notionally inert ^{51}Cr fed simultaneously to a zooplankter (Bricelj et al., 1984). This technique is based on the dual assumptions that the two radiotracers pass through the gut at approximately the same rate and that the inert tracer is not absorbed to any significant extent. Both are incorporated into the same food items together to foster identical movement through the animal's gut.[3] (As with all such uses of radiotracers, the **specific activity concept** is central to accurate implementation: The radionuclide used to reflect movement of the stable nuclide [e.g., ^{14}C for C] must behave identically in chemical and biological processes as its nonradioactive nuclide [e.g., stable C].) In the absence of such an effective pairing of isotopes for exposures, a radioisotope can be fed to an organism, and the difference between the amount of that isotope ingested and egested over a time course can be used to estimate assimilation efficiency. The time to pass through the gut is estimated, and the amount remaining after gut evacuation is assumed to be assimilated. The advantage of the pairing with an inert tracer is that it provides a clear indication of when sampling of egested materials can be stopped, i.e., when all or most of the inert tracer has been egested.

The assimilation efficiencies for various foods and members of a food web are pieced together to predict trends in trophic transfer. They are used to complement field surveys and more-complex laboratory experiments, as described below.

II.B Trophic Transfer

II.B.1 Defining Trophic Position

Inaccuracies in identifying trophic status of species occur in the bioaccumulation literature. For this reason, a few paragraphs will be spent defining some methods for identifying trophic structure.

One means of reducing ambiguity is to conduct a laboratory experiment so that the trophic structure is imposed by the experimental design. The disadvantage of this approach is the artificial context and, perhaps, unrealistic depiction of trophic dynamics relative to natural communities. Recognizing this flaw, field surveys are conducted to complement or test conclusions from such highly structured laboratory experiments.

In field surveys, the prevalent method for determining trophic status is extraction and abstraction of information from the natural history literature. This is adequate for interpreting results of field surveys only to a degree. Much uncertainty can remain because species in communities have complex feeding strategies that change with age, time, and community composition. Some of this uncertainty can be further reduced with visual observations of species interactions and analysis of gut content (Kling et al., 1992). Regardless, most such renderings and associated quantification simplify trophic status to a discrete state for specific species, i.e., primary producer, primary consumer, secondary consumer, and so forth. This would be an inadequate description for species that feed at several levels.

---

[3] With sediments, it can be difficult to incorporate dual tracers identically, since the radiotracers can be distributed differently among sediment fractions. In such cases, it then becomes important to account for feeding selectivity, as some fractions (e.g., fine, organic carbon-rich particles) may be processed differently from others (e.g., large particles).

Another, more accurate method of quantifying trophic status has become readily available during the last 15 years. **Isotopic discrimination**[4] of light elements such as C, N, and S occurs during trophic transfers, providing an opportunity for quantifying trophic status in natural communities (Fry, 1988; Hesslein et al., 1991). Isotopic discrimination tends to reduce the amount of lighter isotopes (^{12}C, ^{14}N, or ^{32}S) in organisms relative to the heavier isotopes (^{13}C, ^{15}N, or ^{34}S) during trophic exchange. The reason for this is that the lighter isotopes are eliminated from the organisms more readily than the heavy isotopes. Carbon isotopic changes with trophic transfer are clear but smaller in magnitude than those of nitrogen isotopes (Hesslein et al., 1991; Cabana and Rasmussen, 1994). Moreover, carbon isotopic ratios vary more within a trophic level, making them a less effective indicator of trophic status than nitrogen isotopes (Hesslein et al., 1991). However, carbon isotopes do provide information on carbon sources (C3 and C4 photosynthetic pathways) as well as trophic structure. Sulfur isotopic ratios may change only slightly (Hesslein et al., 1991) or not at all (Fry, 1988) with trophic exchange, but they change significantly with sulfur source, making this a better indicator of sulfur source than trophic status. For example, Hesslein et al. (1991) used this sulfur isotope technique to identify food bases for broad whitefish (*Coregonus nasus*) and lake whitefish (*Coregonus clupeaformis*). Changes in stable nitrogen isotope composition seem to be the best indicator of trophic status (e.g., Rau, 1981; Fry, 1988; Hesslein et al., 1991; Kling et al., 1992).

Changes in nitrogen isotopes (^{14}N and ^{15}N) in organisms within a trophic web are quantified relative to the isotopic ratio of air (Equation 5.1):

$$\delta^{15}N = 1,000 \left[\frac{\left(^{15}N_{sample}\right) / \left(^{14}N_{sample}\right)}{\left(^{15}N_{air}\right) / \left(^{14}N_{air}\right)} - 1 \right] \tag{5.1}$$

The units for $\delta^{15}N$ are ‰ ("per mill" or "per millage"). The $\delta^{15}N$ increases with each trophic exchange because the lighter isotope (^{14}N) is more readily excreted. Isotopic discrimination of nitrogen is not associated with uptake, metabolic breakdown to amino acids, or deamination (Minagawa and Wada, 1984). Nor is there any evidence of discrimination in the urea cycle or in the formation of uric acid: The discrimination is relatively independent of the diverse processes of urine formation (Minagawa and Wada, 1984).

Changes in $\delta^{15}N$ at each trophic exchange can range widely. Minagawa and Wada (1984) describe a range from 1.3 to 5.3‰ per trophic exchange. However, the average increase is 3.4‰ per trophic exchange (Minagawa and Wada, 1984; Cabana and Rasmussen, 1994). The $\delta^{15}N$ can also change with animal age (Minagawa and Wada, 1984). Despite these considerations, $\delta^{15}N$ can readily reveal trophic structure of diverse field communities (Figure 5.1) and enhance interpretation of contaminant movement in ecological communities. Indeed, it can also be used with human populations to identify diet, e.g., $\delta^{15}N$ for human populations with different but unquantified amounts of diet coming from marine and terrestrial systems (Scheninger et al., 1983). Finally, this approach has the advantage of quantifying intermediate trophic positions. The simplistic assignment of a species to a single trophic level can be avoided. This approach can be augmented with a dual-isotope ($\delta^{15}N$ and $\delta^{13}C$) approach to also imply major vegetable sources of biomass (Fry, 1991) to members of the ecological community.

---

[4] Isotopic discrimination is the rate of or extent to which participation in some biological or chemical process depends on the mass of the isotope. Isotopic discrimination, or the isotope effect, results from differences in the kinetic energy of associated molecules with masses that are slightly different because they contain different isotopes of an element, e.g., ^{14}N instead of ^{15}N. Differences can also result from distinct vibrational and rotational qualities of molecules (Wang et al., 1975). Discrimination between isotopes is measured as a **discrimination ratio**, with a ratio of 1 if no discrimination were occurring.

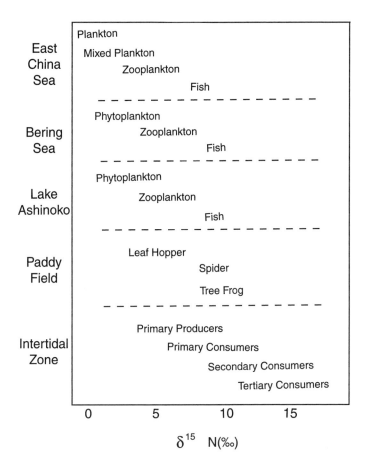

Figure 5.1   The change in δ^{15}N (%) with increase in trophic level for the diverse communities of the East China Sea (an oligotrophic sea with significant amounts of nitrogen fixation at the time of phytoplankton sampling, which brought the δ^{15}N to near atmospheric levels); the Bering Sea (North Atlantic Ocean); Lake Ashinoko (freshwater lake); a paddy field (terrestrial system) in Konosu, Japan; and an intertidal zone. (A composite and modification of Figures 1 and 2 in Minagawa, M. and E. Wada, *Geochim. Cosmochim. Acta*, 48, 1135–1140, 1984.)

II.B.2   Estimating Trophic Transfer

Perhaps the simplest way to quantify biomagnification is to divide the contaminant concentration at trophic level n (C_n) by that at the next lowest trophic level (C_{n-1}) (e.g., Bruggeman et al., 1981; Laskowski, 1991). This **biomagnification factor** (B) may involve individual organisms of a known or assumed trophic status.

$$B = \frac{C_n}{C_{n-1}} \tag{5.2}$$

Such a biomagnification factor is based on the assumption that concentrations have reached steady state in the sampled individuals and that the capacity is the same for the sampled individuals (e.g., individuals have the same lipid content if lipophilic compounds are being studied). This B can also be expressed in terms of the rate constant-based bioaccumulation model.

$$B = \frac{C_n}{C_{n-1}} = \frac{\alpha f}{k_e} \tag{5.3}$$

where α = assimilation efficiency for the ingested species, f = the feeding rate (mass of food $\times$ mass of individual^{-1} $\times$ time^{-1}), and k_e = the elimination rate constant (Bruggeman et al., 1981). Note that assimilation efficiency often diminishes as feeding rate increases (Clark and Mackay, 1991), so this parameterization is conditional.

Biomagnification factors can be estimated for samples of many organisms from two trophic levels by using a body-mass-weighted mean concentration for the two trophic levels:

$$B' = \frac{\left(\sum_{i=1}^{x} C_{n.i}\, w_{n.i}\right)\left(\sum_{j=1}^{z} w_{n-1.j}\right)}{\left(\sum_{j=1}^{z} C_{n-1.j}\, w_{n-1.j}\right)\left(\sum_{i=1}^{x} w_{n.i}\right)} \tag{5.4}$$

where w = the weight of individuals sampled from the n or n − 1 trophic levels.

Definitions for some bioaccumulation indices acknowledge that water may also contribute to differences in concentrations measured for individuals at different trophic levels. This is necessary in field surveys that do not isolate water and food sources to individuals. The **bioaccumulation factor** (BF)[5] is such an index that has the same form as Equation 5.2 except the source is not necessarily food alone:

$$BF = \frac{C_{organism}}{C_{source}} \tag{5.5}$$

where $C_{organism}$ = concentration in the organism resulting from uptake from food and water sources, and C_{source} = concentration in the reference source of contaminant.

For lipophilic organic compounds, concentrations can be expressed as a mass per mass of lipid basis. If the BF is greater than 1, biomagnification may be occurring. If it is less than 1, trophic dilution is suggested, although other factors such as allometric processes or growth dilution may be contributing to the changes. Experimental designs can employ tandem exposures of two subsets of individuals to either contaminant in water alone or contaminant in both food and water. The differences in the treatment results (BCF—bioconcentration factor—derived from the water-alone treatment and BF derived from the food-plus-water treatment) can then be used to assess the significance of trophic exchange relative to accumulation from water alone (Bruggeman et al., 1981). Assuming that uptake from water is significant at the primary-producer level only and that there is a distinct layering of trophic levels with minimal overlap, the transfer up the trophic structure can be calculated with knowledge of the concentration at the lowest trophic level and estimates of assimilation efficiencies and feeding rates at each trophic level (Ramade, 1987; Newman, 1995).

Commonly, the concentrations at trophic levels are referenced to that of the primary or lowest defined source of contaminant. For example, a **concentration factor** (CF) can be estimated for all trophic levels, with the concentration in the water used in the denominator: $CF = C_n/C_{water}$. Reichle and Van Hook (1970) took this approach to estimate concentration factors for radionuclides in a terrestrial food chain with concentrations in plant leaves as the reference concentration. The increase or decrease in concentration relative to that of the source is expressed as a multiple of the source concentration.

---

[5] Note that bioaccumulation factor designated BSAF is used in a more restricted sense in this book to mean the ratio of concentration in an organism and concentration in sediments (see Chapter 3). Also, BAF is the same as BF throughout the literature. For the sake of clarity, only BF will be used here to designate the ratio derived under the assumption that both water and food sources may be contributing to body concentrations of contaminants.

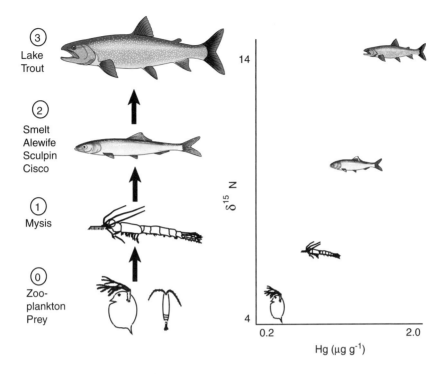

Figure 5.2   The increase in mercury from zooplankton to a zooplanktivorous shrimp (*Mysis relicta*) to pelagic forage fish to lake trout (*Salvelinus namaycush*). The trophic structure is quantified with δ^{15}N, its value increasing roughly 3.4‰ at each trophic exchange. A clear relationship is evident between mercury concentration and trophic status of individuals, indicating biomagnification of mercury. (Constructed using data and Figure 2 in Cabana, G. and J.B. Rasmussen, *Nature*, 372, 255–257, 1994.)

The methods described to this point require assignment of species to discrete trophic levels, but this is often an unrealistic simplification of trophic dynamics. It is more effective to quantify trophic transfer by relating concentrations in members of the community to corresponding δ^{15}N values. Figure 5.2 depicts the results of such an exercise documenting the increase in mercury from zooplankton → a zooplanktivorous shrimp (*Mysis relicta*) → pelagic forage fish → lake trout (*Salvelinus namaycush*) of Canadian lakes (Cabana and Rasmussen, 1994; Cabana et al., 1994). In a survey of seven Canadian Shield lakes with various trophic structures leading to lake trout (Figure 5.3), data for mercury concentration and δ^{15}N fell between the predicted values for a discrete trophic structure. The diagram on the left side of Figure 5.3 shows the expected values for δ^{15}N based on three possible, discrete structures. (Structure is denoted by the number of levels above zooplankton that the lake trout occupy.)

Regression models can be developed if a continuous variable such as δ^{15}N was used to quantify trophic transfer. Figures 5.2 and 5.3 suggest that a simple linear-regression technique could be used to model the increase in mercury with trophic transfer. Studies of bioaccumulation of organic compounds suggested to Broman et al. (1992) and Rolff et al. (1993) that an exponential model may be more appropriate in some cases:

$$C = a\, e^{b(\delta^{15}N)} \tag{5.6}$$

where C = concentration in the organism within the food web, and a and b are estimated parameters. The b is the **biomagnification power**. A positive b indicates a proportional increase in concentration with increase in position within a trophic web (biomagnification), and a negative b indicates a

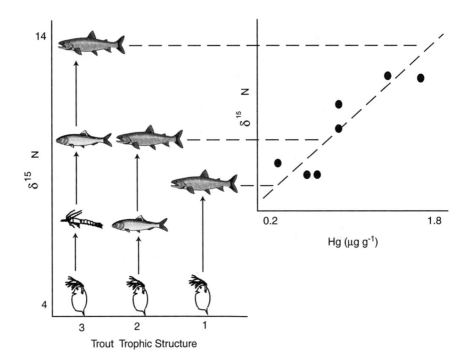

Figure 5.3   Correlation between mercury concentration and trophic structure (δ^{15}N) for seven Canadian lakes with different trophic structure relative to lake trout (*S. namaycush*). In some lakes, shrimp and/or forage fish are missing, giving rise to various trophic structures relative to lake trout. The chart uses a discrete assignment of trophic structure: 3 = same structure as shown in Figure 5.2 (zooplankton → shrimp → forage fish → lake trout), 2 = shortened structure without shrimp (zooplankton → forage fish → lake trout), and 1 = a simple structure with both shrimp and forage fish being insignificant. [Forage fish were predominantly alewife (*Alosa pseudoharengus*), cisco (*Coregonus artedii*), whitefish (*Coregonus clupeaformis, Prosopium cyclindraceum*), sculpin (*Myoxocephalus thompsoni*), and smelt (*Osmerus mordax*).] The panel on the upper right shows the actual mercury concentrations and δ^{15}Ns for lake trout from seven Canadian lakes. Note that trout from these lakes are shifted slightly to positions intermediate between the three discrete trophic structure levels. (Composite and modification of Figures 2 and 3 of Cabana, G. and J.B. Rasmussen, *Nature,* 372, 255–257, 1994.)

proportional decrease in concentration with increase in position (trophic dilution). Broman et al. (1992) found that decreases and increases in concentrations of various polychlorinated dibenzo-p-dioxins and -dibenzofurans in a pelagic food web could be modeled with this relationship.

III INORGANIC CONTAMINANTS

III.A   Metals and Metalloids

Assimilation studies of metals and metalloids have used the twin-tracer technique and single-isotope differences between ingested and egested element. Bricelj et al. (1984) fed ^{14}C/^{51}Cr-labelled algae to clams (*Mercenaria mercenaria*) for 30 to 45 min and then measured the amount of both tracers in feces after the clams were transferred to water containing unlabelled algae. The amounts lost after nearly total recovery of the notionally inert ^{51}Cr label in the feces were used to estimate ^{14}C assimilation. Chromium was assimilated in very small amounts, but carbon assimilation was approximately 80%. With a similar design (^{60}Co used as an inert tracer for ^{65}Zn assimilation), assimilation of zinc from tissue of the macroalga, *Laminaria digitata*, fed to the amphipod, *Orchestia gammarellus*, was estimated to be nearly 100% (Weeks and Rainbow, 1990). Reinfelder

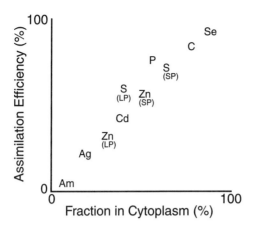

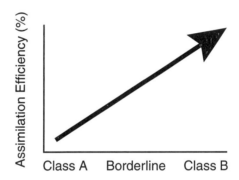

Figure 5.4 Assimilation efficiencies of elements contained in diatoms (*Thalassiosira pseudonana*) fed to zooplankton (*Acartia tonsa, A. hudsonica,* or *Temora longicornis*) plotted against the amount of each element associated with the cytoplasm of the diatoms (top panel). SP represents data derived using stationary-phase (senescent) diatom cultures, and LP represents results derived from diatom cultures in the log phase of growth. The bottom panel depicts the general trend in assimilation efficiency hypothesized relative to the Class B–borderline–Class A classification of metals (see Chapter 3). (Top panel is modified from Figure 1 in Reinfelder, J.R. and N.S. Fisher, *Science,* 251, 794–796, 1991.)

and Fisher (1991) used this design to determine assimilation efficiencies for a sequence of elements when calanoid copepods were fed labeled diatoms. For ^{14}C assimilation, ^{51}Cr was used as the inert tracer. For ^{75}Se assimilation, ^{241}Am was the inert tracer in the twin-tracer technique. Assimilation of other elements (^{110m}Ag, ^{109}Cd, ^{32}P, ^{35}S, and ^{65}Zn) was calculated without an inert isotope marker by dividing the amount retained after complete (estimated) gut evacuation by the amount ingested. Assimilation was estimated for sulfur and zinc using algae from cultures in the log phase or stationary phases of growth (Figure 5.4). Given the short residence time in the gut, the elemental content of the algal cytosol was assumed to reflect that available for "liquid" digestion by the zooplankton. This assumption was supported by the clear correlation between assimilation efficiencies for the various elements and the fraction of each present in the algal cytoplasm. Based on these results and earlier work with thorium, lead, uranium, and radium (Fisher et al., 1987), Reinfelder and Fisher (1991) suggested that Class B metals and those borderline metals that have greater affinities for sulfur than nitrogen or oxygen (e.g., cadmium, silver, and zinc) will be present in the cytoplasm in higher proportions than Class A metals (e.g., americium, plutonium, and thorium), which have greater affinities for oxygen than nitrogen and sulfur. Assimilation efficiencies are generally higher for the Class B and borderline metals than for Class A metals because of the former's higher availability in the cytoplasm for "liquid" digestion during rapid passage through the zooplankton gut.

Additional factors can influence assimilation efficiencies of metals in other feeding interactions. As mentioned, molluscs can sequester metals in granules and lower their potential for damage. The winkle, *Littorina littorea*, does so with zinc. When tissue of this gastropod is fed to the predatory gastropod, *Nassarius reticulatus*, only a small proportion of the zinc in the associated granules is available for assimilation in the predator (Nott and Nicolaidou, 1993). In contrast, there are considerable decreases in magnesium, phosphorus, and potassium in granules during passage through the predator's gut. The detoxification of zinc in the winkle tissue by incorporation into granules decreases the assimilation efficiency of zinc during predation. The assimilation of other metals associated with invertebrate intracellular granules probably is influenced also. Despite the widespread detoxification of metals via sequestration in granules, much work remains to be done regarding their influence on the availability of various classes of metals.

Trophic-transfer studies suggest that biomagnification can be more the exception than the rule for metals and metalloids. Wren et al. (1983) studied a series of elements (aluminum, barium, beryllium, boron, calcium, cadmium, cobalt, iron, lead, mercury, magnesium, manganese, molybdenum, nickel, phosphorus, sulfur, strontium, titanium, vanadium, and zinc) at various trophic levels of a Canadian Precambrian Shield lake and found evidence of biomagnification for only mercury. Cushing and Watson (1971) found no evidence for zinc biomagnification in a food chain of water → periphyton → carp. Terhaar et al. (1977) found that mercury, but not silver, showed evidence of biomagnification in an aquatic food chain. Using *in situ* enclosures placed in an aquatic system, Gächter and Geiger (1979) found no evidence for biomagnification of inorganic mercury, copper, cadmium, lead, or zinc. They speculated that alkylated mercury, but not inorganic mercury, increased with increasing trophic status of a species. Numerous authors report biomagnification of mercury in aquatic systems (Wren and MacCrimmon, 1986; Cabana et al. 1994; Kidd et al., 1995a), although Huckabee et al. (1979) cautioned that, even with mercury, other factors correlated with trophic status of an organism might confound the identification of biomagnification. Mance (1987) reviewed the literature and found only a few cases suggesting metal or metalloid biomagnification in aquatic systems. In addition to mercury, there was an occasional report of arsenic increase in food webs.

Similarly, there appear to be few clear examples of biomagnification of metals or metalloids in terrestrial systems. Beyer (1986) and Laskowski (1991) concluded independently that biomagnification in terrestrial systems is more the exception than the rule. Zinc, an essential and internally regulated metal, might increase with trophic status in terrestrial ecosystems deficient in this element (Beyer, 1986). Wu et al. (1995) suggested that biomagnification occurred in a selenium-contaminated area of California. Selenium concentration factors relative to water-extractable soil concentrations were noted to increase in a soil → plant → grasshopper (*Dissosteria pictipennis*) → praying mantis (*Litaneutria minor*) trophic sequence. However, it is difficult to determine the accuracy of this conclusion because there was not a consistent pattern of increase with each transfer, and covariates such as species longevity and allometric effects were not considered. Only the grasshopper-to-praying mantis concentration factor indicated an increase in selenium concentration, and according to Table 9 of Wu et al. (1995), the increase was quite variable among sample sites.

Vignette 5.1   Dietary Exposure of Piscivorous Birds to Mercury

Lawrence Bryan Jr. and Charles H. Jagoe
Savannah River Ecology Laboratory, Aiken, SC

Mercury is emitted to the environment by both natural and anthropogenic sources, with the latter growing in importance since the early 20th century (Fitzgerald et al., 1998). Anthropogenic emissions include both localized (e.g., industrial effluents) and geographically dispersed sources (e.g., atmospheric deposition from fossil fuel combustion and waste incineration). In general, the former results in more local acute exposure, while the latter results in more widespread chronic exposure. In either case, physical and chemical environmental conditions can affect mercury speciation and bioavailability. These effects in turn influence trophic transfers through food webs that include fish and piscivorous birds.

While most mercury is emitted to the environment in inorganic form, methyl mercury is the dominant form found in fish, accounting for 95% or more of the total burden (Bloom, 1992). The factors controlling mercury methylation are therefore of critical importance in determining exposure and risk to piscivores. Methylation occurs in wetland sediments and flooded soils, as well as within reservoir systems (Rudd, 1995), and is strongly associated with microbial activity, particularly sulfate-reducing bacteria (Gilmour et al., 1992). Chemical parameters including low pH and high DOC (dissolved organic compounds) are associated with enhanced methyl mercury production and bioavailability to fish in freshwaters (Driscoll et al., 1995; Haines et al., 1995). Wetland type and area within catchments affect methyl mercury concentration in northern lakes (St. Louis et al., 1996), demonstrating the importance of wetlands to methyl mercury production. Repeated flooding of wetlands can also enhance the methylation process and mercury cycling in wetland sediments (Porvari and Verta, 1995; Snodgrass et al., 2000). In marine and estuarine systems, sulfur dynamics, particularly the sulfate:sulfide ratio and redox conditions, are key factors controlling mercury methylation and bioavailability (Baeyens et al., 1998; King et al., 1999).

Primary producers and primary consumers are critical intermediaries in the biomagnification of mercury (Hill et al., 1996). The relationship between trophic position and mercury concentration in freshwater fish has been well documented (Kidd et al., 1995a; Watras et al., 1998). Among fish likely to be consumed by piscivorous birds in freshwater wetlands in the southeastern United States, predatory fish generally have higher levels of mercury than herbivores or omnivores (Snodgrass et al., 2000; Brant et al., 2002).

Mercury concentrations also increase with trophic level in estuarine systems (Gardner et al., 1978). However, estuarine prey typically contain less mercury than freshwater prey, and this results in lower mercury concentrations in coastal wood stork (*Mycteria americana*) and bald eagle (*Haliaeetus leucocephalis*) nestlings compared with conspecifics in inland nests (Welch, 1994; Gariboldi et al., 1998). Both of these studies reported that coastal nestlings fed on a combination of fresh- and saltwater prey, while inland nestlings ate only freshwater prey. The ratio of freshwater to estuarine prey is critical in controlling mercury exposure in coastal birds. Gariboldi et al. (2001) reported that in wet years, when temporary freshwater wetlands produced more abundant prey, mercury in coastal wood stork nestlings increased, reflecting the increased fraction of freshwater prey in their diets.

The age of the prey consumed can also affect exposure. Older prey tends to accumulate more mercury than younger and smaller prey, particularly in long-lived predatory species like largemouth bass and pickerel (Lange et al., 1994). Similarly, size of prey is often linked to age and may be directly correlated with mercury concentration (Frederick et al., 1999). Wading birds that consumed larger fish had higher mercury concentrations than birds that ate smaller fish (Sundlof et al., 1994; Beyer et al., 1997). Moreover, environmental conditions such as drought or wetland drainage that make larger fish available as prey can result in higher mercury exposure to predators consuming them.

The trophic level of the birds themselves can also affect exposure. For example, birds that specialize in herbivores or detritivores such as carp would be expected to have lower dietary concentrations of mercury than birds such as eagles that regularly prey on predators such as bass or pickerel. Average mercury concentrations in nestling wood storks that were fed freshwater fish and other aquatic prey by their parents reflected the trophic position of prey items (Gariboldi et al., 2001; Romanek et al., 2000).

Stable isotope ratios, particularly $\delta^{15}N$, have been used to trace food webs and infer trophic position. The relationship between trophic level measured by this method and mercury concentration in tissues is strong in freshwater systems (Kidd et al., 1995a). Bearhop et al. (2000) demonstrated that trophic status, measured by $\delta^{15}N$, influenced mercury concentrations in a pelagic seabird (great skua, *Catharacta skua*). As an example of how foraging behavior can affect mercury exposure, Monteiro et al. (1998) found that mercury concentrations in feathers and prey were fourfold higher in seabirds foraging on mesopelagic prey (those with daytime depths of >200 m that undergo diurnal migrations to near-surface waters) than in those foraging on epipelagic prey (those with daytime depths of <200 m). Mercury methylation occurs in the open ocean, particularly below the thermocline (>200 m; Mason and Fitzgerald, 1990), so mesopelagic items are exposed to greater concentrations of methyl mercury in the deeper waters. By specializing in these items, some seabirds are exposed to higher concentrations of methyl mercury in their prey, even though their trophic position is similar to those seabirds that concentrate on epipelagic items.

Some birds shift their diets during breeding, either in response to physiological needs (such as more calcium for egg formation) or the growth requirements of their offspring (high-energy foods to support growth, or minerals for skeletal development). This shift is not typical among piscivorous birds, but it does occur. Coastal-nesting laughing gulls (*Larus atricilla*) and white ibis (*Eudocimus albus*) shift from a diet of primarily salt- or brackish-water prey to freshwater prey due to reduced salt tolerance by their nestlings (Johnston and Bildstein, 1990; Dosch, 1997). Dietary shifts to fauna or freshwater prey could strongly enhance exposure of young birds to contaminants.

Eggs can be contaminated by chemicals consumed by females during egg development and by chemicals stored in maternal tissues. Mercury elimination from female birds during egg formation ranges from negligible to perhaps 20% of total body burden (Monteiro and Furness, 1995). Mercury in chick down is correlated to concentrations in eggs and may be a good measure of contamination where the mother fed during egg formation (Becker, 1989). Thus, egg and possibly down concentrations can be poor indicators of contamination near the nest site, considering the far-ranging movements of many piscivorous birds.

Chicks can be exposed to some mercury from maternal deposition into the egg, but exposure from consumption of mercury in prey captured near their nest is probably greater. After hatching, nestlings of most piscivorous birds grow at a prodigious rate; a wood stork can increase its weight 25-fold by three weeks after hatching, and eagles can weigh 45 times as much at two months of age as at hatching. Obviously, this requires large amounts of food, and nestlings can consume a relatively large amount of mercury during early life simply because their intake requirements are so high. Daily mercury dose, expressed as μg Hg consumed in diet per kg body weight per day, is a function of dietary intake, body mass, and mercury concentration of prey. The ratio of food intake to body mass changes rapidly during early growth, so even if mercury in prey was relatively constant, maximum daily dose occurs soon after hatching and declines through fledging (Frederick et al., 1999).

Nestlings are able to excrete large amounts of Hg into their growing feathers. In fledgling Cory's shearwater (*Calonetris diomedea*), 80% of the total body burden was contained in feathers (Monteiro and Furness, 1995). Mercury in blood binds with the keratin in developing feathers and is removed from the circulation, reducing exposure to other tissues in growing birds. Young birds have greater risk from dietary mercury during the fledging period, when feather growth has stopped and this excretory pathway is shut down (Frederick et al., 1999; Brant et al., 2002). After feather growth ends, mercury again accumulates within the body tissues, particularly liver and muscle (Braune and Gaskin 1987), and the birds may be at risk of mercury toxicosis (Spalding et al., 2000; Sepulvada et al., 1999).

Older birds generally have higher mercury concentrations in their tissues than younger birds (Burger et al., 1993; Beyer et al., 1997; Bowerman et al., 1994; Bryan et al., 2001; Furness et al., 1990). Unlike nestlings growing complete sets of feathers, postfledging and older birds accumulate mercury in their soft tissues and must await molt and regrowth of feathers to utilize feathers as an excretory pathway. In adult birds, males tend to have higher mercury concentrations because females can deposit up to one fifth of their soft-tissue mercury burden into eggs (Lewis et al., 1993).

As with terrestrial birds, fish-eating sea birds accumulate more mercury than sea birds consuming other prey (Honda et al., 1990). Birds from the order Procellariiformes (e.g., albatrosses) accumulate especially high concentrations of mercury. Concentrations greater than 200 μg Hg kg^{-1} dry weight in liver have been measured in Procillariiforms from remote sites in the North and South Atlantic (Muirhead and Furness, 1988; Thompson and Furness, 1989), and mean concentrations over 75 μg Hg kg^{-1} dry weight were found in livers of Procillariiforms in the Mediterranean (Renzoni et al., 1986).

These pelagic seabirds appear to be able to tolerate considerably higher tissue mercury concentrations than other types of birds (Kim et al., 1996; Monteiro and Furness, 1995), and there is little evidence that these concentrations cause them harm (Monteiro and Furness, 1995). Far lower concentrations are associated with behavioral, physiological, and reproductive effects, as well as mortality, in other bird species (Sundlof et al., 1994; Wolfe et al., 1998; Bouton et al., 1999). Speciation studies suggest that organic mercury species are converted to inorganic forms and stored in seabird livers (Norheim, 1987; Thompson and Furness, 1989). Although it has been hypothesized that demethylation may occur in livers of other bird species (Spalding et al., 2000), pelagic seabirds appear to have evolved enhanced demethylation and inorganic mercury storage capabilities (Monteiro and Furness, 1995; Kim et al., 1996). Many of the species with this capability are considered slow molting, and this process may be a response to the marine environment by those species that cannot excrete via the feather-growth pathway as frequently as other species.

While the accumulation of mercury in piscivorous birds is well documented, the effects of chronic dietary mercury exposure remain unclear. Most studies have focused on effects in nonpiscivorous model species, although some recent studies have looked at fish-eating birds. These works are in general agreement that sublethal dietary exposure results in poor reproduction, behavioral changes, and increased risk of disease (Heinz, 1974; 1979; Heinz and Hoffman, 1998; Bouton et al., 1999; Spalding et al., 2000). However, the dietary concentrations that produce these effects appear to vary widely among species, genera, and families.

Acknowledgments

The authors were aided by financial assistance award DE-FC09–96-SR18546 from the U.S. Department of Energy to the Savannah River Ecology Laboratory.

III.B Radionuclides

This separate consideration of radionuclides is admittedly arbitrary and results in some overlap with the previous discussion of metals and metalloids (III.A). However, the unique sources and effects of radioactive contaminants provide some justification for this separation and will facilitate a more comfortable transition in later chapters.

Assimilation efficiencies of radionuclides vary widely, e.g., 80% for ^{134}Cs (algae $\rightarrow$ carp) to nearly 0% for the radionuclide ^{144}Ce (food $\rightarrow$ fish), which becomes unavailable after being incorporated into structural tissue of prey (Reichle et al., 1970). Amphipods fed brine shrimp had assimilation efficiencies of 6.2%, 9.4%, and 55% for ^{144}Ce, ^{46}Sc, and ^{65}Zn, respectively. Cesium-137 and ^{134}Cs, widespread fission products with long half-lives (circa 30 years for ^{137}Cs and 2 years for ^{134}Cs), have high assimilation efficiencies (65 to 94%) from a wide range of food sources in terrestrial and aquatic systems. Similarly, ^{47}Ca assimilation efficiencies are quite high in diverse systems (69 to 98%). Almost 100% of ^{86}Rb and ^{187}W were assimilated by grazers of plant foliage (Reichle et al., 1970). Reichle et al. (1970) and Blaylock (1982) summarize this type of data for radionuclides in aquatic and terrestrial systems.

Some (radionuclides of nitrogen, phosphorus, potassium, and sodium) increase with passage up the food web but others (^{47}Ca) are diluted (Reichle and Van Hook, 1970). The increase or decrease is a consequence of their availability in the environment relative to the physiological need for each. Davis and Foster (1958) observed that the most common trend for many radionuclides was accumulation to relatively high concentrations at the lowest trophic level (primary producers), with subsequent diminution at each transfer thereafter. For example, ^{32}P displayed this behavior in aquatic systems examined by Kahn and Turgeon (1984). Polikarpov (1966) summarized and described processes resulting in concentration factors for many radionuclides in aquatic biota.

Essential elements in short supply and their radioactive analogs tend to increase during trophic exchange (Reichle and Van Hook, 1970). (An elemental **analog** is an element that behaves like, but not necessarily identical to, another element in biological processes, e.g., cesium is an analog of potassium, and strontium is an analog of calcium.) However, ^{90}Sr does not increase with trophic exchange because it is incorporated into bone or other structural tissues and thus has a very low bioavailability to predators. The consequence is trophic dilution for ^{90}Sr (Woodwell, 1967). Reichle et al. (1970) identified **calcium sinks** such as arthropod cuticles as a mechanism for trophic dilution for calcium and its analogs in terrestrial communities.

The radiocesium isotopes, ^{134}Cs and ^{137}Cs, are potassium analogs that have received much-deserved attention due to their release from fission-related processes and their relatively long half-lives. Radiocesium and potassium are taken up with similar high efficiencies, but radiocesium tends to be eliminated more slowly than potassium (McNeill and Trojan, 1960).[6] Consequently, the potential exists for a net increase in cesium with each trophic exchange. Radiocesium (^{137}Cs) from fallout did biomagnify from plant tissue to mule deer (*Odocoileus h. hemionus*) to cougar (*Felis concolor hippolestes*), suggesting that biomagnification could also increase ^{137}Cs activities in humans who consume tainted mule deer (Pendleton et al., 1964). There was a 3.4-fold increase of ^{137}Cs from deer to cougar. Similar increases have been described for atmospheric fallout-related ^{137}Cs in a food chain leading from lichens to caribou to Alaskan Eskimos, wolves, and foxes (French, 1967; Woodwell, 1967). Further evidence for the increase in ^{137}Cs levels with trophic transfer is the observation that ^{137}Cs activities are often higher in piscivorous fish than nonpiscivorous fish (Blaylock, 1982; Rowan and Rasmussen, 1994).

IV ORGANIC COMPOUNDS

A wide range of behaviors is exhibited by organic contaminants relative to trophic transfer, but in general, biomagnification of compounds not subject to metabolism seems to be predictable with the log K_{OW}. Connolly and Pedersen (1988) suggested that biomagnification is possible if log K_{OW} values were above approximately four; Gobas et al. (1993) suggested log K_{OW} values greater than six as an approximate upper limit for biomagnification. Thomann (1989) indicated a range from five to seven for compounds prone to biomagnification. Above seven, biomagnification is hampered by diminished assimilation efficiencies and BCF for lower trophic levels such as phytoplankton. Below five, decreased uptake and increased elimination rates will limit the capacity for biomagnification. Russell et al. (1995) found that PCBs (polychlorinated biphenyls) with log K_{OW} values of 6.1 or above were subject to biomagnification when white bass (*Morone chrysops*) consumed emerald shiners (*Notropis atherinoides*) (Figure 5.5).

Polychlorinated biphenyls often exhibit biomagnification, although Clayton et al. (1977) argued otherwise for zooplankton. Using nitrogen isotopes to quantify trophic status, PCB concentrations in lake trout from 83 Canadian lakes were shown to be a function of biomagnification processes and species lipid content (Rasmussen et al., 1990). Using the lake trout trophic classification discussed earlier for mercury, concentrations of PCBs were shown to increase 3.5-fold for every trophic exchange.[7] Lake Michigan food-web studies also demonstrated PCB biomagnification (Evans et al., 1991). Clark and Mackay (1991) found the PCB, 2,2′,3,3′,4,4′,5,6′-octachlorophenyl,[8] to biomagnify in guppies (*Poecilia reticulata*) exposed in the laboratory. Bruggeman et al. (1981) fed PCBs to goldfish (*Carassius auratus*) and found that biomagnification occurred, seemingly as a consequence of water solubility-correlated elimination rates.

DDT, DDE, DDD, or the sum of these pesticides may increase with each trophic transfer in field communities (Evans et al., 1991; Kidd et al., 1995b). (These pesticides can be studied individually or together as "ΣDDT.") Toxaphene can display biomagnification (Sanborn et al., 1976; Kidd et al., 1995), but to a lesser extent than DDT (Evans et al., 1991). Dibenzo-p-dioxins and -dibenzofurans

---

[6] Like isotopic discrimination described earlier, a discrimination factor can be estimated for analogs. Discrimination between an element and its analog can be measured as a **discrimination factor or ratio** (e.g., $[Cs]_{food}/[K]_{food}$ divided by $[Cs]_{body}/[K]_{body}$). The discrimination ratio for cesium and potassium in humans is approximately 0.33 (McNeill and Trojan, 1960).

[7] Interestingly, these authors make the point that fisheries management strategies commonly thought to enhance sports fishing can do damage instead. Forage species are added to enhance the trophic base of a lake and produce more trout. This is seen as a positive action. But, each addition to the trophic structure enhances accumulation of PCBs in trout. If the addition of forage species brings trout tissue concentration above a certain level, a fishing warning or ban can result. The practice of stocking forage fish can damage the sports fishery in some cases.

[8] PCBs are manufactured and used as mixtures of chlorinated biphenyls with different numbers and positions of associated Cl atoms. The individual PCBs that share a common form but have different numbers of Cl atoms at different positions are called **congeners**. The numbering of Cl atom positions on the biphenyl common structure of PCB congeners is shown in Figure 5.5.

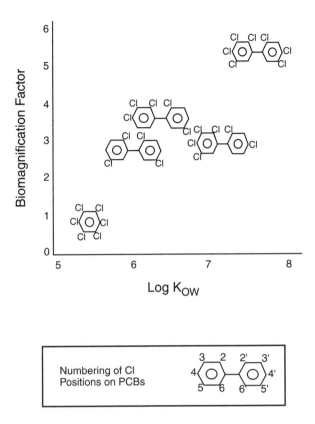

Figure 5.5 The increase in biomagnification factor (concentration in white bass/concentration in emerald shiner) with increasing log K_{OW} values for hexachlorobenzene (bottom left) and four polychlorinated biphenyls (PCBs): 2,2′,5,5′-tetrachlorobiphenyl, 2,2′,3,4,5′-pentachlorophenyl, 2,2′,3,4,4′,5-hexachlorobiphenyl, and 2,2′,3,4,4′,5,5′-heptachlorobiphenyl. The key at the bottom of this figure shows the numbering of chloride atoms attached to the biphenyl rings. The prefixes—tetra-, penta-, hexa-, and hepta-—refer to the total number of chloride atoms in the PCB molecule. (Modified from Figure 2 in Russell, R.W. et al., *Environ. Toxicol. Chem.*, 14, 719–724, 1995.)

also were found to increase with trophic position in a Northern Baltic Sea food web (Broman et al., 1992).

All isomers and congeners do not behave similarly during trophic exchanges. For example, PCB congeners behaved distinctly in Arctic trophic levels, Arctic cod (*Boreogadus saida*) → ringed seal (*Phoca hispida*) → polar bear (*Urus maritimus*) (Muir et al., 1988). The tri- and tetrachloro PCBs were the dominant congeners in cod. Penta- and hexachloro PCB congeners dominated in the ringed seal, and hexa- and heptochloro PCBs were the dominant congeners in polar bears. Older seals tended to have more of the highly chlorinated PCBs than young seals. The reason for this age difference was likely a combination of (1) the slower elimination of the more highly chlorinated congeners and (2) the difference in diet between young and adult seals. Young seals tend to eat more amphipods than do adults. In studies of trophic transfer of dibenzo-p-dioxins and -dibenzofurans in a Baltic food web, the total concentration of these contaminants did not increase, but the most toxic isomers did exhibit biomagnification to high levels in eider ducks (*Somateria mollissima*) (Broman et al., 1992; Rolff et al., 1993). This study underscores the importance of considering individual as well as summed concentrations of isomers or congeners. Catfish (*Ictalurus punctatus*) preferentially accumulate *cis* to *trans* chlordane because the *cis* isomer appears most resistant to metabolism (Murphy and Gooch, 1995). Likely for this same reason, the *cis* form of chlordane increased relative to the *trans* form in a freshwater trophic chain (Sanborn et al., 1976).

V SUMMARY

This chapter discusses the transfer of contaminants during trophic interactions. Methods of estimating such transfer were provided for discrete and continuous trophic structures. Conditions conducive to and examples of biomagnification and trophic dilution were outlined for metals, metalloids, radionuclides, and organic compounds.

This chapter ends Section Two of the book (Chapters 3 to 5), which provides descriptions of the mechanisms leading to bioaccumulation of contaminants. Section Three (Chapters 6 to 12) details the consequences of such bioaccumulation and discusses the effects from the biochemical to the global level of ecological organization.

SELECTED READINGS

Broman, D., C. Näf, C. Rolff, Y. Zebühr, B. Fry, and J. Hobbie, Using ratios of stable nitrogen to estimate bioaccumulation and flux of polychlorinated dibenzo-p-dioxins (PCDDs) and dibenzofurans (PCDFs) in two food chains from the Northern Baltic, *Environ. Toxicol. Chem.,* 11, 331–345, 1992.

Cabana, G. and J.B. Rasmussen, Modelling food chain structure and contaminant bioaccumulation using stable nitrogen isotopes, *Nature,* 372, 255–257, 1994.

McNeill, K.G. and G.A.D. Trojan, The cesium-potassium discrimination ratio, *Health Phys.,* 4, 109–112, 1960.

Murphy, D.L. and J.W. Gooch, Accumulation of *cis* and *trans* chlordane by channel catfish during dietary exposure, *Arch. Environ. Contam. Toxicol.,* 29, 297–301, 1995.

Petersen, B.J. and B. Fry, Stable isotopes in ecosystem studies, *Ann. Rev. Ecol. Syst.,* 18, 293–320, 1987.

Reinfelder, J.R. and N.S. Fisher, The assimilation of elements ingested by marine copepods, *Science,* 251, 794–796, 1991.

Rowan, D.J. and J.B. Rasmussen, Bioaccumulation of radiocesium by fish: the influence of physicochemical factors and trophic structure, *Can. J. Fish. Aquatic Sci.,* 51, 2388–2410, 1994.

Weeks, J.M. and P.S. Rainbow, A dual-labelling technique to measure the relative assimilation efficiencies of invertebrates taking up trace metals from food, *Functional Ecol.,* 4, 711–717, 1990.

SECTION THREE

Toxicant Effects

CHAPTER 6

Molecular Effects and Biomarkers

Pollutants affect biological systems at many levels, but all chemical pollutants must initially act by changing structural and/or functional properties of molecules essential to cellular activities.

Jagoe (1996)

... if effects on the ecosystem are to be predicted and understood, it is necessary to identify the effects of a toxicant on lower levels of biological organization, such as the subcellular level. Specific and sensitive biochemical methods can, therefore, serve as early warning indicators and adverse effects on the ecosystem can be avoided by taking protective measures.

Haux and Förlin (1988)

I INTRODUCTION

Two important concepts relative to biochemical effects of toxicants are exemplified by the above quotes. First, all toxicant effects begin by interacting with biomolecules. Effects then cascade through the biochemical → subcellular → cellular → tissue → organ → individual → population → community → ecosystem → landscape → biosphere levels of organization. Consequently, an understanding of effects at the biochemical level may provide some insight into the root cause of effects seen at the next few higher levels. Also, by understanding biochemical mechanisms, we can better predict effects of untested contaminants based on similarity of biochemical mode of action to well-understood contaminants. Further, if several contaminants are present, specific biochemical changes can provide valuable clues about which contaminant is having an effect. In some histochemical methods, tissue localization of biochemical changes can also provide information relative to exposure at target organs or sites. For example, high Phase I enzyme activity (e.g., monooxygenase activity) in a specific tissue can be measured after animals are exposed in the laboratory to a toxicant[1] known to be metabolized to a potent carcinogen. This finding can be linked to results of surveys indicating high incidence of cancers in the same tissue of individuals from a contaminated site. So, understanding the biochemical mode of action enhances our grasp of causal structure and our ability to predict effects at higher levels of biological organization.

---

[1] A compound that is converted to a carcinogen is a **procarcinogen**. For example, 2-acetylaminofluorene is a procarcinogen that is converted to a carcinogen by Phase I enzymes (N-hydroxylation by monooxygenases) (Stegeman and Hahn, 1994). Similarly, cytochrome P-450 monooxygenases activate benzo[a]pyrene to potent carcinogens. Remember from Chapter 3 that these are cases of activation.

Second, a technical advantage is gained by understanding toxicant effects on biomolecules and molecular responses to contaminants. Changes in biomolecules or suites of biomolecules can indicate exposure to bioavailable contaminants in field situations. The biochemical quality that is changing can be used as a biomarker. As discussed in Chapter 1, a biomarker is a biochemical, physiological, morphological, or histological quality used to imply exposure to or effect of a toxicant. Biomarkers indicate that sufficient toxicant was available for enough time to elicit a response or effect (Melancon, 1995). Because biochemical changes generally are detectable before adverse effects are seen at higher levels of biological organization, the biochemical marker approach is often an early warning or proactive tool. This is a great advantage because responses at higher levels such as the ecosystem are usually measurable only after significant or permanent damage has occurred. Biochemical markers are also useful to monitor the shift back to a normal state after cleanup of a contaminated site. Regardless of their proactive or retroactive utility, the ecological realism (ability to accurately reflect an ecologically meaningful effect or response) is lower for biomarkers than for indicators based on higher-level changes such as species richness or reproductive failure.

A biochemical change should have the following qualities to be useful as a biomarker. First, it should be measurable before any adverse, significant consequences occur at higher levels of biological organization (Haux and Förlin, 1988; Campbell and Tessier, 1996). This enhances its value as a proactive tool. Second, measurement of the ideal biomarker should be rapid, inexpensive, and sufficiently easy so as to be amenable to widespread use by ecotoxicologists. Third, its measurement should accommodate standard quality control/quality assurance practices. Fourth, the ideal biomarker should be specific to a single toxicant or class of toxicants (Haux and Förlin, 1988; Campbell and Tessier, 1996), although nonspecific biomarkers have value too. Fifth, a clear concentration-effect relationship must exist for the toxicant and biomarker (Haux and Förlin, 1988). Sixth, the ideal biomarker should be applicable to a broad range of **sentinel species** (feral, caged, or endemic species used in measuring and indicating the level of contaminant effect during a biomonitoring exercise) so that it might have the widest possible application (Sanders, 1990). Seventh, established linkage of biomarker changes with some toxicant-related decrease in individual fitness is desirable (Sanders, 1990) but not always necessary. This enhances ecological relevance in any discussions of change in a biomarker. Finally, the system should be sufficiently well understood so that other qualities of the organism or its environment that influence the biomarker can be accommodated in the experimental design and data interpretation (Campbell and Tessier, 1996). For example, ambient temperature, age, and sex can influence monooxygenase activity and should be considered in the design of biomonitoring studies using this biomarker (Kleinow et al., 1987).

The remainder of this chapter examines biochemical aspects of ecotoxicology with the dual goals of understanding the mode of action of toxicants and describing current molecular biomarkers. Enzyme activities, conjugates, and products of Phase I and II breakdown of organic compounds are discussed first. Focus then shifts to metallothioneins and stress proteins. Enzymes and products associated with oxidative stress, toxicant effects on nucleic acids, and symptoms of enzyme dysfunction are detailed. Aspects of these topics are explored again during discussions of cancers (see Chapter 7).

II ORGANIC COMPOUND DETOXIFICATION

II.A   Phase I

As will be detailed shortly, Phase I reactions involve hydrolysis, reduction, and oxidation of a xenobiotic. Functional groups are produced or exposed during Phase I reactions, rendering the compound slightly more water soluble (Parkinson, 1996) and producing a biotransformation product that can be acted on in Phase II detoxification reactions. However, some xenobiotics are eliminated effectively without further Phase II biotransformation.

Biochemical shifts associated with **cytochrome P-450 monooxygenase** induction are often used to document a response to organic contaminants such as polycyclic aromatic hydrocarbons

(PAH) (Van Veld et al., 1990; Wirgin et al., 1994), chlorinated hydrocarbons (Walker et al., 1987), organic compounds in pulp and paper mill effluents (Soimasuo et al., 1995), polychlorinated biphenyls (PCB) (Brumley et al., 1995), hydrocarbons (Goksøyr and Förlin, 1992), dioxins (Goksøyr and Förlin, 1992), and dibenzofurans (Goksøyr and Förlin, 1992). These monooxygenases are involved in the metabolism of a wide range of xenobiotics as well as fatty acids, cholesterol, and steroid hormones. The cytochrome P-450 monooxygenase system is often called the **mixed-function oxidase** or MFO system (Di Giulio et al., 1995). Although most Phase I oxidations are associated with the cytochrome P-450 monooxygenase system, the cytochrome b_5 and NADH-cytochrome b_5 reductase system is also important in xenobiotic metabolism (Di Giulio et al., 1995).

The cytochrome P-450 monooxygenase system is often assayed in the microsomal fraction, a cell fraction composed of membrane vesicles (microsomes) derived from the endoplasmic reticulum (ER) during routine separations by tissue homogenization followed by ultracentrifugation. The name P-450 is derived from P for pigment and 450 nm, the wavelength at which they have maximum light absorption when bound to CO (Haux and Förlin, 1988; Di Giulio et al., 1995).

The cytochrome P-450 monooxygenases are hemoproteins associated with membranes, especially in the ER. More specifically, the cytochrome P-450 monooxygenase system is an assemblage of **isoenzymes** (i.e., different forms of the same enzyme that are coded by different gene loci). Each isoenzyme has a molecular weight of approximately 45 to 60 kDa (Goksøyr and Förlin, 1992) and contains an iron protoporphyrin IX prosthetic group (Timbrell, 2000). In the membrane, the isoenzymes are associated with the 78-kDa NADPH cytochrome P-450 reductase that transfers electrons to the isoenzyme assemblage. The isoenzymes, reductase, and membrane phospholipid form the unit responsible for the major portion of Phase I oxidations. This system is most often associated with the conversion:

$$RH + NADPH + O_2 + H^+ \rightarrow ROH + NADP^+ + H_2O \qquad (6.1)$$

where RH is an organic compound undergoing hydroxylation. This reaction involves two enzymes (cytochrome P-450 isoenzymes and NADPH-cytochrome P-450 reductase), NADPH, and molecular oxygen. Although hydroxylation is shown here, cytochrome P-450 monooxygenases also catalyze epoxidation, (N-, O-, and S-) dealkylation, oxidative deamination, (S-, P-, and N-) oxidation, desulfuration reactions, and oxidative and reductive dehalogenation (Stegeman and Hahn, 1994; Di Giulio et al., 1995). Timbrell (2000) describes these reactions in detail, using toxicant and drug transformations as examples.

There is a diversity of nomenclature that must be understood in order to make sense of studies of cytochrome P-450 monooxygenases. The genes coding for cytochrome P-450 monooxygenases are grouped into gene families (circa 27 families) and subfamilies based on similarities in DNA sequences. Names of genes and gene products are based on the root CYP (**Cy**tochrome **P**-450). Numbers and letters are added to the CYP root to designate the particular family (e.g., CYP1 or CYP2), subfamily (e.g., CYP1A or CYP1B), and gene (e.g., CYP1A1 or CYP1A2) (Goksøyr and Förlin, 1992). A standard denotation is used for the protein, mRNA, and DNA associated with any particular gene. The protein and mRNA are designated as detailed above, e.g., CYP1A1. Sometimes the protein can be designated as P-450 1A1 instead. The DNA designation is italicized, e.g., *CYP1A1* (Goksøyr and Förlin, 1992). This is important to keep in mind because protein, mRNA, and DNA are all employed in studies of these important Phase I enzymes. Proteins are often assayed with specific antibodies, and the mRNA is assayed with DNA probes. For example, Haasch et al. (1993) measured immunoreactive CYP1A1 protein and hybridizable CYP1A1 mRNA to suggest liver cytochrome P-450 response in fish to PAH and PCB contamination.

Additional nomenclature appears if enzymatic activities are measured. Activity can be measured by aryl hydrocarbon hydroxylase hydroxylation of benzo[a]pyrene. Results are expressed as **AHH** (aryl hydrocarbon hydroxylase) activity. Another common assay involves O-deethylation of ethoxyresorufin by ethoxyresorufin O-deethylase. Activity is expressed as **EROD** (ethoxyresorufin O-deethylase) activity (Goksøyr and Förlin, 1992). Cytochrome P-450 monooxygenase activities

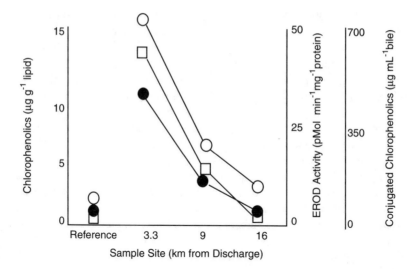

Figure 6.1   The response of biomarkers in juvenile whitefish (*Coregonus lavaretus*) to pulp and paper mill effluents. The concentration of total chlorophenolics in the gut lipids of this sentinel species (●) were elevated in whitefish taken 3.3 km below the discharge relative to the reference samples and rapidly decreased with distance from the discharge. Similarly, two biomarkers—EROD activity (○) and conjugated chlorophenolics in the bile (□)—were highest near the discharge and decreased with distance from the paper and pulp mill effluent. (Generated by combining information from Figures 3, 4, and 5 of Soimasuo, R. et al., *Aquatic Toxicol.*, 31, 329–345, 1995.)

measured as AHH or EROD are common biomarkers reflecting response (induction) to significant amounts of bioavailable xenobiotic. For example, Soimasuo et al. (1995) used EROD activity in whitefish (*Coregonus lavaretus*) liver to measure CYP1A response to pulp and paper mill effluent (Figure 6.1). Haasch et al. (1993) used EROD activity in tandem with assays of CYP1A1 protein and mRNA to monitor fish (catfish, *Ictalurus punctatus*; largemouth bass, *Micropterus salmoides*; and killifish, *Fundulus heteroclitus*) response to PAH and PCB exposure. Tomcod (*Microgadus tomcod*) response to xenobiotics was monitored along the North American coast with CYP1A mRNA (Wirgin et al., 1994). Other enzyme assays that also reflect P-450 activity include ethoxycoumarin O-deethylase (ECOD), aflatoxin B1 2,3-epoxidase (AFBI), lauric acid ω-1 hydrolase (LA), testosterone hydroxylase (TH), and phenanthrene hydroxylase (AH) assays (Goksøyr and Förlin, 1992; Brumley et al., 1995).

Induction of cytochrome P-450 is influenced by numerous factors, including ambient temperature (Kleinow et al., 1987), the particular toxicant and species being studied (Haux and Förlin, 1988), body weight (Parke, 1981), animal sex (Haux and Förlin, 1988), hormone titers (Haux and Förlin, 1988), and tissue oxygen tension (Parke, 1981). When these modifying factors are taken into account, cytochrome P-450 induction has proved to be a reliable biomarker. As discussed already, it was used by Haasch et al. (1993), Wirgin et al. (1994), and Soimasuo et al. (1995) as an indicator for significant exposure to a variety of contaminants. Van Veld et al. (1990) measured cytochrome P-450 protein (immunoassay) and activity (EROD) in spot (*Leiostomus xanthurus*) to indicate PAH contamination in the Chesapeake Bay region. Brumley et al. (1995) used induction of these same biomarkers in sand flatheads (*Platycephalus bassensis*) as indicators of response to PCB exposure.

II.B   Phase II

Phase II enzymes might also be induced by xenobiotics and used in biomarker studies. They involve **conjugation**, "the addition to foreign compounds of endogenous groups which are generally polar and readily available in vivo" (Timbrell, 2000). The readily available compounds that bond to the toxicants can be carbohydrate derivatives, amino acids, glutathione, or sulfate (Timbrell, 2000).

As mentioned already, Phase I biotransformations render a toxicant less lipophilic and, consequently, more readily eliminated. Conjugation produces a more polar molecule from the original toxicant or, very often, from the product of a Phase I reaction. The resulting increase in water solubility and decreased lipid solubility make the molecule more prone to elimination. However, this is not always the case. The biotransformation of a highly lipophilic, volatile organic compound that is usually eliminated via exhalation will diminish the rate at which it is eliminated from the body (Parkinson, 1996).

Glutathione S-transferase (GST), which attaches **glutathione** (GSH, a tripeptide made of cysteine, glutamate, and glycine, glu-cys-gly) to the xenobiotic or its metabolites, is one Phase II enzyme in the cytosol and microsomes. Another is **sulfotransferase**, an enzyme in the cytosol that conjugates sulfate with the compound. **Uridinediphospho glucuronosyltransferase** (UDP-glucuronosyltransferase, UDP-GT) is another. This enzyme catalyzes the transfer of **glucuronic acid** from uridine diphosphate glucuronic acid to electrophilic xenobiotics or their metabolites (Di Giulio et al., 1995). For example, Phase II reactions convert morphine, heroin, and codeine to morphine-3-glucuronide. Morphine is transformed directly without an initial Phase I biotransformation, but the other two drugs must first be acted on by Phase I processes (Parkinson, 1996). Uridinediphospho glucuronosyltransferase also binds covalently with electrophilic compounds such as PAHs (George, 1994).

Both the induced activities of these enzymes and concentrations of conjugated products have been explored as biomarkers, although Di Giulio et al. (1995) suggested that, in general, these enzymes were less valuable biomarkers than Phase I enzymes. As discussed above, Soimasuo et al. (1995) examined GST and UDP-GT induction, concentrations of conjugates in bile, and EROD in whitefish. Induction of EROD was clearly demonstrated for these fish during exposures to pulp and paper mill effluent; however, induction of GST and UDP-GT was not as clear. Conjugated metabolites were sensitive biomarkers that had an obvious trend with distance from discharge (Figure 6.1). Activity of UDP-GT as well as conjugates with glucuronic acid and sulfate did increase after laboratory exposure of sand flatheads to PCBs, suggesting to Brumley et al. (1995) that these qualities might be acceptable biomarkers of PCB exposure.

III METALLOTHIONEINS

Metallothioneins were first described by Margoshes and Vallee (1957) as cadmium-binding proteins in horse kidneys. Now known to be produced in bacteria, invertebrates, and vertebrates, they are found in highest concentrations in the liver, hepatopancreas, kidneys, gills, and intestines (Roesijadi, 1992).

Metallothioneins function in the uptake, internal compartmentalization, sequestration, and excretion of essential (e.g., Cu, Zn) and nonessential (e.g., Ag, Cd, Hg) metals. Their involvement in the normal homeostasis of essential metals results in basal levels of metallothionein being present in the absence of toxic metal exposure. Fluctuations in metallothionein levels associated with processes such as molting and reproduction (Roesijadi, 1992) or changes in substances such as glucocorticoids (Karin and Herschman, 1981) can also occur. Reflecting their roles in metal detoxification and sequestration, they are induced by elevated levels of the above metals. This increases the capacity of the organism to bind and effectively sequester toxic metals away from molecular sites of toxic action. Enhanced levels of metallothioneins have been linked to enhanced fitness during metal exposure of individuals (Bouquegneau, 1979; Hobson and Birge, 1989; Sanders et al., 1983). Further, enhanced capacity to produce metallothionein and the associated lessening of toxic metal effects have been linked to population adaptation to chronic metal exposure. For example, Maroni et al. (1987) detected metallothionein gene duplication in *Drosophila melanogaster* populations, leading to enhanced production of metallothionein and higher survival during copper or cadmium exposure.

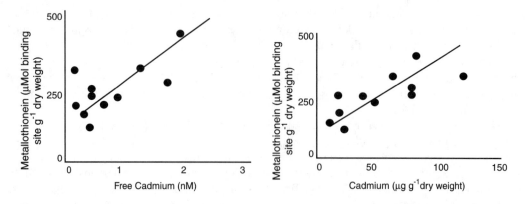

Figure 6.2   The relationships between free cadmium (Cd^{2+}) concentration in lake water (sediment-water interface) and metallothionein concentration in the freshwater mussel, *Anodonta grandis* (left panel),
and cadmium concentration in the entire mussel and metallothionein concentration in this mussel
(right panel). Each point represents samples from a different lake in the Rouyn-Noranda mining
area of Quebec, Canada. (Modified from Figures 3 and 4 of Couillard, Y. et al., *Limnol. Oceanogr.*,
38, 299–313, 1993.)

    These cytosolic proteins have molecular weights of roughly 6 to 7 kDa, and about 25% of their
amino acids are cysteine. They have no aromatic amino acids or histidine (Hamilton and Mehrle,
1986). They also have high heat stability, a quality often used during their isolation (Winge and
Brouwer, 1986). Poorly characterized proteins or proteins not conforming precisely with the above
qualities have been called **metallothionein-like proteins** (Hamilton and Mehrle, 1986). Early in
the study of metal-binding proteins of nonmammalian species, proteins were characterized from
incompletely purified preparations, resulting in confusion. Roesijadi (1992) recently produced the
following, more-general list of metallothionein qualities that seems to fit all relevant biomolecules
that were properly characterized:

1. Low molecular weight with a high metal content
2. High cysteine content and no aromatic amino acids or histidine
3. Unique amino acid sequence, especially regarding cysteine placement
4. Metal-thiolate clusters.

Roesijadi (1992) includes phytochelatin[2] as a metallothionein in this scheme. As discussed in
Chapter 3, phytochelatins are inducible, metal-binding peptides isolated from plants.

    Metallothioneins have been used as specific biomarkers for metal contamination under a
variety of exposure scenarios. For example, metallothionein levels in whelk (*Bullia digitalis*)
consuming metal-tainted grasses reflected site contamination (Hennig, 1986). Hepatic metallothionein levels in juvenile trout (*Salmo gairdneri*) (Roch et al., 1982) and metallothionein in
freshwater mussels *Anodonta grandis* (Couillard et al., 1993; Figure 6.2) were effective biomarkers
of metal contamination.

    Such biomarker studies can be extended even further based on the spillover hypothesis (Camp-
bell and Tessier, 1996; Hamilton and Mehrle, 1986). Under the assumption that binding by
metallothionein sequesters toxic metals away from sites of adverse action, the **spillover hypothesis**
states that toxic effects will begin to emerge after exceeding the capacity of metallothionein to
bind metals. The unbound metals then "spill over" to interact at sites of adverse action. The amount
of metal in cells in excess of that bound to metallothionein is correlated with some measure of

---

[2] If free of metals, phytochelatin has the form of (γ-glutamic acid-cysteine)$_n$-glycine, with n being 3, 5, 6, or 7 (Grill et al., 1985).

adverse effect. One example of such use of the spillover hypothesis is the study by Klaverkamp et al. (1991) of white sucker (*Catostomus commersoni*) inhabiting lakes around the Flin Flon smelter (Manitoba, Canada).

Vignette 6.1   Metallothioneins

G. Roesijadi
Florida Atlantic University, Boca Raton, FL

General

Metallothioneins (MT), a family of low-molecular-weight, metal-binding proteins discovered and initially characterized approximately four decades ago, have since been shown to be ubiquitously distributed in diverse taxa. MTs are known to be involved in regulation of essential metals, specifically of zinc and copper; protection against the toxicity of these and nonessential metals such as cadmium, silver, and mercury; and protection against free-radical toxicity (Cherian and Chan, 1993). Although there is still debate over assignment of specific functions, metal detoxification—particularly of cadmium—remains the central function of this protein (Klaassen et al., 1999). The first reports of the occurrence of MTs in species such as fish and mollusks in the mid-1970s had ecotoxicological contexts, and these reports were closely followed by proposals for the use of MTs in pollution monitoring, a focus that prevails to the present time.

Biochemistry and Molecular Regulation

The amino acid sequence of MT is characterized by the abundance and location of cysteine residues (Kojima et al., 1999), which form cys-cys, cys-X-X-cys, and cys-X-cys motifs, where X represents another amino acid. Metals bind cooperatively to the sulfhydryl groups of the cysteines, resulting in the formation of characteristic metal clusters, typically two, in diverse taxa ranging from mammals (Braun et al., 1992) to invertebrates such as the sea urchin (Wang et al., 1995), crustaceans (Otvos et al., 1982), snail (Dallinger et al., 2001), and nematode (You et al., 1999).

MTs are present at low, basal levels and can be induced by a variety of substances and physical-chemical conditions (Kägi, 1991). For this reason, they have been considered by some to be general stress-related proteins (Ryan and Hightower, 1996). Zinc is usually associated with MTs under basal conditions, and the most thoroughly studied signal transduction pathway for expression of MT involves transcriptional regulation via the zinc-dependent **transcription factor** (or **MTF-1**) and its binding to metal-response elements (MRE) in the promoter of the MT gene (Andrews, 2000). This pathway for MT expression is conserved in organisms as evolutionarily distant as the mouse, human, fish, and fruit fly (Maur et al., 1999; Andrews, 2001; Zhang et al., 2001). The mechanism for induction by metals other than zinc is not completely understood. According to one recent hypothesis, such induction is indirect and results from displacement of zinc from zinc-binding sites and its subsequent interaction with MTF-1 (Palmiter, 1994). However, other metals known to induce MT use signal-transduction pathways that do not involve MTF-1. Induction by copper in yeast occurs via the copper-specific transcription factor for MT induction ACE1 (Szczypka and Thiele, 1989), and induction of cadmium in cells of higher animals occurs by way of the upstream stimulatory factor (USF) and its interaction with a composite USF/antioxidant response element (Andrews, 2000). The existence of direct pathways for these metals strengthen arguments that metal detoxification and protection against metal toxicity are specific functions of MT.

Metals are not the sole inducers of MTs, however, and the MT gene also contains **promoter elements** that are responsive to other signals (Hamer, 1986). Glucocorticoid-response elements and antioxidant-response elements, for example, exist in the promoter sequences of MTs in diverse organisms, enabling free radicals and steroid hormones to induce MT.

MTs and Toxicology

Overexpression of MT is believed to be the basis for increased metal resistance. Induction of MT is associated with increased metal resistance, as is overexpression that results from gene amplification (Beach and Palmiter, 1981; Koropatnick, 1988) and gene duplication (Otto et al., 1986). **Transgenic** mice that overexpress MT also exhibit greater resistance to cadmium (Liu et al., 1995).

Loss-of-function and gain-of-function experiments represent the most convincing examples for a protective function of MT against metal toxicity. Deletion of the MT gene in yeast and mice, for example, results in loss of protection against cadmium, copper, mercury, or zinc toxicity (Hamer et al., 1985; Michalska and Choo, 1993; Masters et al., 1994; Kelly et al., 1996; Satoh et al., 1997). In oyster blood cells, disruption of MT expression with **antisense oligonucleotides** results in susceptibility to cadmium toxicity at concentrations not otherwise toxic (Butler and Roesijadi, 2001). The loss of copper resistance in MT-disabled yeast is restored by insertion of a human MT gene (Thiele et al., 1986).

Several types of intracellular interactions can be envisioned for the role of MT in protection against metal toxicity. These include binding of newly taken-up metal ions (Roesijadi and Klerks, 1989), metal abstraction from structures that have bound toxic metal ions (Huang, 1993), and metal-exchange with metal ions that have replaced zinc in zinc-binding sites (Roesijadi et al., 1998; Ejnik et al., 1999) (Figure 6.3). In these model interactions, the interception of newly taken-up metal ions by MT forestalls toxic interactions (Figure 6.3A), and molecules whose structure and function have been compromised by binding toxic metals are rescued by MT (Figure 6.3B, C). The latter recognizes an active role for MT in repairing structures already affected by binding certain toxic metals. Ideally, detoxification by MT would preclude detection of significant toxicity. If MT expression is not sufficient to keep up with exposure, the type of interactions described in the model interactions can be compromised and result in toxicity. Studies on the intracellular dynamics of metal binding to MT depicted in these models are limited, however, and represent a topic that would benefit from additional research.

That factors other than metals can induce MTs introduces a broader toxicological context for MT. The presence of antioxidant-response elements in the MT gene and the ability of MT to scavenge free radicals imply a specific function in protection against oxidative stress (Andrews, 2000). Additionally, induction of MT through pathways targeting glucocorticoid-response elements in promoter sequences implicates MT

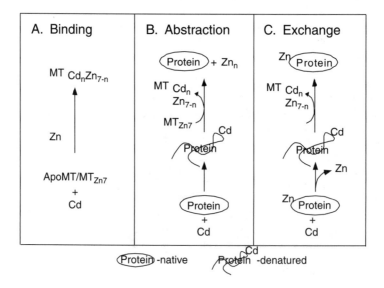

Figure 6.3   Cellular interactions of metallothionein (MT) with toxic metals and targets of metal toxicity. Cadmium is used as an example of a toxic metal. A: Metal-free apoMT or ZnMT binds newly taken up cadmium, thereby intercepting the cadmium and forestalling toxic interactions; B: MT abstracts cadmium from a cadmium-impaired target protein and restores structure and function to the target; C: MT serves as a zinc donor to a cadmium-impaired zinc-metalloprotein through a zinc-for-cadmium exchange, thereby restoring structure.

expression in hormonally mediated processes that are part of general stress responses. The ability to respond to other stressors has implications for natural populations that are responding to complex and variable environmental conditions, because metals alone are not responsible for modulating MT gene expression. Induction of MT is not, by itself, diagnostic of metal exposure.

DNA microarrays are currently being used to elucidate the functional relationship between MT gene expression and expression of other genes in response to metal and other stressors (Frueh et al., 2001; Liu et al., 2001; Yamada and Koizumi, 2002). The ability to screen for expression of a multitude of genes simultaneously is greatly facilitating the analysis of complex interactions, and applications in ecotoxicology will be of value in understanding mechanisms of response in natural environments impacted by multiple chemical agents. Arrays for mammalian and fish species used as biomedical models are already commercially available and applicable to related species of ecological interest. Development of expressed sequence tags (EST) in molluscan species like the snail (*Biomphalaria glabrata*) (Knoght et al., 1998) and the American oyster (*Crassostrea virginica*) (Jenny et al., 2002) represents a contribution to multigene analysis of invertebrates important in ecotoxicology.

MTs and Ecotoxicology

Sensitive and sophisticated techniques to measure MT, its expression, or its metal-binding status are now routine. Analyses directed to different steps in the gene-expression pathway that leads to formation of metallated MT can provide complementary information for evaluating response in natural environments:

1. Metallothionein mRNA as an indicator of MT gene expression (estimated from Northern blot, ribonuclease protection assay, reverse transcription polymerase chain reaction).
2. Metallothionein as an indicator of mobilization of detoxification pathways (estimated from metal content, differential pulse polarography, radionuclide substitution assay, immunoassay, electrophoresis with fluorescent probes, capillary electrophoresis)
3. Metals bound to MT as an indicator of detoxified toxic metals (estimated from metal analysis of isolated MT or MT-containing intracellular fractions by atomic absorption spectrometry and inductively coupled plasma-mass spectrometry).

In general, exposure to certain metals—notably cadmium, zinc, or copper—results in elevated levels of MT mRNA, MT, and MT-bound metals. This has been amply demonstrated in laboratory experiments and has been extended to natural populations inhabiting metal-contaminated environments. However, of these measures, only that for MT-bound metals provides a direct reflection of the contribution that MT detoxification makes to metal bioaccumulation. While MT mRNA and MT should, in principle, reflect changes in levels of inducing metals, actual measurements can be confounded or obscured by simultaneous response to other endogenous and exogenous factors that can modulate MT gene expression or induce MT. Factors such as season, temperature, reproductive condition, and sex can affect levels of either MT or MT mRNA (Hylland et al., 1998; Mouneyrac et al., 2001; Van Cleef-Toedt et al., 2001). Furthermore, elevations in MT or the pool of MT-bound metals do not by themselves signify induction, because increases in the amount of MT and its bound metals can occur in response to metal exposure without induction of MT. For example, cadmium can displace zinc from MT or compete with zinc in binding to newly synthesized MT, resulting in increased stability of MT, reduced turnover of both MT and cadmium, and accumulation of MT and its bound metals without benefit of induction. This appears to have been the case in a population of cadmium-exposed oysters in which MT-bound cadmium levels were elevated as a result of environmental exposure while MT mRNA levels were not (Roesijadi, 1999). Analysis of the different components in the pathway of MT gene expression and accumulation of metallated MT listed above are needed to provide information on the underlying processes associated with changes in levels of MT and its capacity to sequester metals. From the perspective of developing specific biomarkers based on responses associated with MT, however, a focus on response variables that exhibit the least ambiguous relationship to metal exposure and bioaccumulation would be the most desirable.

The existence of metal-specific isoforms of MT that respond preferentially to specific toxic metals would be of value as biomarkers in hazard assessment. To date, a copper-specific MT in yeast (Fürst et al., 1988) and a cadmium-specific form in an earthworm (Sturzenbaum et al., 2001) exhibit such properties.

Metallothionein isoforms with similar promise have recently been reported for an oyster (Tanguy and Moraga, 2001) and mussel (Geret and Cosson, 2002). These might have utility where the environmental effects of the specific metals are of interest. One of the objectives in the use of biomarkers in ecotoxicology is to assess exposure or toxicity to specific contaminants from an environment characterized by multiple anthropogenic and natural stressors. The existence of metal-specific isoforms of MT represents such a potential.

Our current understanding of MT in the diverse species inhabiting natural environments does not allow universal generalizations regarding its utility as a biomarker applicable to all species. In some species, other pathways and biological compartments are quantitatively more important for sequestering metals, and changes in the response of MT are difficult to detect (Ritterhoff et al., 1996). It is clear, though, that natural populations in metal-enriched environments have responded in many instances through increased mobilization of MT. In some cases, such populations have adapted and evolved resistance to metal toxicity (Maroni et al., 1987; Klerks and Levinton, 1989). It had been speculated for some time that such events can have ecological implications apart from changes in fitness of adapted population, since detoxification would enable organisms to accumulate higher concentrations of metals and facilitate trophic transfer. Recent studies demonstrate that (1) the cell fraction containing MT in prey is the most significant contributor to the transfer of cadmium to predators (Wallace and Lopez, 1997) and (2) this cadmium transfer can result in toxic effects in the predator (Wallace et al., 1998; 2000). Thus, it appears that the binding of toxic metals to MT has a direct relationship with transfer of metals and their toxic effects up the food chain. Although the evolution of metal-resistance can provide advantage to the resistant organisms by enhancing their viability in metal-contaminated environments, broader ecological costs may be associated with such adaptation.

Summary

The progress made in understanding the biochemistry, molecular biology, physiology, and toxicology of MTs has greatly increased our understanding of its role in toxicological functions. This progress has been instrumental in refining approaches to the study of MT in natural populations. An understanding of MT function will improve our understanding of ecotoxicological relationships, as well as cellular and organismal function.

Suggested Supplemental Readings

Andrews, G.K., Regulation of metallothionein gene expression by oxidative stress and metal ions, *Biochem. Pharmacol.,* 59, 95–104, 2000.

Cosson, R.P., Bivalve metallothionein as a biomarker of aquatic ecosystem pollution by trace metals: limits and perspectives, *Cell. Mol. Biol.,* 46, 295–309, 2000.

Dallinger, R., Invertebrate organisms as biological indicators of heavy metal pollution, *Appl. Biochem. Biotechnol.,* 48, 27–31, 1994.

Kägi, J.H.R. and Y. Kojima, Chemistry and biochemistry of metallothioneins, in *Metallothionein II,* J.H.R. Kägi and Kojima, Y., Eds., Birkhauser-Verlag, Basel, 1987, pp. 25–61.

Klaassen, C.D., J. Liu, and S. Choudhuri, Metallothionein: an intracellular protein to protect against cadmium toxicity, *Annu. Rev. Pharmacol. Toxicol.,* 39, 267–294, 1999.

Lichtlen, P. and W. Schaffner, Putting its fingers on stressful situations: the heavy metal regulatory transcription factor MTF-1, *Bioessays,* 23, 1010–1017, 2001.

Moffatt, P. and F. Denizeau, Metallothionein in physiological and physiopathological processes, *Drug Metab. Rev.,* 29, 261–307, 1997.

Roesijadi, G., Metallothioneins in metal regulation and toxicity in aquatic animals, *Aquatic Toxicol.,* 22, 81–114, 1992.

Viarengo, A., B. Burlando, N. Cerratto, and I. Panfoli, Antioxidant role of metallothioneins: a comparative overview, *Cell. Mol. Biol.,* 46, 407–417, 2000.

Viarengo, A., B. Burlando, F. Dondero, A. Marro, and R. Fabbri, Metallothionein as a tool in biomonitoring programmes, *Biomarkers,* 4, 455–466, 1999.

IV STRESS PROTEINS

The **cellular stress response** is an "orchestrated induction of key proteins that form the basis for the cell's protein protection and recycling system" (Sanders and Dyer, 1994).[3] A cellular stress response can be elicited under the influence of heat, anoxia, some metals, some xenobiotics, ethanol, sodium arsenate, or UV (ultraviolet) radiation (Craig, 1985; Thomas, 1990; Di Giulio et al., 1995). Stress-induced proteins were first studied after organisms experienced abrupt changes in temperature (5 to 15°C). These **heat shock proteins** (hsp) were distinguished from one another according to their molecular weights. Respectively, there are 90-, 70-, 60-, 16–24-, and 7-kDa groupings designated as hsp90, hsp70, hsp60, low molecular weight (LMW), and ubiquitin. These proteins are now known to be induced by other stressors and, consequently, have been renamed **stress proteins** to reflect this more general role. The terminology based on heat stress has been modified. Now the stress-protein groupings are also called stress90, stress70, chaperon 60 (cpn60), LMW, and ubiquitin. The term **chaperon**, used collectively for stress90, stress70, and cpn60, reflects their role of associating with and directing the proper folding and coming together of proteins. They also protect proteins from denaturation and aggregation, and they enhance refolding of damaged protein to a functional conformation. Their final role is in the transport of proteins to their intercellular location, where they then are folded to a functional conformation (Craig, 1993).

Physical and chemical agents that have significant **proteotoxicity** (toxicity due to protein damage) (Hightower, 1991) induce stress proteins. Enhanced production of stress90, stress70, cpn60, LMW, and ubiquitin is initiated by denatured protein to protect, repair, or vector for breakdown the proteins of the cell. Some stress proteins are always present (e.g., stress70 [Welch, 1990]), but others (LMW) appear only under stressful conditions (Di Giulio et al., 1995). Stress90 is present at high concentrations under unstressed conditions and is induced to even higher levels by stress. Stress70 is present at lower concentrations during normal conditions and is induced by stress. Cpn60, which is found in the mitochondria and facilitates protein movement and folding, is present at low levels under normal conditions and is inducible. Stress70 and cpn60 are excellent candidates as biomarkers because they are inducible by stress, are highly conserved proteins, and are not normally present at high levels, as is stress90 (Sanders, 1990). The LMW proteins are more variable in structure and in inducibility among species, qualities that detract from the utility of LMW as biomarkers. Ubiquitin is induced by stress, and its structure is evolutionarily conservative, making it a potential biomarker (Sanders, 1990).

Stress proteins recognize and bind to exposed regions of denatured proteins, which are rich in hydrophobic peptides (Agard, 1993). With the aid of stress70 and stress90, denatured or aggregated proteins are unfolded and refolded properly to restore their function. Proteins damaged beyond repair are bound by ubiquitin, which helps move them to lysosomes for final breakdown (Di Giulio et al., 1995).

Increased concentrations of the various stress proteins are used as biomarkers of general cellular stress response. For example, Sanders and Martin (1993) correlated general contamination with cpn60 and stress70 in archived tissues of marine species. Because different stressors induce the various stress proteins to different degrees, Sanders and Dyer (1994) suggested that the patterns of stress-protein induction could be used to suggest the particular toxicant inducing the response. Patterns from field samples can be compared with those obtained with single-candidate toxicants in the laboratory. They referred to this approach as **stress-protein fingerprinting**. Because the stress proteins have evolved in a very conservative manner and the cellular stress response is so universal, Sanders and Dyer (1994) suggested that this approach has more universal application than many other biomarker methods.

---

[3] This is one of the most often discussed characteristics of the cellular stress response; however, Sanders (1990) also discussed as part of this response the induction of **glucose-regulated proteins** (grp) under low glucose or oxygen conditions. The grps are structurally similar to hsp, are present at basal levels in unstressed cells, and are induced in glucose- or oxygen-deficient cells exposed to toxicants that modify calcium metabolism, e.g., lead (Sanders, 1990).

V OXIDATIVE STRESS AND ANTIOXIDANT RESPONSE

Combining the reduction of molecular oxygen with energy generation during aerobic metabolism (Figure 6.4) creates the potential for **oxidative stress**, i.e., damage to biomolecules from free oxyradicals. Oxyradical-generating compounds such as hydrogen peroxide (H_2O_2) are also produced by aerobic metabolism and other processes, contributing to oxidative stress. Free radicals[4] such as the superoxide radical ($O_2^{\cdot-}$) and hydroxyl radical ($^{\cdot}OH$) can damage proteins, lipids, DNA, and other biomolecules. Oxyradicals are produced during electron-transport reactions in mitochondria and microsomes; photosynthetic electron transport; phagocytosis; and normal catalysis by prostaglandin synthase, guanyl cyclase, and glucose oxidase (Di Giulio et al., 1989).

Organisms cope with this situation in two ways. They produce antioxidants that react with oxyradicals. These antioxidants include Vitamin E, Vitamin C, β-carotene, catecholamines, glutathione, and uric acid (Winston and Di Giulio, 1991). (Note that glutathione has a dual role as a substrate for Phase II conjugation and as an antioxidant.) In addition to these antioxidants, enzymes that reduce the amount of oxyradicals present at any instant are involved in avoiding oxidative damage. **Superoxide dismutase** (SOD) decreases the amount of superoxide radical in the cell by catalyzing the reaction shown in Equation 6.2. **Catalase** (CAT) and **glutathione peroxidase** reduce levels of hydrogen peroxide via reactions shown in Equation 6.3 and 6.4, respectively.

$$2\ O_2^{\cdot-} + 2\ H^+ \rightarrow H_2O_2 + O_2 \qquad (6.2)$$

$$2\ H_2O_2 \rightarrow H_2O + O_2 \qquad (6.3)$$

$$2\ \text{Reduced Glutathione} +\ H_2O_2 \rightarrow \text{Oxidized Glutathione} + H_2O \qquad (6.4)$$

Xenobiotics can cause oxidative damage indirectly by interfering with these mechanisms of coping with oxidative stress. They can also participate directly in reactions leading to oxidative stress. Xenobiotics can form oxyradicals such as **alkoxyradicals** ($RO^{\cdot}$) and **peroxyradicals** ($ROO^{\cdot}$) and, in so doing, cause damage (Di Giulio et al., 1995). For example, carbon tetrachloride can undergo the following reaction to generate a trichloromethyl radical, $CCL_4 + e^- \rightarrow CCl_3^{\cdot} + Cl^-$. This reaction, involving the NADPH-cytochrome P-450 system, contributes to carbon tetrachloride damage (necrosis, cancers) to the human liver, where the free radical reacts with lipids, proteins, and DNA (Slater, 1984). Xenobiotics can also form free radicals during interactions with oxyradicals. For example, promethazine reacts with the hydroxyl radical, produces OH^-, and becomes a free radical itself. Polycyclic aromatic hydrocarbons (PAH) are activated by biotransformation to their free radical forms (Slater, 1984). Still other contaminants (quinones, aromatic nitro compounds, aromatic hydroxylamines, bipyridyls such as paraquat and diquat, and some chelated metals [Di Giulio et al., 1989]) are reduced to radicals and then undergo **redox cycling** to produce superoxide radicals from molecular oxygen. The initial reduction of the contaminant to a free radical can be facilitated by one of several reductases (Di Giulio et al., 1989). Because the contaminant enters the redox cycle as a free radical but exits in its original form at the end of a cycle, it is available to recycle many times and produce considerable amounts of oxyradicals.

Because contaminants can interfere with the normal mechanisms of coping with oxidative stress and can contribute to free-radical concentrations themselves, considerable effort has been spent in studying contaminant-related changes in oxidative stress. Elevated levels of superoxide dismutase, catalase, or glutathione peroxidase in exposed individuals may suggest elevated oxidative stress. Changes in these enzymes might be observed for individuals in contaminated sites (Regoli and Principato, 1995) or for those exposed to contaminants in the laboratory (Víg and Nemcsók, 1989).

---

[4] A **free radical** is a molecule having an unshared electron. The unshared electron is usually designated by a dot, •. Free radicals are extremely reactive. **Oxyradicals** are free radicals with an unshared electron of oxygen, e.g., $RO^{\cdot}$, where R is some oxygen-containing compound.

O$_2$ Reduction to Water

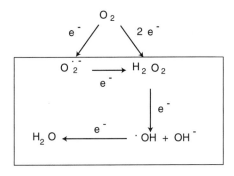

Catalyzed Haber-Weiss Reaction

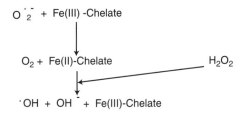

Figure 6.4   Reduction of O$_2$ to water during aerobic respiration (O$_2$ + 4e$^-$ → H$_2$O). Molecular oxygen can be converted to the superoxide radical (O$_2^{\cdot-}$) by adding one electron (e$^-$) or to hydrogen peroxide (H$_2$O$_2$) by adding two electrons. The superoxide radical can be converted to hydrogen peroxide by the addition of another electron. Hydrogen peroxide can then be reduced to the hydroxyl radical ($^\cdot$OH), producing a hydroxyl anion (OH$^-$). Water is produced with the addition of a fourth electron. Two highly reactive, free radicals are generated along this reaction sequence. Hydrogen peroxide is also produced and, through the **Haber-Weiss reaction** (O$_2^{\cdot-}$ + H$_2$O$_2$ → $^\cdot$OH + OH$^-$, as shown in the top panel in the box), generates oxyradicals. The **catalyzed Haber-Weiss reaction** is a greatly accelerated Haber-Weiss reaction catalyzed by metal chelates, as shown in the bottom panel.

Antioxidant pools can also be examined. For example, mussels exposed to metals (Regoli and Principato, 1995) or paraquat (Wenning et al., 1988) show significant changes in glutathione concentrations.

Increased concentrations of free radicals can cause membrane dysfunction due to **lipid peroxidation** (oxidation of polyunsaturated lipids). **Malondialdehyde**, a breakdown product from lipid peroxidation, is indicative of oxidative damage of lipids from a variety of toxicants (Thomas, 1990). For example, Atlantic croaker (*Micropogonias undulatus*) exposed to PCBs (Aroclor 1254) or cadmium had elevated concentrations of microsomal malondialdehyde (Wofford and Thomas, 1988).

Free radicals can form covalent bonds with a variety of biomolecules; free radicals covalently bond to membrane components such as enzymes and receptors to change their structures and functions (Slater, 1984); and free radicals can cause damage by reacting with sulfhydryl groups of proteins and other biomolecules and, in so doing, influence their function (Slater, 1984). There is a final, important reason for the interest paid to contaminant-influenced oxidative stress: The formation of free radicals near DNA can lead to mutations and, as a consequence, increased risk of cancer or other genotoxic consequences.[5] Malins (1993) found DNA (guanine and adenine) lesions induced

---

[5] The formation of free radicals is key to understanding radiation effects too. Free radicals formed from water and molecular oxygen during irradiation cause damage to cells. However, free radicals can also be used to our advantage in radiation treatments of cancers, where our intent is to kill cancer cells. Damage to cancer cells can be enhanced during radiation treatment by administering a radiosensitizer such as derivatives of nitro imidazoles (Slater, 1984). **Radiosensitizers** enhance the production of free radicals in the area receiving radiation, which enhances destruction of cancer cells.

by the hydroxyl radical in fish exposed to carcinogens. He suggested that these types of lesions in fish and humans lead to increased misreading of the DNA template and increased cancer risk.

VI DNA MODIFICATION

As suggested in the paragraph above, toxicants and their products can be discussed in terms of **genotoxicity**, i.e., the damage by a physical or chemical agent to genetic materials such as chromosomes or DNA. At the molecular level, agents such as free radicals produce breaks in one or both sides of the DNA molecule (Figure 6.5). Oxyradicals also oxidize bases. Xenobiotics and their metabolites may bond covalently to a base or, less frequently, to another portion of the DNA molecule to form an **adduct** (Shugart, 1995).

Metals can bind to phosphate groups and heterocyclic bases of DNA and, in so doing, change the stability and normal functioning of the DNA (Eichhorn et al., 1970). Magnesium bonds to phosphate groups in the DNA backbone and stabilizes the DNA structure, but copper bonds between bases, competes with the normal hydrogen binding, and destabilizes the DNA structure (Eichhorn, 1975). Because the matching of specific base pairs is a matter of degree of attraction for complementary versus noncomplementary pairs, the modification of hydrogen bonding by metals can contribute to base mispairing. For example, mercury forms strong crosslinks between the strands

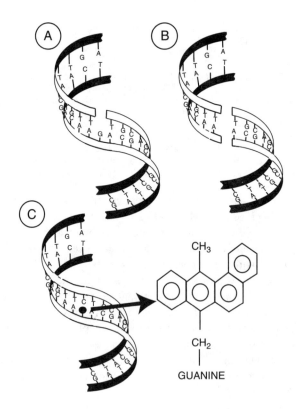

Figure 6.5   Various types of damage can occur to DNA due to toxicants. Toxicants or free radicals can produce single (A) or double (B) strand breaks. Xenobiotics or their metabolites can react with bases to form adducts (C). Here, metabolism of 7,12-dimethyl-benz[a]anthracene leads to covalent bonding of its metabolite to guanine to form a DNA adduct. Interactions such as those with free radicals can also modify bases, i.e., oxidize bases such as thymine and guanine to thymine glycol and 8-hydroxyguanine, respectively (Di Giulio et al., 1995).

of the DNA molecule. Normal DNA repair mechanisms are overwhelmed if alterations occur excessively often, and high mutation rates can result.

The expected diploid chromosome number (2N) can be disrupted due to chromosome breakage, resulting in a deviation from the usual number of chromosomes (**aneuploidy**) or structural aberrations in chromosomes. Agents that cause chromosome damage in living cells are classified as **clastogenic**. All of these genotoxic effects can have **mutagenic** (causing mutations), **carcinogenic** (causing cancers), and **teratogenic** (causing developmental malformations) consequences (Jones and Parry, 1992).

Various methods are used to assess genotoxicity. Chromosome damage can be determined by microscopic examination after appropriate staining. Aneuploidy and other clastogenic effects can be quantified using flow cytometry after staining the DNA in cells with a fluorescent dye. With flow cytometry, fluorescence is used to measure the amount of DNA in individual cells of a sample as each cell flows through an excitation light beam. The distribution of DNA concentrations in the population of cells is examined for significant numbers of cells with atypical amounts of DNA (Shugart, 1995). Clastogenic activity can also be reflected in the number of **micronuclei** (membrane-bound masses of chromatin) in cells. The presence of many micronuclei suggests impairment of the cell's ability to divide properly (Jones and Parry, 1992).

DNA adducts can be assayed using a ^{32}P-labelling method (Jones and Parry, 1992). Deoxyribonucleoside 3′-monophosphates are produced by hydrolysis of the DNA molecule. The DNA molecule is broken down into deoxyribonucleoside 3′-monophosphates and then labeled with [32]P. Most of the deoxyribonucleoside 3′-monophosphates contain one of the four normal bases in DNA, but some will have bases covalently bound to an adduct. The labeled deoxyribonucleoside 3′-monophosphates are separated chromatographically to the four normal bisphosphates of adenosine, cytidine, guanosine, and thymidine, plus bases with adducts. The amount of radioactivity associated with the base-adduct fraction reflects the number of DNA adducts and the potential for genotoxic effects. The occurrence of adducts has been correlated with cancer risk (Gaylor et al., 1992).

The amount of DNA breakage can also be estimated and correlated with exposure to genotoxic contaminants. Shugart (1988) described an alkaline unwinding assay to estimate the degree of single-strand breakage. After a specified time of exposure to alkaline conditions, the DNA strands from samples unwind to different degrees, depending on the relative amounts of double- and single-stranded DNA in the sample. Differences in fluorescence intensity of single- and double-stranded DNA are used to measure the relative amounts of each in the samples. Identical amounts of isolated DNA from control and exposed individuals are placed under alkaline conditions, allowed a set time to unwind, and then the fraction of DNA that is double-stranded in each sample is estimated via fluorescence. This fraction is used to imply the relative amount of single-strand breaks in the various samples. For example, this fraction dropped within ten days of bluegill sunfish (*Lepomis macrochirus*) exposure to benzo[a]pyrene (Shugart, 1988), suggesting an increase in single-strand breaks with exposure. Meyers-Schöne et al. (1993) correlated environmental contamination with single-strand breaks in DNA of freshwater turtles using this alkaline unwinding assay.

VII ENZYME DYSFUNCTION AND SUBSTRATE POOL SHIFTS

Contaminants can have significant effects on enzymes, and these effects are used routinely as biomarkers. Metals can influence protein-mediated catalysis, transport, and gas exchange by modifying protein structure (secondary, tertiary, or quaternary) and stability (Ulmer, 1970). Metals bind to a wide range of electron donor groups of proteins such as imidazole, sulfhydryl, hydroxyl, carboxyl, amino, quanidinium, and peptide groups (Eichhorn, 1975). Changes in secondary and tertiary structure can lead to lowered or elevated enzyme activities. Normally, metals stabilize quaternary structure of many proteins: Substitution of another metal for the usual stabilizing metal can interfere with the coming together of peptide chains to form stable and functional oligomers,

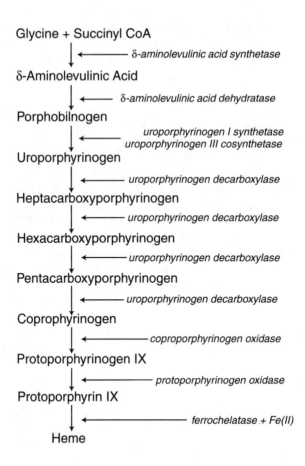

Figure 6.6 Steps in the synthesis of heme. Enzymes catalyzing each conversion are italicized and placed to the right of the reaction sequence. (Modified from Figure 1 in Woods, J.S. et al., *J. Toxicol. Environ. Health,* 40, 235–246, 1993.)

e.g., multimeric enzymes. Some enzymes are stabilized by metals (e.g., lysozyme by Mg^{2+}), and substitution of other metals for these stabilizing metals can enhance denaturation (Ulmer, 1970). Metals are also present at active sites of biomolecules such as carboxypeptidase, alkaline phosphatase, carbonic anhydrase, cytochrome c, and hemoglobin (Eichhorn, 1975). Displacement of the appropriate metal by another can change the functioning of these proteins.

Several steps in the synthesis of heme (Figure 6.6) are modified by toxicants, and associated enzyme activities and substrate pools are used as biomarkers to reflect sublethal poisoning. **Porphyrins** are produced as intermediates in heme synthesis. Porphyrins with four to eight carboxyl groups tend to be produced in excess and are excreted in urine. The relative amounts of each of these excreted porphyrins tend to be consistent among individuals. However, mercury poisoning interferes with normal heme synthesis, promotes oxidation of reduced porphyrins, and shifts porphyrin pools in the urine as a consequence. For example, rats exposed to mercury have increased levels of porphyrins with four and five carboxyl groups (Woods et al., 1993). Woods et al. (1993) used such shifts to indicate sublethal exposure to mercury in dentists who use mercury in silver-mercury amalgam fillings. Male dentists categorized as having either no mercury or mercury levels greater than 20 µg l^{-1} in their urine had pentacarboxylporphyrin concentrations of 0.76 and 3.07 µg l^{-1}, respectively.

Polyhalogenated aromatic hydrocarbons such as PCBs may also interfere with heme synthesis, possibly by inhibiting uroporphyrinogen decarboxylase or by P-450 generation of oxyradicals that oxidize porphyrinogens (Peakall, 1992). Hepatic porphyrins have been used as biomarkers of polyhalogenated aromatic hydrocarbon exposure to species such as pike (*Esox lucius*) (Koss et al., 1986) and herring gulls (*Larus argentatus*) (Fox et al., 1988).

Other biomarkers can indicate effect to the respiratory pigments of animals. Cadmium exposure of flounder (*Pleuronectes flesus*) depressed blood hematocrit, hemoglobin titers, and red blood cell counts (Johansson-Sjöbeck and Larsson, 1978). Lead depresses the activity of δ-**aminolevulinic acid dehydratase** (δ-ALAD or ALAD), an enzyme catalyzing the conversion of δ-aminolevulinic acid to porphobilnogen during heme synthesis (Figure 6.6). Exposure of rainbow trout (*Salmo gairdneri*) to lead lowered ALAD activities and caused anemia (Johansson-Sjöbeck and Larsson, 1979). Fish exposed to lead-containing mine drainage also had low blood ALAD activities (Dwyer et al., 1988; Schmitt et al., 1993). This effect of lead can be modified by the presence of metals such as cadmium and zinc (Berglind, 1986; Schmitt et al., 1993).

Because of their importance in animal osmoregulation and cell-water regulation, ATPases (adenosine triphosphatases) have been the subject of considerable study (Thomas, 1990). Their presence on gill surfaces makes them particularly vulnerable to contact with toxicants. Yap et al. (1971) exposed bluegill sunfish (*Lepomis machrochirus*) to PCBs and noted depressed Mg^{2+}- and Na^+, K^+-ATPase activities. Rock crab (*Cancer irroratus*) gills exposed to cadmium or lead also showed depressed Na^+, K^+-ATPase activity.

VIII SUMMARY

Understanding molecular effects enhances our ability to assign causal linkage to effects at higher levels of organization and to predict effects of untested chemicals based on similar molecular interactions with biomolecules. It also provides biomarkers as tools to proactively measure effects in field situations. Some biomarkers are quite specific (lead's effect on ALAD activity), specific to a class of toxicants (metallothioneins), or general (stress proteins). All have value if researchers can avoid the temptation to attach too much ecological relevance to a particular biochemical response.

SELECTED READINGS

Di Giulio, R.T., W.H. Benson, B.M. Sanders, and P.A. Van Veld, Biochemical mechanisms: metabolism, adaptation, and toxicity, in *Fundamentals of Aquatic Toxicology: Effects, Environmental Fate, and Risk Assessment*, 2nd ed., Rand, G.M., Ed., Taylor & Francis, Washington, D.C., 1995.

Di Giulio, R.T., P.C. Washburn, R.J. Wenning, G.W. Winston, and C.S. Jewell, Biochemical responses in aquatic animals: a review of determinants of oxidative stress, *Environ. Toxicol. Chem.*, 8, 1103–1123, 1989.

Goksøyr, A. and L. Förlin, The cytochrome P-450 system in fish, aquatic toxicology and environmental monitoring, *Aquat. Toxicol.*, 22, 287–312, 1992.

McCarthy, J.F. and L.R. Shugart, Eds., *Biomarkers of Environmental Contamination*, Lewis Publishers, Chelsea, MI, 1990, p. 457.

Peakall, D., *Animal Biomarkers as Pollution Indicators*, Chapman & Hall, London, 1992, p. 291.

Shugart, L.R., Environmental genotoxicology, in *Fundamentals of Aquatic Toxicology: Effects, Environmental Fate, and Risk Assessment*, 2nd ed., Rand, G.M., Ed., Taylor & Francis, Washington, D.C., 1995.

Shugart, L.R., Molecular markers to toxic agents, in *Ecotoxicology: A Hierarchical Treatment*, Newman, M.C. and Jagoe, C.H., Eds., CRC Press, Boca Raton, FL, 1996.

Thomas, P., Molecular and biochemical responses of fish to stressors and their potential use in environmental monitoring, *Amer. Fish. Soc. Symp.*, 8, 9–28, 1990.

Timbrell, J., *Principles of Biochemical Toxicology*, 3rd ed., Taylor & Francis, Philadelphia, 2000.

CHAPTER 7

Cells, Tissues, and Organs

...the problem took the form of habitat pollution $\rightarrow$ DDE accumulation in prey species $\rightarrow$ DDE in predators $\rightarrow$ decline in brood size $\rightarrow$ potential extermination. The same phenomenon can [be written as] lipid-soluble toxicant $\rightarrow$ bioaccumulation in organisms with poor detoxification systems ... $\rightarrow$ vulnerable target organ [shell gland] $\rightarrow$ inhibition of membrane-bound ATPases ... $\rightarrow$ potential extermination. Ecologists would claim a decline in population recruitment, biochemists an inhibition of membrane enzymes.

Simkiss (1996)

I INTRODUCTION

There are two superficially opposing views of how to deal with hierarchical subjects. At one extreme, the reductionist approach attempts to understand the behavior of the simplest units at the lowest levels of organization and then to use this understanding to explain phenomena at all higher levels. By analogy, a clock is understood completely if one understands the workings of all of its parts. In apparent contrast, the holistic approach holds that, because unique properties emerge at higher levels of organization and complex interactions among parts beget complex dynamics, an understanding of higher-order processes is much more useful than building a causal structure from the lowest or most fundamental level to the highest. In reality, both approaches are successful only when used in a mutually supportive fashion (see Newman [2001] for more details). Alone, neither approach works beyond a limited scope. Prediction is limited if one does not understand the components of a system but, instead, only describes phenomena at the highest level of organization. Limited prediction makes it impossible to move effectively toward the scientific goal of ecotoxicology. The reductionist approach alone is also impractical because our limited knowledge makes prediction of some emergent properties impossible based solely on mechanisms at lower levels. Also, causal structure does not always proceed up the artificial, hierarchical structure of biological organization. Effects at the tissue level can be influenced by physiological (one level up) in addition to biochemical (one level down) mechanisms. Arguably, the physiological causes could be traced back to their biochemical mechanisms. However, our incomplete understanding of cascading events, the network of intermeshed cause-effect phenomena, and the uncertainty associated with the magnitude of each phenomenon in the causal web condemns such an exercise to failure beyond a limited number of links.

In some cases, causal structure can be defined clearly for several levels of organization. One such causal sequence from the biochemical $\rightarrow$ cell $\rightarrow$ individual that can be quickly evoked is activation of a xenobiotic by monooxygenases $\rightarrow$ production of oxyradicals and consequent DNA damage $\rightarrow$ increased risk of cancer in the liver $\rightarrow$ increased chance of death to an individual.

However, extrapolation of this sequence to the population level would be difficult, and taking it to the level of an ecological community level would be highly speculative. Causal linkage fails almost immediately for other cases, such as that between ALAD (aminolevulinic acid dehydratase) activity in red blood cells and individual fitness. Future work will eventually allow clearer definition for some of these linkages, but it is unlikely that clear linkage will ever be made in all instances.

Success in ecotoxicology is enhanced by keeping holistic goals and limitations in mind while applying reductionistic methods to questions. Regardless of the level being examined, one should use the reductionist theme of studying the "simplest system you think has the properties you are interested in" (Levinthal quoted in Platt, 1964). This should be done in full anticipation that important properties may emerge at higher levels.

As discussed in Chapter 1 and reflected again in the quote above, conceptual coherency is maintained by understanding that "processes at one level take their mechanisms from the level below and find their consequences at the level above. ... Recognizing this principle makes it clear that there are no truly 'fundamental' explanations,[1] and makes it possible to move smoothly up and down the levels of the hierarchical system without falling into the traps of naive reductionism or pseudo-scientific holism" (Caswell, 1996; also, Bartholomew, 1964). Thus coherency is maintained in a conceptual "relay race" in which all legs (levels of biological organization) are equally crucial to achieving the goal (Newman, 1995). Attempts to fulfill the goal of ecotoxicology using either the reductionist or holistic approaches alone are equally absurd attempts to "swallow the ocean." Whether attempted in one impossibly large gulp or in an impossibly large number of small gulps, the ocean will not be swallowed.

II GENERAL CYTOTOXICITY AND HISTOPATHOLOGY

The study of effects at the cell, tissue, and organ level of organization is invaluable for several reasons. In the context of the hierarchical "relay race" just described, these effects are consequences of biochemical mechanics that provide interpretative power about effects to individuals. They integrate damage done at the molecular level (Hinton and Laurén, 1990a). These biomarkers can be used as an early warning system for potential effects at the level of individual and, sometimes, population. **Histopathology**, the study of change in cells and tissues associated with communicable or noncommunicable disease, provides a cost-effective way to verify toxicant effect as well as exposure (Hinton, 1994).

There are two shortcomings of histopathological biomarkers. First, the normal histology and variations in normal histology with season, diet, reproductive cycle, and other processes may be poorly understood for sentinel species (Hinton and Laurén, 1990a). Much more descriptive work (normal science) is required in this area to improve the effectiveness of histopathological biomarkers. Second, although methods now exist to routinely quantify effects, most histopathological studies are qualitative, and quantitation is needlessly neglected (Jagoe, 1996). Both of these limitations can be overcome through the use of more normal science and a change in emphasis to quantitative methods.

II.A   Necrosis

A wide range of **lesions** (pathological alterations of cells, tissues, or organs) indicate exposure to toxicants and suggest mechanisms of action. Often, lesions are associated with a specific **target organ** as a result of preferential toxicant transport to, accumulation in, or activation within that organ. As an important example of xenobiotic activation within a specific organ, the high levels of

---

[1] A related incongruity arises as a consequence of not fully appreciating this context for ecotoxicological study. Often, statements are made that a specific level of organization is the most important to understanding ecotoxicological phenomena. Such statements are reminiscent of the Ptolemaic theory (the Earth is the center of the universe) and reflect a false paradigm that can be called the **Ptolemaic incongruity**. This incongruity—that any particular level of biological organization holds the central role in the science of ecotoxicology—leads to confusion, narrowness of vision, and wasteful intransigence.

biotransformation and activation of carbon tetrachloride in the liver is responsible for its hepatoxic nature. Cadmium damage to the kidney proximal tubules is linked to its transport to and association with metallothionein in the kidney during oral cadmium poisoning. Cadmium associated with metallothionein enters the kidney, is filtered out of circulation in the glomerulus, and is reabsorbed in the proximal tubules, where lysosomal activity releases the cadmium (Goldstein and Schnellmann, 1996). The cadmium—free of the metallothionein—then damages kidney cells.

Cytotoxicity (toxicity causing cell death[2]) can be reflected in a tissue or **target organ** as **necrosis** (cell death from disease or injury). Pyknosis, the most obvious evidence of necrosis, involves the cell nucleus. With **pyknosis (pycnosis)**, the distribution of chromatin in the nucleus changes, with the material condensing into a strongly staining mass. The nucleus stains intensely basophilic and becomes irregular in shape (Sparks, 1972). Sometimes the nucleus disintegrates (**karyolysis**). The cytoplasm of necrotic cells tends to be more acidophilic[3] (eosinophilic) than that of viable cells. Mitochondria swell and more cytoplasmic granules may appear. Necrosis can also be indicated by displacement or separation of the cell from its normal location in a tissue (Meyers and Hendricks, 1985), e.g., cell sloughing from gill epithelium or arterial wall.

Different types of necrosis are characteristic of various insults. **Coagulation necrosis** is characterized by extensive coagulation of cytoplasmic protein, making the cell appear opaque. The cell outline and position within its tissue are retained for some time after cell death (Sparks, 1972; Hinton, 1994). Coagulation necrosis occurs in the alimentary tract of mammals after ingestion of phenol (Sparks, 1972). It can also result from acute exposure to inorganic mercury because mercury denatures and precipitates proteins (Sparks, 1972). Renal failure can result from accumulation and eventual spillover of metallothionein-associated metals in kidneys (the target organ) and subsequent coagulation necrosis (Hinton and Laurén, 1990a) of kidney cells. This type of necrosis can also occur with an abrupt cessation of blood flow, and for this reason, Hinton and Laurén (1990b) suggest that it is a good biomarker for cytotoxicity. **Liquefactive (cytolytic) necrosis** occurs with rapid breakdown of the cell as a consequence of the release of cellular enzymes. Many cells undergoing liquefactive necrosis in a tissue, especially a tissue with much enzymatic activity, can produce fluid-filled, necrotic spaces in the tissue (Meyers and Hendricks, 1985). Hinton and Laurén (1990b) suggest that liquefactive necrosis is less useful as a biomarker than coagulation necrosis because liquefactive necrosis is often associated with infection. Several other forms of necrosis have been described. With **caseous necrosis**, cells disintegrate to form a mass of fat and protein. **Gangrenous necrosis** is a combination of coagulation and liquefactive necrosis (Sparks, 1972) often resulting from puncture (with an associated lack of blood supply to the damaged tissue[4]) and subsequent infection. Meyers and Hendricks (1985) describe **fat necrosis**, which involves deposits of saponified fats in dead fat cells. **Zenker's necrosis** occurs only in skeletal muscle and is similar to coagulation necrosis. All reflect cell death, but coagulation necrosis seems to best reflect toxicant effect.

II.B Inflammation

Inflammation can be useful as a biomarker of toxicant effect. It is a response to cell injury or necrosis that isolates and destroys the offending agent or damaged cells (Sparks, 1972). As such, inflammation often is associated with toxicant-induced necrosis. For example, hepatic necrosis due to toxicant action can be accompanied by inflammation (Hinton and Laurén, 1990b). The process

---

[2] In everyday use, terms such as *death* lack the clear distinctions that are important in ecotoxicology. The distinction between cell death and somatic death is a good example. Cell death or necrosis occurs in living as well as dead individuals. Death of an individual is **somatic death**. As we have already seen, *stress* is another term requiring unusual attention in its use. In the last chapter, we discussed cellular and oxidative stress. Later, we will discuss Selyean (individual) and ecosystem stresses. Each describes a different phenomenon.
[3] An **acidophilic component** of a cell (e.g., the cytoplasm) is one that is readily stained by an acidic dye, and a **basophilic component** (e.g., the nucleus) is one that is readily stained by a basic dye. In general preparations such as those shown in this chapter, hematoxylin and eosin are used to stain the nucleus and cytoplasm components, respectively.
[4] The resulting inadequacy of blood supply to surrounding tissues is called **ischemia**.

of inflammation continues through to lesion healing. The net result of inflammation is healed tissue with damaged cells being replaced by functioning cells (La Via and Hill, 1971).

The four **cardinal signs of inflammation** are heat, redness, swelling, and pain, although heat is irrelevant for poikilotherms, and redness is relevant to red-blooded animals. Blood vessels in the damaged region dilate to increase blood flow to the area, causing redness and heat. Pain is caused by the pressure of tissue swelling. Swelling is a result of fluid from the blood passing through blood vessel walls and into tissues of the inflamed area. Leucocytes can leave the blood vessels and enter the area of damaged tissue. Consequently, the infiltration of such cells (neutrophilic granulocytes or neutrophils, mononuclear cells, and lymphocytes) also indicates inflammation. Later in the process, small blood vessels begin to form, and connective tissue begins to grow in a mass called the **granulation tissue** (La Via and Hill, 1971). Scar tissue may be formed due to fibroblast and collagen infiltration of the damaged tissue (Sparks, 1972). If inflammation continues for too long as a consequence of chronic damage or infection, dense collections of collagenous connective tissue form and can produce tissue dysfunction. Consequently, such accumulations can be evidence of chronic inflammation.

Although the inflammatory response is common to all metazoan phyla, details vary. Students interested in studying nonmammalian species are advised to consult books describing the pathology and immunology of invertebrates and lower vertebrates (e.g., Sparks, 1972; Cooper, 1976) rather than relaying solely on mammalian pathology books.

II.C   Other General Effects

Several other general effects of toxicants are observed in cells and tissues. Swelling of mitochondria has been correlated with metal contamination (Aloj Totaro et al., 1986) or poisoning (Squibb and Fowler, 1981). **Lipofuscin** accumulation is another biomarker of toxicant effect. Lipofuscin is a degradation product of lipid peroxidation that accumulates in cell vacuoles called **residual bodies** (La Via and Hill, 1971). Lipofuscin accumulates with age in some cell types (e.g., neurons), giving it the name **age pigment** (La Via and Hill, 1971; Aloj Totaro et al., 1986). It also increases with exposure to some toxicants. For example, lipofuscin granules in squid (*Torpedo marmorata*) neurons increased with copper exposure (Aloj Totaro et al., 1985; 1986), reflecting enhanced lipid peroxidation due to oxyradical generation by copper (see Chapter 6). Interestingly, that portion of the squid nervous system with the most lipofuscin accumulation (the electric lobe) had the lowest activities of superoxide dismutase of all central nervous system tissues examined. Metal-rich granules (Figure 7.1) are another biomarker that has been discussed previously. Found in almost all animal phyla (Simkiss, 1981), their presence in unusually high densities in tissues such as the hepatopancreas, arthropod midgut, and kidney suggests a detoxifying response to metal exposure (Brown, 1978; Hopkin and Nott, 1979; Simkiss and Taylor, 1981; Mason et al., 1984; Hopkin et al., 1989).

III GENE AND CHROMOSOME DAMAGE

Contaminants can produce changes in DNA, and such changes can affect cells and tissues, increasing both somatic and genetic risk. **Somatic risk** is risk of an adverse effect to the exposed individual resulting from genetic damage to somatic cells (e.g., damage leading to cancer), and **genetic risk** is the risk to the progeny of an exposed individual as a consequence of heritable genetic damage (e.g., damage to germ cells or gametes, leading to a nonviable fetus or an offspring with a birth defect). These effects are manifested in a variety of cellular structures or processes (Figure 7.2). Several effects are routinely used as biomarkers and provide a mechanism for consequences at the individual level, such as mutation or cancer.

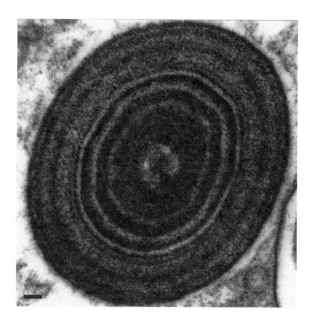

Figure 7.1 A transmission electron micrograph of a Type A granule (cross section) from the digestive gland of the spider, *Dysdera crocata*. Note the characteristic concentric layering in the granule. The granule section is approximately 2 μm in diameter. (Courtesy of S.P. Hopkin, University of Reading. With permission of Chapman & Hall.)

Toxicant-mediated chromatid breakage can be correlated with the incidence of **sister chromatid exchange** (SCE)[5] of DNA (Figure 7.2A). Dixon and Clarke (1982) showed a clear dose-SCE response for mutagen[6] (mitomycin C) exposure of the blue mussel, *Mytilus edulis*. A similar relationship was found for *M. edulis* larvae exposed to mutagens (Harrison and Jones, 1982). Consequently, sister chromatid exchange has been proposed as an assay of DNA damage (Tucker et al., 1993). In this assay, one chromatid in the pair is stained with 5-bromodeoxyuridine. Then, after two cycles of cell division occur, tissue samples are examined for DNA exchange between chromatids of labeled segments. With no exchange, chromatids would remain either completely stained or unstained after cell division. When exchange between chromatids has occurred, chromatids with both unstained and stained segments would be produced. The number of SCEs per metaphase or SCEs per chromosome is used as a measure of DNA damage.

Structural **chromosomal aberrations** include breakage and loss of segments of DNA or chromosomal rearrangements. Chromosomal breaks involve double-strand breaks, as shown in Figure 7.2B (left). Fragments of chromosomes may also be apparent. Also visible may be "gaps" in chromosomes, small discontinuities in the chromosome. A **chromatid aberration** occurs if only one strand (chromatid) is broken (Figure 7.2B, right).

There have been numerous studies using aberrations as biomarkers. Chromosomal aberrations in bone marrow cells were higher in mice (*Peromyscus leucopus*) and cotton rats (*Sigmodon hispidus*) from a petroleum, heavy metal, and PCB (polychlorinated biphenyl)-contaminated site than in mice and rats from uncontaminated sites (McBee et al., 1987). Similarly, chromosomal aberrations in blood lymphocytes of petroleum refinery workers (0.023 to 0.037 breaks per lymphocyte) were elevated relative to those of control populations (0.015 to 0.021 breaks per lymphocyte)

---

[5] Before cells divide, the DNA in each chromosome is duplicated to produce two identical chromatids. As the chromosomes condense prior to division, each appears as a pair of **chromatids** connected at a common centromere. At the metaphase plate, **sister chromatids** come together with those of the other homologous chromosome to form a **tetrad**. The four sister chromatids (two associated with each chromosome) can exchange segments of homologous DNA by the breaking and rejoining with crossing over of DNA segments.

[6] A **mutagen** is a physical or chemical entity capable of producing mutations.

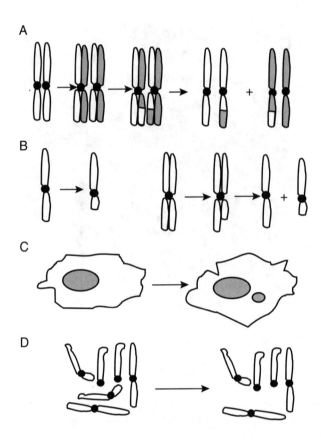

Figure 7.2   Damage to genetic materials can be assessed with a variety of cellular qualities. Increased rates
of sister-chromatid exchange (A) can suggest the rate of DNA damage, although the exchange
itself is not injurious. Beginning as a pair of homologous chromosomes (a **homolog**), four chromatids are formed. Sister chromatids in this **tetrad** (four chromatids paired as two homologs at
metaphase) may exchange material. In the sister-chromatid exchange assay, DNA is stained in
the first round of cell division with 5-bromodeoxyuridine, and the cells then synthesize unstained
DNA in subsequent divisions. This results in stained and unstained chromatids that can exchange
DNA in future cell divisions. This exchange is seen as chromatids with both stained and unstained
segments. The rate of exchange is correlated with the frequency of DNA breakage. Chromosomes
can be damaged (B, left), producing a chromosomal aberration. Viewed under the light microscope,
these aberrations may appear as breaks or gaps in chromosomes. If only one chromatid in a
homolog is broken, a chromatid aberration occurs (B, right). With a chromatid aberration, only one
chromosome in one of the two daughter cells would be damaged. Failure of mitotic processes can
produce micronuclei (C) or aneuploidy, a deviation from the usual ploidy (e.g., the usual 2N diploid
number of chromosomes) (D).

(Khalil, 1995). Lead poisoning increased the incidence of chromosomal aberrations in mouse bone
marrow cells (Forni, 1980). Chromosomal and chromatid aberrations in workers at a lead oxide
factory were elevated by sublethal lead exposure (18.7% versus 5.1% abnormal metaphases per
lymphocyte for factory workers and controls, respectively) (Forni, 1980).

Anomalies during cell division, such as chromosome damage and/or spindle dysfunction, can
produce micronuclei, nuclear segments isolated in the cytoplasm from the nucleus (Nikinmaa, 1992)
(Figure 7.2C). Red blood cells of fish exposed to toxicants have been reported to have higher numbers
of micronuclei (Nikinmaa, 1992). Softshell clams (*Mya arenaria*) exposed to PCB-contaminated
sediments of the New Bedford Harbor in Massachusetts had a threefold higher incidence of micronuclei in blood cells than clams from a clean site (Martha's Vineyard, MA) (Dopp et al., 1996).
Further, the number of leukemic cells per milliliter of hemolymph was correlated with the number

of blood cells with micronuclei, suggesting a common mechanism between production of micronuclei and severity of leukemia in clams from contaminated sites.

Failure of chromosomes to properly segregate during cell division can result in aneuploidy (Figure 7.2D). With a chromosome preparation, such a condition would appear as an atypical number of chromosomes in the cell, e.g., 2N − 1 or 2N + 1 chromosomes. As discussed in Chapter 6, flow cytometry has been used to quantify aneuploidy in cells taken from individuals exposed to genotoxic agents. For example, Lamb et al. (1991) used flow cytometric methods to document aneuploidy in the red blood cells of turtles (*Trachemys scripta*) from a reservoir with elevated levels of radiocesium (^{137}Cs) and radiostrontium (^{90}Sr).

Vignette 7.1   Chromosome Damage

Karen McBee
Oklahoma State University, Stillwater, OK

Chromosomes have been analyzed for evidence of structural damage that results from exposure to clastogenic compounds since early in the 20th century, but effects of environmental contaminants on chromosome structure have been investigated in field settings only for about 20 years. Almost all routine laboratory protocols and techniques have been adapted for use with a wide array of organisms, including plants (Al-Sabti and Kurelec, 1985a), invertebrates (Al-Sabti and Kurelec, 1985b), fish (Al-Sabti, 1985; Al-Sabti and Hardig, 1990), and mammals (Peakall and McBee, 2001). The most commonly used techniques are metaphase chromosome aberration analysis (CA), sister chromatid exchange (SCE), and micronucleus formation (MN).

Metaphase Chromosome Aberration Analysis

Visualization of lesions in metaphase spreads (Figure 7.3) might be the most commonly used and intuitively obvious way to document chromosome damage. Assays to detect induction of chromosomal aberrations (CA) typically have involved microscopic examination of chromosomal spreads extracted from blood, bone marrow, spleen, or testis. Cells are scored for the presence of several categories of chromosomal

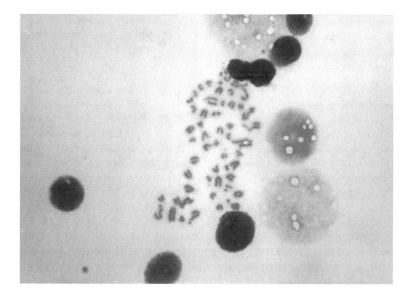

Figure 7.3   Metaphase chromosome spread from *Sigmodon hispidus* (hispid cotton rat) showing multiple chromatid and chromosome breaks. (Photo by Karen McBee.)

aberrations, including breaks, deletions, gaps, and translocations of fragments of one chromosome onto one or more nonhomologous chromosomes.

Chromosome and chromatid aberrations in cotton rats (*Sigmodon hispidus*) living close to two hazardous-waste dumps were compared with animals from a control site by Thompson et al. (1988). After all classes of aberrations were lumped together, the frequency of occurrence was higher in animals from hazardous-waste sites than in animals from the control site. When individual classes of aberrations, such as chromatid breaks, were considered separately, animals from hazardous-waste sites showed higher frequencies, but these differences were not statistically significant.

Chromosomal aberrations were examined in bone marrow of white-footed mice (*Peromyscus leucopus*) and cotton rats (*S. hispidus*) collected over a two-year period at a site contaminated with a complex mixture of petrochemical refinery wastes, diesel fuel, heavy metals, and PCBs. Animals of both species from the contaminated site showed significant increases in number of aberrant cells per individual and mean number of lesions per cell compared with animals from two reference sites. However, there were no statistical differences for either parameter when animals from the two reference sites were compared (McBee et al., 1987). *Peromyscus* from the contaminated site had significantly higher levels of lead and chromium but not cadmium and zinc in muscle tissue than did animals from the reference sites. However, there were only weak correlations (r = 0.12 to 0.48) between level of lead or chromium and number of aberrant cells or lesions per cell, suggesting that something other than metals was responsible for the induced genetic lesions (Tull-Singleton et al., 1994).

Shaw-Allen and McBee (1993) examined bone marrow metaphase chromosomes from *P. leucopus, S. hispidus,* and fulvous harvest mice (*Reithrodontomys fulvescens*) collected from a site contaminated with the PCB mixture Aroclor 1254 and from two uncontaminated reference sites. They found no statistical differences for aberrant cells per individual or number of lesions per cell for any species among the three sites.

Eckl and Riegler (1997) collected feral *Rattus rattus* and *R. norvegicus* in the vicinity of a waste-disposal site near Salzburg, Austria, and found that the level of chromosomal damage in hepatocytes was directly related to age (as determined by weight) for *R. rattus* but not for *R. norvegicus*.

Bueno et al. (1992) collected individuals of two genera (*Akodon* and *Oryzomys*) of South American rodents from an uncontaminated island near Florianpolis and from three sites in the Sango River Basin, which receives wastes from "coal washing" processes that result in the area being heavily contaminated with mercury, lead, cadmium, copper, and zinc. Both *Oryzomys* and *Akodon* showed statistically significant increases in percent cells with chromosomal aberrations at the sites along the Sango River compared with the site near Florianpolis.

Mexican free-tailed bats (*Tadarida brasiliensis*) from Carlsbad Caverns, NM, and from a maternity colony located in a cave in northwestern Oklahoma were examined for levels of chromosome aberrations in bone marrow. No statistically significant differences were observed between males and females or between individuals from the two caves over a four-month period, even though animals from the Carlsbad Caverns colony consistently had higher levels of DDT and its clastogenic metabolites in their tissues than did animals from Oklahoma (Thies et al., 1996).

Sister Chromatid Exchange

Sister chromatid exchange (SCE) is the reciprocal interchange of DNA at homologous loci between sister chromatids. Chromosomes that have undergone SCE are not damaged in the conventional sense in that they are structurally intact, but increased strand breakage during S-phase leads to increased SCE.

Nayak and Petras (1985) examined SCE levels in house mice (*Mus musculus*) from several locations in southern Ontario. They found significant negative correlations between SCE level and both distance to the Windsor-Detroit urban complex and distance to the nearest industrial complex.

Tice et al. (1987) found that *P. leucopus* collected from an EPA Superfund site had significantly higher levels of SCE in bone marrow compared with animals from uncontaminated sites in the northeastern United States. Furthermore, animals collected at the Superfund site during the winter had significantly fewer SCEs compared with animals collected during the summer, even though levels of SCEs in animals collected from the uncontaminated sites were not different between the two collecting periods.

Ellenton and McPherson (1983) examined SCE in herring gull (*Larus argentatus*) eggs collected from five breeding colonies in polluted areas around the Great Lakes. They found no difference in levels of SCEs compared with those in a colony from a pristine area on the Atlantic coast.

Micronucleus Formation

Micronuclei (MN) are cytoplasmic nuclear bodies that are formed when whole or fragmented chromosomes are not incorporated into the nuclei of daughter cells or when small fragments of chromatin are retained from polychromatic erythrocytes after expulsion of the nucleus in the process of erythrocyte maturation in mammals.

Al-Sabti and Hardig (1990) showed that the frequency of MN formation in perch (*Perca fluviatilis*) decreased with increasing distance from wastewater discharge points on the Baltic Sea. Tice et al. (1987) showed increased levels of MN in animals from a Superfund site compared with pristine sites in the northeastern United States. Cristaldi et al. (1990) demonstrated increased levels of MN formation in house mice (*M. musculus*) collected in northern Italy 6 to 12 months after the nuclear incident at Chernobyl, Ukraine, compared with animals collected at the same sites 5 years before the incident. However, Rodgers and Baker (2000) examined bank voles (*Clethrionomys glareolus*) from near the Chernobyl reactor and found no difference in frequencies of MN in animals from inside the exclusion zone, where radiation doses were estimated to be as high as 15 to 20 rads d^{-1}, compared with animals from a reference population outside the exclusion zone, where radiation doses were negligible. Theodorakis et al. (2001) used MN to investigate induction of chromosome damage in kangaroo rats (*Dipodomys merriami*) inhabiting two atomic-blast sites at the Nevada test site, but they were not able to demonstrate significant differences among the blast sites and two reference sites.

Interpretation of Test Data

Although, these assays are relatively easy and time- and cost-effective to perform, we still know little about their meaning in terms of the overall health of a population or community or even their relation to each other. Husby et al. (1999) found no difference in levels of CA in *Peromyscus* from abandoned strip-mine sites and reference sites in eastern Oklahoma, but they did find significantly greater levels of DNA strand breaks (Husby and McBee, 1999) in the same animals. Hausbeck (1995) found only weak correlations between levels of heavy metal bioaccumulation in tissues of these same animals and levels of chromosome damage found by Husby et al. (1999). Tull-Singleton et al. (1994) found only weak relationships between levels of CA and lesions measured by flow cytometry and levels of contamination in tissues. Tice et al. (1987) suggest potential for seasonal variability in sensitivity to or level of exposure to environmental genotoxicants. A series of studies of CA, tissue accumulation, population demographics, immunological response, and enzyme induction (McMurry, 1993; McBee and Lochmiller, 1996; Peakall and McBee, 2001) conducted on *S. hispidus* at three locations within a large Superfund site and three matched reference sites resulted in evidence of significant responses in several endpoints among the six sites, but the pattern of responses was not consistent across sites, seasons, or endpoints. Together, these studies indicate that, just as in a controlled laboratory setting, contaminants can induce chromosomal damage in wild populations, but accurate interpretation of effects of chromosomal damage and its relationships with other toxic effects may be far more difficult than its documentation.

IV CANCER

Normal cells have the capacity to multiply and increase in tissues. This increase in the numbers of cells in a tissue or organ is called **hyperplasia**. Hyperplasia in response to a variety of normal stimuli such as in the tissue repair process described above is called **physiologic hyperplasia** (La Via and Hill, 1971). There is also pathologic or neoplastic hyperplasia. Sometimes, excessive hyperplasia occurs during response to injury or irritation, resulting in a pathologic hyperplasia called **compensatory hyperplasia**. **Neoplastic hyperplasia** results from a hereditary change in a cell such that it no longer responds properly to chemical signals that would normally control cell growth, providing mechanism to cancerous growth. Such **neoplasia** is "hyperplasia which is caused, at least in part, by an intrinsic heritable abnormality in the involved cells" (La Via and Hill, 1971). Neoplasias that tend to remain differentiated in their morphology and grow slowly are not as invasive of tissues as other neoplasias and are termed benign. Those that take on undifferentiated forms, tend to grow rapidly, and invade other tissues are called malignant. Malignant cancers are more life-threatening than benign cancers. Often pieces of the original cancerous growth can dislodge, move to another tissue via the circulatory or lymphatic system, and establish other foci of cancerous growth. This process (**metastasis**) leads to the spread of a malignant cancer from the site of origin throughout the body.

Cancer is a result of heritable change in cells. A chemical (e.g., numerous carcinogens), physical (e.g., ionizing radiation), or biological (e.g., a retrovirus) agent modifies a gene or its normal relation to other genes, which results in neoplasia. This can occur through point mutations, deletions, additions, rearrangements as described above, or by gene insertion by a virus. A retrovirus might insert DNA into a cell chromosome and transform the cell so that it responds improperly to growth-regulating signals. A chemical or physical agent may interact with the cell's genome such that a gene involved with the normal growth and differentiation of cells (**proto-oncogene**) is changed to an **oncogene**, a gene causing cancer. Change to an oncogene results in loss of normal growth dynamics and/or differentiation of the cell. Cancer results from the inappropriate activation or activity of a gene that would normally be involved in cellular growth and/or differentiation (Moolgavkar, 1986). Other genes called **suppressor genes** may function normally to suppress cell growth and might inhibit abnormal growth. Like proto-oncogenes, suppressor genes are thought to function in the process of growth and differentiation in normal cells: The proto-oncogenes are associated with stimulatory and suppressor genes with inhibitory aspects of normal cell growth. Cancer can also develop if some agent or event affects the functioning of the suppressor gene.

Some agents initiate the neoplastic process and some promote the development of the tumor (La Via and Hill, 1971). This fact has led to the classification of two groups of agents: **initiators** that convert normal cells to latent tumor cells, and **promoters** that enhance the growth and continued expansion of latent tumor cells (Figure 7.4). Cancers are thought to pass through an initiation stage

(a)

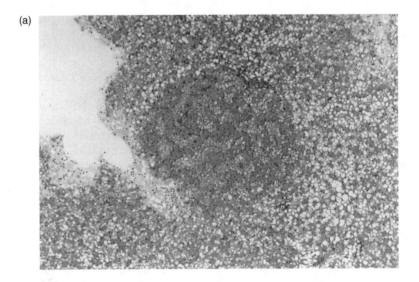

Figure 7.4   Progression of neoplasms in the medaka (*Oryias latipes*) initiated with the carcinogen, diethylnitrosamine. Fish had more rapid development of foci—and shorter times between exposure and realized effect (latent period)—if they were fed the promoter, 17-β-estradiol in their diet. Micrograph (a) shows a basophilic focus of cellular alteration in the liver 25 weeks after exposure to diethylnitrosamine and provision of a diet containing estradiol (Hematoxylin and eosin, ×150. Courtesy of Janis Brencher Cooke, University of California – Davis). Micrograph (b) shows a solid basophilic adenoma in medaka liver exposed as already described for micrograph (a). (Hematoxylin and eosin, ×75. Courtesy of Janis Brencher Cooke, University of California – Davis). Micrograph (c) shows a hepatocellular carcinoma in medaka liver at 12 weeks of exposure. Note the distinct architecture of the cancer relative to the surrounding tissue. Arrows indicate the locations of numerous mitotic figures in the carcinoma. (Hematoxylin and eosin; ×125. Courtesy of Swee The, University of California – Davis.)

(b)

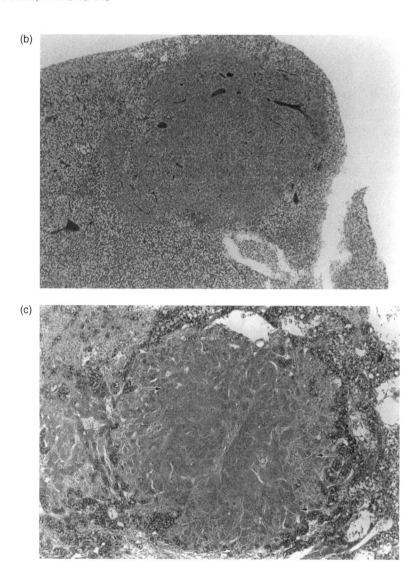

(c)

Figure 7.4   (*Continued*).

and then a promotional stage. Cancer can be promoted by lesions with consequent cell proliferation (e.g., those associated with asbestos fiber in the lung), high levels of hormones with associated hyperplasia (**hormonal oncogenesis** [La Via and Hill, 1971]), and chemical agents. Often, inappropriate inactivation of suppressor genes is also associated with the promotion process (Aust, 1991), resulting in unregulated growth (Van Beneden and Ostrander, 1994). After initiation and promotion, the final step of carcinogenesis occurs: cancer progression. **Cancer progression** is the change in the biological attributes of neoplastic cells over time that leads to a malignancy (La Via and Hill, 1971). Obviously, there can be a long **latent (latency) period** between exposure to a carcinogenic agent and the clinical appearance of cancer. This makes assignment of cause and effect very difficult, because the exposure may have occurred years prior to the appearance of the effect.

The latent period and multistage process leading to cancer also make prediction of the exact shape of the dose-cancer response curve difficult. Two theories exist and are supported by different data. The **threshold theory** assumes that there is no effect below a certain low dose (e.g., Downs

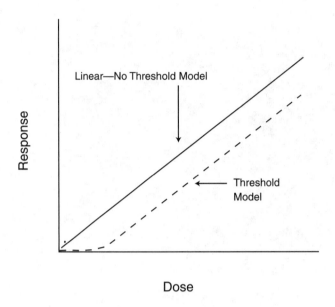

Figure 7.5   The linear no-threshold and threshold theories describe two dose-response models differing from each other at lower doses. The linear no-threshold model is described by a straight line, and the threshold model is described by a "hockey stick" curve.

and Frankowski, 1982; Cohen, 1990). Above the threshold, the slope of the response-vs.-dose curve increases rapidly. The dose-response curve takes on the appearance of a hockey stick (Figure 7.5). The **linear no-threshold theory** is based on several radiation-induced cancer studies where no apparent threshold exists for effect. This theory assumes that any lack of cancers below a certain dose reflects our inability to measure low incidences of cancers at these exposure levels. This dose-response model describes a straight line with no threshold.

Chemical and physical agents can act by initiation or promotion of the process leading to cancer. Relevant mutagenic mechanisms are those described previously for alterations of DNA and chromosomes. Further, agents that inhibit DNA repair may also increase the probably of cancer. Differences in DNA **repair fidelity** (accuracy in repairing and returning the DNA to its original state after damage) associated with various agents can also result in differences in carcinogenicity. For example, Robison et al. (1984) suggested that chromium as chromate, nickel, and mercury cause damage to DNA but that there is a difference in the damage and its repair. Generally, mercury produces more single-strand breaks, while nickel and chromium cause more protein-DNA crosslinking. The difference in the fidelity of repair between these two types of damage leads to nickel and chromium (lower repair fidelity) being more carcinogenic than mercury (higher repair fidelity).

The presence of tumors and various tumor cell qualities associated with gene products are used as biomarkers for environmental carcinogens. In the study of micronuclei in softshell clams discussed above (Dopp et al., 1996), leukemic cells were also measured using an antibody assay that recognized surface alterations of the cells. Enzymatic and histochemical alterations associated with cancer cells can also be used (e.g., Moore and Myers, 1994). More often, the incidence of cancers in populations is correlated with the level of contaminant. For example, liver neoplasms in sole (*Parophrys vetulus*) from Puget Sound were correlated to sediment contamination by polycyclic aromatic hydrocarbons (PAH) (Stein et al., 1990). The incidence of cervical neoplasia in Czech women increased two years (latent period) after the Chernobyl release of radionuclides (Borovec, 1995).

Vignette 7.2   Polycyclic Aromatic Hydrocarbons and Liver Carcinogenesis in Fish

Wolfgang K. Vogelbein

Virginia Institute of Marine Science, Gloucester Point, VA

Polycyclic aromatic hydrocarbons (PAH) are ubiquitous environmental contaminants derived largely from the incomplete combustion of organic matter. These large, complex organic molecules composed of fused aromatic ring structures are often divided into two broad classes: the low-molecular-weight (LAH: 1 to 3 benzene rings) and high-molecular-weight (HAH: 4 to 6 benzene rings) compounds. Major sources of environmental contamination include industrial dischargers such as metal-smelting facilities, wood-treatment plants using creosote, oil-refining operations, and accidental spills of fuel oil, crude oils, and other petroleum products. Atmospheric emissions from incineration and internal combustion engines (e.g., automobiles) and their associated deposition and urban runoff represent another major source of PAH contamination. In heavily industrialized sites, PAHs often co-occur with a variety of other chemical pollutants, including heavy metals, pesticides, and polychlorinated biphenyls (PCB).

Polycyclic aromatic hydrocarbons, especially the high-molecular-weight compounds, are extremely hydrophobic and tend to sorb to organic and inorganic matter. Thus, in aquatic environments, PAHs tend to bind detritus and become immobilized in bottom sediments, and concentrations in surface waters are generally very low. Despite extensive and tenacious sorption to sediments, PAHs appear to be readily bioavailable to aquatic organisms. Fish are exposed to PAH from their diet, by aqueous exposure, and by direct contact with sediments (Meador et al., 1995a). Many benthic invertebrates, especially those that ingest sediment, can accumulate PAH and, therefore, can constitute an important dietary source of PAH exposure to fish (Meador et al., 1995b).

Although sediment PAHs are generally bioavailable and tend to accumulate in some aquatic invertebrates, they do not normally accumulate to any appreciable degree within tissues of aquatic vertebrates such as teleosts. Additionally, there does not seem to be any significant biomagnification of PAH across trophic levels in fish (Suedel et al., 1994). Thus, measured PAH concentrations are generally very low in fish tissues and are not a useful method of assessing environmental exposure. The primary reason for the low tissue concentrations in the higher animals is that vertebrates rapidly and efficiently transform these hydrophobic contaminants into polar, more-water-soluble forms that can then be more easily excreted. Xenobiotic-metabolizing enzyme systems such as the cytochrome P-450 monooxygenases and glutathione-s-transferases are responsible for biotransformation of these chemicals. Thus these enzymes serve an important detoxification function and provide a mechanism whereby large hydrophobic and potentially toxic xenobiotics are eliminated from the tissues and the body. In vertebrates, the primary organ responsible for the metabolic detoxification of hydrophobic organic contaminants such as PAH is the liver.

Although they do not generally accumulate in vertebrate tissues, PAHs have been documented to cause significant adverse health effects in exposed organisms. While metabolic conversion of PAH serves mainly in detoxification, some of the intermediate metabolites produced during this process exhibit potent cytotoxic, immunotoxic, mutagenic, and carcinogenic properties. Recent investigations have shown that some PAHs, including benzo[a]pyrene, benzfluoranthene, benz[a]anthracene, dibenz[a,h]anthracene, indeno[1,2,3-c,d]pyrene and dibenzo[a,l]pyrene, are highly carcinogenic in laboratory rodents (e.g., National Toxicology Program, 1999). Carcinogenic potential of some of these PAHs has been confirmed in fish (e.g., Hawkins et al., 1990; 1995; Hendricks et al., 1985; Reddy et al., 1999). Additionally, most of these carcinogens have been documented at biologically significant concentrations in PAH-contaminated aquatic sediments (Reddy et al., 1999).

Over the past 25 years, numerous studies have been conducted on the adverse health effects of contaminant exposure in wild fish. These studies have shown that some fish populations inhabiting highly industrialized environments exhibit many of the same health impacts documented in PAH exposures of laboratory rodents and fish. An important focus of these studies has been on the relationship between chemical exposure, in particular, to the PAH, and development of liver cancer in wild fish. In North America, cancer epizootics have been documented in fish from over 40 freshwater and estuarine waterways (Clark and Harshbarger, 1990). These include liver cancers and associated lesions in winter flounder, *Pleuronectes americanus*, from Boston Harbor, MA (Moore and Stegeman, 1994; Moore et al., 1996); brown bullhead, *Ictalurus nebulosus*, from tributaries of the southern Great Lakes region (Baumann, 1989; Baumann et al., 1991); English sole, *Pleuronectes vetulus*, and other bottom-fish species from Puget Sound, WA (Myers et al., 1987; 1998); and mummichogs, *Fundulus heteroclitus*, from the Elizabeth River, VA (Vogelbein et al., 1990; 1997). Polycyclic aromatic hydrocarbons are implicated as causative agents in all of these epizootics.

However, the most thorough investigations to date are those conducted on the English sole in the Pacific Northwest. In general, the prevalence of fish liver disease has been found to increase with increasing industrialization and urbanization. Liver lesion prevalences as high as 90% have been documented in fish inhabiting waterways where sediment PAH concentrations are very high, while lesion prevalences have

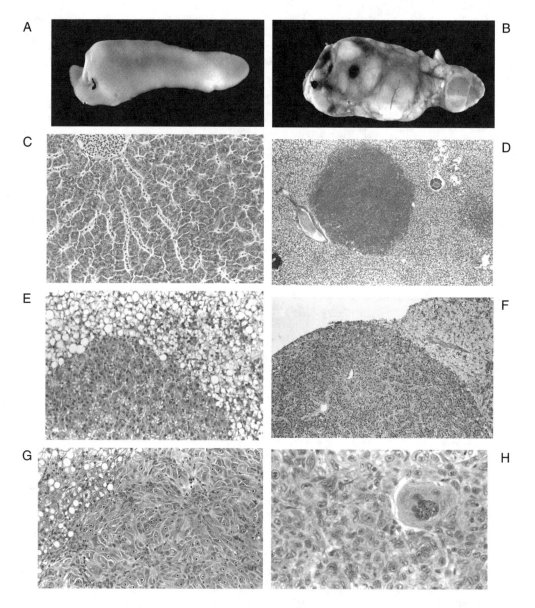

Figure 7.6   Gross and histopathologic anatomy of the liver of mummichogs, *Fundulus heteroclitus*, inhabiting uncontaminated and PAH-contaminated environments in the Elizabeth River, VA. A: Normal, healthy liver of mummichog from uncontaminated environment. B: Liver of mummichog from a PAH-contaminated habitat exhibiting multiple focal lesions, tumorous nodules, and cystic lesions. C: Histologic section of normal healthy mummichog liver showing typical tubulosinusoidal architecture. D: Altered hepatocellular focus (eosinophilic focus) in liver of exposed mummichog. E: High magnification of eosinophilic focus in panel D showing subtle blending of the lesion border with surrounding normal tissue. F: Large hepatocellular adenoma showing swelling of liver capsule and well-demarcated border with normal liver tissue. G: Hepatocellular carcinoma showing sharp, locally invasive border with normal liver tissue and cellular and nuclear pleomorphism of the tumor cells. H: High magnification of a hepatocellular carcinoma showing cellular and nuclear pleomorphism typical of poorly differentiated, highly malignant neoplasms. (Photos by Wolfgang W. Vogelbein.)

been very low or inconsequential (<1%) in fish from relatively uncontaminated habitats. Further, PAH levels in benthic prey organisms and fish gut contents (Malins et al., 1984; Myers et al., 1987; 1993), induction of cytochrome-P-450-mediated xenobiotic metabolizing enzymes (Van Veld et al., 1992; 1997), DNA adduction in fish tissues (Varanasi et al., 1986), and PAH metabolite levels in fish bile (Krahn et al., 1984) all are reported to be much higher in fish from PAH-contaminated environments than in fish from relatively uncontaminated habitats. These findings strongly support the view that PAHs are bioavailable and that they are the causative agents responsible for the development of liver disease in wild fish. However, direct experimental evidence that sediment PAH causes liver cancer in wild fish is scant. To date, the most direct evidence for this association has been provided by Metcalfe et al. (1988), who induced hepatocellular carcinomas in rainbow trout, *Oncorhynchus mykiss*, by microinjecting fry with sediment extracts from Hamilton Harbor, Ontario, and by Schieve et al. (1991), who induced altered hepatocellular foci in English sole exposed to an extract of a contaminated marine sediment.

Because the effects of exposure to toxic chemicals are expressed as characteristic tissue lesions, histopathology has been applied to many of the above-mentioned investigations and has also gained acceptance in coastal pollution-monitoring programs. The U.S. EPA has selected fish liver histopathology as one of several indicators of biological impacts for marine dischargers holding 301(h)-modified NPDES permits. Both the EPA Environmental Management and Assessment Program (EMAP) and NOAA National Benthic Surveillance Project (e.g., Moore and Myers, 1994) have used fish liver histopathology in their assessments of environmental pollution in coastal habitats.

The adverse effects of PAH exposure in wild fish generally comprise a complex spectrum of liver lesions, represented here by the mummichog from a creosote-contaminated site in the Elizabeth River, VA. This small nonmigratory fish is abundant in most tidal marshes along the eastern United States and is found in several PAH-contaminated portions of the Elizabeth River, where it exhibits severe liver disease. Whereas livers of fish from uncontaminated reference sites appear texturally homogeneous (Figure 7.6, panel A), livers of fish from the creosote-contaminated sites exhibit multiple small focal lesions, pigment foci, large tumorous nodules, and fluid-filled cysts (panel B). Histologically, livers of fish from uncontaminated sites exhibited normal teleost tissue structure (panel C). In contrast, mummichogs from the creosote-contaminated site exhibited a spectrum of lesions that included cytotoxic and specific degenerative cellular changes, and a suite of lesions representative of hepatocarcinogenicity, including preneoplastic altered hepatocellular foci (panels D, E), large benign neoplasms called hepatocellular adenomas (panel F), and hepatocellular carcinomas (panels G, H) (Vogelbein et al., 1990; 1999). Additionally, these fish also exhibited extrahepatic neoplasms, including tumors of the exocrine pancreas, bile ducts, vascular system, kidney, and lymphoid tissues (Fournie and Vogelbein, 1994; Vogelbein and Fournie, 1994; Vogelbein et al., 1997). To date, neoplastic changes in five different organ systems and eight specific cell types have been documented in mummichogs from this site. Several observations suggest that this epizootic is caused by chronic exposure of the fish to xenobiotics: (1) sediments at this site are highly contaminated with wood preservatives (creosote and pentachlorophenol); (2) chemical analyses of sediments confirm the presence of high concentrations of PAHs, several of which are potent carcinogens; and (3) fish captured directly across the river and less than 600 m away from this facility, as well as from other uncontaminated sites, exhibit no cancerous liver lesions.

V GILLS AS AN EXAMPLE

Gills are often damaged or changed by exposure to toxicants. Gills are used here as an example for integrating many of the nongenetic cellular effects described in this chapter. (The consequences of genetic damage are already illustrated relative to cancer.) Gills also provide good examples of unique changes not easily predicted from biochemical mechanisms alone. The consequences of gill changes to the individual are also easily demonstrated.

The fish gill (Figure 7.7, top panel) is composed of **primary lamellae (filaments)** that extend outward at right angles from the branchial arches. On the dorsal and ventral sides of each primary lamella are parallel rows of **secondary (respiratory) lamellae**, which are the major sites of gas exchange. The epithelium covering the lamellae is a double layer of cells with intercellular lymphoid spaces between the two cell layers. **Chloride cells** (Figure 7.8), which function in ion regulation, are found predominantly on the epithelium of primary lamellae but also on the secondary lamellae.

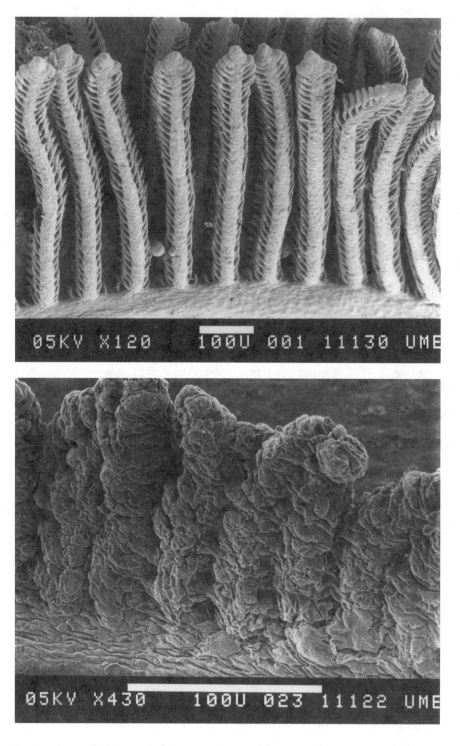

Figure 7.7   Electron micrographs of gills from Atlantic salmon (*Salmo salar*) fry. The top panel shows the normal gill morphology with primary lamellae extending (vertically here) from the branchial arch. Perpendicular to and on both sides of the main axis of each primary lamella are the secondary lamellae. The bottom micrograph shows the gills of salmon fry after 30 days exposure to 300 µg l^{-1} of aluminum. Note the extensive fusion of the secondary lamellae. (Courtesy of C.H. Jagoe, University of Georgia. With permission.)

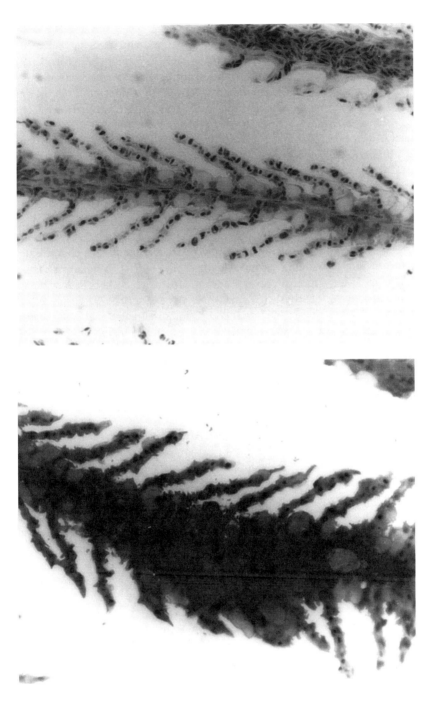

Figure 7.8   Micrographs of mosquitofish (*Gambusia holbrooki*) gills in cross section. The top panel shows the normal gill with secondary lamellae extending outward from the primary lamellae. Note the large chloride cells on the primary lamellae between the secondary lamellae. After exposure for 14 days to 60 μg l^{-1} of inorganic mercury (bottom panel; see Jagoe et al., 1996), the primary lamellar epithelium began filling in the spaces between adjacent secondary lamellae. The chloride cells (large cells with lightly stained cytoplasm) were involved to a large extent in this hyperplasia, becoming larger (hypotrophy) and more abundant on the primary lamellae. The secondary lamellae appeared to shorten or disappear as a consequence. Although reported elsewhere in response to gill irritation with toxicants (e.g., Tuurala and Soivio, 1982), no necrosis, separation of epithelium from the secondary lamellae, or inflammation were noted. (Toluidine blue; distance between secondary lamellae at their base is circa 20 μm; Courtesy of C.H. Jagoe, University of Georgia. With permission.)

Upon exposure to such diverse toxicants as metals, detergents, elevated H^+, or nitrophenols, the outer epithelium of the secondary lamellae often lifts away and leaves an enlarged, fluid-filled space between itself and the inner epithelial layer of cells (Skidmore and Tovell, 1972; Mallatt, 1985; Evans, 1987). Granulocyte densities in this space can rise, suggesting inflammation. With acute exposure to agents such as zinc, chloride cells can begin to separate from the epithelium, or ruptures may occur in the outer epithelial layer. Cells may appear swollen with distended mitochondria, i.e., necrosis occurs.

The number of chloride cells on the primary lamellae increased with the exposure of rainbow trout (*Salmo gairdneri*) to dehydroabietic acid, a component of kraft mill effluent (Tuurala and Soivio, 1982). Increased number of chloride cells also occurred in mosquitofish (*Gambusia holbrooki*) exposed to inorganic mercury (Jagoe et al., 1996). Also, the size of chloride cells of mosquitofish gills increased with inorganic mercury exposure. This increase is an example of **hypertrophy**, an increase in cell size (and function) resulting from an increase in the mass of cellular components, often as a compensatory response (La Via and Hill, 1971; Meyers and Hendricks, 1985). The hyperplasia and hypertrophy of chloride cells is thought to be a compensatory response to ion imbalance associated with gill damage (Jagoe et al., 1996). Individually or together, hyperplasia and epithelial lifting often produce a fusion of secondary lamellae, with a consequent reduction in the capacity for gaseous exchange across the gill surface. Exposure of salmon fry to aluminum (Figure 7.7, bottom) produced extensive fusion of lamellae. The hyperplasia and hypertrophy of chloride cells of the secondary lamellae caused similar fusion of lamellae as the spaces between primary lamellae were filled with cells from the primary lamellae (Figure 7.8, bottom). The consequences at the physiological and individual levels were a reduction in oxygen exchange across gill surfaces and death at lethal exposures.

VI SUMMARY

This chapter describes the effects at the cellular, tissue, and organ level from exposure to toxicants. The effects included necrosis, inflammation, and specific changes, such as increased numbers of lipofuscin deposits and metal-laden granules. Damage to genes and chromosomes was described along with several associated assays for damage. Cancer development as a consequence of exposure to chemical and physical agents was discussed briefly and linked to damage due to somatic mutations. Cancer is a clear example of consequences to individuals of cellular damage. Finally, gill response to toxicants was used as an example integrating several of the cellular changes discussed in the chapter. This example linked changes at the cellular level to consequences at the physiological (ion imbalance and decreased oxygen diffusion) and individual (somatic death) levels of biological organization.

SELECTED READINGS

Hinton, D.E., Cells, cellular responses, and their markers in chronic toxicity of fishes, in *Aquatic Toxicology: Molecular, Biochemical and Cellular Perspectives*, Malins, D.C. and Ostrander, G.K., Eds., CRC Press, Boca Raton, FL, 1994.

Li, A.P. and R.H. Heflich, *Genetic Toxicology*, CRC Press, Boca Raton, FL, 1991, p. 493.

Malins, D.C. and G.K. Ostrander, *Aquatic Toxicology: Molecular, Biochemical, and Cellular Perspectives*, CRC Press, Boca Raton, FL, 1994, p. 539.

Meyers, T.R. and J.D. Hendricks, Histopathology, in *Fundamentals of Aquatic Toxicology: Methods and Applications*, Rand, G.M. and Petrocelli, S.R., Eds., Hemisphere Publishing Corp., Washington, D.C., 1985.

CHAPTER 8

Sublethal Effects to Individuals

Our notions of law and harmony are commonly confined to those instances which we detect; but the harmony which results from a far greater number of seemingly conflicting, but really concurring laws, which we have not detected, is still more wonderful.

Thoreau (1854)

I GENERAL

In the last chapter (Chapter 7), we discussed effects at the cellular to organ levels including cell death. Death of individuals (somatic death) will be detailed in the next chapter (Chapter 9). Sandwiched between these two categories of toxicant effects are sublethal effects to individuals. The emphasis in the last chapter was on use of sublethal effects to indicate mode of action. The emphasis in this chapter is on adverse impacts. Although these sublethal effects are often more difficult to measure than lethal effects, they are likely as, or more, important in determining the ultimate consequences of many pollution scenarios.

Sublethal effects are effects occurring at concentrations or doses below those producing direct somatic death (Rand, 1985). They are most often recognized as some change in an important physiological process, growth, reproduction, behavior, development, or a similar quality. Nearly always, they are adverse or putatively adverse effects that lower an individual's fitness.

The concept of sublethal effects was first developed from a pharmacological or mammalian toxicological vantage. Although also useful and widely applied, the meaning of the term *sublethal effect* is more ambiguous in ecotoxicology. Some sublethal effects will have lethal consequences in an ecological context, i.e., an arena in which the individual must successfully compete with other species, avoid predation, find a mate, and cope with multiple stressors. For example, an individual may be able to survive a sublethal exposure but have reduced ability to evade predators or effectively forage for food. Thus, the result of a sublethal exposure may be death in a natural setting. The concept of **ecological mortality (death)** is used to describe the toxicant-related diminution of fitness of an individual, functioning in an ecosystem context, that is of a magnitude sufficient to be equivalent to somatic death (Newman, 1995). It is important to remain open to the very real possibility of ecological mortality at exposures that are judged sublethal based on laboratory assays. Also, a sublethal effect that results in an individual that is incapable of producing viable offspring could be considered a lethal effect because the individual's fitness (i.e., ability to contribute offspring to the next generation) could be equivalent to that of a dead individual

(Rand and Petrocelli, 1985). On the other hand, complex behaviors such as avoidance of contaminated areas of the habitat may ameliorate sublethal consequences of contamination.

II SELYEAN STRESS

The concept of stress is frequently invoked when dealing with sublethal effects. The implication is that sublethal exposure causes individuals to function suboptimally. The concept of stress to individuals must be clarified here because it does not always have the same meaning to different people. When applied in a medical or scientific sense to individuals, stress has a precise meaning that does not correspond precisely with its general meaning. This confusion of the scientific/medical and general meaning of stress in the ecotoxicology literature has led to many problems.

Hans Selye formulated the medical concept of stress as applied to individuals about 60 years ago. As a medical student, he noted a nonspecific response of the human body if extraordinary demands were made upon it (Selye, 1973). He referred to this nonspecific response as stress and defined it as "the state manifested by a specific syndrome which consists of all the nonspecifically induced changes within a biological system" (Selye, 1956).[1] Selyean stress is a specific suite of changes constituting the body's attempt to reestablish or maintain homeostasis while under the influence of a stressor (Adams, 1990).

The **general adaptation syndrome** (GAS) associated with Selyean stress has three phases: the alarm reaction, adaptation or resistance, and exhaustion phases (Figure 8.1). All phases of the GAS serve to resist deviation from, or to regain, homeostasis (Newman, 1995). In the alarm component, blood pressure and heart rate increase as short-term responses to compensate for an immediate stressor. Because these short-term, energy-intensive mechanisms cannot be maintained indefinitely, the organism may enter the second stage if stress persists. Responses that enhance tissue-level compensation, such as enlargement of the adrenal cortex, are typical of the adaptation stage; however, adverse responses such as shrinkage of the thymus, lymph nodes, and spleen, and appearance of gastric ulcers can also occur for mammals stressed to this phase (Selye, 1973). If exposure to enough stressor continues for sufficient time, the individual's ability to resist change is exceeded and it enters an exhaustion phase. In this last stage, the individual slowly fails in its efforts to compensate for the effects of the stressor and will eventually die if exposure continues.

Selyean stress is a specific syndrome. Phenomena such as necrotic damage or metallothionein induction were not included in the stress concept as originally proposed by Selye. The term *Selyean stress* should not be used for these effects. Necrosis constitutes damage, not a response to a stressor. In many cases, metallothionein induction is a specific response to a class of chemicals, i.e., cadmium, copper, mercury, silver, and zinc. Induction of cytochrome P-450 monooxygenase by a polycyclic aromatic hydrocarbon (PAH) would not be part of a stress response, but the cellular stress response to a stressor as described in Chapter 6 would seem to fit into the Selyean stress concept. It is important to keep clear these distinctions among Selyean stress, damage, and specific responses to and effects of toxicants. In the remainder of this chapter, most of the sublethal effects described are not examples of Selyean stress, yet they might be discussed as such elsewhere in the ecotoxicological literature.

III GROWTH

To measure sublethal effects, growth is often chosen as the response variable. Not only is it easy to measure, but it also integrates a suite of biochemical and physiological effects (Mehrle and Mayer, 1985) into one quality that is often associated with individual fitness. For example, growth is reduced for fish under acidic conditions or for green tree frog (*Hyla cinerea*) tadpoles exposed to a combination

---

[1] We will further qualify stress to an individual as **Selyean stress** so as to distinguish it from other applications of the term in this book and throughout the literature.

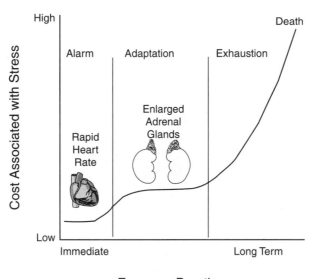

Figure 8.1 The three stages of the general adaptation syndrome (GAS). In the initial alarm phase, immediate responses to stress include increases in blood pressure and heart pumping rate. Cells of the adrenal cortex also discharge their granules into the blood. With long durations of exposure, adrenal enlargement occurs, and cells of the adrenal cortex can become rich in granules again (Selye, 1973). Finally, after sufficient exposure, the ability of the body to compensate for the stressor's effects is exceeded, and the individual slowly becomes exhausted. If the stress continues, the individual will die.

of low pH and elevated aluminum concentrations, conditions that are often found in water bodies impacted by acid rain (Jung and Jagoe, 1995). Woltering et al. (1978) measured reduced growth of largemouth bass (*Micropterus salmoides*) exposed to ammonia and attributed the reduced growth to decreased feeding rate. Growth was reduced for redbreast sunfish (*Lepomis auritus*) inhabiting a stream contaminated with mixed waste (Adams et al., 1992). In contrast, white sucker (*Catostomus commersoni*) from a lake contaminated with metals had higher growth rates but reduced longevity relative to white suckers from a nearby, uncontaminated lake (McFarlane and Franzin, 1978). Some toxicants reduce growth in fish, but differences in body size between exposed and nonexposed individuals may disappear later due to compensatory growth (Sprague, 1971).

The dose-response relationship for toxicant-influenced growth often conforms to a threshold model (Figure 7.5), but this is not always the case. Sometimes, a stimulatory effect is exhibited with exposure to low, subinhibitory levels of toxicants or physical agents. Such **hormesis** is not usually a toxicant-specific response. Instead, it is a general phenomenon that appears during exposure to a variety of stressors. For example, peppermint (*Mentha piperita*) exposed to a series of concentrations of the growth inhibitor, phosfon, grew fastest at the lower concentrations of 2.5×10^{-12} to 2.5×10^{-8} M (Calabrese et al., 1987). Faster growth of peppermint at these concentrations than in the control group yields a biphasic dose response. The **biphasic dose-effect model** is shaped like the threshold model in Figure 7.5, but the curve dips down from the controls before increasing with dose. In this case, the downward dip would reflect enhanced growth of plants exposed to low concentrations, but in other cases, it might reflect decreased mortality or increased fecundity at low concentrations relative to controls. Stebbing (1982) and Sagan (1987) suggest that the mechanism for hormesis may be regulatory overcompensation or an over response of organisms after subinhibitory challenge. Regardless, hormesis has been documented for survival, growth, seed germination, cancer incidence, antibody titer, and fertility in individuals exposed to a variety of agents including toxicants and radiation (Doust et al., 1994; Robohm, 1986; Stebbing, 1982; Sagan, 1987; Wolff, 1989).

Figure 8.2   A pharmacy display window in Barcelona, Spain, advertising homeopathic medication ("homeopa-
tia" = homeopathy). Because homeopathic medicine does not focus on the causal mechanisms of
disease, it is judged to be an untenable practice by many professionals. Perhaps the concept of
hormesis has not been incorporated as fully as warranted into the science of ecotoxicology because
of its association with this controversial branch of medicine.

Interest in hormesis is not restricted to nonhuman species. Indeed, hormesis is a foundation
concept of **homeopathic medicine**, a branch of medicine not given full attention by many physicians
in North America but practiced to varying degrees throughout the world (Figure 8.2). Homeopathic
medicine, founded by Samuel Hahnemann, is based on the **law of similars**: a drug that induces
symptoms similar to those of the disease will aid the body in defending itself by stimulating the
body's natural responses. By stimulating these responses (symptoms), a drug is thought to enhance
the processes that the individual uses in resisting the disease.

IV DEVELOPMENT

IV.A   Developmental Toxicity and Teratology

Some contaminants can adversely impact the developing fetus or embryo. This can result in
death (embryo lethality), anatomical malformation, functional deficiencies, or slowing of growth.
Often there are critical periods in embryo or fetus development when a particular developmental
effect can occur. The *in utero* effect of mercury on humans was mentioned briefly in our discussion

of the Minamata disease. *In utero* effects of mercury include macrocephali (abnormally large head), asymmetrical skull, depressed optical region of the skull, lowered IQ, poor muscular coordination, hearing loss or impairment, poor speech, poor walking skills, and mental retardation (Khera, 1979). Any physical or chemical agent such as mercury that is capable of causing developmental malformations is called a **teratogen**. **Teratology** is the science of fetal or embryonic abnormal development of anatomical structures. Notice that some of the qualities mentioned above for mercury involve functional deficiencies, e.g., lowered IQ, that may not be considered in classic teratology. Also, some contaminants can slow the growth of the developing organism without producing an anatomical abnormality. The broader term **developmental toxicity** is often used to include altered growth and functional deficiencies in addition to teratogenic effects (Weis and Weis, 1989a).

Teratogenic effects often have a threshold: a critical amount or concentration of teratogen is needed before an effect is manifest (Weis and Weis, 1989a). This is due to "the high restorative growth potential of the ... embryo, cellular homeostatic mechanisms, and maternal metabolic defenses" (Rogers and Kavlock, 1996). Although some teratogens, such as the infamous thalidomide,[2] are specific in their action, most teratogenic contaminants are believed to be relatively nonspecific. According to **Karnofsky's law**, any agent will be teratogenic if it is present at concentrations or intensities producing cell toxicity (Bantle, 1995). Teratogens act by disrupting mitosis, interfering with transcription and translation, disturbing metabolism, and producing nutritional deficits (Weis and Weis, 1987). Consequences of these disruptions include abnormal cell interactions, migration, and growth as well as excessive or inadequate cell death. In general, effects early in development tend to be more deleterious than those occurring later, because early damage affects cells that will go on to differentiate and to become involved in a wider range of organs and tissues (Bantle, 1995).

Most developmental toxicology focuses on effects occurring during or after exposure of the egg to contaminants. Exposure can occur across the placenta, from contaminants deposited in the yolk, from egg exposure before fertilization, between egg shedding and elevation of the chorion, or after elevation of the chorion (Weis and Weis, 1989a). However, evidence suggests that some congenital diseases and birth defects in humans may also be linked to male exposure to contaminants (Gardner et al., 1990; Stone, 1992). For example, men working at the Sellafield nuclear fuel processing plant in England had a higher incidence of children with leukemia than control groups (Gardner et al., 1990), notionally due to a chromosomal translocation that activated proto-oncogenes (Evans, 1990; Kondo, 1993). Previously, such **male-mediated toxicity** (disease and birth defects produced by a father's exposure to a physical or chemical agent) had been judged to be insignificant based on epidemiological studies of atomic bomb survivors of Hiroshima. However, Stone (1992) reports that males in some professions (painters, mechanics, and farmers) may be at higher risk of fathering children with teratogenic problems.

A wide range of developmental problems has been described for organisms exposed to contaminants. For fish, the most common are those of the skeletal system and associated musculature, circulatory system, optical system, and retardation of growth (Weis and Weis, 1989a). Skeletal system problems include **scoliosis** (the lateral curvature of the spine) and **lordosis** (the extreme, forward curvature of the spine) (Figure 8.3). Slowing of growth may increase the time a developing individual remains in a critical stage and, consequently, the likelihood of a problem becoming manifest (Weis and Weis, 1989a). Birds exposed to contaminants produced embryos with a variety of eye, limb, beak, heart, and brain abnormalities (selenium exposure) (Ohlendorf et al., 1986) or had increased incidence of egg failure (mercury, selenium, DDE, and DDT exposure) (Henny and Herron, 1989).

---

[2] In the early 1960s, the sedative, thalidomide was given to pregnant European women to treat nausea. Babies were born without limbs (**amelia**) or with limbs having reduced bone lengths (**phocomelia**). It soon became apparent that abnormalities occurred if it was taken during a critical two-week period of active limb morphogenesis (Lenz 1968; 1996; McBride, 1961; La Via and Hill, 1971). Roughly 10,000 children were born with severely malformed limbs. This event lead to revision of drug-testing procedures. Although it has become symbolic of the tragic consequences of drug use prior to extensive testing, thalidomide is prescribed more judiciously today as an immunosuppressant and is being considered for treatment of some symptoms of AIDS, including associated cancers (Richardson et al., 2002).

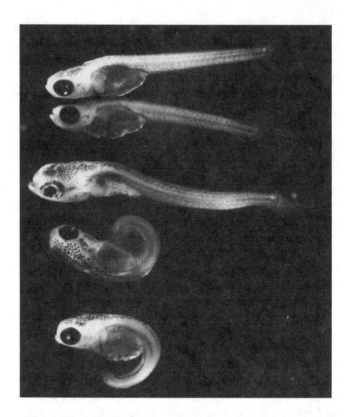

Figure 8.3   Fish (*Fundulus heteroclitus*) exposed to no Pb^{2+} (top three individuals) or 10 mg l^{-1} of Pb^{2+} (bottom two individuals) during development. The effect threshold is typically 1 mg l^{-1} of Pb^{2+}. Note the failure of the exposed individuals to uncurl after hatching. (Courtesy of P. Weis, UMDNJ – New Jersey Medical School. With permission.)

Recently, Weis and Weis (1995) demonstrated that **behavioral teratology** (behavioral abnormalities in otherwise normal-appearing individuals arising after exposure to an agent as an embryo) can also be important. They showed that exposure of mummichog (*Fundulus heteroclitus*) embryos to methyl mercury lowered the ability of this fish to capture prey after hatching.

There are standard assays for measuring developmental effects of environmental contaminants. The widely accepted **FETAX** (frog embryo teratogenesis assay—*Xenopus*) uses embryos of the clawed frog, *Xenopus laevis*. Although the test species is not native to North America or Europe, the convenience of producing ample amounts of eggs and sperm, well-established procedures for the assay (e.g., Bantle and Sabourin, 1991; ASTM, 1993), and the large database for this species make it an appealing tool (Bantle, 1995). In the assay, eggs are exposed to different concentrations of the contaminant for a set time, e.g., 96 h. The proportions of exposed eggs showing mortality and the proportion of living embryos with developmental abnormalities are scored for each treatment concentration. Using formal methods described in Chapter 9, the concentrations producing 50% mortality of eggs (LC50) and producing 50% of embryos with abnormalities (EC50 or TC50) are calculated. (EC50 is the median effective concentration, and TC50 is the **median teratogenic concentration**.) A **teratogenic index** (TI), calculated as the LC50 divided by the EC50, reflects the developmental hazard of a contaminant, with higher values indicating increased developmental hazard of the tested contaminant. Bantle (1995) believes that TI values less than 1.5 indicate little developmental hazard.

Osano et al. (2002) provide a good example of applying the FETAX assay to understanding the environmental risks of chloroacetanilide herbicides and their aniline degradation products. Atypically, *Xenopus laevis* was endemic to the study area about which the risk statements were being made, i.e., agricultural areas of Kenya. (This coincidence inspired the atypically droll line in the author's acknowledgments, "And thank you God for providing the *Xenopus* frogs at my doorstep!") The agricultural chemicals, alachlor and metolachlor, are widely applied in Kenya. Alachlor and its degradation product, 2,6-diethylaniline, and metolachlor and its degradation product, 2-ethyl-6-methylaniline, were used in assays. Embryotoxicity, growth, and teratogenicity were noted after 96 h of exposure (midblastula to early gastrula stages). Although alachlor and metolachlor had TI of 1.7 and 0.2, respectively, their degradation products had TI of 2.1 (2,6-diethylaniline) and 2.7 (2-ethyl-6-methylaniline). The results for metolachlor illustrated that a chemical with no teratogenic risk can be the source of a degradation product that poses a significant teratogenic risk. Curiously, agricultural use of alachlor in North America has been replaced by metolachlor application.

IV.B Sexual Characteristics

Estrogenic chemicals (xenobiotic estrogens or xenoestrogens) mimic estrogen and can cause changes in the sexual characteristics of individuals. Like estrogen, these chemicals regulate the activity of estrogen-responsive genes by binding to estrogen receptors (Jobling et al., 1996). They disrupt hormonal systems, affecting sex organ development, behavior, and fertility. For example, affected male sea gulls might ignore nesting colonies, and females may pair and nest together as a consequence of modified behavior by estrogenic chemicals (Hunt and Hunt, 1977; Luoma, 1992). Indeed, gull egg treatment with DDT results in feminization of the reproductive system of hatched males (Fry and Toone, 1981). Agonistic (estrogenic) substances bind to and have an effect at estrogen receptors. Males might develop female traits, or female endocrine-related features might change. Antagonistic or antiestrogenic substances block receptors, preventing normal binding of estradiol (Leung et al., 2002). Estrogenic chemicals include xenobiotics such as DDT, DDE, dioxin, polychlorinated biphenyls (PCB), alkylphenols (e.g., p-nonyl-phenol and the surfactant, alkylphenol polyethoxylate), and pharmaceuticals released in sewage effluent. Some xenobiotic estrogens bind to hormone receptors and induce an abnormal elevated response. Others with minimal estrogenic activity bind to the receptor, thus blocking the normal hormone's action (McLachlan, 1993). Kelce et al. (1995) found that some estrogenic chemicals such as DDE can also block androgen-receptor-mediated processes and, in doing so, act as **androgen receptor antagonists**.

Bergeron et al. (1994) provided a straightforward demonstration of PCB acting as xenobiotic estrogens. Eggs of the red-eared slider turtle (*Trachemys scripta*) were exposed to either: no PCB, different concentrations of PCB, or the hormone, estradiol-17β (Figure 8.4). Because this species displays temperature-dependent sex determination, researchers could manipulate the sexes of hatchlings by controlling incubation temperatures. In the experiment, eggs were incubated at a temperature (26 to 28°C) that would produce all males. The application of 200 μg of 2′,4′,6′-trichloro-4-biphenylol to the eggshell surface resulted in 100% of the hatchlings emerging as females despite the incubation temperature. The estrogenic activity of a second PCB (2′,3′,4′,5′-tetrachloro-4-biphenylol) was much lower than that of 2′,4′,6′-trichloro-4-biphenylol.

Masculinization of females can also result from exposure to contaminants. **Imposex** (the imposition of male characteristics on females, e.g., a penis or vas deferens) occurred for mosquitofish (*Gambusia* sp.) exposed to kraft mill effluent (Howell et al., 1980). Females displayed male reproductive behavior and developed gonopodia, a modified anal fin that functions as an intromittent organ in males. Bortone et al. (1989) suggested that mosquitofish exposed to kraft mill effluent take in sterols (stigmastanol and β-sitosterol) that have been modified by *Mycobacterium smegmatis* to compounds having androgenic effects. Tributyltin (TBT) compounds used in antifouling paints have also been implicated in high incidence of imposex in marine snails inhabiting coastal regions of North America and England (Bryan and Gibbs, 1991; Saavedra Alvarez and Ellis, 1990). In areas of heavy boat traffic, the occurrence of a large proportion of reproductively incompetent individuals resulted in decimated populations of ecologically important whelk species and consequent shifts in the composition of the ecological community (Bryan and Gibbs, 1991).

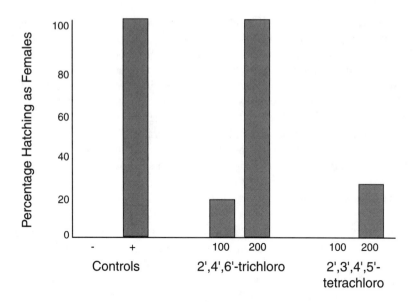

Figure 8.4   The percentage of hatchling turtles that are females after exposure to PCBs. Eggs were incubated at 26 to 28°C, which will cause all hatchlings to be males (negative control). Eggs in the negative control treatment (–) were spotted on their surfaces only with the solvent used for the other treatments (95% ethanol), and those in the positive control treatment (+) were spotted with the hormone, estradiol-17β. The two PCBs spotted onto eggs were 2´,4´,6´-trichloro-4-biphenylol (labeled 2´,4´,6´-trichloro) and 2´,3´,4´,5´-tetrachloro-4-biphenylol (labeled 2´,3´,4´,5´-tetrachloro). Both were dosed at 100 and 200 μg per egg, as indicated above the PBC labels. (Modified from Figure 1 of Bergeron, J.M. et al., *Environ. Health Perspect.,* 102, 780–781, 1994.)

Vignette 8.1   Environmental Estrogens: Occurrence of Ethynylestradiol and Adverse Effects on Fish Reproduction

Irvin Schultz
Battelle Marine Science Laboratory, Sequim, WA

Introduction

During the past several decades, there has been growing concern that many environmental pollutants can interfere with the normal functioning of human and animal endocrine systems. This type of toxicant interaction is now called endocrine disruption, and pollutants that disturb the endocrine system in some manner are called **endocrine disruptors**. The **endocrine system** is composed of several tissues and is broadly defined as any tissue or cells that release a chemical messenger (hormone) that signals or induces a physiological response in some target tissue (Thomas and Thomas, 2001). Due to the dispersed nature of the endocrine system and its integration with many biological functions, a wide range of toxicological effects might be expected from exposure to endocrine disruptors. Greatest scrutiny has been placed on sexual differentiation, sexual development, and reproduction as targets of endocrine disruptors because of the critical role of sex hormones in directing these processes. This is a particular concern with fish, as a number of compelling studies have linked disturbances in normal reproductive function with chemical exposure (Jobling et al., 1998; Howell et al., 1980).

The awareness that endocrine disruptors can adversely affect fish reproduction has prompted research into understanding how pollutants interact with male and female sex-hormone-signaling pathways. It is generally recognized that endocrine disruptors can act as either sex hormone agonists or antagonists or perhaps as mixed agonist-antagonists. Considerably more research has been performed using sex hormone agonists, and it is unclear at present whether alternative mechanisms of action represent a significant threat to fish reproduction. A comparatively large number of environmental pollutants are hypothesized to act as estrogen agonists. In females, estradiol is the principle female reproductive hormone, and it is synthesized primarily in the ovarian follicle surrounding the oocyte. Male fish typically have very low levels of circulating estradiol (Jensen et al., 2001) and are in many ways considered to be more sensitive to exposure to estrogenic chemicals. Although numerous contaminants have been identified to bind to fish estrogen receptors, the affinity is typically several orders of magnitude below that of estradiol and synthetic steroid analogues (Kloas et al., 2000). One of the more potent synthetic estrogens is 17 α-ethynylestradiol (EE2), a biologically persistent analogue of estradiol that is widely used in oral contraceptives. The presence of EE2 in surface waters is attributed to its incomplete removal from municipal sewage discharges (Larsson et al., 1999). In the remaining portion of this vignette, I will briefly review reports of environmental concentrations of EE2 and highlight important aspects of its disposition in fish in addition to describing its consequences on reproduction.

Occurrence of Ethynylestradiol and Disposition in Fish

Until recently, environmental levels of EE2 were considered low and insignificant. During the past decade, however, improvements in analytical detection methods have gradually made it possible to detect EE2 in sewage effluents and occasionally in surface waters (Roefer et al., 2000). Reported values for EE2 concentrations in surface waters range from less than 0.05 ng l^{-1} up to 9 ng l^{-1} (Ternes et al., 1999; Larsson et al., 1999). The disparity in reported EE2 concentrations is undoubtedly related to the site of sample collection and EE2 detection methods, but also to variable removal by sewage treatment facilities and annual hydrologic fluctuations. Despite the erratic reports of EE2 in the environment, it has been implicated as an important contaminant (along with other steroidal estrogens) that contributes to the estrogenic activity in surface waters from both the U.K. and the United States (Desbrow et al., 1998; Snyder et al., 1999).

An important consideration in evaluating the occurrence of EE2 in surface waters and its estrogenic potency in fish is the formation of glucuronide conjugates of EE2 that occurs in both humans and fish. In humans, EE2 is extensively glucuronidated and excreted into urine, which serves as the elimination pathway in women who ingest EE2 as an oral contraceptive (Ranney, 1977). Besides facilitating elimination of the steroid, glucuronidation is also considered to be an inactivation process, as estrogen glucuronide conjugates have low biological activity in fish (Tilton et al., 2001). Interestingly, it appears that little of the excreted EE2-glucuronide conjugate persists in sewage, as monitoring of both raw and treated effluents indicates that most of the EE2 present is the unconjugated form (Larsson et al., 1999; Belfroid et al., 1999). This would suggest that glucuronide conjugates of EE2 are environmentally unstable. Many bacteria possess β-glucuronidase, an enzyme that can hydrolyze the conjugate and, in so doing, regenerate the parent EE2. Considerable β-glucuronidase activity can also be present in bacteria found in the intestinal tract. This can be important in animals that secrete the EE2-glucuronide conjugate into bile. When bile is subsequently released into the gut, EE2 can be reformed (via hydrolysis of the glucuronide conjugate) and reabsorbed, entering a cycle called enterohepatic recirculation. A description of the conjugative reaction and conceptual model of the process of enterohepatic recirculation is depicted in Figure 8.5. This phenomenon occurs in fish, as demonstrated by a recent study in rainbow trout (*Oncorhynchus mykiss*) showing that EE2 is extensively conjugated to glucuronide and secreted into the bile (Schultz et al., 2001). An important question to ask is whether trout are like humans with respect to elimination of the EE2 conjugate. Is the conjugate hydrolyzed and reabsorbed, or is it eliminated? To answer this question, an experiment was performed that measured both the parent EE2 and its glucuronide conjugate that was eliminated by trout after intravascular injection of EE2. In this experiment, trout were placed in 300 l of water and injected with 1 mg kg^{-1} EE2 via a cannula made of Teflon tubing inserted into the dorsal aorta of the fish. Periodically after the injection, aliquots of tank water were removed, and the concentrations of EE2 and its glucuronide conjugate were measured. The results of this experiment are shown in Figure 8.6 and indicate that, unlike in humans, trout only eliminate a minor fraction of the EE2 glucuronide conjugate. Apparently, the EE2-glucuronide conjugate is deconjugated by β-glucuronidase activity present in the gut microflora and reabsorbed into the bloodstream. The importance of enterohepatic recirculation in extending the persistence of EE2 was demonstrated in trout, where pharmacokinetic model predictions of the biological half-life of EE2 increased from 6 hours to more than 30 hours when enterohepatic recirculation was occurring (Schultz et al., 2001).

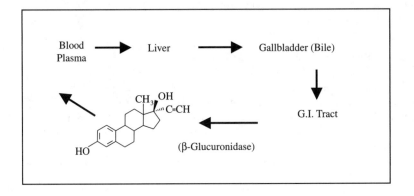

Figure 8.5   Depiction of EE2 and its conjugation with glucuronide.

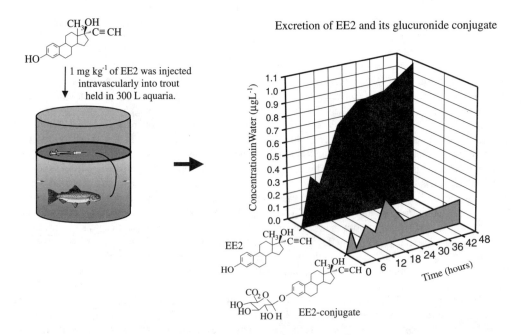

Figure 8.6   Excretion of EE2 after intra-arterial injection of a 1.0 mg kg^{-1} dose. Trout were maintained in a static water-exposure tank (300 l) after injection, and the exposure water was assayed for EE2 (black) and conjugated EE2 (grey). Mean ± SD ($n = 3$) is shown.

Effects on Sexual Differentiation and Reproduction

Gender reversal is perhaps the most profound effect on reproduction caused by exposure to potent estrogens such as EE2. Most fish species are gonochoristic, which means the differentiation of the gonads is under genetic control and is constant (irreversible) after development into ovaries or testes (Redding and Patino, 1993). However, it has been known for some time that exposure to potent sex steroids prior to sexual differentiation can cause sex reversal (Hunter and Donaldson, 1983). In fact, intentional sex reversal of male fish is now routinely practiced in aquaculture to allow cultivation of all-female fish populations, which can benefit the fish farm through increased productivity (Piferrer, 2001). Although well characterized from an aquacultural perspective, the environmental significance of sex reversal is less clear. In theory, environmental estrogens such as EE2 could produce a similar effect on wild fish populations if exposure above some threshold occurred during a vulnerable life stage. Unfortunately, determination of gender reversal is not always a simple matter because identification of genetic sex can be problematic. One fish species used in laboratory studies that can allow easy determination of genetic sex is the Japanese medaka (*Oryzias latipes*). In this species, a mutant strain known as the d-rR medaka has been demonstrated to possess sex-linked pigmentation, with males having an orange-red coloration and females having white coloration. Several laboratory studies utilizing the d-rR medaka have exploited this feature to demonstrate that exposure to EE2 can induce sex reversal in genetic males. For example, when medaka embryos were nano-injected with 0.5 or 2.5 ng EE2 and cultured to sexual maturity, 25% and 80% of the genetically male fish had undergone sex reversal and were now phenotypically female (Papoulias et al., 2000). A similar finding was reported by Scholz and Gutzeit (2000), who exposed newly hatched medaka larvae for two months in various EE2 concentrations up to 0.1 μg l^{-1}, followed by a 6-week grow-out period. At completion of the study, a complete shift in phenotype from male to female had occurred in all genetic male fish exposed to 0.1 μg l^{-1} of EE2 (Scholz and Gutzeit, 2000). These studies indicate that if embryonic or larval fish are exposed to sufficiently high concentrations of EE2, genetically male fish can develop into phenotypic females.

In addition to gender reversal caused by embryonic and larval exposure to EE2, deleterious effects on reproduction can also occur if sexually mature fish are exposed to EE2. For example, two recent studies of reproduction in the zebra fish (*Danio rerio*) and rainbow trout have demonstrated decreased fertility in these species after short-term exposures to EE2 concentrations as low as 5 ng l^{-1}. In zebra fish, reproductive performance of breeding pairs was evaluated after a 21-day exposure to EE2 concentrations of 5, 10, and 25 ng l^{-1} (Van den Belt et al., 2001). Gender-specific effects on reproductive performance were also evaluated by mating exposed fish with control fish of the opposite sex. The results indicated that exposure to 10 and 25 ng l^{-1} significantly decreased the percentage of spawning females (defined as the number of breeding pairs producing viable offspring) (Van den Belt et al., 2001). This study also reported that mating of exposed male zebra fish to control females reduced fertilization success to below 70% (the median value for control fish reported by the authors) at all EE2 exposure levels (Van den Belt et al., 2001). More-precise determinations of fertility can be made using fish species such as the rainbow trout and the use of controlled-fertilization trials. A controlled-fertilization trial involves the *in vitro* mixing of known quantities of sperm and eggs, and then assessing fertilization after allowing sufficient time to pass before fertilization becomes apparent. In salmonid species, fertilization can often be determined after 12 to 24 hours or delayed for 3 to 4 weeks, at which point the "eyed" stage of development can be used to assess fertilization and successful embryonic development (see Figure 8.7). This latter approach was recently used to assess the effect of a 62-day exposure to 10 and 100 ng l^{-1} EE2 on reproduction in male rainbow trout (Schultz et al., 2002). In this study, semen was collected from exposed trout and used to fertilize eggs obtained from a single control female donor. The results indicated that exposure to either EE2 concentration reduced the fertilization capacity of semen harvested from exposed trout by approximately 50% when compared with control trout.

Conclusion

Exposure to EE2 at dose levels closely approaching those described in the environment has been shown to alter normal phenotypic development and reduce fertilization success in fish. Therefore, both juvenile and sexually mature fish can be considered to be a vulnerable life stage to EE2 exposures. These findings are particularly disturbing given the increasing reports of EE2, estradiol, and other estrogens in surface waters. A worrisome aspect of these compounds is that many of the adverse effects on reproduction at low

"eyed"

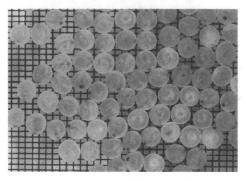

Eggs fertilized with
control semen

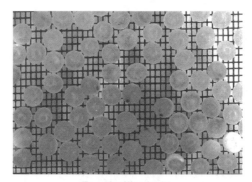

Eggs fertilized with
semen from EE2
exposed trout

Figure 8.7   Determination of the eyed stage of embryonic development. A known quantity of sperm collected
from control and EE2-exposed trout is incubated with eggs from nonexposed trout. After incubation
for 28 days, the number of eyed eggs is counted. (Photos by Irvin Schultz.)

exposure rates are subtle or occur without overt signs of toxicity. Thus, the effects on sexual differentiation
or reproduction might be easily overlooked and could be more widespread than previously thought.

IV.C   Developmental Stability

Developmental stability, the capacity of an organism to develop into a consistent phenotype
in an environment, can also be influenced by and used to suggest the impact of contaminants.
Developmental stability can be correlated with fitness within a particular setting. Individuals that
fail to achieve a consistent phenotype often show lower survivorship or reduced reproductive output.

As will be seen, developmental stability is easily and inexpensively measured in the field or laboratory, and it is relatively universal in its applicability (Zakharov, 1990; Graham et al., 1993a). Thus developmental stability has great advantage as an indicator of contaminant effects.

Developmental stability can be examined by measuring deviations from perfect form. For bilaterally symmetrical organisms, this is often calculated as deviations from perfect symmetry. Bilateral characteristics, such as the lengths of the right and left claws of a crab, are measured and the difference calculated, i.e., d = Length$_{right}$ − Length$_{left}$. Usually many characters are measured or counted. Deviations from perfect bilateral symmetry (**fluctuating asymmetry** [FA]) measured within a population are thought to reflect perturbations of normal developmental processes. If the mean of the d values from a population is zero and the distribution of d is normal, the variance of d within the population is a measure of fluctuating asymmetry. Using the notation of Zakharov (1990), the mean (M_d) and standard deviation (σ_d) of d are as expressed in Equations 8.1 and 8.2:

$$M_d = \frac{\sum d_i}{n} \tag{8.1}$$

$$\sigma_d = \sqrt{\frac{\sum (d - M_d)^2}{n - 1}} \tag{8.2}$$

where n = the number of individuals measured.

Bilateral organisms can also exhibit directional asymmetry or antisymmetry. **Directional asymmetry** exists if the mean is not zero.[3] For example, measurement of d for weights of left and right arms of humans would display directional asymmetry because most humans are right-handed. **Antisymmetry** is indicated if the distribution of d is bimodal. Graham et al. (1993b) give the example of male fiddler crab claws as antisymmetry: some males have larger right claws but others have larger left claws. Although directional asymmetry and antisymmetry can be used to measure contaminant effects on developmental stability, fluctuating asymmetry is most often used.

Minimal fluctuating asymmetry is expected under optimal (benign) conditions. Fluctuating asymmetry is expected to increase as conditions become increasingly different from some optimal range, as in the case of increasing concentrations of a contaminant. This increase in fluctuating asymmetry reflects movement away from developmental stability.

Several field and laboratory studies have shown the value of measuring fluctuating asymmetry to assess toxicant effects. For example, Graham et al. (1993a) measured traits of morning glory (*Convolvulus arvensis*) growing at various distances from an ammonia plant in the Ukraine and found that fluctuating asymmetry was highest for leaves taken nearest the factory. In a laboratory study in which flies (*Drosophila melanogaster*) were fed either lead or benzene, fluctuating asymmetry increased with exposure to these toxicants, suggesting a diminished ability to maintain a consistent phenotype (Graham et al., 1993c).

V REPRODUCTION

Lowered fitness of individuals due to reproductive impairment is arguably among the most useful sublethal effects measured by ecotoxicologists (Sprague, 1971; 1976). In Chapter 1, reproductive failure traced to DDT/DDE inhibition of Ca-dependent ATPase in the eggshell gland (Kolaja and Hinton, 1979) and consequent eggshell thinning was discussed as one of the earliest events leading to our increased awareness of pesticide impacts to nontarget species. In this chapter, we discussed adverse impacts to the seagull reproductive system due to an estrogenic xenobiotic, and

---

[3] According to Zakharov (1990), a simple t-test can be used to determine if M_d is significantly different from zero.

we also examined population failure of whelks experiencing high incidences of imposex as a consequence of TBT exposure. Reproduction remains one of the most frequently measured qualities in both field and laboratory studies of contaminants.

The following field studies further illustrate the value of measuring reproductive variables. White suckers (*Catostomus commersoni*) from a lake near the Flin Flon metal smelters (Manitoba, Canada) had smaller eggs and reduced egg and larval survival than suckers sampled from a clean lake (McFarlane and Franzin, 1978). *Catostomus commersoni* also had a high incidence of reproductive failure in a lake contaminated with metals (Munkittrick and Dixon, 1988). However, this failure was attributed to modification of the food base for the sucker, not to a direct effect of the metals. Western mosquitofish (*Gambusia affinis*) taken from a selenium-contaminated water body showed lowered fry survival and more stillborn fry than mosquitofish from a clean site (Saiki and Ogle, 1995). Starry flounder (*Platichthys stellatus*) from a highly urbanized region of the San Francisco Bay had more previtellogenic oocytes and lower embryo success than flounder sampled from a less urbanized region (Spies et al., 1989). And egg fertilization for purple sea urchins (*Strongylocentrotus purpuratus*) was diminished by oil production effluent (Krause, 1994).

Numerous laboratory experiments have demonstrated the impact of toxicants on reproductive performance, augmenting field studies such as those just mentioned. In one such study, reproductive impairment by metals was estimated by using a target 16% reduction in the number of young produced per female *Daphnia magna* (Biesinger and Christensen, 1972). Mosquitofish (*G. holbrooki*) in mercury-spiked mesocosms had low numbers of late-stage embryos (Mulvey et al., 1995). Standard laboratory methods also are used for measuring various reproductive qualities as affected by contaminants. Weber et al. (1989) outline these methods for reproduction of zooplankton species and fathead minnow (*Pimephales promelas*). Chapman (1995) describes a standard method for quantifying fertilization success of purple urchin eggs.

All of these methods are useful for predicting effects on field populations if combined with an adequate understanding of the reproductive biology and ecology of the species in question. Without such information, the effectiveness of identifying consequences of the sublethal effect is uncertain. For example, a delay in the onset of reproduction can have trivial consequences for one species but serious consequences for another (Newman, 1995).

VI PHYSIOLOGY

Deviations from homeostasis associated with sublethal exposure often reflect physiological alterations. Such physiological alterations can be used to infer a mode of action of the toxicant as well as to document a lowered capacity to maintain homeostasis or normal functioning (Brouwer et al., 1990). For example, acetylcholinesterase inhibitors[4]—notably many of the organophosphate and carbamate insecticides—affect feeding, respiration, swimming, and social interactions of nontarget as well as target species by impairing the senses and neuromuscular activities (Mehrle and Mayer, 1985). While some toxicants like acetylcholinesterase inhibitors are straightforward in their mode of action, others can involve a complex series of causal linkages. For example, contaminant-modified functioning of the endocrine system can impact numerous processes under hormonal control (Brouwer et al., 1990).

Some physiological effects involve threshold concentrations. Ammonia toxicity to rainbow trout (*Oncorhynchus mykiss*) is a good example of a physiological effect with a threshold (Lloyd and Orr, 1969). As ammonia concentration increases, it causes increased water flux into trout that is counterbalanced by an increase in urine production. Ammonia only becomes lethal above a threshold reflecting the maximum rate of urine production.

---

[4] **Acetylcholinesterase inhibitors** are compounds such as some insecticides that inhibit the normal functioning of acetylcholinesterase, an enzyme that breaks down the neurotransmitter, acetylcholine. After release from the presynaptic neuron, acetylcholine diffuses across the synapsis to bind at a receptor on the postsynaptic neuron (or on the neuromuscular junction), resulting in nerve impulse transmission. At the receptor site, it must then be hydrolyzed by acetylcholinesterase to choline and acetic acid in order to facilitate subsequent normal transmission of nerve impulses.

Physiological changes often studied for nonhuman animals include impaired performance (e.g., swimming speed or stamina), respiration, excretion, ion regulation, osmoregulation, and bioenergetics (e.g., food conversion efficiency) (Sprague, 1971). Infrequently, immunological capabilities or disease resistance are examined (Anderson, 1990), although these responses have considerable potential for understanding effects of toxicants (Newman 2001). Common sublethal effects to plants include water status, stomatal function, root growth, respiration, transpiration, nitrogen or carbon fixation, chlorophyll content, and photosynthesis (Baker and Walker, 1989). Most of these sublethal effects are assumed to compromise important physiological functions for which diminished efficiency implies lowered fitness of the affected individual.

Studies of toxicant effects on respiration describe oxygen consumption directly under normal, resting, or maximum exertion regimes. Also used is the **scope of activity** or metabolic scope, the difference between the rates of oxygen consumption under maximum and minimum activity levels. The scope of activity reflects the respiratory capacity or amount of energy available for the diverse demands on or activities of an organism. The oxygen consumption rate can be combined with the nitrogen (ammonia) excretion rate (O:N ratio) to suggest the relative dependence of respiration on carbohydrate and lipid resources versus the deamination of amino acids. For example, exposure of white mullet (*Mugil curema*) to benzene and of shrimp (*Macrobranchium carcinus*) to metals produced significant shifts in the O:N ratio (Correa, 1987; Correa and Garcia, 1990).

More-complicated indicators of metabolism may be needed to address specific problems. For example, energetic analysis can reveal a disruption of energy balance associated with toxicant exposure (Dillon and Lynch, 1981). The **scope of growth** can be calculated from an energy budget of exposed individuals. The scope of growth (P = production) is the amount of energy taken into the organism in its food (A) minus the energy used for respiration (R) and excretion (U): P = A – R – U (Cockerham and Shane, 1994). The value of P reflects the energy available for growth and production of young. For organisms exposed to toxicants, energy is expended in response to the toxicant, and the prediction is that scope of growth will be decreased.

The **adenylate energy charge** (AEC) (Equation 8.3) can also reflect the balance of energy transfer between catabolic and anabolic processes:

$$AEC = \frac{ATP + 1/2\ ADP}{ATP + ADP + AMP} \tag{8.3}$$

where ATP, ADP, and AMP = concentrations of adenosine tri-, di-, and monophosphate, respectively. Giesy et al. (1981) reported a drop in AEC in crayfish (*Procambarus pubescens*) and freshwater shrimp (*Palaeomonetes paludosis*) exposed to cadmium, suggesting a diminished energy status for these exposed crustaceans.

Sublethal effects on respiratory activity or respiratory organs can be determined by examining movements associated with respiratory organs. In a common assay, fish are placed into a chamber receiving a solution of toxicant or an effluent suspected of containing a toxicant. The chamber is equipped with electrodes that measure the change in low-voltage electrical fields produced by respiratory movement. In the absence of contaminant, changes in potential over time (Figure 8.8, Panel A) show regular and uniform respiratory movements with occasional **coughs or gill purges**. The coughs are abrupt, periodic reversals of water flow over the gills that function to dislodge and eliminate excess mucus from the gill surfaces. Cough frequency may increase during exposure to various contaminants, often because the irritant causes excess mucus production on the respiratory surfaces (Figure 8.8, Panel B). For example, cough rates increased for bluegill sunfish (*Lepomis macrochirus*) exposed to heavy metals or chlorinated hydrocarbons (Bishop and McIntosh, 1981; Diamond et al., 1990) and for brook trout (*Salvelinus fontinalis*) exposed to very low concentrations of methyl or inorganic mercury (Drummond et al., 1974). The frequency and amplitude of the signals from ventilation are also sensitive indicators of sublethal effect. Ventilation frequency increased with exposures to cadmium (Bishop and McIntosh, 1981), copper (Thompson et al., 1983), zinc (Thompson et al., 1983),

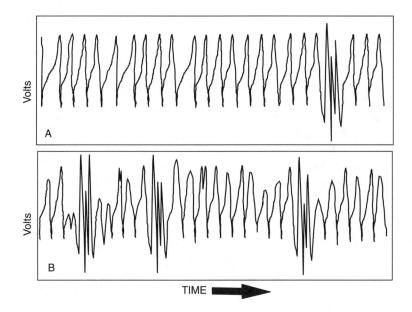

Figure 8.8   Patterns for ventilation and coughing for fish. In panel A, ventilation (small, evenly spaced peaks)
and coughing (strong, rapid cluster of peaks toward the right-hand side) are traced through time
for a control individual. Ventilation frequency increases, amplitude decreases, and cough frequency
increases with exposure to contaminants (Panel B).

chlorinated hydrocarbons (Diamond et al., 1990), and other organic compounds (Kaiser et al, 1995).
Diamond et al. (1990) and Kaiser et al. (1995) demonstrated that the amplitude of the ventilation
signal can also decrease with exposure to sublethal concentrations of toxicants.

Osmoregulation and ion-regulation capacities can also decline during contaminant exposure.
Atlantic salmon (*Salmo salar*) in ammonia-spiked seawater experienced an increase in plasma
osmolality (Knoph and Olsen, 1994). Eel (*Anguilla rostrata*) osmoregulation was disrupted by DDT
inhibition of Na^+,K^+- and Mg^{2+}-ATPase activity in the intestine, where the flux of water is linked
with the flux of these ions (Janicki and Kinter, 1971). Ion balance was disrupted in Atlantic salmon
(*S. salar*) exposed to ammonia, in flounder (*Platichthys flesus*) exposed to cadmium, and in rainbow
trout (*O. mykiss*) exposed to copper (Larsson et al., 1981; Laurén and McDonald, 1985; Knoph
and Olsen, 1994). Acid conditions, alone or in combination with elevated aluminum concentrations,
also altered ion regulation by fish (Fromm, 1980; Witters, 1986).

A wide range of physiological qualities is also measured for plants under the influence of
contaminants. Metals (cobalt, nickel, and zinc) modify the water balance, stomatal closure, and
leaf orientation of bean (*Phaseolus vulgaris*) seedlings (Rauser and Dumbroff, 1981). Exposure to
PCBs (Doust et al., 1994) and heavy metals (Baker and Walker, 1989) can reduce photosynthetic
activity of plants. Air pollutants such as ozone and sulfur dioxide (SO_2) cause leaf **chlorosis**, a
blanching of green color due to the lack of chlorophyll production or the increased destruction of
chlorophyll (Landis and Yu, 1995). Heavy metals can affect plant respiration, alter carbon and
nitrogen fixation, and inhibit photosynthesis (Baker and Walker, 1989).

VII BEHAVIOR

Various animal activities are studied in **behavioral toxicology**, the science of abnormal behaviors produced by exposure to chemical or physical agents. In addition to the behaviors already
discussed relative to ventilation movement and teratology, behavioral abnormalities include changes
in preference or avoidance, activity level, feeding, performance, learning, predation, competition,

Table 8.1   Behaviors Commonly Used to Reflect Sublethal Effects of Contaminants

Behavior	Examples
1. Preference or avoidance	Change in response to light, temperature, salinity, or current; may avoid or move toward a stimulus differently after toxicant exposure; salmon exposed to DDT shift their preferred water temperature from 19.1 to 23.4°C (Ogilvie and Miller, 1976)
2. Activity level	Fatigue (lethargy) of workers with chronic lead poisoning (Bornschein and Kuang, 1990); hyperactivity of fiddler crabs (*Uca pugilator*) exposed to tributyltin (Weis and Perlmutter, 1987) or Arctic char exposed to chlorine (Jones and Hara, 1988)
3. Feeding	Cessation of or diminished feeding by fish after exposure (Jones and Hara, 1988); deviation from predictions of optimal foraging theory (Sandheinrich and Atchison, 1990)
4. Performance	Ability to swim against a current or maintain proper orientation to a current (rheotaxis) (Little and Finger, 1990); critical swimming speed of fish lowered by exposure (Schreck, 1990)
5. Learning	Memory impairment of humans exposed to excess metals or metalloids, and memory loss due to mercury poisoning (Bornschein and Kuang, 1990); poorer response and higher error rate of exposed mammals in "lever pulling" learned behavior experiments (Gad, 1982)
6. Respiratory activity	See examples given in Section VI on physiological effects of sublethal exposures
7. Predation	Lowered ability to avoid predator (largemouth bass) by mosquitofish (*G. holbrooki*) exposed to radiation or mercury (Kania and O'Hara, 1974; Goodyear, 1972); suboptimal predator foraging or prey-switching (Atchison et al., 1996)
8. Competition	Zooplankton species grazing and filtration rates modified by toxicants (general details discussed by Atchison et al., 1996)
9. Reproductive behavior	Decreased libido after occupational lead exposure (Bornschein and Kuang, 1990); masculinization of mosquitofish discussed earlier in this chapter
10. Social interactions	Grooming in mammals after exposure (Gad, 1982); emotional lability of humans exposed to excess manganese, or increased irritability or depression associated with lead poisoning of humans (Bornschein and Kuang, 1990)

reproduction, and a variety of social interactions such as aggression or mutual grooming (Table 8.1) (Rand, 1985; Henry and Atchison, 1991). Most often, these effects are measured in a laboratory setting, but some studies measure *in situ* changes in behavior (e.g., Gray, 1990). Unfortunately, behavioral effects are underutilized in assessments of risk for three reasons (Giattina and Garton, 1983): (1) it is difficult to objectively score some behaviors, thus leaving open the possibility of generating biased information, (2) considerable variability can exist in behavioral data, and (3) it is often difficult to extrapolate accurately from highly structured laboratory experiments of behavior to behavior in field situations. However, the first two problems can be minimized by careful design and execution of experiments; the third point is no truer of behavioral assays than for many assays used today in risk assessment. Regardless, the results of behavioral studies are most effectively used to assess contaminant impact when combined with results from a suite of other lethal- and sublethal-effects studies (Atchison et al., 1987).

Vignette 8.2   The Role of Behavior in Ecotoxicology

Mark Sandheinrich

University of Wisconsin – La Crosse, La Crosse, WI

Behavior integrates genetic, biochemical, physiological, and environmental attributes that influence the evolutionary fitness of organisms. Consequently, behavior transcends single levels of biological organization and provides a link between subcellular processes that can be measured in the laboratory and ecological responses to contaminants observed in the field. Weis et al. (2001) demonstrated that altered levels of brain neurotransmitters and thyroid hormones of killifish (*Fundulus heteroclitus*) from a contaminated site were correlated with altered behavior, specifically reduced locomotor activity, impaired feeding, and increased vulnerability to predators. These altered behaviors explained reduced growth, condition, and longevity of the fish in the field as well as population changes in their major prey, the grass shrimp (*Palaemonetes pugio*). The role of behavior in ecotoxicology has been defined primarily by studies of (1) the sublethal effects of chemicals on organisms, (2) behavior as a modifier of organism exposure and chemical toxicity, (3) the use of behavior in identifying toxic modes of action, and (4) the incorporation of behavior into routine monitoring of water quality.

Numerous reviews have concluded that behavior is an ecologically important and sensitive indicator of toxicant stress in aquatic organisms (Atchison et al., 1987; Beitinger, 1990; Birge et al., 1993; Henry and Atchison, 1991; Little et al., 1993; Westlake, 1984). Contaminants affect a variety of behaviors (Table 8.1), many that influence growth, reproduction, and survival—the classic endpoints of standard acute and chronic toxicity tests. For example, Little and Finger (1990) reported that swimming behavior of fish exposed to a variety of chemicals was altered at concentrations as low as 0.7 to 5% of the median lethal concentration. Swimming activity was frequently affected at concentrations that subsequently reduced growth. Sandheinrich and Atchison (1990) noted that feeding behavior of fish was disrupted by toxicants at or near concentrations that decreased growth and that altered behavior was often observed within hours of exposure.

Behavior, however, is not routinely used in hazard assessment or in establishment of water quality criteria. Lack of test standardization and field verification of behavioral responses are major challenges that have limited the acceptance of behavior as regulatory endpoints (Little, 1990). However, avoidance responses to contaminants, particularly by fish, do occur in the field and are ecologically important because the response can alter aquatic communities through emigration of organisms (Atchison et al., 1987). Sprague and colleagues demonstrated that Atlantic salmon (*Salmo salar*) avoided low concentrations of copper and zinc in the laboratory (Sprague, 1964) and subsequently reported that Atlantic salmon migrating upstream during spawning runs avoided areas contaminated with a mixture of zinc and copper (Sprague et al., 1965; Saunders and Sprague, 1967). Geckler et al. (1976) reported that production of fish within a stream was reduced by avoidance of copper. Standardized acute and chronic toxicity tests conducted on site failed to predict this response. Avoidance of contaminated areas has been accepted as legal evidence of injury for Natural Resource Damage Assessments under the Comprehensive Environmental Response, Compensation, and Liability Act of 1980 (Natural Resource Damage Assessments, 1986; Little et al., 1993).

Although little studied, one of the more intriguing areas of ecotoxicology is the effect of behavior on rate of uptake and sensitivity of organisms to toxicants. Feeding, avoidance or attraction to contaminated areas, and social behaviors may alter exposure of fish to contaminants. For example, hierarchical positions within social groups of freshwater fish, such as salmonids and centrarchids, are established through aggressive or agonistic behaviors that may include nudging, biting, chasing, and flaring fins. Consequently, subordinate individuals within the group often respond by producing elevated levels of the stress hormone, cortisol (Fox et al., 1997), which subsequently alters osmoregulation and transport of ions (e.g., metals) by the gills (Jobling, 1995). Sloman et al. (2002) found that the position of a fish within a social hierarchy affected uptake of copper. Sublethal concentrations of copper did not alter established hierarchies within laboratory populations of rainbow trout (*Oncorhynchus mykiss*), but subordinate fish had significantly higher concentrations of copper in gills and liver than dominant fish due to increased uptake of copper from water. Behaviorally altered rates of toxicant uptake may influence susceptibility to toxicants. Sparks et al. (1972) found that the subordinate fish of a pair of bluegill (*Lepomis macrochirus*) succumbed more quickly to a lethal concentration of zinc than the dominant fish. When provided with a shelter, the frequency of aggressive interactions decreased between the fish, and there was no difference in their sensitivity to zinc.

Behavior of aquatic organisms can also be used to identify the mode of action of chemicals. For example, Diamond et al. (1990) demonstrated that the amplitude and frequency of ventilation, number and type of gill purges, and frequency of erratic movement in bluegill could be used to differentiate between different groups of chemicals. The heavy metals, zinc and cadmium, decreased the ventilatory amplitude and increased the frequency of gill purges within minutes of exposure. The chlorinated hydrocarbons, dieldrin and trichloroethylene, increased erratic movement, number of gill purges, and ventilatory frequency of the bluegill.

Drummond and coworkers used behavioral and morphological changes in 30-day-old fathead minnows (*Pimephales promelas*) to classify organic chemicals according to toxic modes of action (Drummond et al., 1986;

Drummond and Russom, 1990). They initially developed a system for visual observation of 40 different behavioral and morphological characteristics from 10 general categories (Drummond et al., 1986). These categories included equilibrium, locomotor activities, schooling and social behavior, body movement and coloration, ventilatory patterns, and general pathology and mortality. Fathead minnows were observed during acute exposure to 139 single chemicals. Loss of schooling was usually the first major symptom to occur and was caused by 96% of the chemicals. Discriminant function analysis, a statistical method for identification of groups and patterns, demonstrated that behavioral data could be used to identify and separate chemicals into four categories based on probable cause of action. Narcosis-inducing chemicals (e.g., alcohols, ethers, ketones, phthalates) caused loss of equilibrium, depressed swimming activity, altered ventilation, and darkened body color. Chemicals that disrupt metabolism (e.g., benzenes, phenols) induced hyperactivity and increased ventilatory rates and amplitude. Neurotoxins (e.g., carbamate and organophosphate insecticides) depressed swimming activity but increased ventilation, tetany, deformities, and sensitivity to outside stimulation. Skin irritants (e.g., acrolein) produced symptoms similar to disrupters of metabolism, but fish were extremely hyperactive and exhibited agonistic and other unusual behaviors. Drummond and Russom (1990) refined these categories and classified more than 300 organic compounds based on three behavioral toxicity syndromes that represent a different mode of toxic action: (1) hypoactivity syndrome (narcosis), (2) hyperactivity syndrome (metabolic dysfunction), and (3) physical deformity syndrome (neurological dysfunction). Subsequently, Rice et al. (1997) used similar methods to assess sublethal toxicity to Japanese medaka (*Oryzias latipes*) of five chemicals with different modes of action. Behavioral toxicity syndromes may be useful, in conjunction with other endpoints, for predicting modes of actions of unknown compounds, identifying toxicants in complex effluents, and for testing wastewater prior to discharge into the environment.

One of the most practical uses of behavior is for the continuous monitoring of water quality at wastewater or drinking-water treatment facilities and for providing a method of detection of toxicants from episodic events, such as stormwater runoff or accidental chemical spills. Instrumentation is commercially available to identify changes in behavior of fish, mussels, and water fleas (*Daphnia*) due to a toxic event and to signal an alarm to facility personnel. Water from lakes, rivers, or treatment plants is continuously passed through single or multiple chambers containing aquatic organisms, and their behavior is monitored electronically. Biomonitoring systems are used throughout Western Europe to protect surface waters but have received only limited use in North America.

The amplitude and frequency of rhythmic ventilatory movement, gill purges, and erratic locomotor behaviors of captive fish are sensitive indicators of environmental stress (Diamond et al., 1990). Neuromuscular activity of the fish generates a microvolt bioelectric signal. Electrodes within a chamber can capture the signals, which are subsequently amplified and transmitted to a computer system. Significant departures from baseline activity due to abnormal behavior are indicative of a toxic condition.

Monitoring systems that measure changes in position of the valves (i.e., gape) of mussels have also been developed (e.g., Sluyts et al., 1996). Small sensors are attached to the shell of the mussel and measurements of shell gape are obtained using an electromagnetic induction technique. Shell gape and the number of valve movements change under stressful conditions, such as those elicited by exposure to chemicals.

Video imaging software and complex algorithms are used to evaluate changes in the swimming behavior of groups of *Daphnia*. Measurements of mean swimming speed, distribution of swimming speeds among individuals, turns and circling movements, height of the *Daphnia* in the water column, distance between organisms, and number of moving *Daphnia* are combined to calculate a toxic index. Statistically significant changes in the toxic index trigger an alarm (Lechelt et al., 2000).

In conclusion, behavior is a sensitive indicator of sublethal toxicant stress and provides an ecologically relevant measure of contaminant effects. In developing methods for measuring chemically altered behavior, Little et al. (1993) proposed that endpoints should be evaluated based on (1) the sensitivity and ability of the behavioral response to measure effect or injury from exposure, (2) the utility of the response in identifying the chemical causing toxicity, and (3) the capacity of the behavioral response to increase predictability of the ecological consequences of toxicant exposure. Behavioral endpoints of greatest use in ecotoxicology will be those that provide linkages between the endpoints of growth, reproduction, and survival derived in the laboratory and the changes in population and community structure predicted to occur in the field.

VIII DETECTING SUBLETHAL EFFECTS

Sublethal and chronic sublethal effects are detected and quantified in several ways. For regulatory testing, data are often generated with experimental designs similar to that shown at the top of Figure 8.9. Effects are measured at one or several time intervals in replicates receiving various concentrations of toxicant. For example, fathead minnow (*Pimephales promelas*) larvae can be exposed to a series of toxicant concentrations (four replicate tanks containing 10 larvae each per exposure concentration) and the growth of the larvae measured after 7 days. The results are then analyzed using either a hypothesis-testing or regression method.

Hypothesis testing most often involves a one-way analysis of variance (ANOVA) and postanalysis of variance approach (Figure 8.10), although other tests are also applicable. Generally, in **analysis of variance** (ANOVA), the total variance (total sum of squares) is broken down into the variance among and within treatments, e.g., among the different concentration treatments and within replicates for each concentration treatment. The variance within treatments (mean sum of squares$_{within}$) is assumed to reflect the sampling or error variance, and that among treatments (sum of squares$_{among}$)

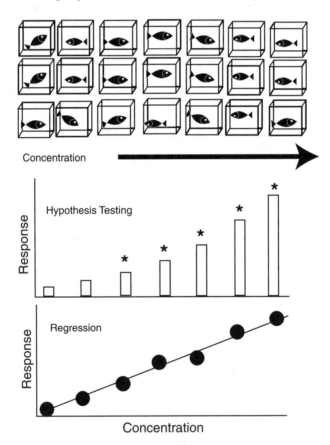

Figure 8.9 For regulatory testing, sublethal effects are typically analyzed using hypothesis-testing methods. A series of tanks (seven sets of triplicates are depicted in the figure) receiving increasing concentrations of contaminant are used to estimate the response for each concentration, including a control set of tanks. Data are then tested to determine if the response at each concentration was significantly different from that of the control. In the middle panel of the figure, asterisks are used to indicate exposure concentrations with effects that are significantly different from the control. Here, the responses of the five highest concentrations were significantly different from the control: The lowest exposure concentration was not significantly different from the control. These data can also be used in regression models to develop a predictive relationship between concentration and effect (bottom panel).

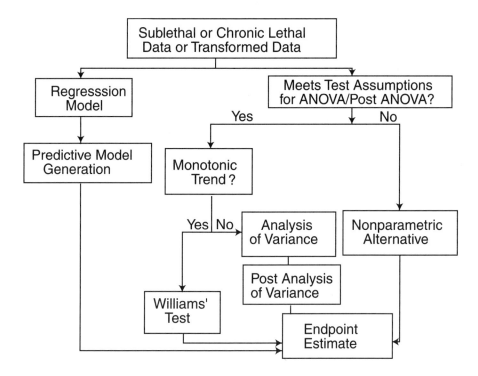

Figure 8.10 A flow diagram of methods for analyzing sublethal and chronic lethal effects.

is an estimate of error variance plus any additional variance associated with the treatment. If there were no differences among treatment means, then these two measures of variance would be equal. This fact leads to a useful statistic for testing the null hypothesis of equal means among treatments. The ratio of these two variance estimates ($F = $ mean sum of squares$_{among}$/mean sum of squares$_{within}$) is compared with tabulated F statistics to test for significant deviation from the null hypothesis of no difference among treatments.

The ANOVA approach has two statistical assumptions: equal variances among treatments and normally distributed data. These two assumptions are assessed before performing the ANOVA.[5] Normality is conveniently tested with a formal method such as the **Shapiro-Wilk's test**, although graphical methods are also adequate (Newman, 1995). The assumption of homogeneity of variance is tested using one of several tests (**Bartlett's test** is the most commonly applied) for data that, according to some test such as the Shapiro-Wilk's test, already satisfy the assumption of normality. The ANOVA methods are relatively robust to violations of these two formal assumptions (Miller, 1986; Salsburg, 1986). Newman (1995) states that "probabilities derived from ANOVA are close to the real probabilities if the underlying distribution is at least symmetrical and the variances for the treatments are within three-fold of each other."

Often transformations of the data aid in meeting these formal requirements for ANOVA. The most common transformation is the **arcsine square root transformation**, which is particularly helpful in meeting the assumption of homogeneous variances for proportions of exposed individuals responding:

$$\text{Transform } P = \arcsin \sqrt{P} \qquad (8.4)$$

where $P = $ the measured effect, such as the proportion of the exposed organisms responding.

---

[5] A third and very important assumption of ANOVA is that observations are independent. It is usually satisfied with rigorous experimental design in which subjects are randomly assigned to treatments.

If ANOVA leads to the rejection of the null hypothesis, we know that treatment (i.e., exposure concentration) had a significant influence on mean response, but we do not know which means were significantly different. Extending the fathead minnow growth example, we know that mean growth rate was different among the different exposure concentrations, but we do not know which concentrations had mean growth rates that were significantly higher or lower than the others. A series of post-ANOVA methods can identify the specific treatments that are different from each other. Some test for significant difference between all possible pairs with the control (control mean vs. each treatment-concentration mean). For example (Figure 8.9), six pairs of means are tested because there was a control treatment and six exposure concentration treatments: one pair for each concentration tested against the control. Other methods can also be applied, as will be discussed (Chapter 9).

If the data or transformations of the data grossly violate one or both of the assumptions of ANOVA, it is possible to use nonparametric, post-ANOVA tests at the cost of sacrificing some statistical power. **Steel's many-one rank test** and the **Wilcoxon rank sum test with Bonferroni's adjustment** are the most commonly employed (Weber et al., 1989). Steel's many-one test is often recommended if there are equal numbers of observations for all treatments (e.g., triplicate tanks of fish for all concentrations, including the control treatment). The Wilcoxon rank sum with Bonferroni adjustment is recommended often for unequal number of observations (Weber et al., 1989), although the slightly more powerful Steel's many-one rank test can also be used in designs with unequal observation numbers (Newman, 1995).

If the data or their transformations are acceptable for ANOVA, several parametric methods are available that have more statistical power than the nonparametric methods just described. **Dunnett's test** or **t-tests with a Bonferroni adjustment** can be applied, although Dunnett's test is the slightly more powerful. Newman (1995) suggests that, although widely used, the t-test with a Bonferroni adjustment is also slightly less powerful than the equally convenient but seldom applied **t-test with a Dunn-Šidák adjustment**.

If one can assume that a monotonic trend (i.e., a consistent increase or decrease in response) will occur with increasing toxicant concentration, the even more powerful **Williams's test** can be applied. Unlike the tests described above, the alternate hypothesis is no longer inequality of means between treatment pairs. Instead, the alternate hypothesis becomes "there is a monotonic trend in effect with treatment concentration." The test is done in two steps. In the first, significant deviation from the null hypothesis (equal mean responses among all treatments) is tested. If there is a significant deviation, a second step of the test is completed in which the lowest concentration having a significantly different mean from the control is identified. Obviously, Williams's test would not be appropriate if one suspected hormesis in the data, because the underlying assumption of monotonicity would not be met.

Any of these methods could produce a data summary as depicted in the middle panel of Figure 8.9. A mean response of the control (leftmost bar) is compared with mean responses for the treatment concentrations, and those significantly different from the control response are identified. In Figure 8.9, an asterisk was placed above treatment means differing significantly from the control.

With results from analyses of the kind just described, biological effects of various concentrations of chemicals are predicted for sublethal and chronic lethal effects. However, the process of using tests of statistical significance to make projections about biological significance is more difficult than it may first appear. Statistical hypothesis tests only demonstrate that something is present in the data set that differs significantly from the null hypothesis: they say nothing about the biological significance of that deviation. Considerable judgment must be applied to statistical results in order to successfully predict ecotoxicological consequences.

To assist in the extrapolation of statistical results to ecotoxicological consequences, a number of descriptive concepts and terms have been developed. The **no-observed-effect concentration** (or level) (NOEC or NOEL) is the highest test concentration for which there was no statistically significant difference from the control response. To emphasize that the effect is an adverse one, the term can sometimes be expanded to no-observed-adverse-effect concentration (NOAEC). In Figure 8.9, the lowest experimental concentration would be the NOEC (second bar from the left). The **lowest-observed-effect concentration** (or level) (LOEC or LOEL) is the lowest concentration in a test with a statistically significant difference from the control response (third bar from the left). Again, the word *adverse* might be added to the term (LOAEC). The **maximum acceptable toxicant concentration** (MATC) is "an undetermined concentration within the interval bounded by the NOEC and LOEC that is presumed safe by virtue of the fact that no statistically significant adverse effect was observed" (Weber et al., 1989). Notice that the boundaries for statistical significance are represented by NOEC and LOEC, and biological "safety" is defined as being within those boundaries. Weber et al. (1989) define the term **safe concentration** to mean "the highest concentration of toxicant that will permit normal propagation of fish and other aquatic life in receiving waters." The concept of a "safe concentration" is a biological concept, whereas the "no observed effect concentration" is a statistically defined concentration.

The above definitions and their implied applications in assessing risk have several shortcomings. First, the values that the NOEC and LOEC can take are totally dependent on the particular concentrations chosen for the experiment: They are tied as much to the experimental design as to any toxicological reality. Next, the process produces higher than optimal NOEC and LOEC values if one uses a suboptimal experimental design (low statistical power) or poor technique (high error variance). Consequently, inferior design and technique can be rewarded with higher NOEC and LOEC values than would be calculated with superior design and technique; the concentrations identified as having an effect would be higher with inferior methods. The presumption of a "safe" MATC between the NOEC and LOEC cannot be extended beyond the design, species, and exposure durations used in the specific test without much additional supportive information. The MATC has no statistical confidence interval because the LOEC and NOEC are used to define it. Finally, such data may have dubious predictive value for estimating a safe concentration (i.e., "that [permitting] propagation of fish and other aquatic life in receiving waters"). A statistically significant reduction in reproduction (e.g., 50%) may be much higher than that which will eventually lead to local extinction of some species populations (e.g., 20%) but not others. Regardless, these types of data, augmented with supportive data and professional judgment, are used extensively in ecological risk assessments today.[6]

Another approach to analyzing concentration-response data is to use regression methods (Figure 8.9, bottom panel) (Stephan and Rogers, 1985; Hoekstra and Van Ewijk, 1993). Data can be fit to a specific concentration-effect model by least-squares or maximum-likelihood methods. Concentrations (and their associated confidence intervals) having some biologically significant effect, such as a 10% reduction in fecundity, are calculated via interpolation with the model. The ability to extrapolate downward from results is another potential advantage of this approach if one is confident in the shape of the concentration-response model. However, if one incorrectly assumes a linear model when a threshold concentration exists, predictions from regression models will lead to false conclusions. Also, if there is so much variation in the data that a good model cannot be identified, the regression approach becomes compromised and the ANOVA approach may be required (Stephan and Rogers, 1985).

IX SUMMARY

This chapter describes sublethal effects to individual organisms, including the general adaptation syndrome (GAS) and effects to growth, development, reproduction, physiology, and behavior. Many sublethal effects overlap these artificial categories, in particular behavioral teratology. The ambiguity associated with whether or not such "sublethal" effects may result in death in an ecological arena

---

[6] Statistical significance is often used directly to assign ecological significance in ecological risk assessment. This unwise practice will be labeled the **maulstick incongruity**. (A maulstick is a stick used by artists to simply steady the brush hand while painting.) Just as an inferior painting would be expected from an artist who used the maulstick instead of the brush to apply paint to canvas, the use of statistical methods alone—instead of biological data supported by statistical methods—to determine biological significance leads to an inferior decision. The misuse of an otherwise effective tool leads to an inferior product and misinformation.

was emphasized. Statistical methods used to detect, model, and predict sublethal response to toxicants were described, along with the difficulties of using the associated results to predict ecotoxicological impact or risk. Regardless of the difficulties in prediction, measures of sublethal effects are likely to be as important, or more important, than the measures of acute or chronic lethal effect (described in Chapter 9) to accurately assess the consequences of contamination.

SELECTED READINGS

Atchison, G.J., M.G. Henry, and M.B. Sandheinrich, Effects of metals on fish behavior: a review, *Environ. Biol. Fishes,* 18, 11–25, 1987.

McLachlan, J.A., Functional toxicology: a new approach to detect biologically active xenobiotics, *Environ. Health Perspect.,* 101, 386–387, 1993.

Newman, M.C., Hypothesis tests for detection of chronic lethal and sublethal stress, in *Quantitative Methods in Aquatic Ecotoxicology,* Newman, M.C., Ed., CRC Press, Boca Raton, FL, 1995.

Palmer, A.R., Waltzing with asymmetry, *Bioscience,* 46, 518–532, 1996.

Selye, H., The evolution of the stress concept, *Am. Sci.,* 61, 692–699, 1973.

Sprague, J.B., Measurement of pollutant toxicity to fish, III: sublethal effects and "safe" concentrations, *Water Res.,* 5, 245–266, 1971.

Stebbing, A.R.D., Hormesis: the stimulation of growth by low levels of inhibitors, *Sci. Total Environ.,* 22, 213–234, 1982.

Weis, J.S. and P. Weis, Effects of environmental pollutants on early fish development, *CRC Crit. Rev. Aquatic Sci.,* 1, 45–73, 1989.

CHAPTER 9

Acute and Chronic Lethal Effects to Individuals

Pollutants matter because of their effects on populations, and so, indirectly, on communities too, but pollutants act by their effects on individual organisms.

Moriarty (1983)

I GENERAL

I.A   Overview

Most methods used by ecotoxicologists to determine lethality have their origins in mammalian toxicology. In the developing field of ecotoxicology, this transplantation allowed very rapid, initial advancement, since established techniques and concepts could be quickly incorporated. It also led to some inconsistencies that must be resolved if ecotoxicology is to progress further. Also, some very worthwhile techniques in mammalian toxicology have yet to be assimilated into ecotoxicology. This chapter explores these borrowed concepts and techniques, discusses those remaining under-exploited, and identifies inconsistencies between application in mammalian toxicology and ecotoxicology.

I.B   Acute, Chronic, and Life-Stage Lethality

Early in ecotoxicology, a crude distinction was made between acute and chronic lethality. **Acute lethality** refers to death following a brief, and often intense, exposure. The duration of an acute exposure in toxicity testing is generally 96 or fewer hours (Sprague, 1969). Although death is assumed to occur within that short time, there are exceptions in which acute exposure results in death over a long period of time. For example, a brief exposure to high concentrations of beryllium may produce an effect that becomes apparent only after a period has transpired. Small amounts of beryllium that remain lodged in the human lungs can cause a cell-based immune response after environmental exposure has ended (Burns et al., 1996). **Chronic lethality** refers to death resulting from a more prolonged exposure. By recent convention in ecotoxicology, a chronic test should be at least 10% of the duration of the species' life span (Suter, 1993), but this is not always the case. Sometimes a test of shorter duration is discussed as a chronic test. The distinction between acute and chronic is often blurred.

Another important distinction can be made for lethality testing based on life stages. An elaborate **life cycle study** might determine lethality, growth, reproduction, development, or other important qualities at all stages of a species' life (e.g., Mount and Stephan, 1967). **Critical life stage testing** focuses on a particular life stage such as neonates. Often, the most critical life stage is, or is assumed to be, an early life stage, leading to the development of **early life stage (ELS) tests** (McKim, 1985; Weber et al., 1989). The critical life stage approach is based on the sound assumption that protection of the most sensitive stage will ensure protection of all life stages; the most sensitive stage of an individual's life cycle will determine its fate under lethal challenge. A dubious extension of this concept (**weakest link incongruity**) is often made in which one assumes that exposure of field populations to concentrations identified in testing as causing significant mortality at a critical life stage will also result in significant impact on the field population. However, loss of individuals from certain life stages may or may not have much bearing on population demographics or the risk of local extinction for species populations.[1] For example, a 10% reduction in the number of larvae during an oyster spawn due to toxic effect may have minimal impact on the likelihood of an oyster population becoming extinct. Very high mortality is expected for the planktonic larvae under normal conditions, and oyster populations accommodate widely varying annual recruitment. It is important to keep in mind that the term *critical life stage* refers to the life stage of an individual most sensitive to poisoning, not necessarily the life stage most critical to population viability. More will be added to this point in Chapter 10.

I.C   Test Types

Tests used to quantify lethality vary, depending on the medium of concern, i.e., water, food, sediment, or soil. The nature of the toxicant or toxicant mixture can also influence the test design. For example, exposure solutions might not be aerated for tests involving volatile compounds. For tributyltin (TBT)-based antifouling paints, dosing can be done by placing painted surfaces (e.g., discs or rods) into the feed-water flows to the different exposure tanks (Bryan and Gibbs, 1991). Different amounts of TBT leach from these surfaces into the tanks, producing a range of exposure concentrations. Exposure duration, test species, resources, and expendable time also influence methods. A chronic exposure of a suite of endemic species in the laboratory might be highly desirable yet impractical. Instead, a representative species that is easily cultured in the laboratory might be exposed for only the most sensitive time in its life cycle. The results can then be used to imply the risk to endemic species over their entire life cycles.

A series of exposure designs has been established for tests quantifying lethal effects of toxicants in waters. In **static toxicity tests**, individuals are placed into one of a series of exposure concentrations. The exposure water is not changed during the test. The advantage of this design is that it is easy to perform and inexpensive. Also, minimal volumes of toxic solutions are generated (Peltier and Weber, 1985). But toxicant concentrations can change during exposures due to sorption to the container walls and other solid phases, volatilization, bacterial transformation, photolysis, and many other processes. Waste products of the test organisms may build up during the test, and oxygen concentrations may drop to undesirable levels. For these reasons, most static tests are used to measure acute lethality, not chronic lethality. A **static-renewal test** can minimize some of these problems. Test solutions are completely or partially replaced with new solutions periodically during exposures, or organisms are periodically transferred to new solutions. A **flow-through test** uses continuous flow or intermittent flow of the toxicant solutions through the exposure tanks. The flow-through design eliminates or greatly minimizes the problems just discussed for static tests. However, flow-through tests produce large volumes of toxicant solutions that must be treated. They also

---

[1] This situation results from the difference in emphasis between medical/mammalian toxicologists and ecotoxicologists. For a toxicologist dealing with humans, the individual is justifiably the focus of decisions, concepts, and associated methodologies. The ecotoxicologist tends to focus on population and community viability instead. This distracting inconsistency is a consequence of the rapid infusion of methods from mammalian toxicology into ecotoxicology.

require more time, space, and expense (Peltier and Weber, 1985). Although individual containers of various toxicant concentrations can be used as sources of test waters, often a special apparatus called a **proportional diluter** mixes and then delivers a series of dilutions of the contaminant solution to the test tanks. The solution being diluted can be a toxicant solution or an effluent suspected of having an adverse effect on aquatic biota. With effluent testing, lethal effects are expressed as percentages of the total exposure water volume made up of the effluent (e.g., 45% effluent blended with 55% diluent by volume) resulting in the toxic response.

There are several methods associated with solid phase testing. For example, organisms can be placed into spiked or contaminated soils. This type of test has been used with important soil invertebrates such as nematodes (Donkin and Dusenbery, 1994) and earthworms (Gibbs et al., 1996). Similarly, sediment toxicity tests can involve spiked or contaminated sediments. The **spiked bioassay approach** (SB) generates a concentration-response model for, or tests hypotheses regarding, effects to individuals placed into sediments spiked with different amounts of toxicant (Giesy and Hoke, 1990). Concentration may be based on total concentration in the sediment, interstitial water concentration, or the concentration in some notionally bioavailable fraction of the sediments. Sediment toxicity can also be implied using an elutriate test with a nonbenthic test species. In an **elutriate test**, a nonbenthic species such as *Daphnia magna* is exposed to an elutriate produced by mixing the test sediment with water and then centrifuging the mixture to remove solids (McIntosh, 1991). Exposure to various dilutions of the elutriate allows an amount-response analysis of associated data, as described above for effluent tests. The results of such a test would seem most appropriately used to assess lethality of plumes produced during sediment dredging activities.

II DOSE-RESPONSE

II.A Basis for Dose-Response Models

More as a consequence of the early history of ecotoxicology than through careful comparison to alternatives, most lethality tests in ecotoxicology involve the dose-response[2] approach. This approach and associated quantitative methods were taken directly from mammalian toxicology and remain the cornerstone of toxicity testing in ecotoxicology. The time-response models described later are utilized much less frequently.

With the **concentration-response approach**, a series of toxicant concentrations is delivered to containers, as illustrated earlier in Figure 8.9. There are replicate containers for each treatment (concentration) to allow estimation of variation within each treatment and at least one control treatment receiving no toxicant.[3] Individuals are randomly placed into the containers until a predetermined number of individuals is present in each, e.g., ten fish per tank. Mortality is tallied in each tank as the number dying of the total number exposed after a certain time or series of times such as 24, 48, 72, and 96 h. The paired data (proportion dying and exposure concentration) for the tanks are used to calculate lethality.

The predominant methods for analyzing dose- or concentration-response data are based on the **individual effective dose (IED)** or **individual tolerance concept**. According to this concept, there exists a smallest dose (or concentration) needed to kill any particular individual, and this IED is a characteristic of that individual. This concept and its applicability to dose-response data analysis is illustrated in Figure 9.1. The top panel shows the skewed, log-normal distribution of

---

[2] To improve readability, dose-response and concentration-response are used interchangeably here. The analyses of these two models are identical: only the delivery of toxicant differs. A dose-response model is associated with a delivered amount of toxicant, e.g., 5 mg of toxicant is injected into the organism. A concentration-response model is associated with exposure to a concentration in the organism's environment, e.g., 5 mg l^{-1} of toxicant is present in the water into which a fish is placed.
[3] A solvent may be used to produce the toxicant solution as required for some organic compounds. In such a case, a "solvent control" treatment is included too and methods for quantifying toxic effect are modified slightly to include this control in the experimental design. The modifications are made to ensure that any effect of the solvent on response does not unintentionally bias the results.

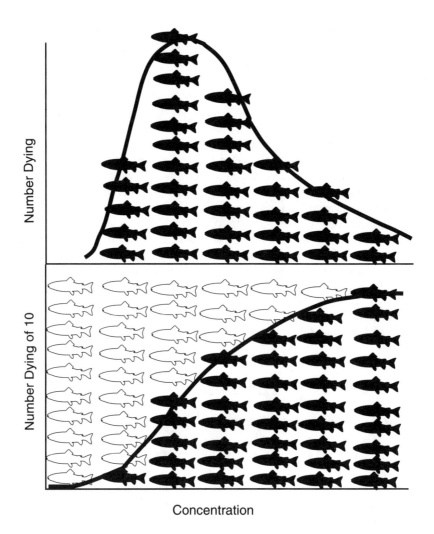

Figure 9.1   The IED concept and the analysis of dose-response lethality data. The top panel depicts the distribution of individual effective doses among 35 fish taken randomly from a population. Each individual's response was placed into a category (column) based on whether its IED fell into one of six different ranges of IED values. The distribution of such IED values within a population is thought to be log normal, as evidenced by an asymmetric curve with a few very tolerant individuals. The bottom panel shows the results of exposing sets of 10 individuals from this same population to a series of doses for a set time. Surviving fish (white) had IED values greater than the exposure dose, and dead fish (black) had IED values less than or equal to the exposure dose. The result is a typical, sigmoidal dose-response curve.

IED values thought to be typical of populations. A sigmoidal dose-response curve would be produced if seven random samples of ten fish, each from this same population, were given doses corresponding to the six IED groupings in the top panel and a control dose. This presumptive log-normal curve is the basis for the probit method, the most common approach to analyzing dose-response data.

Although the IED concept is presented almost exclusively as the foundation of dose-response models, other concepts are invoked to support various methods of analysis. The log-logistic model has been suggested for years as an alternative to the log-normal model, leading to a protracted controversy about the relative values of logit versus probit methods. The log-logistic model also predicts a sigmoidal curve like that shown in Figure 9.1. The foundation for the logistic model is its linkage to processes such as enzyme kinetics, autocatalysis, and adsorption phenomena (Berkson, 1951).

Berkson (1951) questioned the IED concept and advocated the use of the log-logistic model instead of the log-normal model. He based his argument on an experimental screening of combat pilots for tolerance to high-altitude conditions. Candidate aviators were placed into decompression chambers and their individual tolerances measured. Assuming the IED concept was correct, those failing to exhibit symptoms of the "bends" to a certain critical pressure passed the test, and low-tolerance individuals were rejected from further consideration as pilots. Berkson broke from the standard test protocol and asked that a set of candidates be retested to see if individual responses remained the same between tests. The results showed poor agreement between repeated tests: The IED concept had failed to explain this dose-response phenomenon.

As a consequence, a counterargument to the IED concept has been made based on the idea that individuals do not have unique tolerances. The argument is made that the probability of death for any individual is a consequence of a random process or set of processes that may conform to a log-normal, log-logistic, or another model. Berkson (1951) and Finney (1971) argued that some processes, such as those described in Chapter 7 for cancer risk, are based on probabilities that a specific sequence of events will occur and lead to death or cancer. The distribution of differences in occurrence of the event (i.e., appearance of a clinical cancer or death) is not related to differences among unique individuals. Rather, the distribution is a consequence of probabilities associated with events taking place in all individuals. In this model, which particular individual responds quickly ("sensitive") or slowly ("tolerant") is a matter of chance alone. Individual response is a consequence of random events that are described by probability distributions. Certainly, the observation that the log-normal model seems to work for tests with microbes and zooplankton composed of cloned individuals casts some doubt on the IED concept as the sole underlying explanation for the log-normal model. Gaddum (1953) suggested that a random process in which several "hits" are required to produce death could also form the basis for the log-normal model. Remarkably, whether one or both of these concepts is the basis for the majority of dose-response relationships remains poorly tested. Only recently, Newman and McCloskey (2000) formally tested these concepts and found that neither was universally valid.

Although this point may appear trivial, the consequences of repeated exposure of a population are different under these two concepts (Newman, 2001). This can be illustrated with a thought problem in which all covariates affecting lethal impact such as animal sex, size, and age are identical for all individuals. If the IED concept were correct, survivors of a first exposure would be the most tolerant, and the impact on the population of survivors would be less in a subsequent exposure. In the second case, where probability of death is the same for all individuals, the survivors will not be inherently more tolerant, and the impact of a second exposure would be as large as that of the first.

A Weibull model also provides good fit for the dose-response curve but is seldom used (Christensen, 1984; Christensen and Nyholm, 1984; Newman, 1995; Newman and Dixon, 1996). The Weibull model can describe a multistage process such as that discussed in Chapter 7 for carcinogenesis. When applied, it seems to fit dose-response data as well as the generally accepted log-normal and log-logistic models.

Many dose-response relationships for lethality have threshold concentrations below which no discernible increase in mortality occurs. Most of the models described here can and should be modified to include lethal thresholds if required. Further, with chronic exposures, natural or spontaneous mortality may be occurring simultaneously with toxicant-induced mortality. Such spontaneous mortality can also be included in dose-response models if necessary.

II.B Fitting Data to Dose-Response Models

Methods have been developed to analyze dose-response data based on the concepts and models just discussed. Data (proportions, doses, or concentrations) might be used directly or after transformation. Often, the objective of transformations is to make linear the relationship between dose and response. Measurements or transformations of measurements used in the analysis of biological tests are termed **metameters**. Both dose or concentration metameters and effect metameters can be used for dose-response data. The most common dose or concentration metameter is the logarithm of dose or concentration. Which effect metameter is appropriate depends on whether the log-normal, log-logistic, or another model is assumed. For example, the log-normal model is assumed if the log-dose or log-concentration transformation is paired with the probit metameter of the proportion dying. The log-logistic model is assumed if the log dose or concentration is paired with the logit metameter of the proportion dying.

The probit transformation is derived from the **normal equivalent deviation** (NED), the proportion dying expressed in units of standard deviations from the mean of a normal curve. For example, a proportion corresponding with the mean (50% of exposed individuals are dead) would have an NED of zero; a proportion below the mean by one standard deviation (16% of exposed individuals are dead) would have an NED of –1. In introducing the probit method, Bliss (1935) viewed negative NED values as inconvenient and added five to NED values to avoid negative numbers.[4] The resulting metameter is the **probit**. Probit analysis (log-normal model) is performed using the log dose or log concentration versus probit of the proportion dead:

$$\text{Probit}(P) = \text{NED}(P) + 5 \qquad (9.1)$$

where P = proportion of exposed individuals that died by the end of the exposure, and NED = the normal equivalent deviation.

The **logit** or "log odds" metameter is based on the log-logistic model and has the form:

$$\text{Logit}(P) = \ln\left[\frac{P}{1-P}\right] \qquad (9.2)$$

A transformed logit is more commonly employed than that calculated by Equation 9.2 because values of this transformed logit are nearly the same as probit values, except for proportions at the extreme ends of the curves.

$$\text{Transformed Logit} = \left[\frac{\text{Logit}(P)}{2}\right] + 5 \qquad (9.3)$$

where logit(P) = logit value estimated by Equation 9.2.

Other effect metameters are used much less commonly, as discussed in Newman (1995). The Weibull transformation is one metameter that could be used more often (Christensen, 1984; Christensen and Nyholm, 1984). Equation 9.4 gives the form of the **Weibull metameter**:

$$U(P) = \ln(-\ln(1-P)) \qquad (9.4)$$

All of these metameters will generate a straight line for appropriate dose-response data (Figure 9.2). Except at the tails of these lines, the probit and logit metameters have nearly identical values for any set of data.

The **median lethal dose** (LD50) and its associated confidence limits are the most common statistics derived from dose-response models. The LD50 is the dose resulting in death of 50% of exposed individuals by a predetermined time, e.g., 96 h. Similarly, a **median lethal concentration** (LC50)[5] and its confidence limits are commonly derived from concentration-effect data. It is the concentration resulting in death of 50% of exposed individuals by a predetermined time. Median

---

[4] The need to avoid negative numbers can and has been questioned. Analysis with the NED instead of the probit produces the same results. The NED used in this way is often called the **normit** metameter, and the associated method is normit analysis.

[5] For sublethal or ambiguously lethal effects, the term **median effective concentration** (EC50) can be used instead of LC50. It is often applied to species such as invertebrates for which death is difficult to score and events such as cessation of ventilation or general movement are scored instead. It is also the term used if sublethal events are being analyzed.

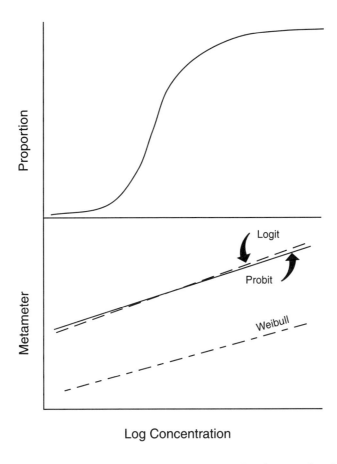

Figure 9.2 The typical, sigmoidal curve for concentration-response data (top panel) and lines resulting from the probit, logit, and Weibull transformations (bottom panel).

values were adopted as benchmarks instead of low values (e.g., 1% or 5%) because medians tend to be more consistent and to have narrower confidence intervals than lower percentiles. Median values also have an advantage because models such as the probit and logit produce very similar results toward the median. Although these advantages are quite real, it is important to keep in mind that the median was not chosen because it has any particular biological significance. In many cases, concentrations killing lower proportions would be much more meaningful for the ecotoxicologist attempting to determine risk upon toxicant release to the environment.

Many ecotoxicologists estimate values other than the median. Such LCX statistics (e.g., 96 h LC5 = the concentration killing 5% of exposed individuals after 96 h) are more helpful in assessing adverse effects, but they have generally wider confidence intervals, and results are more model dependent. This last point is often forgotten because of the misconception, based on our preoccupation on the median, that all models give similar statistics. Model independence is true for practical purposes at the median, but it becomes progressively less so toward the tails of the model. Careful model selection (e.g., probit versus logit analysis) becomes important as attention moves away from the LC50. Methods of comparing candidate models will be discussed toward the end of this chapter section.

Numerous methods for estimating the LC50 are available (Table 9.1). Indeed, interpolation from a simple line produced with different sets of metameters, such as the probit of P vs. log of concentration (Figure 9.3), could be used to graphically estimate the LC50. However, the 0% and 100% mortality treatments would not easily be plotted on such a graph, and visual fitting would

Table 9.1   Established Methods for Estimation of LC50

Method	Advantages	Disadvantages	References
Litchfield-Wilcoxon	Quick, semigraphical method	Results are dependent on the individual who is fitting the data "by eye"	Litchfield and Wilcoxon (1949); Stephan (1977); Peltier and Weber (1985); Newman (1995)
Maximum likelihood estimation (MLE) or χ^2 fitting of specific model	Powerful, parametric method that can use any of a series of possible models	Requires a specific model such as the log-normal or log-logistic models; iterative method that is tedious to do manually; MLE results are slightly biased; MLE methods may not converge properly	Armitage and Allen (1950); Berkson (1955); Stephan (1977); Peltier and Weber (1985); Newman (1995)
Spearman-Karber	A robust, nonparametric method not requiring a specific model; can trim data to minimize the undue influence of extreme values	Toxicity curve must be symmetrical; not as powerful as parametric methods	Hamilton et al. (1977); Stephan (1977); Peltier and Weber (1985); Newman (1995)
Binomial	Can be used for data with no partial kills	Estimation of confidence interval ignores sampling error (Newman, 1995)	Stephan (1977); Peltier and Weber (1985); Newman (1995)
Moving average	Easily implemented	Simple equations for this method require specific progression of toxicant concentrations and numbers of replicates	Stephan (1977); Peltier and Weber (1985)

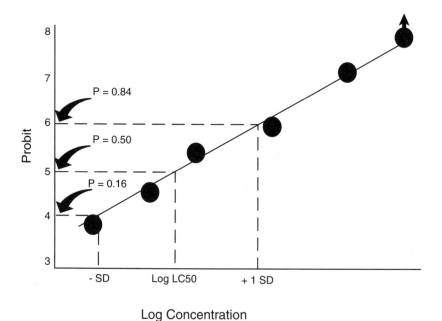

Figure 9.3   Based on a log-normal model, probit methods can be used to generate a line from concentration-effect data. An estimate of the log LC50 ± one standard deviation can be grossly approximated from such a graph. Note that a point with 100% mortality is often indicated by an arrow attached to the data point. This indicates that 100% mortality would likely have occurred prior to the time endpoint, and some lower concentration likely would have produced 100% at exactly that time endpoint. Other, more-rigorous methods discussed in the text provide better estimates.

be subjective: For these reasons, more-formal methods are applied for estimation of the LC50, its 95% confidence interval, and the slope of the concentration-effect line. These last two parameters are as important as the LC50 itself. The LC50 has little utility without some measure of confidence in its calculated value. Also, the slope of the line provides valuable information, as can be illustrated easily with Figure 9.3. Imagine a second line intersecting the drawn line at the LC50, but give this second line a much steeper slope. Although the LC50 would be the same for both lines, a small change in concentration has much more of an effect with one toxicant (steep slope) than the other (shallow slope). This is an important piece of information.

The easiest, but most subjective, method for estimation of an LC50 is the **Litchfield-Wilcoxon method**. It is a semigraphical method in which data are graphed first as in Figure 9.3, and the points on the line corresponding to proportions 16%, 50%, and 84% mortality are used to estimate the LC50 and its 95% confidence interval. The antilogarithm of the log concentration corresponding with the p = 0.50 is taken as the LC50. A slope factor (S) is calculated from the concentrations corresponding with 16% (LC16), 50% (LC50), and 84% (LC84) mortalities:

$$S = \frac{\dfrac{LC84}{LC50} + \dfrac{LC50}{LC16}}{2} \tag{9.5}$$

This S and the total number of animals exposed in treatments within the range of LC16 and LC84 (N′) are used to generate the upper and lower 95% confidence limits:

$$f_{LC50} = S^{\frac{2.77}{\sqrt{N'}}} \tag{9.6}$$

$$\text{Upper Limit} = LC50 \cdot f_{LC50} \tag{9.7}$$

$$\text{Lower Limit} = LC50 / f_{LC50} \tag{9.8}$$

As easy as this method is, it is not a consistent tool because visual fitting of the line will vary among individuals.

Maximum likelihood estimation (MLE) for the log-normal, log-logistic, or other models is a parametric method that avoids subjectivity but takes on the assumption of a specific model. Probit, logit, and other approaches are most often applied with MLE methods if there were two or more **partial kills** (treatments in which some, but not all, exposed individuals are killed). Maximum likelihood estimation results have very good precision relative to those of other methods. The MLE estimates can be slightly biased for small sample sizes; however, this bias is often within the range of that of the other methods. The MLE method is an iterative process and, for this reason, it is most often done with a computer. Data are fit to the model, a goodness-of-fit statistic is calculated, and then the estimated parameters are changed slightly and the process is repeated. This process is repeated until the goodness-of-fit statistic indicates that the maximum likelihood method has found ("converged on") a set of parameter estimates that best fit the data to the model. Occasionally, the MLE method will fail to converge after many iterations or will converge on an inappropriate ("local") solution.

Because various models can be fit to data, questions arise about goodness-of-fit among candidate models. The χ^2 values estimated for each model fit to the data can be used for this purpose. Because the χ^2 value will decrease as fit improves, the ratio of χ^2 values for the different models will reflect relative goodness-of-fit for each model to the data. If the χ^2 **ratio** is less than 1, the model whose χ^2 value is in the numerator (e.g., χ^2_{probit} for $\chi^2_{probit}/\chi^2_{logit}$) fits the data better than the model whose χ^2 value is in the denominator. More formal tests can be carried out with the χ^2 ratio to determine if the fit for one model is significantly better than that of the other.

If it is difficult or unnecessary to assume a specific model for the dose- or concentration-effect data, the nonparametric **Spearman-Karber method** is available to estimate the LC50. The technique only requires a symmetrical toxicity curve. The technique has many steps but can be applied with a hand calculator if necessary. During application of this method, values at the extreme tails may or may not be trimmed to minimize undue influence of extreme values on the estimate. Trimming rules are provided by Hamilton et al. (1977). If trimming is done, the influence of anomalous values will be reduced; however, the standard error of the estimate will increase.

Two other methods are commonly discussed in ecotoxicology. The **binomial method** allows estimation of the LC50 if there were no partial kills, although sampling error is ignored with this method (Newman, 1995). The **moving average method** can be implemented with straightforward equations if the toxicant concentrations are set in a geometric series and there are equal numbers of individuals exposed in each treatment.

Occasionally, a model cannot be unambiguously fit to data for chronic toxicity tests. An ANOVA design as described previously can then be applied. The lowest concentration at which mortality is significantly higher than the control is determined via hypothesis testing.

II.C Incipiency

Incipiency, when applied to lethality of contaminants, is the lowest concentration (or dose) at which an increase in toxicant concentration (or dose) begins to produce an increase in the measured effect. It is often measured with the **incipient median lethal concentration** (incipient LC50), the concentration below which 50% of exposed individuals will live indefinitely relative to the lethal effects of the toxicant. This concentration is also called the asymptotic, ultimate, or threshold LC50 by various authors. It can be determined graphically in several ways. Figure 9.4 shows one approach taken by van den Heuvel et al. (1991) for rainbow trout (*Oncorhynchus mykiss*) exposed to pentachlorophenol. The LC50 values were determined for a series of exposure times, and the reciprocals of the LC50 values were plotted against duration of exposure. Alternatively, a double logarithm plot can be used (Newman, 1995). The concentration at which the curve becomes parallel to the

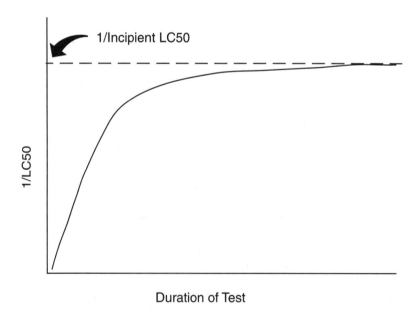

Duration of Test

Figure 9.4   The incipient lethal concentration is estimated as the point at which the curve of 1/LC50 vs. duration of exposure begins to run parallel to the *x*-axis.

x-axis is an estimate of the incipient LC50. Although this method is widely used and is convenient to estimate incipiency, it is difficult to assign ecotoxicological significance to this incipiency measure. Recall that the median value has ambiguous meaning relative to the continued viability of an exposed population. It follows that any measure of incipiency based on the median also carries the same ambiguity relative to ecological significance. Statistical limitations also exist for this graphical approach (Chew and Hamilton, 1985). For instance, concentration is set in the design yet treated as an independent variable in subsequent analysis. Applying regression models under such conditions is statistically inappropriate.

II.D Mixture Models

Many contaminants of concern are introduced to the environment as mixtures (see Chapter 2), and estimation of lethal effects is more difficult than has been described to this point for exposure to single toxicants. Regardless of the difficulty, it is important to understand joint effects of chemicals in mixture. Such understanding begins by distinguishing among four different situations that can arise with mixtures: potentiation, additivity, synergism, and antagonism.

Potentiation might occur if one chemical, not toxic itself at the exposure concentration or dose, enhances the toxicity of a second chemical in a mixture.[6] For example, sublethal concentrations of isopropanol greatly enhance the toxic effects of carbon tetrachloride to the mammalian liver (Klaassen et al., 1987). Disulfiram (also called tetraethylthiuram disulfide and Antabuse) greatly enhances ethanol toxicity in humans (Timbrell, 2000). The fact that this potentiation occurs at nontoxic doses of disulfiram has led to its use as a medication to reinforce drink abstinence by alcoholics. After consumption, ethanol is converted by alcohol dehydrogenase to acetaldehyde, which is then converted to acetate by aldehyde dehydrogenase. Disulfiram inhibits acetaldehyde breakdown, producing an acetaldehyde buildup. The resulting acetaldehyde syndrome includes headache, nausea, vomiting, and dizziness. Potentiation is also used to our advantage to improve the effectiveness of some insecticide formulations. Piperonyl butoxide, added to insecticide formulations, potentiates pesticide action by inhibiting pesticide breakdown by the cytochrome P-450 system. Thus less pesticide need be released to the environment during application.

Joint effects of chemical mixtures can also result in effect additivity, synergism, or antagonism. Simply put, **additivity** exists if the measured mixture effect was simply the sum of the expected effects for the individual toxicants. **Synergism** occurs if the observed effect level of the mixture was higher than the sum of the predicted effects for the individual toxicants in the mixture. Toxicant **antagonism** occurs if the observed effect level of the mixture was lower than that predicted by summing the predicted effects for the individual toxicants in the mixture.

Combined effects of chemicals have often been illustrated using the simplified concept of **toxic units** (TU), amounts or concentrations of different toxicants expressed in units of lethality such as units of LD50 or LC50. At the onset of its use, the toxic unit was most often expressed as a fraction of the incipient median lethal concentration. For example, if toxic units were based on the incipient LC50, chemical A with an incipient LC50 of 20 mg l^{-1} would be present at 0.5 TU in a 10 mg l^{-1} solution. Similarly, chemical B (LC50 = 100 mg l^{-1}) would be present as 0.5 TU in a 50 mg l^{-1} solution. If the toxicity of two toxicants in combination were (concentration) additive, the simple sum of the toxic units of the two toxicants would equal the actual toxicity measured for the mixture, e.g., 0.5 TU of A + 0.5 TU of B should equal 1.0 TU of effect when combined as a mixture. In the above example, a mixture of 10 mg l^{-1} of A plus 50 mg l^{-1} of B should result in 50% of the exposed individuals dying. If the mixture actually results in less than 1 TU of lethality, chemicals A and B are said to be less than additive or antagonistic. They are synergistic if their combined effect is more than additive. This concept can also be used to illustrate the **isobole approach** to mixture analysis (Figure 9.5). (See Chen and Pounds (1998) for further discussion.) In Figure 9.5,

---

[6] The term, *potentiation*, is also used by some authors (e.g., Thompson, 1996) in the context of synergism, as discussed below.

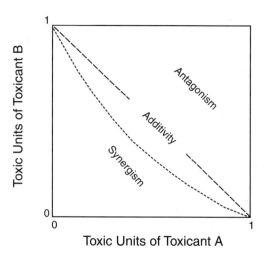

Figure 9.5   An illustration of the isobole approach. The line connecting 1.0 on both axes reflects additivity and deviation from additivity if a point lies to the lower left (synergism) or upper right (antagonism) of the line. The dashed line represents a hypothetical hyperbole for two toxicants A and B displaying synergistic joint action.

the straight line connecting 1.0 on both axes reflects additivity and deviation from additivity if a point lies to the lower left (synergism) or upper right (antagonism) of the figure. The dashed line represents a hypothetical hyperbole for two toxicants displaying synergistic joint action. In isobole analysis, the observed line produced for different proportions of the toxicants is compared with the theoretical line for strict additivity.

Antagonism can be broken down based on the underlying mechanism (Klaassen et al., 1987). **Functional antagonism** results from two chemicals eliciting opposite physiological effects and, as a consequence, counterbalancing each other. With **chemical antagonism**, two toxicants react with one another to produce a less toxic product. For example, cyanide and a toxic metal may combine in mixture to form a less toxic complex. **Dispositional antagonism** involves the uptake, movement within the organism, deposition at specific sites, and elimination of the toxicants. The presence of the two toxicants together shifts one or more of these processes to lower the impact of the toxicants on the site(s) of action or target organ(s). This might involve lowered chances of interaction with a target, e.g., less time available to interact or lowered concentration available to interact. For example, ethanol enhances mercury elimination in mammals (Hursh et al., 1980; Khayat and Shaikh, 1982) and could modify the toxic effect of mercury as a consequence. The last type, **receptor antagonism**, occurs where two or more toxicants bind to the same receptor, and each toxicant blocks the other from fully expressing its toxicity. Klaassen et al. (1987) give the example of using O_2 to counter the effects of carbon monoxide poisoning based on receptor antagonism. Newman and McCloskey (1996a) found evidence of receptor antagonism for binary mixtures of metals on the response of the **Microtox**[®] **assay**, a rapid, bacterial assay in which a decrease in bioluminescence is thought to reflect toxic action.

The toxic unit approach just described is based on the concept of **concentration additivity**: After adjustment for relative potency, concentrations of toxicants can be added together to predict effects under the assumption of additivity. However, as we saw earlier in this chapter, toxicant concentration (or dose) is often related to lethal effect by a sigmoid, not a linear, relationship. Some relationships are pseudolinear over a range of concentrations of interest, and concentration additivity can be used to approximate effects. But summation of potency-adjusted concentrations or doses to predict combined effect is not always valid. (The interested reader is directed to Berenbaum (1985) for a mathematically explicit explanation of this point that the **summation rule** based on concentration or dose is not generally valid.) Instead, to explore additivity, effect levels predicted with models such as the probit or logistic model for each toxicant in a mixture should be added together, not the toxicant concentrations in the mixture.[7] For example, the effect predicted for a certain exposure concentration of toxicant A is predicted from a probit model and added to the effect predicted from another probit model for a certain concentration of toxicant B. The sum of the predicted effects could then be compared with the observed effect of the actual combination of the two concentrations of A and B in order to assess conformity to or deviations from effect additivity. For example, the separate effects of two toxicants on the proportion of exposed individuals dying (P_A and P_B) are estimated with the following equations:

$$\text{Probit}(P_A) = \text{Intercept}_A + \text{Slope}_A(\log \text{Concentration}_A) \qquad (9.9)$$

$$\text{Probit}(P_B) = \text{Intercept}_B + \text{Slope}_B(\log \text{Concentration}_B) \qquad (9.10)$$

To progress from this approach to a more involved treatment of mixtures, the distinction must be made between toxicants that act independently or similarly. Toxicants can have **similar joint action**, in which case they act by the same mechanism (i.e., have identical modes of action), and "one component can be substituted at a constant proportion for the other. … [T]oxicity of a mixture is predictable directly from that of the constituents if their relative proportions are known" (Finney, 1947). **Independent joint action** of toxicants exists if each toxicant produces an effect independent of the other and by a different mode of action (Finney, 1947). Relative to quantifying joint action of such chemicals, Finney (1947) makes an important distinction, "In mixtures whose constituents act similarly any quantity of one constituent can be replaced by proportionate amount of any other without disturbing the potency, but for mixtures whose constituents act independently the mortalities, not the doses, are additive." This distinction is important to keep in mind in the following discussions.

A group of individuals is exposed to a mixture of Concentration$_A$ and Concentration$_B$, and the proportion dying (P_{A+B}) measured. The three effect proportions (P_A, P_B, and P_{A+B}) can then be used to estimate any deviation from predictions of simple independent joint action. The predicted effect level for the mixture would be the following if the mixed toxicants A and B acted independently in producing the effect (e.g., death) (Finney, 1947):

$$\text{Predicted } P_{A+B} = P_A + P_B(1 - P_A) \qquad (9.11)$$

This can be, and often is, rearranged to the following (e.g., Berenbaum, 1985):

$$\text{Predicted } P_{A+B} = P_A + P_B - P_A P_B \qquad (9.12)$$

Using Equation 9.11 or 9.12, the predicted P_{A+B} can be compared with the observed proportion dying after exposure to the mixture of A and B in this case of independent action of the toxicants in the mixture. They should be equivalent if the toxicants display simple independent joint action.

Likely, it is not obvious to the reader why the right side of the equation is not simply $P_A + P_B$. The reason that the right side of Equation 9.11 contains the $(1 - P_A)$ term can be understood if one envisions the proportion responding as being the probability of an exposed individual dying. If the independently acting toxicants A and B are mixed at the specified concentrations, the probability of dying from A is estimated as P_A. However, an individual must have survived the effects of A in order be available to die from B. The probability of surviving the exposure to A is simply $1 - P_A$,

---

[7] The details provided here are only the most rudimentary for understanding this complex and important topic of joint mixture effects. For a fuller understanding, the reader is urged to review Finney (1947), Plackett and Hewlett (1952), Berenbaum (1985), Chen and Pounds (1998), Eide and Johnsen (1998), Groen et al. (1998), and Mumtaz et al. (1998).

so the P_B must be multiplied by $1 - P_A$ instead of the implied 1 to accommodate for this fact. This relationship can also be rearranged to the following:

$$P_{A+B} = 1 - (1 - P_A)(1 - P_B) \tag{9.13}$$

Finney (1947) expands this equation to calculate mixture effects for more than two toxicants with independent action to the following:

$$P_{A+B+C+\cdots} = 1 - (1 - P_A)(1 - P_B)(1 - P_C)\cdots \tag{9.14}$$

A slightly different approach is used for toxicants that have similar action. Toxicants with similar action often show parallel slopes in their probit models, so a simple measure of relative potency can be calculated to predict the effect of two similarly acting toxicants in mixture (Finney, 1947). Using Finney's equations converted to the form shown above:

$$\text{Probit}(P_A) = \text{Intercept}_A + \text{Slope}(\log \text{Concentration}_A) \tag{9.15}$$

$$\text{Probit}(P_B) = \text{Intercept}_B + \text{Slope}(\log \text{Concentration}_B) \tag{9.16}$$

The log of the relative potency can be calculated as the following:

$$\log \rho_B = \frac{(\text{Intercept}_B - \text{Intercept}_A)}{\text{Slope}} \tag{9.17}$$

And the predicted effect for the mixture of A and B can be calculated with Equation 9.18:

$$\text{Probit}(P_A + P_B) = \text{Intercept}_A + \text{Slope}(\log \text{Concentration}_A + \rho_B(\log \text{Concentration}_B)) \tag{9.18}$$

Other approaches can be used. As a more involved example, Carter et al. (1988) applied the regression methods for the logistic isobologram model:

$$\log\left[\frac{P}{1-P}\right] = \beta_0 + \beta_A C_A + \beta_B C_B + \beta_{AB} C_A C_B \tag{9.19}$$

where P = proportion of exposed individuals dying, C_A = concentration metameter for toxicant A, C_B = concentration metameter for toxicant B, β_0 = the y-intercept, β_A = regression coefficient for the effect of toxicant A, β_B = regression coefficient for the effect of toxicant B, and β_{AB} = regression coefficient for the interaction term, $C_A C_B$. Deviations from additivity would be suggested by the β_{AB} coefficient. The equivalent probit model would be the following:

$$\text{Probit}(P) = \beta_0 + \beta_A C_A + \beta_B C_B + \beta_{AB} C_A C_B \tag{9.20}$$

Perhaps the best example of coping with mixtures of similarly acting toxicants is the approach taken for mixtures of compounds having aryl hydrocarbon (AH) receptor interaction as the first stage of their toxic action. Compounds with very similar modes of action of this kind include dioxins, dibenzofurans, and dioxin-like polychlorinated biphenyls (PCB). These chemicals are often present in mixtures, and assessment of their net effect must be made. Based on empirical evidence gathered in the laboratory, **toxic equivalency factors** (TEF) are calculated for each such compound in the mixture, and the concentrations of individual chemicals in the mixture are added together after multiplication of each by its toxic equivalency factor. One of the most toxic AH binding compounds 2,3,7,8-tetrachlorodibenzo-p-dioxin (abbreviated TCDD) is given a value of 1, and other similarly acting compounds are given experimentally derived TEF scaled to the TCDD TEF of 1. For example, a moderately active PCB might be given a TEF of 0.001. The **toxic equivalence** (TEQ) of this PCB would be its concentration multiplied by its TEF, i.e., its effect scaled to that of TCDD. Assuming that these compounds are similarly acting, the joint effect of a mixture can be expressed as the sum of the TEQs for all of the constituent dioxins, dibenzofurans, and dioxin-like PCBs (Van den Berg et al., 1998).

Let us return for a moment to assessments of mixture additivity. Many mixture studies in ecotoxicology use the simple additive index approach to determine whether additive or nonadditive action is occurring. Marking and Dawson (1975) generated an **additive index** for assessing the joint action of toxicants in mixtures. Letting A_m and B_m = the toxicity (e.g., incipient LC50) of toxicants A and B when present in mixture, and A_i and B_i = toxicity of A and B when they are tested separately, Equation 9.21 can be used to assess the mixture interaction:

$$\frac{A_m}{A_i} + \frac{A_m}{B_i} = S \qquad (9.21)$$

The toxicants are antagonistic if $S < 1$, synergistic if $S > 1$, or additive if $S = 1$ (Figure 9.6, "sum of toxic contributions" scale at the top of the figure). Although shown here for a binary mixture, several toxicants can be added to Equation 9.21 if desired. Unfortunately, the units are not linear to the right and left sides of 1 on this scale. A change from +1 to +2 is not of the same magnitude as a change of −1 to −2. Units can be made linear by making values to the left of additivity equal to −S + 1 and values to the right of additivity equal to 1/S − 1, as seen in Equations 9.22 and 9.23, respectively:

$$\text{Additive Index (AI)} = -S + 1 \quad \text{for } S > 1.0 \qquad (9.22)$$

$$\text{Additive Index (AI)} = \frac{1}{S} - 1 \quad \text{for } S \le 1.0 \qquad (9.23)$$

This additive index (AI) is linear on both sides of additivity (zero) (Figure 9.6, "corrected sum of toxic contributions" at the bottom of the figure). Negative and positive numbers indicate less than additivity and greater than additivity, respectively.

In common ecotoxicology practice, the toxic unit and additive index are the most widely used tools for quantifying joint toxicant effects. Although the mathematical models provided above for independent- and similarly acting toxicants are more inclusive, Marking's additive index (Marking and Dawson, 1975; Marking, 1985) is a straightforward means of visualizing combined effects of mixtures. The concept of concentration additivity also forms the foundation for estimation of combined effects of contaminants in solutions or solids. For example, Di Toro et al. (1990) assumed additivity in estimating the combined effects of metals in contaminated sediments and performed mathematical modeling accordingly. The current, extensive use of the toxic unit and additive index is a consequence of their ease of application and historical role in ecotoxicology, not their power to quantify and test for significant deviations from additivity. More thorough analysis of mixtures can be done with the general linear-model approach including interaction terms (e.g., Equations 9.11–9.14 and 9.18–9.20; see Neter et al. (1990) for details of regression methods). There are now many convenient software packages for personal computers that implement such general linear-modeling procedures, which should promote more widespread implementation of these methods.

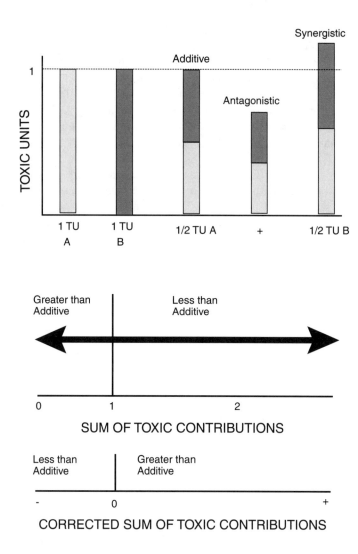

Figure 9.6   The combined effect of toxicants can be quantified by expressing toxicant concentrations in mixtures as toxic units (top panel). If the realized effect expressed in terms of TU is less than the calculated sum of TU for both toxicants A and B in a mixture, the chemicals are said to be antagonistic. Their effect is synergistic if the realized effect is greater than the calculated effect based on their individual actions. The two scales include the nonlinear scaling of Marking and Dawson (1975) (middle) and an additive index (bottom), which is a linear scaling of combined toxicant effect.

Vignette 9.1   The Use of Lethal Effects Data in the Control of Toxicant Discharge

Daniel Fisher
University of Maryland, Queenstown, MD

Acute and chronic lethality tests have formed the backbone of chemical regulation for the past few decades. As described in this chapter these tests have the benefit of being relatively inexpensive and easy to conduct. Data can thus be obtained quickly. Because of this, a large database exists on the acute and chronic toxicity of many chemicals. This vignette presents a few examples on how these tests are used in the regulatory arena today.

One major use of these tests is the development of Water Quality Criteria for specific chemicals of concern. The Federal Water Pollution Control Act Amendments (PL 92-500), known as the Clean Water

Act (CWA), were passed in 1972. The objective of the CWA was to restore and maintain the chemical, physical, and biological integrity of the nation's surface waters. The national goal was to eliminate the discharge of pollutants to navigable waters by 1984. Initially, in carrying out the mandates of the CWA, the U.S. EPA focused on control of conventional pollutants such as BOD and suspended solids. The failure of the U.S. EPA and states to implement toxic substance control provisions in the early 1970s prompted a series of lawsuits by the Natural Resources Defense Council (NRDC). The landmark lawsuit, resolved in a 1976 consent decree, required the U.S. EPA to develop water quality criteria (WQC) for 129 specific toxic substances (priority pollutants). The WQC are a set of numeric criteria that set the maximum level of discharge that should protect against acute and chronic effects. To date, WQC for approximately 200 toxic substances have been issued.

The methods and rationale for deriving these criteria are presented in the U.S. EPA "Guidelines for Deriving Numerical National Water Quality Criteria for the Protection of Aquatic Organisms and their Uses" (Stephan et al., 1985) and in supporting documentation by Erickson and Stephan (1988). For protection from acute toxicity, these guidelines require the calculation of a final acute value (FAV) based on acceptable simple laboratory acute lethality toxicity data from at least eight different taxonomic families. The method for calculating the FAV puts more emphasis on the toxicity values for the four most sensitive species and is designed, by extrapolation, to protect 95% of the species in an ecosystem. The guidelines provide specific requirements for determining the families that must be included in the data set. The acute criterion, called the *criterion maximum concentration* (CMC), is set at one-half the FAV. The criterion would then read, "except possibly where a locally important species is very sensitive, aquatic organisms and their uses should not be affected unacceptably if the 1-h average concentration does not exceed the CMC more than once every three years." The guidelines also outline procedures for protecting against chronic toxicity. Besides lethality data, other more sensitive endpoints such as growth and reproduction are used in this calculation.

Although this WQC approach has worked well at setting criteria for approximately 200 priority chemicals, it has several drawbacks if the final goal is to protect the environment from toxic chemicals. What about the tens of thousands of chemicals discharged each year that do not have criteria? What about the interaction of chemicals with WQC in a waste stream? These interactions can cause unpredictable changes in the toxicity of a waste stream and thus possible damage to the environment. These concerns led to the establishment of the whole effluent toxicity (WET) testing program within the National Pollutant Discharge Elimination System (NPDES) permit program. The CWA prohibits the discharge of any wastewater to surface waters without an NPDES permit. Thus, it is the administrative mechanism for regulating the discharge of wastewater from point sources. The CWA requires the elimination of discharge of "toxic substances in toxic amounts," regardless of whether there are existing criteria for all of the chemicals in a waste stream.

The WET testing program is the cornerstone of the effort to eliminate the toxicity of discharges from point sources in the United States. The program involves the use of simple laboratory acute and chronic tests on whole effluents collected before their discharge to receiving waters. Representative freshwater and estuarine organisms are used nationwide to allow for ease of culture and comparison of results. These tests range from short-term 48- to 96-h static-renewal acute lethality tests to 7-day short-term chronic tests measuring effluent effect on survival in addition to more-sensitive sublethal effects to growth and reproduction. The species most often used in these tests are the early live stages of the fathead minnow (*Pimephales promelas*) and sheepshead minnow (*Cyprinodon variegatus*) and the crustaceans, *Daphnia magna*, *Ceriodaphnia dubia*, and *Americamysis bahia* (formerly *Mysidopsis bahia*). The crustaceans allow for the measurement of a sensitive reproductive effect. In fact, the 7-day *C. dubia* test represents a full life-cycle test. If one examines the sensitivity of various species to chemicals that have published WQC, *C. dubia* is consistently one of the most sensitive species tested. These WET tests have been used extensively in the United States since the 1980s for regulation of point-source discharges.

How well has this program worked? Fisher et al. (1998) found a significant decrease in the incidence and severity of acute toxicity over an 8-year period for effluents in the State of Maryland. Of the 59 facilities in the state required to decrease their WET toxicity during that time, 84.9% (200 million gallons per day or Mgd of discharge) eliminated acute toxicity. Their paper shows that a long-term point-source biomonitoring program with emphasis on simple acute and chronic toxicity tests can lead to substantial reductions in the acute and chronic toxicity of effluents discharged to surface waters.

These simple tests have also been used on a case-specific basis to identify and correct environmental problems. Fisher et al. (1995) used simple 48-h static-renewal bioassays to characterize the toxicity of storm water runoff from an international airport during deicing storm events and at other times of the year. Each storm event was tested with two species, *Daphnia magna* and *Pimephales promelas*. Runoff samples were collected from the airport during the events with a composite sampler. The samples were returned to the laboratory, where the tests were conducted. At the time of sampling, all runoff from the airport runways and deicing/anti-icing areas flowed directly through concrete pipes to nearby streams. The major components of the deicing/anti-icing fluids were propylene glycol and ethylene glycol. The stormwater runoff from this airport was regulated through the NPDES permit system and was thus subject to effluent toxicity limitations. Samples from winter storm events caused acute toxicity to both species, with LC50 values as low as 1.0 to 2.0% effluent. Samples from the summer storm events when deicing/anti-icing was not occurring were not toxic, suggesting that the toxicity during the winter events was related to the deicing/anti-icing activities. High oxygen demands and elevated total nitrogen levels are other potential problems during deicing/anti-icing activities. Because of this study, a new discharge permit was issued for this airport requiring the collection and recycling or disposal of the deicer/anti-icer mixtures. The airport has since constructed contained deicing/anti-icing stations near the ends of the runways. A system of pumps, piping, and storage tanks was constructed for the used fluid. The fluids are slowly fed into the local sewage treatment plant, where the propylene glycol and ethylene glycol are rapidly broken down. Thus, results from these simple acute-toxicity tests hastened the elimination of toxic discharges to local receiving streams.

Acute and chronic lethality tests are also used in testing sediments. Sediments are collected from the field and transported to the laboratory. Laboratory-cultured organisms such as amphipods (*Hyalella azteca, Leptocheirus plumulosus,* and *Ampelisca abdita*) and midges (*Chironomus tentans*) are added to the sediments, and the toxicity is determined in either 10-day short-term acute tests or 28- to 42-day longer-term chronic tests. The longer-term tests measure more-sensitive sublethal endpoints such as growth and reproduction. Sediment toxicity tests are used for a variety of purposes. The U.S. Army Corp of Engineers uses them in its evaluations of dredged material proposed for discharge in waters of the United States (EPA/U.S. Army Corps of Engineers, 1998). Hall et al. (1997) used sediment tests as part of a suite of tests to investigate ambient toxicity in the Chesapeake Bay, United States. McGee et al. (1999) used the acute 10-day *L. plumulosus* test to assess sediment toxicity and its relationship to sediment contamination and *L. plumulosus* population viability in Baltimore Harbor, MD. They found that sediment toxicity was strongly correlated with impaired populations of the amphipod.

Coupling these simple sediment toxicity tests with benthic community analysis and sediment chemistry data allows investigators to detect impaired sediment areas and the possible causes of the impairment. This is known as the sediment quality triad (Chapman et al., 1987). In the future, these sediment tests will be used in the total maximum daily loads (TMDL) process. The CWA requires states to inventory the quality of their surface waters, to identify waters that are not consistently attaining water quality standards (WQS), and the pollutants causing the problems. States then are required to develop TMDL allocations that are translated into source controls for pollutants responsible for nonattainment of WQS. These simple sediment toxicity tests coupled with the rest of the sediment quality triad are being used to identify areas of concern for the TMDL process because sediments act as sinks for contaminants in the surface water. Again, very simple toxicity tests are being used effectively in regulatory procedures.

Lastly, these lethality tests are used quite extensively in the ecological risk assessment procedure (EPA, 1998). Laboratory-derived acute and chronic toxicity data (e.g., LC50 or LC5 values) and actual measured environmental concentrations of the chemical of concern are used to construct graphs showing the distributions of cumulative exposure and effects (Figure 9.7). When these graphs are overlaid, it is quite easy to discern whether these two distributions overlap and to what degree they overlap. The overlap suggests the degree of risk. For example, if there were no overlap between the exposure and effects curves, then one would predict little or no risk (Figure 9.7, Scenario 1). If there were 50% overlap, one would predict a high probability of risk (Figure 9.7, Scenario 2). One can then set the level of acceptable risk and make decisions based on that level. For example, is it permissible to protect 95% of the species 95% of the time? By using large amounts of simple acute and chronic toxicity data and comparing them with actual measured environmental exposure levels, one can get an excellent idea of possible risks for toxicity or ecological effects in the field. Solomon et al. (1996) used this procedure in conducting an extensive ecological risk assessment of atrazine in North American surface waters.

Although the preceding examples do not fully illustrate the ways these acute and chronic toxicity test data are used, they should give a feeling for the importance of these simple test procedures in the regulation

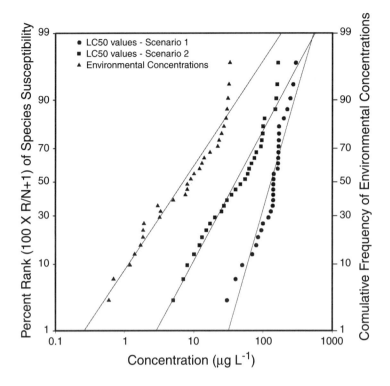

Figure 9.7   Hypothetical species sensitivity distributions vs. environmental concentration.

of toxic compounds. There is some question concerning the extrapolation of findings from these simple laboratory tests to impacts in the field. The general assumption when these tests are used in a regulatory sense is that, if the database is large enough and contains information for a number of families and life stages, the extrapolation from laboratory to field becomes much less troublesome. In addition, some of the above uses, particularly in ecological risk assessment, provide means for calculating levels of uncertainty in the lab to field extrapolations.

Vignette 9.2   Avian Toxicity Testing

Dr. Michael J. Hooper
Texas Tech University, Lubbock, TX

The toxic effects of chemicals in birds were first seriously considered in the 1960s and early 1970s with the discoveries that DDT caused eggshell thinning and reproductive impairment, and that dieldrin, parathion, monocrotophos, and other highly toxic pesticides were causing substantial mortality under standard use conditions. Methods for laboratory studies of chemical toxicity in birds were developed in response. They incorporate assessments of effects from acute and subchronic exposures as well as long-term exposure and the resulting effects on reproduction. These tests and their variants are still performed in the laboratory and the field, in standard test species, and in wildlife of interest. They are now also applied to chemical contaminants as well as pesticides. Findings from standardized assessments, such as single-species LD50, LC50, and reproduction tests, have led to extensive field and semifield studies that incorporate a variety of avian species found on pesticide-treated fields or contaminated waste sites. The results of these tests can help predict the safety of new pesticides and chemicals, drive decision-making on hazardous waste sites, and, once regulations are in place or remediation has occurred, can help evaluate the effectiveness of programs put in place for the protection of avian species.

Basic avian testing includes a suite of standardized toxicity tests performed on established species, providing data similar to those developed in rats for human health assessments. Though methods for these assessments have their roots in pesticide toxicity determinations, harmonization of test methods has led to procedures common for regulation of both pesticides (under the Federal Insecticide, Fungicide and Rodenticide Act) and hazardous chemicals (under the Toxic Substances Control Act). The U.S. EPA Office of Prevention, Pesticides and Toxic Substances (USEPA OPPTS) and the Organization for Economic Cooperation and Development (OECD) are also proceeding toward common methods for countries under these two regulatory bodies. In most cases, recommended test species are an upland game bird such as the northern bobwhite quail (*Colinus virginianus*) in the United States or the Japanese quail (*Coturnix coturnix*) in Europe, and an aquatic bird species, generally the mallard duck (*Anas platyrhynchos*).

Acute Oral Toxicity (OPPTS Method 850.2100)

Mortality following an acute oral chemical exposure is measured as the LD50 and assessed on the basis of milligrams of active ingredient per kilogram of body weight. Following a 14-day acclimation period, test chemical in an appropriate carrier is administered orally by gavage or capsule to birds that have been fasted overnight. The dose range is chosen based on a range-finding test to include at least three doses with partial mortality, preferably with one dose greater and one dose less than 50% mortality. The standard protocol calls for at least five doses and a control group, each with ten individuals (generally five male and five female birds), all at least 16 weeks old. The birds are observed for 14 days or until mortality or until signs of intoxication are not observed for three days. The LD50, the slope of the dose-response curve, and the LD50 95% confidence interval are determined using probit analysis or another appropriate statistical method.

Two other acute oral toxicity methods can be used to decrease the number of animals necessary for the test. The *limit test* can be used if the test compound is thought to have low toxicity. A single dose of 2000 mg per kg of body weight is administered to ten birds, while another ten birds serve as controls. If no mortalities occurred, the LD50 value is reported as "greater than 2000 mg per kg." The *approximate lethal dose*, or ALD, can be determined using an "up-and-down test" that uses sequential dosing of individual birds followed by an observation period of 2 days. The benefit here is the use of five to ten animals in the LD50 estimation. A preliminary dose is chosen based on data on other species dosed with the same chemical or structurally/mechanistically similar compounds. Subsequent doses are based on the results of the initial dose and are modified upward or downward in an effort to identify reversal of survival or mortality outcomes. An LD50 value is calculated using maximum-likelihood methods. The use of ALD determinations allows lethal dose estimates using a fraction of the number of animals in a standard LD50 determination, making testing of nonstandard test species (i.e., field-collected wild species) a more reasonable proposition. The need for additional LD50 values for an acute toxicity characterization of the chemical is discussed below.

Subacute Dietary Toxicity (OPPTS Method 850.2200)

Mortality following ingestion of chemically treated feed is measured in the LC50 test, where the lethal concentration to 50% of a test population is expressed on a part per million (ppm) basis of the chemical in the diet. Considerations of acclimation, range finding, numbers of individuals, and dose spacing are similar to those of the acute oral toxicity test; however, the ages of the test birds are younger (mallards, 5 to 10 days old; bobwhite quail, 10 to 15 days old). Following acclimation, exposure to treated diet occurs for five consecutive days followed by a 3-day observation period (on nondosed feed) that can be extended should mortality or signs of toxicity persist. The LC50 value, the slope of the dose-response curve, and the LC50 95% confidence interval are determined as in the acute oral toxicity test. A limit test, similar to that in the acute oral toxicity test, can be performed, using 5000 ppm as the limit value. Avoidance of the treated diet must be considered in the subacute test, as aversive test materials may lead to cessation of eating and interruption of the dosing process. Food consumption rates and body weights are monitored to ensure avoidance does not invalidate the test. Concern for this issue has recently led to a new test to address aversion.

Avian Reproduction Test (OPPTS Method 850.2300)

Dietary exposures to chemicals are assessed for effects on avian reproduction, both on the production of fertile, viable eggs by adult birds as well as the health of the eggs and resulting nestlings. Following a 2-week acclimation period, pairs of adult bobwhites or mallards are pre-exposed to a chemical-containing diet for 10 weeks under lighting conditions that suppress reproductive behavior (7 to 8 hours of light per day). Light levels are gradually increased to 16 hours per day over the next 2 weeks, at which point breeding and egg laying commence and continue for an additional 10 weeks, all with continued exposure to the dosed feed. Three dose levels and a control are used, with 12 pairs of birds in each group, although 20 pairs of controls provide a better statistical foundation. Doses should reflect anticipated field food-item residue levels and doses that bracket that level by a factor of five. Throughout the egg-laying portion of the test, eggs are collected daily, accumulated for 14 days, and then placed in an incubator to encourage chick development and hatch.

Reproductive impairment associated with adults and egg-laying include direct impacts on adult health or survival, decreases in the numbers of eggs laid or in egg fertility, and increases in occurrence of eggs with thinned or cracked eggshells. During incubation, the number of viable 18-day embryos is determined using egg candling techniques. Hatchability, or the number of eggs hatched as a percent of viable eggs, is determined, and chicks—marked or segregated based on specific breeding pairs—are observed for 14 days post-hatch. A wide range of potential abnormal outcomes is monitored, from physical defects in dead embryos and hatchlings to decreased survival, gross clinical abnormalities, or signs of toxicity (e.g., decreased growth rates) in 14-day surviving chicks. No observed effect concentration (NOEC) and the lowest observed effect concentration (LOEC) are determined based on test results.

Toxicity data alone, however, do not fully explain the potential risk to a bird in the field. Ecological risk assessments predict the overall risk to birds by combining expected environmental concentrations of chemicals in the field with laboratory toxicity test results. Exposure data come from pesticide residue studies in field crops, avian food-item assessments, and sometimes from direct determinations obtained from birds in chemically treated or contaminated areas.

Historically, the findings of the three basic tests were used to determine the need for additional field-level assessments as part of the pesticide registration process. When toxicity levels approached anticipated or documented concentrations of pesticides in avian food items, pesticide manufacturers were required to perform extensive avian field studies under actual pesticide use conditions. These studies, generally requiring sufficient numbers of replicated study fields for statistical significance testing, incorporate analyses of the study chemical in soils, crop, and avian food items, avian field censusing, carcass searching, and reproductive success assessments of breeding birds on the sites. Today, because of new directions in risk assessments, field studies are more focused on capturing data relative to particular endpoints in need of characterization, such as testing chemical levels in environmental matrices.

In addition to the three basic toxicity tests and field assessments, a number of additional tests have been used or are being developed to address other aspects of chemical toxicity to birds. Screening methods for endocrine effects, under development by both OECD and the EPA, will likely bring modifications to the avian reproduction test to better incorporate endocrine effects assessments, perhaps even leading to the addition of a second generation component to the test. Testing for secondary poisoning has been suggested for chemicals such as rodenticides that leave carcasses accessible to birds of prey or scavengers. Semifield studies that assess behavioral responses of dosed birds can demonstrate levels of exposure that can make species more vulnerable to predation or, alternatively, that can make predators less capable of capturing prey. Expansion of existing tests to less common species can help better interpret the relevance of standard test species to the overall class of birds. It is this approach that is currently of interest to regulators due to the incorporation of sensitivity distributions into risk assessments.

The desire to better use toxicity data to protect avian species has led to the development of probabilistic risk assessments that incorporate toxicity distributions. Working with all of the avian toxicity data available for a compound, a distribution of LD50 values (the endpoint usually used) can be created that reflects the variability and extremes of sensitivity to better predict levels of chemical exposure that are protective of the majority of bird species. The dose that protects a chosen proportion of exposed birds (e.g., 95% of avian species) from LD50-level exposures can be extrapolated using these distributions (Figure 9.8). Rich data sets can provide fairly accurate estimates of protective values for both high-toxicity (squares) and lower-toxicity (circles) compounds. Compounds with small numbers of datum and with widely divergent

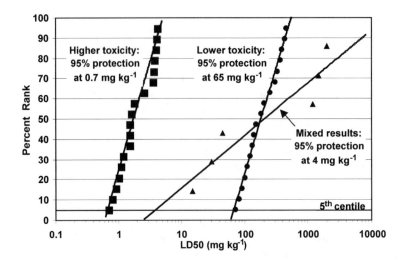

Figure 9.8 Probability distributions of avian LD50 values for three hypothetical chemicals.

toxicities (triangles) demonstrate the need for extrapolation of safety factors, expanded data sets, and understanding of mechanisms leading to such variability in order to better protect sensitive species. The ALD method for LD50 determinations can aid in this situation by providing data on additional species while using only small numbers of wild birds.

A final assessment to ensure the accuracy of laboratory-to-field extrapolation in toxicity prediction is postregistration monitoring. In a regulatory environment that increasingly stresses models rather than field studies for predicting effects, follow-up studies to assure the accuracy and validity of registration decisions are now needed more than ever. This is particularly true for compounds with wide ranges of avian toxicity or those with which we have limited historical experience.

Suggested Supplemental Readings

EPA, Office of Prevention, Pesticides and Toxic Substances (USEPA OPPTS), http://www.epa.gov/docs/OPPTS_Harmonized/, site contains referenced protocols for harmonized avian test guidelines.

Hart, A., D. Balluff, R. Barfknecht, P. Chapman, T. Hawkes, G. Joermann, A. Leopold, and R. Luttik, *Avian Effects Assessment: A Framework for Contaminants Studies*, SETAC Press, Pensacola, FL, 2001, 193 pp.

Mineau, P., A. Baril, B.T. Collins, J. Duffe, G. Hoermann, and R. Luttik, Reference values for comparing the acute toxicity of pesticides to birds, *Rev. Environ. Contam. Toxicol.*, 24, 24, 2001.

Organization for Economic Cooperation and Development, Acute Oral Toxicity: Up-and-Down Procedure, OECD Guideline for Testing of Chemicals, Guideline no. 425, OECD, Paris, 2001, 26 pp.

Solomon, K.R., J.P. Giesy, R.J. Kendall, L.B. Best, J.R. Coats, K.R. Dixon, M.J. Hooper, E.E. Kenga, and S.T. McMurry, Chlorpyrifos: ecotoxicological risk assessment for birds and mammals in corn agroecosystems, *Hum. Ecol. Risk Assessment*, 7, 497–632, 2001.

III SURVIVAL TIME

III.A Basis for Time-Response Models

The **time-response approach** complements the dose- or concentration-response approach. In the dose- or concentration-response approach, exposure time is held constant, although results (numbers dying) are often acquired at a series of time intervals. This dose-response approach provides a gross

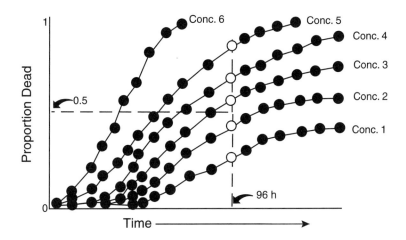

Figure 9.9   An idealized time course of mortality occurring at six different concentrations. Note that, if only data from 96 h were used, only five points would be available to calculate the LC50. In contrast, many more data points (approximately 70 points) would be gathered by 96 h if time-to-death data for individuals were noted instead of the final proportion dying.

indication of the influence of exposure duration. However, full consideration of temporal dynamics is sacrificed in order to generate estimates of chemical toxicity expressed as amounts or concentrations.

Accurate measurement of the effect of exposure duration is also extremely important to assessing ecotoxicological consequences of contamination. To accomplish this, times to death (TTD) are measured in the time-response approach. More data are generated for each exposure tank, resulting in enhanced statistical power. For example, ten TTD values can be generated instead of one proportion from an exposure tank containing ten fish. Also advantageous, because the endpoint (TTD) is associated with an individual, important qualities influencing toxic consequences, such as individual sex or weight, can be included in the analysis. The enhanced power allows more effective inclusion of other qualities such as temperature, toxicant concentration, or nutritional history. The inclusion of concentration in the survival-time approach provides a highly desirable model that predicts effect as a function of both exposure duration and concentration (Figure 9.9). Finally, as we will discuss in Chapter 10, results of time-response analysis can be readily incorporated into demographic analysis.

Despite these clear advantages, ecotoxicologists still underutilize time-response methods. Traditions—including those taught to students and formalized in regulations—seem to have a strong hold on ecotoxicologists in this area. Kooijman (1981), Chew and Hamilton (1985), Dixon and Newman (1991), Newman and Aplin (1992), Newman et al. (1994), Newman (1995), Roy and Campbell (1995), Newman and Dixon (1996), Newman and McCloskey (1996b), and Crane et al. (2002a, 2002b) provide general discussion and examples of time-response methods application in ecotoxicology.

III.B   Fitting Survival-Time Data

The **Litchfield method** is a simple method for analyzing survival-time data (Litchfield, 1949). It was developed during the same year as the Litchfield-Wilcoxon method for estimating LC50. The tests are very similar, differing only in minor details. First, the proportion of exposed individuals responding is tabulated for a series of exposure durations. The probit of the proportion responding is plotted against log of exposure duration to produce a straight line, e.g., Figure 9.10. (Originally log-probability paper was used, but these transformations produce the same results.) A line is fit

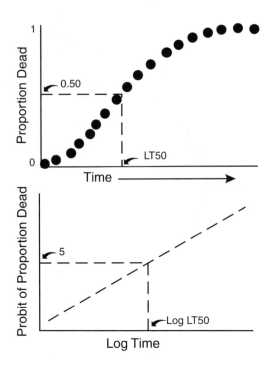

Figure 9.10   Linear transformation of time-to-death data assuming a log-normal model. A sigmoidal curve is generated if the cumulative proportion of the exposed individuals that have died is plotted against exposure time (top panel). Under the assumption of a log-normal model, a straight line is produced (bottom panel) if log time is plotted against the probit of the proportion dead. A curve that is not straight, but instead has a break in its slope, is called a **split probit**. Traditionally, a split probit is assumed to reflect either two distinct mechanisms of toxicity at the beginning and later in the exposure, or two distinct subpopulations of individuals in the exposed group that differ in tolerance to the toxicant. A third possibility—that the log-normal model is an inappropriate model for the data—is rarely assessed.

by eye to these data, and the time corresponding to the probit for 50% mortality is the estimated LT50 (**median lethal time**).[8] To calculate the 95% confidence interval, a slope factor is derived from this line:

$$S = \frac{\dfrac{LT84}{LT50} + \dfrac{LT50}{LT16}}{2} \tag{9.24}$$

where LT16, LT50, and LT84 = time intervals corresponding to when 16%, 50%, and 84% of exposed individuals were dead, respectively. An f_{LT50} is then calculated to estimate the 95% confidence interval for the LT50.

$$f_{LT50} = S^{\frac{1.96}{\sqrt{N}}} \tag{9.25}$$

$$\text{Lower Limit} = LT50/f_{LC50} \tag{9.26}$$

$$\text{Upper Limit} = LT50 \cdot f_{LT50} \tag{9.27}$$

---

[8] As discussed for LC50, a **median effective time** (ET50) can be used instead of LT50 if the effect is sublethal or ambiguously lethal.

Table 9.2   Transformations of Times-to-Death Data Used to Select from among Candidate Models

Candidate Model	Transformation of Mortality	Transformation of Time
Exponential	ln S(t)	t
Weibull	ln [−ln S(t)]	ln t
Normal	Probit [F(t)]	t
Log normal	Probit [F(t)]	ln t
Log logistic	ln [S(t)/F(t)]	ln t

Note: t = duration of exposure; S(t) = cumulative survival to time t, expressed as the proportion of exposed individuals still alive at time t; and F(t) = cumulative mortality to time t, which is 1 − S(t).

where N = total number of individuals exposed during the test if all individuals died during the test. If there were survivors, the N in this equation must be modified slightly, as described originally by Litchfield (1949) or conveniently as described by Newman (1995).

Although ecotoxicology has traditionally considered only this log-normal model, other models for time-to-death data are commonly assumed or explored in other fields. The most commonly employed include the exponential, Weibull, normal, log-normal, and log-logistic models. Fit to these models can be crudely assessed with a series of linear transformations (Table 9.2). The model with a set of transformations that results in a straight line (e.g., Figure 9.10 for the log-normal model) is selected as the most appropriate.

Like dose-response data, time-to-death data can be analyzed with several methods, ranging from the simple Litchfield method to more-involved nonparametric, semiparametric, and fully parametric methods. The nonparametric **product-limit (Kaplan-Meier) methods** do not require a specific model for the survival curve. Product-limit methods allow estimation of survival through time and can be combined with tests for significant differences among treatments. At the other extreme, fully parametric techniques assume a specific model for the survival curve (e.g., log-normal or Weibull model) and a specific function relating survival to covariates (e.g., to exposure concentration or animal size). The underlying distribution of mortality through time can be selected from a series of candidate models such as those in Table 9.2. These models are usually fit using a maximum-likelihood method, as described earlier for dose-response data.

Depending on which model is selected to describe the shape of the survival curve, the fully parametric model takes one of two general forms: proportional hazard or accelerated failure time. Selection of an exponential or Weibull model produces a proportional hazard model. A **proportional hazard model** relates the hazard (proneness to or risk of dying at any time, t) of one group (e.g., smokers) quantitatively to that of a reference group (e.g., nonsmokers). Results of proportional hazard models for human mortality are often expressed as easily understood **relative risks**, risks of one group expressed as a multiple of that of another. For example, the risk of dying from lung cancer may be X times higher for smokers relative to nonsmokers, or the risk of surviving a heart attack is Y times higher if one exercises relative to that for someone who does not exercise regularly. Proportional hazard models have the general form:

$$h(t, x_i) = e^{f(x_i)} h_0(t) \tag{9.28}$$

where h(t, x_i) = hazard at time t as modified by the value (x_i) of the covariate x, $h_0(t)$ = hazard of some reference group or type that is described by a specific model such as a Weibull model, and f(x_i) = some function of the covariate (x) making the hazards proportional. For example, f(x) could be a simple function making hazards proportional among different animal sizes, e.g., f(x) = a + b(log animal weight).

Use of the log-normal, log-logistic, normal, or gamma model results in an **accelerated failure time model**, a model in which the time to death (ln TTD$_i$) of a particular type of individual (e.g., smoker) is changed ("accelerated") as a function of some covariate (e.g., classification relative to smoking habit). For example, an increase in toxicant concentration may accelerate the expected time until death of an individual. The simple form of the accelerated failure-time model is shown in Equation 9.29:

$$\ln \mathrm{TTD}_i = f(x_i) + \varepsilon_i \tag{9.29}$$

The error term (ε) is fit to the assumed model, e.g., log-normal model. Some function of the covariate modifies the ln TTD. It can be any function of the covariate such as that given above for the effect of animal weight, e.g., $f(x) = a + b(\log$ animal weight$)$.

Most applications fitting parametric survival-time models allow parameter estimation so that predictions can be made relative to **median times to death** (MTTD) or some proportion other than 50% mortality. They also allow one to test for significant effects of different covariates on lethal effect.

Often, especially in clinical studies comparing various treatments, it is impossible or unnecessary to model the underlying survival curve. The exact underlying model is much less important than estimating the influence (hazard or relative risk) of some covariate on survival. For example, the focus of a study may be on determining if a postsurgical treatment improves patient survival relative to the standard treatment (reference treatment). In such a study, knowledge of the exact form of the survival curve is not required. A semiparametric method (**Cox proportional-hazard model**) is designed to allow examination of proportional hazards without taking on the assumption of any specific model for the baseline or reference hazard.

III.C Incipiency

The LT50 or MTTD can be used to estimate lethal incipiency, the concentration at which 50% of exposed individuals will live indefinitely relative to the toxicant effects. The point at which the line for toxicant A begins to run parallel to the log MTTD axis is an estimate of this lethal threshold in Figure 9.11. Note that there may be no apparent threshold for some toxicants (e.g., Toxicant B). Also, there may be a minimum time required before an effect can be expressed. Toxicant B illustrates such a **minimal time to response**.

III.D Mixture Models

Mixture effects can also be quantified for time-response approach as done with the dose-response approach. For example, the lethal threshold estimated with the MTTD or some other estimate is used instead of the incipient LC50. However, more-involved treatments become possible because more data can be extracted from a toxicity test with these methods. As an example, Roy and Campbell (1995) took advantage of this enhanced power of survival-time models to quantify the combined effect of aluminum and zinc to young Atlantic salmon (*Salmo salar*).

IV FACTORS INFLUENCING LETHALITY

IV.A Biotic Qualities

Many biological qualities influence toxicant effects. Some have already been discussed, e.g., developmental stage, lipid pools within the individual, feeding behavior, and induction of detoxification mechanisms. Life stage is also important (e.g., Campbell, 1926). Because of their widespread occurrence, two general biological qualities that can influence lethality, acclimation and allometry, will be discussed briefly here.

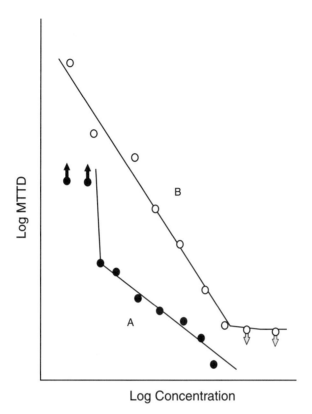

Figure 9.11 Incipiency for the time-to-death approach. Chemical A (closed circles) has a distinct incipient MTTD. Below a certain concentration, 50% of exposed individuals will live indefinitely relative to the effect of toxicant A. Note that the two lowest concentrations resulted in less than 50% mortality regardless of how long the individuals were exposed. Some toxicants (B) may show no evidence of incipiency. There may also be a minimum time to get a response, e.g., individuals can only die so fast regardless of the toxicant concentration. Toxicant B in the diagram has such a threshold. (Arrows attached to a point signify that the "true" value is probably in the indicated direction from the point; e.g., the true log MTTD for the last point on curve B is less than the value of the last point.) (Modified from Figure 5 in Sprague, J.B., *Water Res.,* 3, 793–821, 1969, and Figure 4 in Newman, M.C., *Quantitative Methods in Aquatic Ecotoxicology,* Lewis Publishers, Boca Raton, FL, 1995.)

Acclimation[9] is the modification of biological functions, especially those physiological, or structures to maintain or minimize deviations from homeostasis despite change in some environmental quality such as temperature, salinity, light, radiation, or toxicant concentration. It is an expression of phenotypic plasticity of individuals in response to a sublethal change in some environmental factor. Often the distinction is made in the literature between acclimation (shifts taking place in a controlled or laboratory setting) and **acclimatization** (shifts taking place under natural conditions). It is important to understand that acclimation is not the shift in population qualities, i.e., not a change in genetic composition or demographic structure, in response to a stressor. Such adaptation or genetic change in a population as a consequence of selection will be discussed in Chapter 10.

Acclimation can occur after pre-exposure to sublethal concentrations of toxicants, enhancing survival of individuals during a subsequent, more intense exposure. However, some toxicants cause damage regardless of their concentrations, e.g., cyanide pre-exposure does not enhance survival of

---

[9] Acclimation can also be used to identify the time allowed for an organism, population, community, or ecosystem to stabilize to a set of conditions prior to testing. Acclimation will not be used in this context (relative to an experimental protocol) here.

rainbow trout because of kidney damage during the first exposure (Dixon and Sprague, 1981b). In contrast, pre-exposure of human lymphocytes to low levels of radiation leads to reduced chromatid aberrations during a subsequent exposure to x-rays (Olivieri et al., 1984). Dixon and Sprague (1981a) found that, above a certain concentration, pre-exposure of rainbow trout to copper enhanced their tolerance to otherwise lethal concentrations of copper. It is important to note that the enhanced tolerance increased with copper concentration only to a certain point; beyond that point, pre-exposure caused damage and consequent diminished tolerance in pre-exposed fish. Enhanced tolerance was a complex function of both acclimation concentration and time (Dixon and Sprague, 1981). For example, both strongly influenced lethal consequences to salmon exposed to high temperatures after varying intensities and durations of pre-exposure (Elliott, 1991).

Allometry can also influence toxicant effect to individuals (Newman and Heagler, 1991). The classic work of Bliss (1936) used arsenic intoxication of silkworm larvae to estimate the influence of size on toxic impact. First, he took time-to-death data and transformed these to rate of toxic action (rate of toxic action = 1000/TTD). Next, he performed a multiple regression to generate the model:

$$\text{Rate} = a + b_1 (\log \text{ dose}) + b_2 (\log \text{ weight}) \qquad (9.30)$$

where a, b_1, and b_2 are constants derived during model fitting. This relationship was transformed to generate an adjustment factor (weight^h) for the influence of animal weight.

$$\text{Rate} = a + b_1 \left[\log \frac{\text{dose}}{\text{weight}^h} \right] \qquad (9.31)$$

where $h = b_2/b_1$. This approach has been adopted during the last 60 years as the primary approach to incorporating size effects into time-to-death data. The approach (Bliss, 1936) was later modified to dose- or concentration-effect models (Anderson and Weber, 1975).

$$\log \text{ LC50} = \log a + b \log \text{ weight} \qquad (9.32)$$

or

$$\text{LC50} = a \text{ weight}^b \qquad (9.33)$$

Newman et al. (1994) and Newman (1995) demonstrated that a more general approach exploring the various models described for survival-time models can lead to a better fit to the data.

IV.B Abiotic Qualities

Numerous physical and chemical factors modify the toxic action of contaminants. Ambient and acclimation temperatures modify the impact of some toxicants. For example, toxic impact (LC50) of cadmium was modified by acclimation and ambient temperature during exposures of a snail (*Potamopyrgus antipodarum*) (Møller et al., 1994). Ambient light can also influence the action of photolabile chemicals by changing the rate at which they break down to more-toxic products. **Photo-induced toxicity**, toxicity of a chemical in the presence of light due to the production of toxic photolysis products, was found for bluegill sunfish (*Lepomis macrochirus*) exposed to anthracene, a polycyclic aromatic hydrocarbon (PAH) (Bowling et al., 1983). Some chemicals can also enhance the **photosensitivity** (sensitivity of cutaneous tissues to the effects of light evoked by a chemical) of individuals. Channel catfish (*Ictalurus punctatus*) treated with the antibiotic, oxytetracycline, become extremely sensitive to sunlight, resulting in skin (sunburn) and eye (lesions) damage (Stacell and Huffman, 1994).[10]

[10] This is also the case for humans taking certain antibiotics. Physicians and labels on some prescription antibiotics warn patients not to spend much time in the sun because of increased sensitivity to sunburn.

Organic and inorganic constituents of waters can modify metal toxicity (e.g., Andrew et al., 1977; Bradley and Sprague, 1985; Azenha et al., 1995). Freshwater **hardness** (sum of the concentrations of dissolved calcium and magnesium) is often related to toxic effect of metals such as beryllium (EPA, 1986), cadmium (Carrol et al., 1979), copper (Howarth and Sprague, 1978), lead (Davies et al., 1976), and zinc (EPA, 1987b). Several mechanisms have been proposed, including competition of hardness cations with toxic metals for binding sites on biomolecules, modification of biological processes, and modification of metal speciation in the exposure waters (Newman, 1995). These effects on metal toxicity are predicted in EPA water quality criteria documents (e.g., EPA, 1985e) with a simple, empirical model:

$$\log \text{ Toxicity Endpoint (e.g., LC50)} = \log a + b \, (\log \text{ Hardness}) \tag{9.34}$$

or

$$\text{Toxicity Endpoint} = a \, \text{Hardness}^b \tag{9.35}$$

where log a and b = estimates from linear regression of the intercept and slope. Although Equation 9.35 produces a biased prediction of toxic effect (Newman, 1991; Newman, 1993), it is used extensively in the ecotoxicological literature and water quality regulations.

Sediment and soil qualities also influence toxicity, as described in earlier discussions of factors modifying bioavailability. Further, QSARs (quantitative structure-activity relationships) are readily developed for lethality as well as bioaccumulation. The reader is encouraged to review information covered in Chapter 4, as topics discussed there are relevant to this discussion of abiotic factors influencing toxicity.

V SUMMARY

In this chapter, lethality was described under acute and chronic exposure scenarios. The possibility of differences in toxic impact at different life stages of an individual was explored along with associated implications. Basic toxicity test designs were outlined and their relative advantages and disadvantages highlighted. The predominant approach to measuring toxicity (dose- or concentration-response design) was detailed, and methods of analyzing dose-response data were contrasted. An alternative approach, the survival time design, and methods for analyzing time-to-death data were compared with the dose-response methods. Toxic incipiency and ways of quantifying effects of toxicant mixtures were provided for the dose-response and survival-time designs. Finally, biotic and abiotic factors modifying toxicity were explored only briefly, since many had already been discussed in the context of factors modifying bioavailability (Chapter 4).

SELECTED READINGS

Anderson, P.D. and L.J. Weber, Toxic response as a quantitative function of body size, *Toxicol. Appl. Pharmacol.*, 33, 471–483, 1975.

Christensen, E.R., Dose-response functions in aquatic toxicity testing and the Weibull model, *Water Res.*, 18, 213–221, 1984.

Crane, M., M.C. Newman, P.F. Chapman, and J. Fenlon, *Risk Assessment with Time-to-Event Models*, Lewis Publishers, Boca Raton, FL, 2002, p. 175.

Dixon, D.G. and J.B. Sprague, Acclimation to copper by rainbow trout (*Salmo gairdneri*): a modifying factor in toxicity, *Can. J. Fish. Aquat. Sci.*, 38, 880–888, 1981.

Marking, L.L., Toxicity of chemical mixtures, in *Fundamentals of Aquatic Toxicology*, Rand, G.M. and Petrocelli, S.R., Eds., Hemisphere Publishing Corp., Washington, D.C., 1985.

Newman, M.C. and M.S. Aplin, Enhancing toxicity data interpretation and prediction of ecological risk with survival time modeling: an illustration using sodium chloride toxicity to mosquitofish (*Gambusia holbrooki*), *Aquatic Toxicol.,* 23, 85–96, 1992.

Newman, M.C. and J.T. McCloskey, Time-to-event analyses of ecotoxicology data, *Ecotoxicology,* 5, 187–196, 1996.

Sprague, J.B., Measurement of pollutant toxicity to fish, I: bioassay methods for acute toxicity, *Water Res.,* 3, 793–821, 1969.

Stephan, C.E., Methods for calculating an LC50, in *Aquatic Toxicology and Hazard Evaluation*, ASTM STP 634, Mayer, F.L. and Hamelink, J.L., Eds., American Society for Testing and Materials, Philadelphia, 1977.

CHAPTER 10

Effects on Populations

There is an enormous disparity between the types of data available for assessment and the types of responses of ultimate interest. The toxicological data usually have been obtained from short-term toxicity tests performed using standard protocols and test species. In contrast, the effects of concern to ecologists performing assessments are those of long-term exposures on the persistence, abundance, and/or production of populations.

Barnthouse et al. (1987)

I OVERVIEW

Sometimes, the focus of our efforts is protection of individuals, particularly when those efforts involve effects to humans or endangered species. More often, effects to individuals are measured with the intent to predict consequences to populations. Indeed, the primary goal of most ecotoxicologists is assuring the persistence and vitality of populations within ecological communities. Qualities used to assess population-level effects include many already discussed, but they are interpreted in a slightly different context. For example, cancer can now be examined in the context of relative incidences in populations occupying different microhabitats or having different demographic qualities. Age-dependent effects of toxicants can be woven into explanations of demographic change. Differences in individual effective doses can be considered in the context of selection for tolerant genotypes in populations. Knowledge of activation mechanisms for carcinogens can be sought to support inferences about causal structure in studies of disease prevalence.

This chapter describes the means of determining the status of exposed populations. Also discussed are population responses that could lessen adverse effects. First, approaches are detailed for describing the occurrence of toxicant-related disease in extant populations. Associated epidemiological information helps us to assess the imminence of failure of the afflicted population in addition to estimating the probability of an individual with certain qualities being adversely affected. Second, impacts on population demography are described along with the advantages of generating toxicity test data amenable to demographic analysis. Third, we examine the change in population genetics, including the evolution of tolerance that might occur after long periods of exposure to contaminants.

II EPIDEMIOLOGY

Epidemiology is the science concerned with the cause, incidence,[1] prevalence, and distribution of infectious and noninfectious diseases in populations. Most often, disease is linked through correlation with risk factors[2] such as qualities of individuals and **etiological agents** (an agent responsible for causing, initiating, or promoting the disease [Rench, 1994]). An example of an etiological agent might be the high mercury concentrations in seafood taken from Minamata Bay. Two subdefinitions are relevant to studies of disease resulting from chemical (e.g., pollutants or chemicals in the workplace) and physical (e.g., radiation, UV light, high temperatures, or asbestos fibers) etiological agents. Here, **environmental epidemiology** is defined as that subdiscipline of human epidemiology concerned with diseases caused by chemical or physical agents (Rench, 1994). **Ecological epidemiology**, frequently associated with retrospective ecological risk assessments, is the name often given to epidemiological methods applied to determining the cause, incidence, prevalence, and distribution of adverse effects to nonhuman species inhabiting contaminated sites (Suter, 1993).

A range of straightforward metrics can be generated during epidemiological analyses. Those from life tables, as described shortly, were among the first to be applied to human epidemiology. Disease **incidence rate** (I, expressed in units of individuals or cases per unit of exposure time being considered in the study) for a nonfatal condition can be calculated as the number of individuals with the disease divided by the total time that the population had been exposed, e.g., 10 new cases per 1000 person years:

$$I = N/T \tag{10.1}$$

where N = number of diseased individuals or cases, and T = total time at risk of contracting the disease (Ahlbom, 1993). The T can be expressed as the total number of time units that individuals were exposed to disease risk during the study period, e.g., per 1000 person years of exposure.

Prevalence (P) is simply the incidence rate (I) times the amount of time (t) that individuals were at risk:

$$P = I \times t \tag{10.2}$$

If there were two cases per 1,000 person years, the prevalence in 10,000 person years (e.g., a population of 1,000 people exposed for 10 years) would be (two cases/1,000 person years) × (10,000 person years) or 20 cases.

Occurrence of disease in a population can also be expressed relative to that in another population. Often one is a control or reference population. The simple difference in incidence rates can be used to compare disease in two populations, e.g., 25 more cases per year in population A than in population B. The difference often is expressed in terms of a standard size (10,000 individuals) because the populations will likely differ in size. Also, the ratio of occurrences of the disease in the two populations (**relative risk** or RR) can be expressed as the ratio of incidence rates (**rate ratio**).

$$RR = I_A/I_0 \tag{10.3}$$

---

[1] Incidence and prevalence have slightly different meanings in epidemiology. Disease *incidence* is the number of new individuals scored as having the disease in a certain time interval, e.g., 10 cases per week. *Prevalence* is simply the total number or proportion of individuals with the disease at a particular time, e.g., 157 cases in New York City (Ahlbom, 1993). Often prevalence is expressed as a ratio, e.g., 157 cases per 10,000 people in 1957.

[2] A **risk factor** is any quality of an individual (e.g., age or dietary habits) or an etiological factor (e.g., chronic exposure to high levels of a toxicant) that modifies an individual's risk of developing the disease.

where I_A = incidence rate in population A, and I_0 = incidence rate in the reference or control population. For example, 23 diseased individuals occurring per year in a standard sample size of 10,000 individuals for a heavily industrialized city might be compared with an annual incidence rate of 0.5 individuals per 10,000 individuals in a small town. The relative risk would be expressed as a rate ratio of 46. Note that relative risk calculated with survival time models described earlier can be used in such epidemiological analyses too.

Relative risk can also be expressed as an **odds ratio** in case-control studies. The number of disease cases (individuals) that were (a) or were not (b) exposed to the risk factor, and the number of control individuals free of the disease that were (c) or were not (d) exposed to the risk factor are used to estimate the odds ratio (Ahlbom, 1993):

$$\text{Odds Ratio} = (a/b)/(c/d) = ad/bc \tag{10.4}$$

Say, for example, that 750 cases of a fatal disease were documented, with 500 of them associated with individuals who had been previously exposed to an etiological agent (a) and 250 of them (b) were associated with individuals who had never been exposed to this agent. In another sample of 500 control individuals showing no signs of the disease, 60 had been exposed (c) and 440 (d) had not been exposed to the agent. The odds ratio would be (500)(440)/(250)(60) or roughly 14.7. This odds ratio suggests that exposure does influence proneness to the disease. An example study employing odds ratios is provided by Rench (1994), in which a disproportionately high number of soil conservationists fell victim to non-Hodgkin's lymphoma relative to a control group, notionally because of high pesticide exposures associated with their jobs.

There is a wealth of methods such as proportional odds and logistic regression models (SAS Institute, 1990) for analyzing odds data from epidemiological studies. Easily assessable textbooks such as those written by Ahlbom (1993) and Marubini and Valsecchi (1995) describe these and other statistical methods applicable to epidemiological data. They detail the calculation of confidence intervals for estimates and statistical tests of significance. Although most of these methods focus on human epidemiology and clinical studies, there are no inherent obstacles to their application to ecological epidemiology. Unfortunately, most of these powerful methods remain underexploited in ecotoxicology.

Logical rules have been developed to enhance inferential soundness of epidemiology because most approaches in this field relay heavily on inferentially weak correlations of disease with risk factors. Two such sets of rules—Hill's (1965) nine aspects of noninfectious disease association and Fox's (1991) **rules of practical causal inference**—are commonly applied in ecotoxicology. Only Hill's rules will be described, as both sets of rules are similar. **Nine aspects of disease association** have been identified by Hill (1965) for assigning linkage of a risk factor and disease in environmental epidemiology: strength of association, consistency of association, specificity of association, temporal association, biological gradient (dose-response), biological plausibility, coherence of the association, experimental support, and analogy.

The *strength of association* between some risk factor and disease is important to consider. For example, the 200-fold higher prevalence of scrotal cancer for chimney sweeps relative to men in other occupations added strength to the supposition that the cancer was initiated or promoted by some occupational risk factor (carcinogenic PAHs in soot and tar).

Consistency of association, the consistent observation of the association under numerous, varying conditions, also strengthens conclusions. For example, soundness of the conclusion associating asbestos fibers with lung cancer is enhanced if the incidence of lung cancer is high for either male and female workers in many diverse occupations sharing one common factor: high asbestos fiber densities in workplace air. However, consistency may also be generated by a bias in the database, so caution must be exercised in applying this rule. As an example of potential bias, a certain occupation may have a higher number of individuals in it that are predisposed to a particular condition, e.g., an unusually sedentary group within the population who are prone to cardiovascular disease may be overrepresented in a study group of smokers. In such a case, the disease could possibly be associated incorrectly with a preexisting workplace risk factor, e.g., some chemical used in the occupation.

Identification of association under very specific situations (*specificity of association*) can enhance one's ability to assign causation or linkage. For example, only a very specific type of behavior or occupation may have the associated linkage between the risk factor and disease. Indeed, Hill (1965) observed that it was the lack of specificity in the association between lung cancer and smoking that allowed counter arguments to any causal linkage to persist for so long.

A *temporal association* might be considered in order to reinforce causation. The cause or promoter of the disease should be present before the disease occurs. There is potential for considerable uncertainty and bias here also.

Obviously, a clear *biological gradient (dose-response) in the association* strengthens evidence also for a relationship.

Although the knowledge is often not available, *biological plausibility* (a plausible, underlying mechanism for the association) can enhance the probability of making a correct linkage between a disease and risk factor.

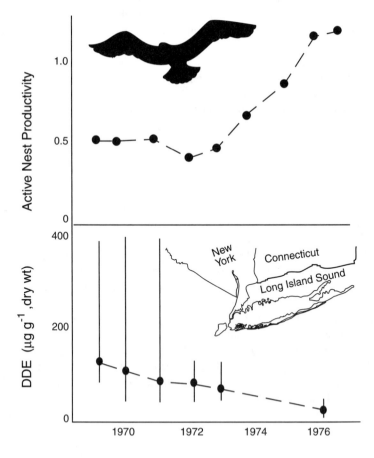

Figure 10.1   The gradual increase in active nest productivity (average number of young fledged per active nest) of Long Island Sound osprey. Nesting success was extremely low prior to the widespread banning of pesticides such as DDE. The nest productivity slowly recovered (top panel) as DDE concentrations decreased in eggs from osprey nests (bottom panel), suggesting that DDE was a significant risk factor in nest failure. (Modified from Figure 1 in Spitzer, P.R. et al., *Science*, 202, 333–335, 1978.)

Coherence of the association with what is already known about the disease is also very helpful. For example, findings of liver lesions after chronic exposure of laboratory rats to a particular toxicant could support inference of an association between human exposure to that toxicant in the workplace and an increased incidence of liver cancer.

Experimental support of association (manipulation of the association with measured change in the disease response) can help, if practical. This type of support in human studies is limited for ethical reasons and, consequently, tends to be more applicable for nonhuman species. However, these data are available for humans under special circumstances. Hill (1965) provides one example of lung and nasal cancer incidences in nickel refinery workers before and after workplace exposure routes were controlled. The incidence of cancer dropped significantly after control measures were taken to reduce worker exposure. Figure 10.1 shows an analogous situation with the effect of DDE on nest productivity of osprey (*Pandion haliaetus*). The nestling production gradually increased after a ban on the application of such pesticides. This recovery supports causal linkage between DDE and reproductive damage. As discussed before, the association between DDE and raptor population decline was also supported by biological plausibility, i.e., eggshell thinning.

Analogy is the final quality of associations enhancing the accuracy with which risk factors are identified. Similarity of the association to another well-documented association fosters accuracy. Hill (1965) suggests that the demonstration of thalidomide's effect on fetuses enhanced the credibility of subsequent suggestions of birth-defect linkage to other drugs taken by pregnant women.

None of these nine factors alone leads to accurate identification of a real association. Rather, the goal is to accumulate enough information with all of the nine factors to identify associations as either sufficiently plausible or implausible. Of course, another possible outcome of an epidemiological study is the conclusion that insufficient evidence exists to judge plausibility.

Obviously, these same nine factors can enhance plausibility of association between a risk factor and some adverse consequence in ecological epidemiology. Suter (1993, Table 10.4) outlined Hill's nine factors, providing a slightly more ecological explanation for each. The only change was associated with a greater capacity to perform and to emphasize controlled experiments with nonhuman species. These and similar logical tools will be discussed again in more detail in Chapter 13, which describes ecological risk assessment techniques.

III POPULATION DYNAMICS AND DEMOGRAPHY

III.A Overview

A **population** is a group of individuals of the same species occupying a defined space at a particular time. As recognized early on and studied extensively in ecology and human demography, external factors influence the size, nature, and distribution of nonhuman[3] and human[4] populations. Growth under limitation in artificial systems (such as agricultural plots) and species tolerance ranges (as determined in laboratory trials) were used to predict presence and size of populations in the field. Ecotoxicologists have extended this traditional ecological approach to predicting population sizes and the probabilities of local extinction of populations exposed to contaminant gradients based on laboratory, mesocosm, small-field-plot, and enclosure studies.

---

[3] Perhaps the best-known examples are **Liebig's law of the minimum** and **Shelford's law of tolerance**. Liebig's law states that a population's size will be limited by some essential factor in the environment that is scarce relative to the amounts of other essential factors, e.g., phosphorus will limit algal growth in many lakes. Shelford's law states that the tolerance of a species along an environmental gradient (or a series of environmental gradients) will determine its realized population distribution and size in the environment. For example, salinity and temperature gradients may define the location and abundance of an oyster species along the east coast of the United States.

[4] The Englishman, Thomas R. Malthus (1766–1834) established a series of assumptions and observations regarding limitations on human populations. **Malthusian theory** was first published in 1798 as the profoundly influential essay, *An Essay on the Principle of Population as it Affects the Future Improvement of Society with Remarks on the Speculation of Mr. Godwin, M. Condorcet and Other Writers.*

Predictions of population viability are often made from results of toxicity tests and ANOVA-derived NOEC and LOEC values. Concentrations that affect such qualities as individual growth, larval and adult survival, and number of young produced per female are pooled to produce these predictions. Decisions may be based on whether or not change was statistically significant. As discussed previously, the major flaw in this approach is that statistical significance does not dictate biological significance. For this and other reasons already discussed, a movement away from such an approach has begun. (See Chapman et al. [1996] as an example.) A more thoughtful approach might be to decide on a magnitude of change in some relevant quality beyond which the population is assumed to be adversely impacted. The rationale is the following: (1) If environmental stewardship is to be enhanced, consensus must be reached about the level of change required to trigger concern or action by regulators, and (2) consensus opinion should focus on a level of biologically meaningful change that one can detect with reasonable confidence in most cases (e.g., Hoekstra and Van Ewijk, 1993). Both of these pragmatic approaches (NOEC-LOEC and magnitude of change for some population quality) are potentially compromised, as they do not directly answer the question of whether or not the population will remain viable despite the presence of the toxicant. A 20% reduction of a particular quality may be catastrophic to one species population but trivial to another species population. Similarly, a 20% change in one quality may be trivial, but a similar change in another quality may lead to imminent extinction of a population. Fortunately, traditional population and demographic analyses can be used to predict the possible outcomes of exposure and their probabilities of occurring. Although most toxicity testing methods do not produce information directly amenable to demographic analysis, more and more ecotoxicologists have begun to design tests and interpret results in this context, e.g., Kammenga and Laskowski (2000) and Newman (2001). Further, current EPA documents describing ecological risk assessment methods clearly recognize and now articulate this inconsistency between traditional toxicity tests and data needs.

During the past two decades, toxicological endpoints (e.g., acute and chronic toxicity) for individual organisms have been the benchmarks for regulations and assessments of adverse ecological effects. ... The question most often asked regarding these data and their use in ecological risk assessments is, "What is the significance of these ecotoxicity data to the integrity of the population?" More important, can we project or predict what happens to a pollutant-stressed population when biotic and abiotic factors are operating simultaneously in the environment?

Protecting populations is an explicitly stated goal of several Congressional and Agency mandates and regulations. Thus it is important that ecological risk assessment guidelines focus upon protection and management at the population, community, and ecosystem levels. ...

(EPA, 1991a)

The focus has begun to shift to population vital rates. This approach, as described here, has much promise for enhancing prediction of population effects of contaminants.

III.B General Population Response

The simplest models of population response treat all individuals identically and predict change in total number or density of individuals over time. Surveys of widespread population trends such as that by Sarokin and Schulkin (1992) often treat populations in this general manner. Changes in total numbers of individuals can be correlated with epidemics of pollutant-linked cancers in the population or with **epizootics** (outbreaks of disease in a large number of individuals) caused by a biological agent in pollution-weakened populations.

Unrestrained exponential growth of a population can be predicted as a function of the population size (N) and its **intrinsic (or Malthusian) rate of increase** (r) with a simple differential equation:

$$dN/dt = rN \tag{10.5}$$

With knowledge of the initial population size (N_0), its size at any time (t) in the future can be predicted:

$$N_t = N_0 e^{rt} \tag{10.6}$$

Doubling times ($t_d = (\ln 2)/r$) can also be estimated for populations if r is known.

A difference equation can be used instead of Equation 10.5 if population size is measured at discrete intervals, such as might be done with a population with nonoverlapping generations. A population of an annual plant or an insect population could have nonoverlapping generations.

$$N_{t+1} = \lambda N_t \tag{10.7}$$

where $\lambda =$ **finite rate of increase**,[5] and N_{t+1} and $N_t =$ population size at times $t + 1$ and t, respectively.

If population size is measured initially (N_0) and at some time in the future (N_t), r can be calculated from λ.

$$\lambda = (N_t/N_0) = e^r \tag{10.8}$$

All three parameters—λ, r, and t_d—can and have been used as meaningful metameters for adverse population effects. For example, Marshall (1962) calculated the effects of γ radiation on the intrinsic rate of increase for *Daphnia pulex*. Rago and Dorazio (1984) detailed means to estimate the influence of toxicants on zooplankton population finite rate of increase. These qualities have the advantage over measures such as a simple percentage reduction in reproduction in that they are easily fit into predictive ecological models. They also are readily incorporated into results from more complex life table methods, as will be discussed.

Obviously, exponential population growth cannot continue indefinitely. In the simplest form, these equations are modified so that growth rate decreases as the population size approaches some **carrying capacity** of the environment, K (the maximum population size expressed as total number of individuals, biomass, or density that a particular environment is capable of sustaining). For populations with overlapping generations, Equation 10.9 is relevant:

$$dN/dt = rN[1 - (N/K)] \tag{10.9}$$

The classic **Ricker model** (Equation 10.10) is relevant to populations with nonoverlapping generations or experimental designs with discrete intervals of population growth:

$$N_{t+1} = N_t e^{r\left(1 - \frac{N_t}{K}\right)} \tag{10.10}$$

Even these simplest of models (Equations 10.9 and 10.10) predict very complex behavior for population dynamics under certain conditions (May, 1974; 1976b). According to these models, populations can increase to and remain at K, oscillate around K with gradual convergence to K,

---

[5] This term is widely used; however, May (1976a) believes that a better term is the **multiplicative growth factor per generation**. For further discussion of this point, please refer to May (1976a) or Newman (1995).

oscillate indefinitely around K, or fluctuate chaotically. Consequently, field observations of population densities other than those near K or of population densities that fluctuate widely do not necessarily reflect an adverse consequence of contamination. Also, recovery of a population after a toxicant-related decrease in population size might not always involve a simple monotonic increase back to K. Oscillations can occur and influence the probability of successful recovery or extinction. For example, Simkiss et al. (1993) found that sublethal exposure of blowflies (*Lucilia sericta*) to cadmium modified population dynamics in a food-limited environment.

III.C Metapopulation Dynamics

Populations living in areas with patches of superior and inferior habitat are often distributed unevenly. Subpopulations within a superior habitat may have high rates of increase, while those within inferior habitats may have negative growth rates (r). Some subpopulations act as a source of individuals because of the surplus of offspring produced, while others act as sinks as surplus individuals move in from outside the patch. This source-sink or patch structure results in dynamics distinct from those predicted by the simple models above (Pulliam and Danielson, 1991). As an example, a habitat so contaminated that reproduction is impossible can still have a high density of individuals if there is a nearby source of individuals. Maurer and Holt (1996) illustrate the relevance of this concept to ecotoxicology with the example of pesticide impact on wildlife inhabiting a patchy habitat.

Interest in the dynamics of populations in source-sink habitats has spawned a new field of population biology. The complex of populations occupying a habitat mosaic within a landscape and exchanging individuals by migration is called a **metapopulation**, and the field of metapopulation ecology has emerged to explain and predict the dynamics of such systems. Equation 10.11 is a simple metapopulation model that includes some of the more relevant aspects of metapopulation dynamics to ecotoxicology. Other relevant information can be found in O'Connor (1996) and Newman (2001),

$$dp/dt = m(1 - p) - ep(1 - p) \qquad (10.11)$$

where p = proportion of available patches that are occupied, e = probability of local extinction in a patch, and m = probability of population reappearance in a vacant patch.

Two important themes are embedded within this model: the rescue effect and the propagule rain effect. The **rescue effect** is simply the increased probability of vacated-patch reoccupation if nearby patches are occupied, with the probability of a patch extinction decreasing as p increases. In contrast, the **propagule rain effect** is independent of the occupation density of nearby patches. The likelihood of a patch population extinction decreases—and that of patch population reappearance increases—if there is a seed bank or dormant stage that acts as a constant source of propagules to the patch, regardless of the proportion of nearby patches that are occupied. Both of these effects are relevant to ecotoxicology. If a patch has been vacated due to the action of a stressor, then patch-density-dependent and -independent mechanisms can reestablish the patch population. Also, the importance of effective migration corridors among patches is also clear for effective recovery of a toxicant-stressed patch.

Two additional facets of metapopulations have ecotoxicological consequences. It is possible that an individual could be exposed in one patch but experience the effect of toxicant exposure only after migrating to another patch. Spromberg et al. (1998) call this spatial separation between exposure and effect the **effect-at-a-distance hypothesis**. Another issue arises from the fact that individuals in different patches have different fitnesses. Some habitat patches can be of such high quality that they are **keystone habitats** essential to maintaining the vitality of the metapopulation. The loss of a hectare of keystone habitat to contamination or to habitat remediation can be much more damaging than the loss of an equivalent area of marginal habitat.

Vignette 10.1   The Impacts of Chemicals on Spatially Explicit Populations

Wayne G. Landis

Institute of Environmental Toxicology, Western Washington University, Bellingham, WA

Clearly one of the frontiers of environmental toxicology is the interaction of toxicants with ecological systems over space and time. Toxicology has been a science of the organism, but environmental toxicology must take into account the ecological ramifications of the impacts of toxicants. The simplest ecological unit is the population. The impacts of chemicals upon populations occur within a landscape.

A population is a group of potentially interacting individuals of the same species. Typically, a population comprises individuals of different ages and different sexes, and only part of the population is reproductively active. The organisms share a great deal of genetic material, enough so that successful breeding can occur. Sharing so much genetic information also means that the members of the population utilize the same resources, invoking intense competition. Competition for resources can be reduced to a degree by segregating the use of environments by gender or age. However, the toughest competitors for reproductive success are ultimately members of one's own population.

Organisms that make up a population are not evenly distributed in the environment, and they often migrate singly or as a group. The remainder of this vignette demonstrates the critical roles of spatial distribution and migration of organisms in understanding the impacts of toxicants on populations and the ecological systems that contain them.

Populations in Space

There are several different spatial arrangements that populations can take (Figure 10.2). Populations may exist as a single patchy population, an isolated population, a metapopulation, or a mainland-island metapopulation. A metapopulation is defined as a "population of populations" (Levins, 1969) connected through immigration and emigration. In a metapopulation, most individuals spend most of their life spans in a single patch, but occasional migration does occur. Not all available habitats that can successfully maintain the species are always occupied. A single patchy population is characterized by frequent migration of individuals between the various habitat patches. In an isolated-population structure no migration takes

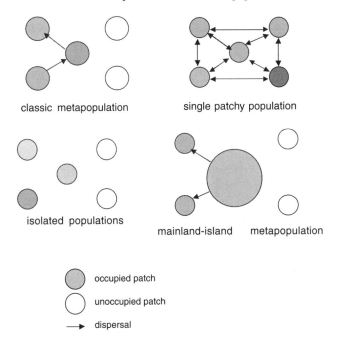

classic metapopulation   single patchy population

isolated populations

mainland-island   metapopulation

occupied patch

unoccupied patch

dispersal

Figure 10.2   Kinds of patch structure in a patchy environment. The models described here generally fit the single patchy population with varying arrangements and distances between patches.

place. A mainland-island pattern is where one population is much larger than the populations in surrounding patches. The properties of the different types of distributions affect the types of impacts that a toxicant can have on one patch and how individual patches affect populations in other patches.

In creating simulation models of metapopulation or patch populations, there are two basic assumptions:

1. There is a minimum viable population (MVP) size below which extinction of the population in the habitat patch will occur.
2. There is a population carrying capacity, i.e., a population size that can be maintained without a tendency to increase or decrease.

There are also some basic definitions:

1. A population in a habitat patch serves as a sink if it is below the MVP and is recruiting emigrants.
2. A subpopulation serves as a source for nearby habitat patches by providing immigrants to them.
3. The rescue effect is where a population that is below the minimum viable population is rescued by immigrating organisms from a source population or patch.
4. If a population within one patch becomes extinct due to a stressor, it can be rescued by other nearby patches.

The Impacts of Chemicals on Spatially Explicit Populations

Using a template from Wu et al. (1993), my colleagues and I have developed a computer model of a generic animal patchy population that has at least one contaminated patch (Spromberg et al., 1998; Landis and McLaughlin, 2000; McLaughlin and Landis, 2000; Landis, 2002). The basic outline of the simulation model is presented in Figure 10.3. The single species patchy population model is based upon deterministic equations for growth of the population, migration between patches, and the fate of the toxicant. In some of the simulations, the toxicant is persistent; in others, the toxicant is degraded. The models use a statistical distribution, the Poisson, in a stochastic (probabilistic) function to estimate exposure of the organism to the toxicant within a habitat patch. The amount of toxicant to which the organisms are exposed in the habitat patch is determined by the persistence of the chemical in the patch and the chance encounter of the organism with the toxicant.

The simulations begin with a standard set of initial conditions: an amount of toxicant in one or more of the patches, an initial population size, and a set distance between the habitat patches. The model is then

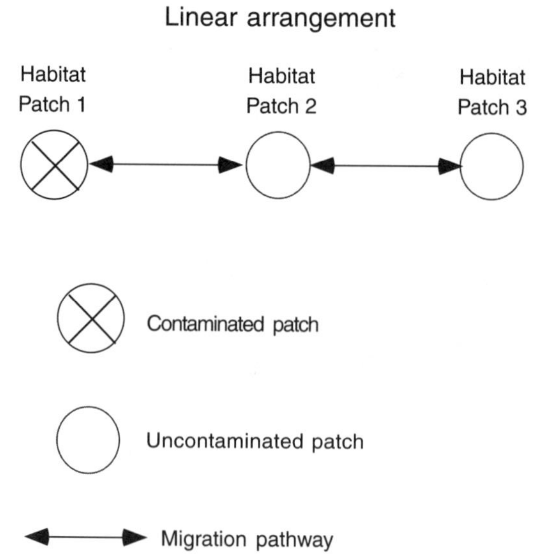

Figure 10.3 Arrangement of habitat patches in the patchy population models used in our studies. The contaminated patch could be in any position within the model landscape. Distances between habitats could also be altered, changing the rate of migration between each habitat patch.

run, and the results of each run are used for the next iteration. The models typically run for 300 iterations, and the sizes of the populations in each habitat patch are plotted.

In our first simulations, a persistent toxicant was placed in the end patch, and all of the patches had the same initial population size. The first finding is that populations in patches removed from the contamination were still affected by the presence of the toxicant (Figure 10.4). The effects were the reduction of the population below carrying capacity and a fluctuation in population size in all of the patches. The reduction in number and the fluctuations in the nondosed patches resulted in population sizes that were equal to those in the dosed habitat patch, even with a dose equivalent to the EC50 (Panel A). In the simulations when the

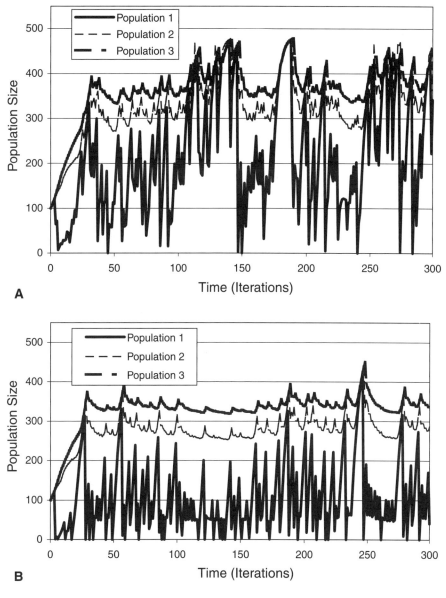

Figure 10.4 Action at a distance: In the three patch simulations where the end contaminated patch has a toxicant concentration equal to an EC50 (Panel A), the dosed patch has wide fluctuations. However, there is an occasional overlap among all three populations. The populations in nondosed habitat patches are below the carrying capacity of 500 and exhibit fluctuations. Even at an EC100 (Panel B) in the dosed patch, organisms are still extant and occasionally reach numbers comparable with the nondosed patches. In both simulations the population in the dosed habitat patch is rescued by the other patches. The initial population size for each habitat patch was 100, and the distance between the patches is equal.

dosed patch was at EC100 (Panel B), organisms could still be found in the dosed patch due to the rescue effect from other patches. Given the stochastic nature of the exposure between the toxicant and the organism, the simulations are repeatable only in type of outcome, not in the specific dynamics.

We next performed simulations with a toxicant that degrades. Our primary finding is that many of these simulations had several discrete outcomes from the same set of initial conditions. The range and types of outcomes depend on the specifics of toxicant concentration, initial population size, and distance between patches. The outcomes can be as varied as all three populations reaching carrying capacity to all three becoming extinct, with associated probabilities of occurrence.

In the first example, the simulation has three patches that are at a specified distance from each other. Only one patch contains the toxicant, which is degraded about halfway through the simulation. With an initial population size of 100 in each of the patches, 80% of the simulations resulted in all three of the populations in the patches reaching the minimum viable population. All three populations reach carrying capacity in 20% of the simulations.

In the second example, the initial populations sizes were increased to 140 for each habitat patch. In no instance did the populations decline to the MVP. In 82% of the simulations, the outcome was all three populations reaching the carrying capacity. However, 18% of the simulations resulted in a stable oscillation or bifurcation of all three populations. This is a very different outcome, yet only the initial population sizes were altered.

In the previous examples all of the habitat patches had the same initial population size. In the next series of simulations, the initial population sizes were set so that one habitat patch had a source population and the others had initial populations at the MVP or lower. The patch that acts as the source was at the end or in the middle patch in the simulations.

The finding from these simulations is that the type of habitat patch dosed is important in determining the potential range of outcomes. In a series of simulations, the source patch on the end had an initial population size of 200, and the other two patches started at 50, i.e., near the MVP. When the source was habitat patch 1 dosed with a nonpersistent toxicant, the four outcomes were: (1) In 50% of the simulations, only the population in patch 2 survived, and it survived at the MVP. (2) In 26% of the runs, the populations in patches 1 and 2 survived at the MVP. (3) In 10% of the simulations, all three populations survived at the MVP. (4) In only 14% of the runs, all three populations reached carrying capacity. Placing the toxicant in nonsource patches 2 and 3 (in the middle and at the far end) with population densities at 200, 50, and 50 for patches 1, 2, and 3, respectively, resulted in all populations reaching carrying capacity.

In another series of simulations, the arrangement was that habitat patch 2, in the middle, was the source (population 200) and dosed with the toxicant, while the other patches had population sizes of 50. In these simulations, there were only three possible outcomes observed: (1) In 56% of the runs, the populations in patches 1 and 3 existed at the MVP. (2) In 28% of the cases, all three patches reached MVP. (3) In 16% of the cases, all three populations reached carrying capacity. As with the first set of simulations, when applying the contaminant to nonsource populations (in this case, at the ends of the landscape), all patches reached carrying capacity.

In both series of simulations, the alteration in possible outcomes and outcome frequency depended on the location of the contaminated patch in the context of that specific landscape arrangement. Dosing of the source population had a much greater impact than dosing the sink populations.

The simulations summarized above clearly demonstrate that a relationship exists between patch arrangement, initial population size, and the placement and amount of the toxicant in determining the number and frequencies of discrete outcomes. Changes can lead to new outcomes and alter the probabilities of each of these outcomes. Johnson (2002) confirmed several of these findings using an individual-based modeling system. The findings of these modeling efforts led Spromberg et al. (1998) to hypothesize "action at a distance."

The basic premise of the action-at-a-distance hypothesis is that the impact of a toxicant can be transmitted to populations in other habitat patches by changes in the rate and direction of migration between patches by the individual organisms. The resultant patterns can be expressed at a variety of biochemical, organismal, and ecological levels. Action at a distance does not require the direct contamination of a patch or of any organism that resides in a patch. The patterns in the resulting dynamics can be varied and nonlinear. The questions remain about whether these results are artifacts of the numerical simulations and whether they can be found in simulated populations and ecological systems.

Partial experimental confirmation of the action-at-a-distance hypothesis has been provided recently by the research of Macovsky (1999). This study created a novel laboratory metapopulation model of a single insect species, *Tribolium castaneum*. Arranged linearly, habitat patches were linked by density-dependent dispersal of the adult. Patches were monitored for the indirect effects on population demographics beyond the patch that received a simulated adulticide over the period of approximately one and one-half egg-to-adult cycles. It was demonstrated that indirect effects do occur in patches beyond the patch where adulticide occurred. The indirect effects were dose related and correlated with distance from the directly disturbed patch.

Significance of Action at a Distance

These findings have importance beyond the realms of the computer and laboratory. First, spatially structured populations exist and are being produced by alterations to the spatial arrangements in ecological systems. Second, this spatial structure has implications on our ability to predict impacts upon ecological systems and the existence of reference or control sites.

There are populations of interest to resource managers that are spatially structured. Thorrold et al. (2001) discovered that the forage fish, *Cynoscion regalis* (weakfish), along the Atlantic Coast of North America exists as metapopulations. Similarly, the Pacific herring of the British Columbia coast has been identified as a metapopulation comprising several patches or subpopulations (Ware et al., 2000).

Alterations of the terrestrial ecological systems by both agriculture and urbanization have fragmented the landscape, producing a variety of habitat patches. Aquatic environments have been modified, altering the movement of organisms by the introduction of dams, modification of the channels, or by contamination of the water and sediment. Clearly, spatially structured populations exist, even if they were continuous in the past. There are four clear implications of the existence of spatial structure in populations.

First, in order to predict the effects of a toxicant upon spatially structured populations, it is important to understand the spatial structure. Impacts on one part of the population can have effects on other parts through changes in migration patterns across the landscape. A change in migration patterns or sources and sinks within a landscape can change the occupancy of the patches. Placement of the same amount of toxicant in different parts of the landscape changes the ranges of potential outcomes for the populations. To understand the dynamics of a population in one habitat patch, the spatial context of that population needs to be clearly understood. Impacts to a distant population may have important effects on a local population.

Second, our simulations often resulted in more than one outcome being realized by the same set of initial conditions. The reason for this is that a probabilistic function was incorporated to describe the dosing of the individuals within a patch. Outcomes as divergent as possible for a population, from extinction to reaching carrying capacity, can be possible from the same set of initial conditions. Only a clear knowledge of the properties of the organisms as individuals, of the distribution of the toxicant, and the spatial arrangement of the populations will allow an accurate prediction of the range of outcomes.

Third, action at a distance can occur: Individuals within a habitat patch need not be exposed to a contaminant for impacts to be realized upon the population. Measurement of contaminant levels in organisms of that patch will not indicate any exposure, although there is an impact due to contamination in another part of the landscape. Exposure and effect need to be understood in a landscape context.

Fourth, if one habitat or patch is connected to another contaminated site by migration, it cannot serve as a reference site. Changes in one part of a landscape can be transmitted throughout. Real impacts can be masked, as in our first simulation example. This brings the question of whether areas clearly not linked by migration can be used. In such a case, there might not be sufficient genetic and community similarity to serve as a reference. The conclusion is that there is no such thing as a reference site when it comes to populations and landscapes.

III.D   Demographic Change

... ten times twelve solar years were the term fixed for the life of man, beyond which the gods themselves had no power to prolong it; that the fates had narrowed the span to thrice thirty years, and that fortune abridged even this period by a variety of chances...

Niebuhr's *History of Rome* as cited in Deevy (1947)

A more comprehensive analysis can be made by separately examining population **vital rates**, rates at which important life-cycle events or processes such as birth, migration, and death occur for individuals in populations. Vital rates can be considered for age classes or life stages to obtain a rich understanding of population qualities such as rate of change and stable age structure. Life tables are constructed with these age-specific vital rates to predict population qualities influencing persistence. As suggested by Bezel and Bolshakov (1990), such information is critical for accurate assessment of population effects of pollutants because irreversible shifts in population structure are more often the cause of population extinction than outright death of individuals.

Life tables may include only survival data similar to that modeled in the previous chapter. The conventional notation includes x as the unit of age and l_x as the number of individuals in a cohort that are alive at x. Such l_x **life tables or schedules** summarize information, just as the survival models described earlier might. However, no specific underlying distribution need be assumed in the l_x schedule (Table 10.1). The **age-specific number of individuals dying** ($d_x = l_x - l_{x+1}$) and the **age-specific death rate** (probability of dying in interval x or $q_x = d_x/l_x$) can also be derived from l_x.

Some ecologically meaningful statistics can be generated from l_x tables alone, as illustrated with Table 10.2. This table is a rendering of the high-dose survival data in Table 10.1; however, l_x is now expressed as the number of survivors. The average years lived (L_x) is estimated as ($l_x + l_{x+1}$)/2 for each age class. The total years lived (T_x) for the age class is estimated by summing the L_x values from the bottom of the chart (e.g., for x = 21 here) up to the pertinent age class (x). The **expected life span** for individuals of age x can then be estimated as $e_x = T_x/l_x$. Bechmann (1994) measured changes in expected life span of a marine copepod exposed to copper and found an increase in life span at low copper concentrations, suggesting a hormetic or compensatory response by the copepod population.

Age-specific birth rates (m_x, the mean number of females born to a female of that age class) can be added to produce an $l_x m_x$ **life table**. Much more information can then be extracted from the population. Newman (1995) gave the fictitious example (Table 10.3) of an $l_x m_x$ table for a population living in a contaminated mesocosm. The expected number of females to be produced during the lifetime of a newborn female (R_0 or **net reproductive rate**) can be estimated from this table as the sum of the $l_x m_x$ products in the table. In this case ($R_0 = 0.846$), each female is not replacing herself, and the population size will likely decline if conditions do not change. The **mean generation time** (T_c) is estimated as the sum of the $x l_x m_x$ column divided by R_0. (The x used here is the midpoint of the age class, e.g., 0.5 for the 0 to 1-year age class.) The mean generation time for this population is 1.435/0.846 or 1.7 years. An approximation of the intrinsic rate of increase is ln R_0/T_c or –0.098, indicating a decline in population size through time. The intrinsic rate of increase is more accurately estimated with the **Euler-Lotka equation**:

$$\sum_{x=0}^{\text{infinity}} l_x m_x e^{-rx} = 1 \tag{10.12}$$

The estimate of r generated above can be used initially in the Euler-Lotka equation. The estimated r is then adjusted upward or downward until the solution to the left side of the equation is sufficiently close to 1. The r resulting in such a condition is deemed adequate for most purposes. Use of the Euler-Lotka equation carries the assumption that the population is stable. If conditions do not change with time, a population with a particular r will eventually establish a stable distribution of individuals among the various age classes. Such a population is called a **stable population**.

For a stable population, the expected contribution of offspring during the life of an individual (**reproductive value** or V_A) is easily calculated for each age class. This V_A can be thought of as

Table 10.1 Survival (l_x) for *D. pulex* Exposed to Zero and 75.9 R h^{-1} of Radiation

Age Class or x (days old)	Control (zero R h^{-1})	High Dose (75.9 R h^{-1})
0	1.00	1.00
1	1.00	0.98
2	1.00	0.96
3	1.00	0.96
4	1.00	0.96
5	1.00	0.96
6	1.00	0.96
7	0.98	0.96
8	0.98	0.96
9	0.98	0.96
10	0.98	0.96
11	0.98	0.96
12	0.98	0.94
13	0.98	0.94
14	0.98	0.94
15	0.98	0.94
16	0.98	0.94
17	0.98	0.81
18	0.98	0.67
19	0.98	0.29
20	0.98	0.17
21	0.98	0.02
22	0.91	0.00
23	0.81	0.00
24	0.58	0.00
25	0.49	0.00
26	0.35	0.00
27	0.28	0.00
28	0.19	0.00
29	0.14	0.00
30	0.14	0.00
31	0.14	0.00
32	0.07	0.00
33	0.05	0.00
34	0.05	0.00
35	0.00	0.00

Note: The proportion of the original cohort surviving in each treatment is shown instead of raw counts of individuals. These proportions are identical to the cumulative survivals or S(t) described for survival data in Chapter 9.

Source: Derived from Table I in Marshall, J.S., *Ecology,* 43, 598–607, 1962.

the reproductive value for the x age class, divided by that for a neonate, i.e., V_x/V_0.

$$V_A = \sum_{x=A}^{infinity} \frac{l_x}{l_A} m_x \qquad (10.13)$$

This V_A can be used to reinforce the point discussed previously that it is inappropriate to estimate the consequences to population viability based solely on reductions in most sensitive life stage survival and NOEC values for lowered reproduction. Reproductive value can be used directly to assess such consequences to populations. Indeed, Kammenga et al. (1996) clearly demonstrated with such an approach that the most sensitive life cycle trait of the nematode *Plectus acuminatus* was not the most critical demographic quality impacted by cadmium exposure. For example, a toxicant that eliminated the age class or set of age classes with the highest reproductive value would

Table 10.2   Estimation of Expected Life Span (e_x) from the High Dose in Table 10.1
·   Assuming 100 Individuals in the Original Cohort

Age (x)	l_x[a]	d_x	q_x	L_x	T_x	e_x (days)
0	100	2	0.02	99.0	1774.0	17.7
1	98	2	0.02	97.0	1675.0	17.1
2	96	0	0.00	96.0	1578.0	16.4
3	96	0	0.00	96.0	1482.0	15.4
4	96	0	0.00	96.0	1386.0	14.4
5	96	0	0.00	96.0	1290.0	13.4
6	96	0	0.00	96.0	1194.0	12.4
7	96	0	0.00	96.0	1098.0	11.4
8	96	0	0.00	96.0	1002.0	10.4
9	96	0	0.00	96.0	906.0	9.4
10	96	0	0.00	96.0	810.0	8.4
11	96	2	0.02	95.0	714.0	7.4
12	94	0	0.00	94.0	619.0	6.6
13	94	0	0.00	94.0	525.0	5.6
14	94	0	0.00	94.0	431.0	4.6
15	94	0	0.00	94.0	337.0	3.5
16	94	13	0.14	87.5	243.0	2.6
17	81	14	0.17	74.0	155.5	1.9
18	67	38	0.57	48.0	81.5	1.2
19	29	12	0.41	23.0	33.5	1.2
20	17	15	0.88	9.5	10.5	1.1
21	2	2	1.00	1.0	1.0	1.0
22	0	—	—	—	—	—

[a] The l_x in this table is l_x in Table 10.1 multiplied by 100.

Table 10.3   A $l_x m_x$ Life Table for a Fictitious Population in a Contaminated Mesocosm

Age Class	Midpoint of Class (x)	l_x	m_x	$l_x m_x$	$x l_x m_x$
0 to <1 year	0.5	1.000	0.000	0.000	0.000
1 to <2 year	1.5	0.312	2.238	0.698	1.047
2 to <3 year	2.5	0.095	1.390	0.132	0.330
3 to <4 year	3.5	0.037	0.410	0.015	0.053
4 to <5 year	4.5	0.003	0.400	0.001	0.005
5 to <6 year	5.5	0.000	—	$\Sigma = 0.846$	$\Sigma = 1.435$

Source: Modified from Newman, M.C., Quantitative Methods in Aquatic Ecotoxicology, Lewis
Publishers, Boca Raton, FL, 1995.

have strong, adverse consequences to the population's viability. Conversely, a toxicant might have
very minor consequences relative to population viability if it eliminated large numbers of individuals
in an age class that had very low reproductive value. Unfortunately, this approach has been largely
ignored for assessments of effect, although Barnthouse et al. (1987) did effectively include reproductive value in their models of fish population response to contaminants. Martínez-Jerónimo et al.
(1993) did estimate V_A for *Daphnia* populations exposed to increasing concentrations of kraft mill
effluent.

Marshall (1962) exposed female *D. pulex* to various levels[6] of ^{60}Cobalt γ radiation (Figure 10.5)
and estimated age-specific changes in birth and death rates. The changes in intrinsic rate of increase
for the exposed populations were then calculated from these vital rates. Stable age structure of

[6] Radiation is expressed here as the exposure dose rate. Units of dose are Roentgen per hour. A **roentgen** (R) is a measure
of the amount of energy deposited in some material by a certain amount of radiation. By convention, it is expressed relative
to energy dissipation in 1 cm^3 of dry air. Use of R to express dose allows one to normalize for the different amounts of
energy that are deposited by different types of radiation. You will remember from Chapter 1 that R was incorporated in the
rem, Roentgen equivalent man. The rem is the measure of radiation dose expressed in units of potential effect to humans.

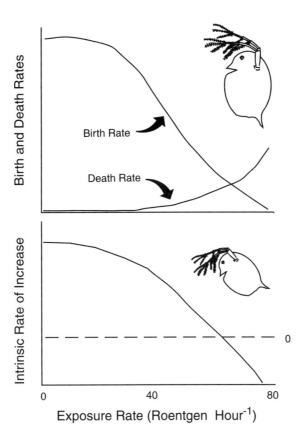

Figure 10.5 Birth, death, and intrinsic rate of increase for *D. pulex* exposed continuously to different amounts of ^{60}Cobalt γ radiation. Age-specific survival (l_x) and fertility (m_x) rates were measured for females exposed to different levels of radiation, and the intrinsic rate of increase was estimated by iterative solution of the Euler-Lotka equation. (Modified from Figures 1 and 2 of Marshall, J.S., *Ecology*, 43, 598–607, 1962.)

populations can be estimated if the rate of increase (r or λ) and l_x values are known, and this information can then be used to suggest contaminant effects. The proportion of all individuals in age class x (C_x) is estimated using Equation 10.14:

$$C_x = \frac{\lambda^{-x} l_x}{\sum_{i=0}^{\text{infinity}} \lambda^{-i} l_i} \tag{10.14}$$

Marshall (1962) applied this approach to predict stable age structures for *D. pulex* exposed to radiation (Figure 10.6). The stable age structure slowly shifted away from a preponderance of neonates at zero R d^{-1} to a structure composed primarily of older individuals at 75.9 R d^{-1}.

The life-table approach to demographic analysis just presented was used because it involves relatively straightforward mathematics. A more powerful approach based on matrix algebra was developed by Leslie (1945, 1948) and has been widely applied to demography (e.g., Caswell, 2001). A **Leslie matrix** can be constructed that has the probability (P_x) of (1) a female alive in period x_i to x_{i+1} being alive in period x_{i+1} to x_{i+2} in the matrix subdiagonal and (2) the number of daughters (F_x) born in the interval t to t + 1 per female of age x to x + 1 in the top row of the matrix. The Leslie matrix is multiplied by a vector of the number of individuals alive at each period, e.g., x_i to x_{i+1}, at some time (t) to estimate the number of individuals expected in each period at some future

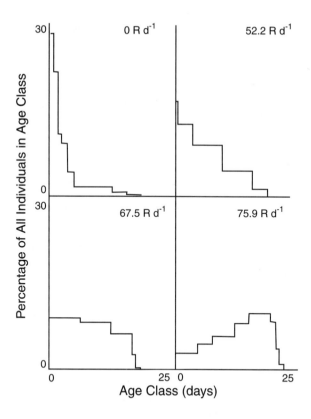

Figure 10.6 Stable age structures of *D. pulex* exposed to 0 to 75.9 Roentgens per hour of ^{60}Cobalt γ radiation. Note the gradual trend toward a structure with few neonates as dose rate increases. (Adapted from Figure 3 of Marshall, J.S., *Ecology*, 43, 598–607, 1962. With permission.)

time, t + 1. This multiplication can be repeated for many time steps to project the population size and structure through time.

$$
\begin{bmatrix}
F_0 & F_1 & F_2 & F_3 & \cdots & F_\omega \\
P_0 & 0 & 0 & 0 & \cdots & 0 \\
0 & P_1 & 0 & 0 & \cdots & 0 \\
0 & 0 & P_2 & 0 & \cdots & 0 \\
\cdots & \cdots & \cdots & \cdots & \cdots & \cdots \\
0 & 0 & 0 & 0 & P_{\omega-1} & 0
\end{bmatrix}
\cdot
\begin{bmatrix}
n_{0,t} \\
n_{1,t} \\
n_{2,t} \\
n_{3,t} \\
\cdots \\
n_{\omega,t}
\end{bmatrix}
=
\begin{bmatrix}
n_{0,t+1} \\
n_{1,t+1} \\
n_{2,t+1} \\
n_{3,t+1} \\
\cdots \\
n_{\omega,t+1}
\end{bmatrix}
\tag{10.15}
$$

Demographic methods are being used to great advantage in an increasing number of studies of copepods (Daniels and Allan, 1981; Bechmann, 1994; Green and Chandler, 1996) and cladocerans (Winner et al., 1977; Schober and Lampert, 1977; Daniels and Allan, 1981; Hatakeyama and Yasuno, 1981; Van Leeuwen et al., 1985; Day and Kaushik, 1987; Wong and Wong, 1990; Martínez-Jerónimo et al., 1993; Koivisto and Ketola, 1995). Sibly (1996) tabulated a literature search of such studies, which included studies primarily of arthropods, but also of algae, gastrotrichs, nematodes, and humans.

Caswell (1996) developed a demographic assay for the effects of pollutants and advocated more attention to demography. Ferson and Akçakaya (1991) produced software that allows risk assessors to estimate population persistence probabilities based on pollution-induced changes in demographic qualities. A significant literature and theory are beginning to accumulate on alterations of life history traits associated with contamination (e.g., Adams et al., 1992; Bezel and Bolshakov, 1990; Holloway et al., 1990; McFarlane and Franzin, 1978; Mulvey et al., 1995; Neuhold, 1987; Postma et al., 1995a, b; Schnute and Richards, 1990; Sibly and Calow, 1989; Sibly, 1996).

Vignette 10.2   Contaminant Effects on Population Demographics

Valery E. Forbes

Roskilde University, Roskilde, Denmark and Peter Calow, The University of Sheffield, Sheffield, U.K.

General Introduction

We shall consider contaminant effects on population demographics and complications associated with them by focusing on population growth rate (pgr), which can be expressed as either r or λ. For each section, there will be a general introduction followed by evidence derived from experimental observations, analytical techniques, and simulations.

Theoretical Ideal versus Practical Possibilities

Introduction

For wildlife, we are most often concerned with protecting populations rather than just the individuals within them. Yet it is rarely possible to carry out sufficiently comprehensive studies on the effects of chemicals on life-table variables to predict effects on pgr directly. Instead, ecotoxicological tests focus on the effects of chemicals on survival in short-term experiments at relatively high concentrations, and on fecundity and developmental rates in longer-term experiments at relatively low concentrations. Thus, the key question is, "To what extent are these measurements on individual-level variables likely to be more or less sensitive to chemicals than is pgr itself?"

Review of Literature

In a review of 41 studies that included a total of 28 species and 44 toxicants, Forbes and Calow (1999) found that out of 99 cases considered, there were only 5 in which chemical effects on pgr were detected at lower exposure concentrations than those resulting in statistically detectable effects on any of the individual demographic variables. In 81.5% of the cases considered (out of a total of 81), the percentage change in r was less than the percentage change in the most sensitive of the individual demographic traits, 2.5% in which the percentage change in r was equal to that of the most sensitive trait, and 16% in which the percentage change in r was greater than the percentage change in the most sensitive demographic trait. Although proportional changes in r were significantly correlated with proportional changes in fecundity and with time to first reproduction, these correlations were weak, and trend analysis indicated that the relationships were nonlinear. Surprisingly, the correlation between proportional reduction in overall survival (i.e., juvenile and adult survival probabilities pooled) and proportional reduction in r was not statistically significant. Overall, there was no consistency in which of the measured individual-level traits was the most sensitive to toxicant exposure, and none of them, considered individually, was a very precise predictor of toxicant effects on r.

Elasticity Analyses

Forbes et al. (2001a) used a generalized form of elasticity analysis on a two-stage life-cycle model to consider the extent to which small percentage changes in individual life-cycle variables result in greater or lesser changes in λ. In the neighborhood of steady state (i.e., $\lambda = 1$), small changes in time to first reproduction, time between reproductive events, juvenile and adult survivorship, and fecundity resulted in, at most, the same percentage changes in λ. However, if λ were allowed to increase above one, elasticities might be such that small changes in life-cycle variables would lead to proportionally greater changes in λ.

Simulations

Following from the analytical approach, Forbes et al. (2001a) used simulation to consider how small-percentage changes in life-cycle variables affected λ in several typical life cycles. These were organized to represent a benthic invertebrate, a fish, a daphnid, and an algal species. For all of the life cycles, the effect of impacting one life-cycle trait at a time by 10% resulted in a less-than-or-equal percentage change in λ. For three of the four life cycles, simultaneously impairing all life-cycle traits by 10% still led to a less than 10% reduction in λ. The exception was the benthic macroinvertebrate in which a simultaneous change of 10% in all life-cycle variables resulted in a 17% reduction in λ. This is due to life-cycle complications that will be discussed further below.

Conclusions

From all the evidence, it would seem that, for most cases, protection limits based on juvenile survival, which is very typically used as an ecotoxicological endpoint, are lower than those based on λ. This conclusion can also be extended to other life-cycle variables. Hence, using individual-level endpoints in ecotoxicology appears to be protective and could be overprotective in certain circumstances. However, Forbes et al. (2001a) also identified conditions in which this was not the case. So where resources allow, full life-table analyses would always be preferable to risk assessments based on individual life-cycle variables only.

Life Cycle as a Complication

Introduction

It is implicitly assumed in ecotoxicological analyses that the relationship between effects of chemicals on individual-level variables means the same for different taxa. Thus 50% mortality (LC50) is treated as implying the same in demographic terms for all species. Yet analyses of demographic models very clearly show that this is not the case and that the demographic consequences of similar changes in individual-level variables can vary widely among species with different life cycles.

Analytical Models

Calow et al. (1997) considered a series of simplified, but plausible, scenarios to explore the complications introduced by life cycles. A number of general conclusions arose out of this analysis. (1) As expected, the effect on pgr of a toxicant that reduces juvenile survivorship or fecundity will be greater for **semelparous** (i.e., species that reproduce once) as compared with **iteroparous** (i.e., species that reproduce more than once) **species**, and the reverse will be the case for effects of toxicants on adult survival. (2) Iteroparous species with life cycles in which the time to first reproduction is shorter than the time between broods will be more susceptible to toxicant impacts on survival or fecundity than will species in which time to first reproduction is longer than the time between broods. (3) Anything that shortens time to first reproduction relative to the time between broods (e.g., increased temperature, increased food availability) is expected to increase the population-level impact of toxicant-caused impairments in survival or fecundity. (4) Lengthening of the time to first reproduction should lessen the population impact of toxicant-caused impairments in survival or fecundity.

Simulation

The simulation work described above is relevant for a consideration of life-cycle issues. The comparison among the four life-cycle types showed that a 10% reduction in juvenile survival (i.e., an LC10) would result in a 10% reduction in λ for a semelparous benthic invertebrate life cycle, a 5% reduction in λ for a green-algal life cycle, a 2% reduction in λ for an iteroparous fish life cycle, and only a 0.6% reduction in λ for a daphnid life cycle (Figure 10.7). Clearly, a chemical having similar effects on juvenile mortality would be expected to have vastly different population-level consequences for these life cycles. Likewise, although the benthic invertebrate might have a higher LC10 value than the daphnid for a given chemical, its population dynamics could nevertheless be more sensitive. The analysis indicated that a 5% reduction in juvenile survival of the benthic invertebrate life cycle would have the same effect on λ as an 80% reduction in juvenile survival of the daphnid life cycle.

Conclusions

An important consequence of this analysis is the realization that life-cycle variability can play an important part in differential sensitivity to chemicals among species. Yet, in general, species sensitivity

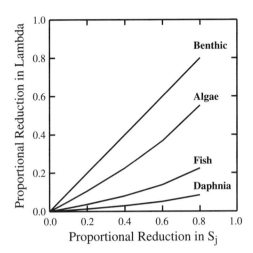

Figure 10.7 The relationship between S and lambda (λ) for four categories of common toxicity testing species.

distributions are more often constructed on the basis of differences in responses of individual-level variables, and this is a reflection of physiological rather than life-cycle variability. In theory, life-cycle variability can magnify, ameliorate, or reverse these kinds of differences, and this can cause problems for carrying out risk assessments using species-sensitivity distributions based only on physiological variability (Forbes and Calow, 2002).

Density as a Complication

Introduction

Most ecotoxicological work is carried out, if not on "growing" populations, certainly in systems designed not to provide restrictions on food supply or space. Similarly, all of the theoretical models described above do not specifically take density effects into account. Yet it is presumed that most natural populations are under some form of regulation. Population growth rate is generally under regulation such that there are negative relationships between population size and pgr (Sibly and Hone, 2002). To what extent is this complication likely to mean that observations/predictions made under nonregulated conditions under- or overestimate responses of density-regulated populations to chemicals?

Analytical Work

Forbes et al. (2001b) investigated the interaction between toxicity and density using a two-stage demographic model. The key determinant of toxicant-density interactions was the shape of the functional relationship between λ and the life-cycle variables contributing to it. Depending on these shapes, various outcomes—in terms of the additivity, less-than-additivity, or more-than-additivity—between density and toxicant effects on λ were possible. Clearly, if we knew enough about these functional relationships, we should be able to predict the responses of populations living under density-dependent control to chemical exposure. However, because the number of factors influencing the outcome is large, this would seem unlikely, and at least for the near future we shall have to approach the problem experimentally.

Review

Forbes et al. (2001b) reviewed experimental studies exploring density × toxicant interactions. They produced mixed results, with some showing additive interactions between density and chemical effects on pgr, whereas others found less-than-additive effects or more-than-additive effects. There is even some indication that the form of the interaction may vary across a chemical concentration gradient, with effects shifting from less-than-additive at low toxicant concentrations to more-than-additive at higher toxicant concentrations (Linke-Gamenick et al., 1999). In addition, it appears that the kind of interaction that is observed is, to an important extent, constrained by the kind of experimental design employed.

Simulation

There have been very few simulation studies carried out to date. One by Grant (1998) explored the consequences of simulated density effects on previously published life-table response data for *Eurytemora affinis* exposed to dieldrin. The other (Hansen et al., 1999) simulated potential influences of competition and predation on population responses of *Capitella capitata* type I to nonylphenol. Both studies suggested that toxicants are likely to have less of an effect on density-regulated populations than on ones growing exponentially.

Conclusions

The main messages from this vignette are that density needs to be taken into account in ecotoxicological assessments, that the effects of density are difficult to predict *a priori*, and that therefore density needs to be introduced more explicitly into ecotoxicological tests, with care being taken to use an appropriate experimental design.

Suggested Supplemental Readings

Caswell, H., *Matrix Population Models: Construction, Analysis, and Interpretation*, 2nd ed., Sinauer Associates, Sunderland, MA, 2001.
Stearns, S.C., *The Evolution of Life Histories*, Oxford University Press, Oxford, U.K., 1992.

III.E   Energy Allocation by Individuals in Populations

Responses of individuals making up a population have been described relative to energy allocation. These responses can produce significant changes in population demographics. Sibly and coworkers (Sibly and Calow, 1989; Holloway et al., 1990; Sibly, 1996) describe an **optimal stress response** for species exposed to toxicants. The optimal stress response involves a shift in the balance in energy allocation between somatic growth rate and longevity (survival) to optimize Darwinian fitness under stressful conditions. Sibly (1996) gives several examples of toxicants influencing these crucial demographic qualities. Kooijman (Kooijman, 1993; Kooijman and Bedaux, 1996) provides a theory-rich approach utilizing energy budgeting for individuals as the central theme through which survival, growth, and reproduction are affected by toxicants. This **dynamic energy budget (DEB) approach** has been formalized into the DEBtox model (Kooijman and Bedaux, 1996).

Toxicant-related effects can be interpreted in the context of the more encompassing **principle of allocation**: There exists a cost or trade-off to every allocation of energy resources. Also discussed as the **concept of strategy**, this principle is used to interpret responses ranging from immediate responses to stress (e.g., early sexual maturity or delayed growth) (Sibly, 1996) to evolutionary responses (e.g., enhanced metal tolerance at the cost of impaired growth rate) (Wilson, 1988) (Figure 10.8). There exists a limited amount of energy available that must be optimally allocated among different processes and functions by an individual in order to enhance Darwinian fitness. Energy spent producing defense proteins (e.g., cytochrome P-450 proteins) cannot be used for reproduction or growth.

Related to this concept is the **limited life span paradigm** that a genetically defined maximum life span is an inherent quality of an individual.[7] Parsons (1995) extends this to the more germane

---

[7] Curtsinger et al. (1992) disagree with this concept.

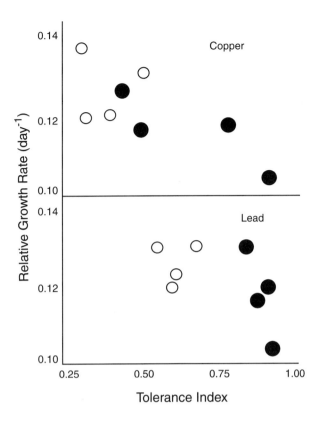

Figure 10.8 The cost in terms of growth rate for plant (*Agrostis capillaris*) strains with varying degrees of heavy metal tolerance. Plants were taken from reference locations (open circles) and areas with long histories of heavy metal mining in Wales (closed circles). Growth was measured in near-optimal media with heavy metal concentrations below those causing stress. Tolerance was measured as [2 (root length in heavy metal solution)]/[root length in heavy metal solution + root length in control solution]. Plants adapted to high metal concentrations (high tolerance index) had generally slower growth under ideal conditions than nontolerant strains, indicating a trade-off associated with tolerance. (Modified from Figures 1 and 2 of Wilson, J.B., *Evolution,* 42, 408–413, 1988.)

rate of living theory of aging, i.e., that the total metabolic expenditure of a genotype is generally fixed, and longevity depends on the rate of energy expenditure. He notes that the immediate response to stress, an increase in metabolic rate, diminishes longevity. He further advocates a **stress theory of aging**, i.e., that selection takes place for resistance to stress, and that as an epiphenomenon, individuals resistant to stress will predominate in extreme age classes of a population. The diminution of homeostasis under stress with age should be lowest in individuals with the highest longevity. He illustrates the concept with populations of *Drosophila*, correlating longevity with reduced metabolic rates under stress and increased antioxidant activity.[8] He refers to the work of Koehn and Bayne (1989), in which high stress resistance was associated with efficient use of metabolic resources. In discussing genetic evidence supporting this stress theory of aging, he notes that different genetically determined forms of the glycolytic enzyme, glucosephosphate isomerase (abbreviated as GPI or PGI), seem to impart an advantage to individuals in a variety of species under stressful conditions. As we will soon see, this observation is reinforced relative to ecotoxicology by recent genetic studies.

---

[8] The observation of increased longevity with increased antioxidant activity could also be used to support the **disposable soma theory of aging**. This theory suggests that aging is a consequence of a gradual accumulation of cellular damage via random molecular defects (Parsons, 1995).

IV POPULATION GENETICS

Kettlewell (1955) provided a telling investigation of **industrial melanism** (the gradual increase to predominance of melanic forms in industrialized regions) in pepper moths (*Biston betularia*) of Great Britain. These moths are active at night but remain still on surfaces during the day in order to avoid the notice of visual predators, especially birds. The fitness of rare dark morphs quickly increased relative to the light morphs as surfaces darkened with soot. Although not thought of as such, this premier example of natural selection in the wild is an equally good example to ecotoxicologists of contaminant effects on population genetics. It also reinforces the important point that physical or chemical contaminants do not have to kill or impair individuals outright in order to influence populations. Rapid shifts in population genetics occurred as a consequence of shifting Darwinian fitness relative to a species interaction, predation.

Vignette 10.3   Industrial Melanism

Bruce Grant

College of William and Mary, Williamsburg, VA

Industrial melanism refers to the evolution of dark body colors in animal species that live in habitats blackened by industrial soot. Kettlewell (1973) and Majerus (1998) list some of approximately 100 examples of industrial melanism that have been reported in a variety of species. Most examples are insects, and the vast majority of these are Lepidoptera, especially night-flying moths that rest by day. By far, the most thoroughly documented example is the peppered moth, *Biston betularia*.

The common name for this moth derives from the appearance of the typical phenotype, a pale moth "peppered" with white and black scales. Melanic phenotypes are effectively solid black and are sometimes called *carbonaria* (Figure 10.9). Intermediates (or *insularia*) also occur. The range of phenotypes is genetically determined by multiple alleles at a single locus (Lees and Creed, 1977). Most population studies focus on temporal and geographic frequency distributions of the melanics as distinct from nonmelanics (typicals plus *insularia*).

The melanic form was first discovered near Manchester, England, in 1848. By 1895 about 98% of the specimens near Manchester were melanic, and this once rare phenotype had spread across the industrial regions of Britain. With only one generation per year, the nearly complete reversal in phenotype frequency was astonishingly rapid.

Figure 10.9   Melanic and typical peppered moths, *Biston betularia*. At rest the wings span 4–5 cm. (Photo by Bruce Grant.)

Tutt (1896) spelled out the basic story of industrial melanism in peppered moths much as it is taught today. The moths are active at night and remain still during daylight hours, resting or hiding on the surfaces of trees. Most insectivorous birds hunt by day and locate prey primarily by vision. Moths that are well concealed against the backgrounds upon which they rest are more likely to escape detection by birds than are conspicuous moths. Tutt proposed that "speckled" peppered moths in undisturbed environments gain protection from predators by their resemblance to lichens. In manufacturing regions where lichens have been destroyed by pollution and the surfaces of trees blackened by soot, the speckled forms fall victim to bird predators, whereas the melanic forms escape detection and pass their traits on to their progeny.

Tutt's ideas were not tested until the 1950s, when Kettlewell (1973) initiated a series of experiments to determine whether or not birds actually ate melanic and typical peppered moths selectively. His experiments included three main components: (1) quantitative rankings of conspicuousness (to the human eye) of typical and melanic phenotypes placed on various backgrounds; (2) direct observations of predation by birds on moths placed onto tree trunks; (3) recapture rates of marked moths released onto trees in polluted and unpolluted woodlands. The design and execution of Kettlewell's experiments have been subjected to considerable scrutiny over the years. Criticisms about densities of the moths, the positions on trees where they were placed, the methods of release, and the mixed origins of moth stocks—all of these inspired additional experiments by other workers. The bulk of that work is in qualitative agreement and corroborates rather than contradicts the basic conclusion that conspicuous moths are more readily eaten by birds (reviewed by Majerus, 1998).

Because reflectance from the surface of tree bark is strongly negatively correlated with atmospheric levels of suspended particles (Creed et al., 1973), the testable prediction is that melanic phenotypes should be more common in sooty, polluted regions than they are in unpolluted regions. This prediction was clearly met by the frequency of melanic peppered moths in populations surveyed across Britain during the 1950s (Kettlewell, 1973).

Several studies have shown that the geographic distribution of melanism in British peppered moths is more strongly correlated with SO_2 concentration than with smoke (Steward, 1977). Such SO_2, as a gas, is more widely dispersed in the atmosphere than particulate matter, which tends to settle locally as soot. Although selection may operate locally, gene flow from the migration of individuals among populations has contributed to more gradual clines in melanism in peppered moths, a relatively mobile species, than in the sedentary *Gonodontis* (= *Odontoptera*) *bidentata*, in which the frequency distribution of melanic phenotypes is sharply subdivided over its range (Bishop et al., 1978). The power of gene flow to obscure genetic adaptation to local conditions resulting from selection was too little appreciated by early workers, who entertained various forms of nonvisual selection in attempts to account for apparent anomalies in clines.

In 1956, Britain initiated the Clean Air Acts to establish so-called smokeless zones in heavily polluted regions. Following significant reductions in atmospheric pollution, particularly SO_2, melanic peppered moths have declined in frequency as the typical form recovered (Clarke et al., 1985). Indeed, the decline in melanism in the latter half of the 20th century is much better documented than the increase in melanism in the latter half of the 19th century. A continuous record begun in 1959 by C.A. Clarke near Liverpool includes over 18,000 specimens that show a drop in the frequency of melanics from 93% to 8% by 1996 (Grant et al., 1998). The most recent national survey taken in 1996 shows that marked declines in melanism have occurred everywhere in Britain (Grant et al., 1998). The predicted correlation between changes in the levels of pollution and in the frequencies of melanic phenotypes in peppered moth populations has been firmly established.

Parallel evolutionary changes have also occurred in the American peppered moth, *B. betularia cognataria* (Grant et al., 1996). Owen (1961) showed that the rise and spread of melanism started in America about 50 years later than in Britain. Museum collections do not include any melanic specimens prior to 1929 in the vicinity of Detroit, MI, but by 1959, when Owen began sampling moth populations there, the frequency of melanics exceeded 90%. In 1963, clean-air legislation was inaugurated in the United States. Records for southeastern Michigan show that atmospheric SO_2 and suspended particulates have declined significantly, and the frequency of melanic peppered moths had fallen below 20% by 1994 (Grant et al., 1996). The decline in melanism continues even now, with only 15 melanic moths (5.3%) observed among 283 specimens collected in 2001 (Grant and Wiseman, 2002). The significance of a sharp decline in melanism in American peppered moths coincident with reductions in atmospheric pollution is complemented by comparisons of SO_2 concentrations from southeastern Michigan to western Virginia, where melanism has not exceeded 5% over the same time interval. The Michigan location recorded higher concentrations of SO_2 than the Virginia location in 23 of the 25 years measurements were taken at both places (Grant and Wiseman, 2002).

The recent decline in melanism in Michigan, and its near absence in Virginia, might be mistaken for *thermal* melanism, a phenomenon not uncommon in the Lepidoptera (Majerus, 1998). However, there is direct evidence against thermal melanism in *Biston betularia*—namely, the absence of latitudinal clines. This is clear from latitudinal variation in melanic frequencies in Britain (Kettlewell, 1973) and in Scandinavia (Douwes et al., 1976).

Both American and British peppered moth populations are now converging on monomorphism for their respective typical phenotypes correlated with reduced levels of atmospheric pollution on both sides of the Atlantic. These phenotypic changes reflect genetic changes as the populations continue to adapt by natural selection to environments modified by human activity.

IV.A   Change in Genetic Qualities

Toxicants can influence population genetics in many ways, although ecotoxicologists have been preoccupied with selection-associated changes. By mechanisms already discussed, chemicals and radionuclides can directly change the genetic qualities of individuals within a population. Baker et al. (1996) describe large changes in DNA (the gene for mitochondrial cytochrome b) in voles (*Microtus arvalis* and *M. rossiaemeridionalis*) living near the damaged Chernobyl reactor.

Toxicants can influence population genetics in less direct ways. A toxicant can reduce the **effective population size** (the number of individuals contributing genes to the next generation) and result in a net loss of genetic variation. A **genetic bottleneck** occurs if there are too few individuals available to ensure that an allele makes it into future generations. Under less severe conditions, reduction of effective population size can accelerate the rate of loss of a rare allele from a population. **Genetic drift** (random change in allele frequencies in a population) is accelerated at low effective population sizes. The net result of a toxicant's effect on a population, even in the absence of selection, could be the loss of a specific allele or an overall decrease in genetic variability. As genetic variation is the raw material for evolutionary change, this could reduce the ability of a population to adapt to and survive future changes in its environment. Murdoch and Hebert (1994) found reduced variability in mitochondrial DNA of the brown bullhead (*Ameiurus nebulosus*) from the Great Lakes and attributed it to pollutant-induced reductions in effective population size. Kopp et al. (1992) demonstrated a reduction in genetic heterozygosity in stressed populations of the central mud minnow (*Umbra limi*).

Under the presumption that individuals possessing the most genetic variation tend to be most robust,[9] the argument has been made that an increase in genetic variability could occur in populations impacted by toxicants (Mulvey and Diamond, 1991). The individuals most able to survive and reproduce in the polluted environment may be the most heterozygous. For example, Kopp et al. (1992) noted for laboratory assays that central mud minnows (*Umbra limi*) with the highest genetic diversity had enhanced abilities to survive stressful conditions (low pH and high aluminum concentrations). Diamond et al. (1989) found that mosquitofish (*Gambusia holbrooki*) with high numbers of heterozygous loci tended to have longer times-to-death during mercury exposure than mosquitofish with fewer heterozygous loci. However, Newman et al. (1989) demonstrated later that the particular heterozygosity effect described by Diamond et al. (1989) was an artifact. Schlueter et al. (1995) found no effect of heterozygosity on time-to-death of fathead minnows (*Pimephales promelas*) exposed to high concentrations of copper. Although plausible mechanisms exist (e.g.,

---

[9] Measures of fitness have been correlated with the number of loci found to be heterozygous in individuals (Samallow and Soule, 1983; Koehn and Gaffney, 1984; Danzmann et al., 1986). The mechanism for the enhanced fitness is often assumed to be **multiple heterosis**, a generally higher fitness as a consequence of combined advantages of being heterozygous for each individual locus (heterosis). (**Heterosis** is the general term used to describe the superior performance of heterozygotes.)

Table 10.4   Qualities Modifying the Rate of Tolerance Acquisition

Quality	Specific Influence on Tolerance Acquisition
1. Genetic Qualities	
Dominance	Most rapid in early generations if tolerance is controlled by a dominant allele
Single gene vs. many genes involved	Most rapid if determined by a single gene
Two or more selection components	Opposing selection components can balance each other or slow tolerance acquisition
Relative differences in fitness	Most rapid if the differences in fitness among tolerant and sensitive individuals is large
2. Reproductive Qualities	
Rate of increase and generation time	Most rapid with high population growth rate and short generation time
Size of population	In general, smaller populations will have less variation than larger populations
3. Ecological Qualities	
Migration	Influx of nontolerant individuals due to immigration could slow tolerance acquisition
Presence of refugia	The presence of refugia such as uncontaminated areas will slow tolerance acquisition
Life stage	Sensitive life stage will have large influence on tolerance acquisition

Source: Modified from Table I in Mulvey, M. and S.A. Diamond, Genetic factors and tolerance acquisition in populations exposed to metals and metalloids, in *Metal Ecotoxicology: Concepts and Applications,* Newman, M.C. and McIntosh, A.W., Eds., Lewis Publishers, Chelsea, MI, 1991.

multiple heterosis, inbreeding depression, or overdominance), presumptions cannot be made at this time regarding any specific field situation that multiple-locus heterozygosity does or does not influence fitness relative to toxicant effects.

IV.B   Acquisition of Tolerance

Toxicants can also act as selective agents for exposed populations. **Natural selection**, the process by which genes from the most fit individuals are overrepresented in the next generation,[10] can result in enhanced tolerance to toxicants, as is amply demonstrated by the adaptation of pests to pesticides. Considerable effort is made to counterbalance tolerance acquisition by insect (e.g., Comins, 1977; Mallet, 1989), rodent (Webb and Horsfall, 1967; Partridge, 1979), and other target species of pesticides. Populations of nontarget species can also adapt and become more tolerant of toxicants. The probability of obtaining, or the rate at which they obtain, enhanced tolerance is influenced by many factors (Table 10.4).

The likely, but sometimes overlooked, consequence of exposure is local extinction of the exposed population, not enhanced tolerance (Klerks and Weis, 1987; Mulvey and Diamond, 1991). Pollution as a selection force is extreme in its rate of change relative to many other environmental factors to which organisms must adapt (Moriarty, 1983). Populations successfully adapting to so rapid a change in environmental conditions are probably exceptional. Therefore, the occasional arguments made to ease regulations—based on the premise that "adapted" field populations will be more tolerant than predicted from laboratory assays using "nonadapted" individuals—are flawed because they ignore this possibility (Klerks and Weis, 1987).

---

[10] Evolution via natural selection carries several assumptions. It is assumed that surplus numbers of individuals are produced by populations. In a particular environment, individuals vary in their abilities to survive and reproduce, i.e., their fitness. All or a portion of these differences in fitness are heritable. The net result is natural selection.

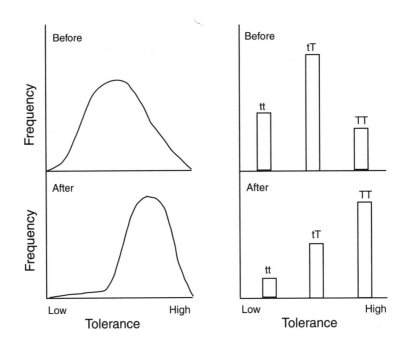

Figure 10.10   Shifts in tolerance expected under polygenic (left side of figure) or monogenic (right side of figure) control. With polygenic control, differences in tolerance will appear continuous as shown in the "before" selection panel on the left side of the figure. With selection, the mean tolerance will shift upward, and the variation about this mean will narrow. With tolerance determined by a single gene (right side), the distribution of genotypes among homozygous for intolerance (tt), homozygous for tolerance (TT), and heterozygous (Tt) will shift to a predominance of the tolerance allele (T). In this illustration, the T allele is dominant to the t allele. (Modified from Figure 1 of Mulvey, M. and S.A. Diamond, in *Metal Ecotoxicology: Concepts and Applications,* Newman, M.C. and McIntosh, A.W., Eds., Lewis Publishers, Chelsea, MI, 1991.)

Differences in tolerance[11] of target and nontarget species can be controlled by a single gene (**monogenic control**) or several genes (**polygenic control**) (Figure 10.10). Monogenic control of tolerance to endrin and other cyclodiene pesticides was found for mosquitofish (*Gambusia affinis*) populations inadvertently exposed during agricultural spraying (Yarbrough et al., 1986; Wise et al., 1986). In contrast, Posthuma et al. (1993) found polygenic control of heavy metal tolerance in populations of the springtail, *Orchesella cincta*. Enhanced tolerance of this soil insect was associated with differences in metal-excretion efficiency among populations.

Tolerance acquisition varies relative to its cost (resource allocation) (Figure 10.8). Hickey and McNeilly (1975) found that metal-tolerant plants are at a disadvantage relative to nontolerant plants if grown in a noncontaminated soil. Postma et al. (1995b) found that midges (*Chironomus riparius*) from cadmium-tolerant populations had poorer survival, growth, and reproductive success than nontolerant populations if reared in low-cadmium conditions. The lowered fitness was attributed to an apparent zinc deficiency in tolerant individuals living under low cadmium conditions.

There may be cross-resistance among toxicants, depending on the mechanism underlying enhanced tolerance. **Cross-resistance** or **co-tolerance** is the condition in which enhanced tolerance to one toxicant also enhances tolerance to another. For example, plants tolerant to one s-triazine herbicide display cross-resistance to other s-triazine herbicides due to elevated levels of a herbicide-binding protein (Erickson et al., 1985). Lead detoxification is higher in isopod populations tolerant

---

[11] Distinction is made by many authors between the terms *tolerance* and *resistance*. **Tolerance** is often reserved for enhanced ability to cope with a factor due to physiological acclimation. **Resistance** is used if the enhanced abilities are associated with genetic adaptation. Following the lead of Weis and Weis (1989b), tolerance is used here for both acclimation and genetic adaptation.

to copper (Brown, 1978). Metallothionein gene duplication as an adaptation to elevated levels of one metal (Lange, 1989) could enhance tolerance to other metals. Enhanced rotenone tolerance imparted by elevated mixed function oxidase activity (Fabacher and Chambers, 1972) would likely enhance tolerance to other pesticides detoxified by this mechanism.

IV.C Measuring and Interpreting Genetic Change

Allozymes (allelic variants of an enzyme coded for by a particular locus) were first introduced in the late 1970s by Nevo and coworkers (Nevo et al., 1977; 1978; 1981; Lavie and Nevo, 1982; 1986; Baker et al., 1985) to reflect changes in population genetics as a consequence of environmental pollution. Now, they are occasionally useful for assessing population response to toxicants (Gillespie, 1996), although recent DNA techniques may supplant them in the near future. Allozymes have a distinct advantage in that they can be rapidly scored in large numbers of individuals using starch gel or some other type of electrophoresis. (During electrophoresis, the different forms of an enzyme notionally coded for at a locus are separated from tissue homogenates of individuals in an electrical field. Biochemical staining is then used to visualize enzyme activity on the electrophoretic medium. The genotype for each sampled individual is then implied from the pattern of staining on that medium.)

Allele and genotype frequencies in field populations have suggested pollutant-related loss of genetic diversity (Kopp et al., 1992) or selection (Battaglia et al., 1980; Gillespie and Guttman, 1989; Heagler et al., 1993). Interpretation of field results is frequently supported by laboratory studies suggesting differential mortality among allozyme genotypes (Battaglia et al., 1980; Diamond et al., 1989; Newman et al., 1989; Heagler et al., 1993; Keklak et al., 1994; Schlueter et al., 1995). For example, Heagler et al. (1993) interpreted changes in the frequency of alleles associated with a glucosephosphate isomerase locus (GPI-2) of a field population of mosquitofish (*G. holbrooki*) using results of survival analysis of GPI-2 genotypes exposed to high concentrations of mercury in the laboratory.

Too often, the enzyme itself is assumed to be responsible for the observed differences in fitness among genotypes without due consideration of alternative explanations. For example, the different allozymes are thought to have different availabilities of sites to bind with metals and, consequently, different susceptibilities to inactivation by the metals. Although this is a reasonable explanation, it is seldom tested rigorously. It was not true in one case in which it was tested. Differences in GPI-2 genotype sensitivity under acute mercury exposure of mosquitofish (*G. holbrooki*) were not a consequence of differential inactivation of allozymes by mercury (Kramer et al., 1992; Kramer and Newman, 1994). Results suggested that differences among genotypes were more readily interpreted in the context of optimal energy resource allocation under general stress. Further, a scored enzyme locus might only be acting as a marker for a closely linked gene that is actually responsible for the difference in tolerance among genotypes. Such **genetic hitchhiking**[12] is very often given inadequate consideration as a mechanism underlying the observations.

Alleles can be unevenly distributed throughout a structured population, e.g., among lineages. If differences in tolerance exist within the structure, this would result in correlations between tolerance and allozyme genotypes that falsely suggest that an allozyme itself is directly linked to tolerance. Lee et al. (1992) reinforced this point by demonstrating a strong family effect relative to mosquitofish (*G. holbrooki*) tolerance to mercury. To our knowledge, no other pollution-related study of allozymes has carefully tested this alternative and equally reasonable explanation for the correlation between allozyme genotype and tolerance. Consequently, conclusions tend to remain

---

[12] Endler (1986) defines genetic hitchhiking as "a situation in which a given allele changes in frequency as a result of linkage or gametic disequilibrium with another selected locus... [it can] give a false impression of selection at a particular locus. ... Similarly, if there is genotypic correlation among quantitative traits, then selection will appear to affect a trait directly, although it is actually only affected through its correlation with another selected character."

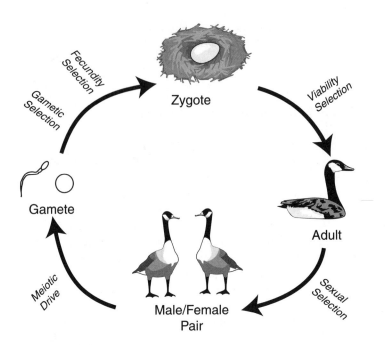

Figure 10.11 Components of the life cycle of an individual in which natural selection (selection components) can occur. Although rarely considered, selection components can be acted upon by contaminants. See the text for a detailed explanation.

ambiguous regarding direct linkage of differential inactivation of allozymes by contaminants to differential tolerance of genotypes.

Population structuring can also produce an apparent deficit of heterozygotes relative to Hardy-Weinberg expectations.[13] These deficits can be mistakenly attributed to selection against heterozygotes. However, a deficit of heterozygotes can arise if two genetically distinct groups of individuals are mixed, as in the case of unknowingly sampling a highly structured population under the assumption of uniformity in the sample. It can also occur if significant amounts of migration have occurred, as in the case of an influx of individuals into a population recently decimated by a pollution event. The **Wahlund effect** predicts that there will be a net deficit of heterozygotes if two populations, each in Hardy-Weinberg equilibrium but possessing different allele frequencies, are mixed and their combined genotype frequencies quantified. This effect has been ignored in most pollution-related studies to date. Woodward et al. (1996) recently demonstrated such a Wahlund effect during a population study of midges (*Chironomus plumosus*) inhabiting a mercury-contaminated lake.

Genotype-related differences in pollutant tolerance are almost exclusively examined relative to survival. However, selection can and does occur at several stages in the life cycle of an individual (Figure 10.11). There are several such **selection components**: viability selection, sexual selection, meiotic drive, gametic selection, and fecundity selection. **Viability selection**, or selection based on differential survival, begins at the zygote and continues throughout the life of the individual. This component can be further broken down to viability at different ages or stages of development. **Sexual selection** involves differential mating success of individuals. It can be associated with females or males, i.e., female or male sexual selection. **Meiotic drive** is the differential production

---

[13] At **Hardy-Weinberg equilibrium**, the frequency of genotypes will remain stable through time. For a two allele locus (i.e., T and t), the frequencies of the genotypes will be q^2 for TT, 2pq for Tt, and p^2 for tt, where q and p are the allele frequencies for T and t, respectively. The following conditions are assumed: (1) the population is large (effectively infinite) and composed of randomly mating, diploid organisms with overlapping generations, (2) no selection is occurring, and (3) mutation and migration are negligible.

of gametes by different heterozygous genotypes. One allele may be underrepresented in the gametes produced by a heterozygous individual. **Gametic selection** involves differential success of gametes produced by heterozygotes. The last component, **fecundity selection**, is the production of more offspring by matings of certain genotype pairs than produced by other genotype pairs. Selection components can act in opposite and balancing directions (Endler, 1986); therefore, measurement of only one, such as viability selection, may result in inaccurate predictions of changes in allele frequencies under selection pressures from pollutants. For example, Mulvey et al. (1995) found this to be the case with mercury-exposed mosquitofish (*G. holbrooki*). The GPI-2 genotype that was at a disadvantage during acute exposure (viability selection) was not the same as that genotype at a disadvantage relative to female sexual selection. Unfortunately, selection-component analysis is ignored in most studies.

V SUMMARY

In this chapter, the importance of assessing effects to populations is emphasized. A brief sketch of epidemiological metrics and logic is provided as applicable to ecotoxicology. Demographic approaches are described that greatly improve our ability to predict the population consequences of toxicant exposure. The increasing use of demographic approaches by ecotoxicologists is demonstrated with several examples. The potential influences of toxicants on population genetics are outlined in addition to the possible consequences of such changes. Acquisition of tolerance, factors influencing the rate of tolerance acquisition, and other related processes are described.

SUGGESTED READINGS

Barnthouse, L.W., Population-level effects, in *Ecological Risk Assessment*, Suter, G.W., II, Ed., Lewis Publisher, Chelsea, MI, 1993.

Daniels, R.E. and J.D. Allan, Life table evaluation of chronic exposure to a pesticide, *Can. J. Fish. Aquatic Sci.*, 38, 485–494, 1981.

Deevey, E.S., Jr., Life tables for natural populations of animals, *Q. Rev. Biol.*, 22, 283–314, 1947.

Klerks, P.L. and J.S. Weis, Genetic adaptation to heavy metals in aquatic organisms: a review, *Environ. Pollut.*, 45, 173–205, 1987.

Mulvey, M. and S.E. Diamond, Genetic factors and tolerance acquisition in populations exposed to metals and metalloids, in *Metal Ecotoxicology: Concepts and Applications*, Newman, M.C. and McIntosh, A.W., Eds., Lewis Publishers, Chelsea, MI, 1991.

Newman, M.C., Effects at the population level, in *Quantitative Methods in Aquatic Ecotoxicology*, Lewis Publishers, Boca Raton, FL, 1995.

Newman, M.C., *Population Ecotoxicology*, John Wiley & Sons, Chichester, U.K., 2001.

Rench, J.D., Environmental epidemiology, in *Basic Environmental Toxicology*, Cockerham, L.G. and Shane, B.S., Eds., CRC Press, Boca Raton, FL, 1994.

Sibly, R.M. and P. Calow, A life-cycle theory of responses to stress, *Biol. J. Linn. Soc.*, 37, 101–116, 1989.

CHAPTER 11

Effects to Communities and Ecosystems

> The accumulation of persistent toxic substances in the ecological cycles of the earth is a problem to which mankind will have to pay increasing attention. ... What has been learned about the dangers in polluting ecological cycles is ample proof that there is no longer safety in the vastness of the earth.
>
> **Woodwell (1967)**

I OVERVIEW

I.A   Definitions and Qualifications

An ecological **community** is "an assemblage of populations living in a prescribed area or physical habitat: it is an organized unit to the extent that it has characteristics additional to its individual and population components. ... [It is] the living part of the ecosystem" (Odum, 1971). The community is made up of species that interact and form an organized unit (Magurran, 1988), although some species interact only loosely. Much of the following material is structured around this abstraction.

The impossibility of studying all species, or even all important species, in any community results in studies that focus on some taxonomic or functional subset of the community, such as the fur bearers of a woodland or the fish in a lake. Pielou (1974) suggests that the term **taxocene** should be used to distinguish these operationally defined subsets from true communities. Magurran (1988) suggests the term **species assemblage** for any operationally defined grouping. Many models and indices are framed in the community context but are applied to species assemblages out of necessity. Although such an approach remains valuable and necessary, interpretation of associated results should be tempered with this understanding.

Similarly, the ecosystem concept is also an abstraction or simplification that should not be confused with reality (Newman, 1995). The **ecosystem** concept combines the biota (community) and abiotic environment into an organized system. (See Golley [1993] for a detailed discussion of the ecosystem concept.) Species interact with each other and loosely interact with their physical environment. Biotic and abiotic components act together to direct the flow of energy and cycling of materials. Obviously, application of this concept to real situations is highly dependent on the scale (time and space), distinctiveness of system boundaries (e.g., a distinct, spring-fed lake versus a diffuse bottomland hardwood ecosystem along a river), and the particular qualities under study (e.g., cation flux from a watershed ecosystem versus oxygen dynamics of a dimictic lake). This concept must be applied intelligently to avoid illogical conclusions regarding qualities of an operationally defined "ecosystem" using ideal characteristics of the ecosystem abstraction.

I.B   Context

Community and ecosystem qualities are affected by abiotic factors including pollutants (Dunson and Travis, 1991). Despite this, effects at these levels are often addressed in less detail than warranted (Taub, 1989). This neglect probably reflects this field's historical roots in mammalian toxicology, a field that emphasizes effects to individuals. In illustration of this fact, Clements and Kiffney (1994) noted that only 12% of 699 environmental toxicology articles published from 1980 to 1982 dealt with populations, communities, or ecosystems. They further noted a disappointingly low percentage (18% of all papers) in a more recent (1992) survey of the journal, *Environmental Toxicology and Chemistry*. Clearly, a better balance is needed in ecotoxicology.

Causal mechanisms for community change are often to be found at the next lower level of organization, i.e., at the population level. For example, change can occur because a particular species population's viability was lowered sufficiently by a toxicant's effect on growth, survival, or reproduction. This scenario is consistent with the approach advocated in Chapter 1 for maintaining conceptual coherency in any hierarchical science. However, **emergent properties** must also be considered carefully at higher levels of organization. In hierarchical systems, not all of the properties can be predicted solely from our limited understanding of a system's parts or components.[1] The counterexample to that just given is the indirect loss of several species because an important keystone species was killed directly by the toxicant. (A **keystone species** is one that influences the community by its activity or role, not its numerical dominance.) A species resistant to the direct action of a toxicant could disappear if another species performing a crucial role in the community were eliminated. Another example is industrial melanism, a situation where community processes (i.e., predator-prey interactions) influence population genetics (i.e., predominance of melanism). Causal structure is reversed with interactions among species populations in the community (higher level) producing an impact to a population (lower level).

This chapter deals primarily with communities, but processes occurring in whole ecosystems are discussed toward the end of the chapter. Many ecosystem topics have already been described in Chapter 2, and some additional properties associated with ecosystems are addressed again in Chapter 12. This chapter begins by describing simple species interactions relative to the influence of toxicant action. Then community qualities, including structure and function, are described in the context of laboratory, mesocosm, and field research. Although much of the ecotoxicological work done with communities has been descriptive, the emphasis in this chapter will be explanatory principles derived primarily from experimental efforts. Field studies and methods are detailed toward the end of this chapter. Most field methods focus on structural changes observed in species assemblages such as soil arthropods or stream macroinvertebrates. Conventional community indices (e.g., species richness) or more specialized indices (e.g., the index of biological integrity) are applied to species assemblages from and around contaminated areas. Less often, community functions are assessed. These functions are discussed briefly toward the end of the chapter.

I.C   General Assessment of Effect

A wide range of experimental approaches has been taken to determine the concentration of toxicants below which the community is protected. The **most sensitive species approach** takes the results for the most sensitive of *all tested species* as an indicator of that concentration most likely to protect *all species in the community*. Despite the great advantage of its simplicity, several difficulties arise with this notionally cost-effective approach (Cairns, 1986). One must make the dubious assumption that the tested species and measured effects truly reflect the most sensitive within the community. The most sensitive species approach might not be cost effective if one considers

[1] This concept is central to the tedious holistic-reductionistic debate in ecology. Time wasted debating this obvious point distracts ecotoxicologists from the real challenge, enhancing the inferential strength of their science regardless of the conceptual vantage taken.

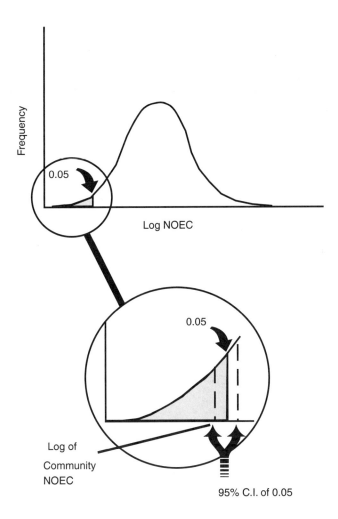

Figure 11.1 One method of estimating a community-level NOEC. The 5% quantile is used to estimate the log concentration at which all but 5% of species would be protected. The concentration corresponding to the 5% quantile or the lower limit of the 95% confidence interval around this estimate could be used as the community-level NOEC (Van Straalen and Denneman, 1989; Wagner and Løkke, 1991).

the high costs of bad management decisions based on flawed approaches (Cairns, 1986). Remember also that the biological significance of the effect remains ambiguous if effect metameters such as an NOEC are used, i.e., the maulstick incongruity.

Recently, a statistical permutation of this most-sensitive-species approach has appeared for determining concentrations protective of a community. A collection of available NOEC values or other measures of effect are pooled for a sample of species to estimate the concentration below which a predetermined percentage of all species (e.g., 5%) are protected (Figure 11.1) (Van Straalen and Denneman, 1989; Wagner and Løkke, 1991; Posthuma et al., 2002). This concentration is thought to be generally protective of the community. However, Hopkin (1993a) questioned the assumption that a 5% loss of all species is always acceptable. Arguing in the context of soil species, he noted that elimination of one keystone species such as the earthworm would dramatically influence the soil community, even if 95% of all species were protected. It is difficult to justify the assumption that the NOEC values used in any such analysis accurately reflect the effect of concentrations within a community or species assemblage (Jagoe and Newman, 1997; Newman et al., 2002). Often, values are derived from those of standard test species and are biased toward certain taxa.

Beyond this bias, it is difficult to know how many NOEC values are needed to effectively capture the differences among species in an entire community or even a species assemblage.

II INTERACTIONS INVOLVING TWO OR A FEW SPECIES

II.A   Predation and Grazing

An adverse effect on predator-prey interactions can lead to local extinction of a species population, even if toxicant concentrations are below those causing diminished growth, reproduction, or survival of individuals in the population. This premise prompted laboratory experiments quantifying such effects. A simple predator-prey arena (Figure 11.2) was used to demonstrate the influence of γ irradiation on the ability of mosquitofish (*Gambusia holbrooki*, formerly *G. affinis holbrooki*) to avoid predation by largemouth bass (*Micropterus salmoides*) (Goodyear, 1972). Mosquitofish were provided with a shallow refuge to simulate normal mosquitofish behavior of avoiding bass predation by staying in the shallows close to the water's edge. The influence of irradiation on predator avoidance over ten days was dose-dependent, with more mosquitofish failing to stay in the refuge as radiation dose increased. This approach was applied again by Kania and O'Hara (1974) to demonstrate that sublethal concentrations of inorganic mercury increase predation in a concentration-dependent fashion. Mosquitofish previously exposed to low concentrations of mercury were incapable of maintaining the most effective orientation relative to predator location in the test chamber. In a slightly more elaborate arena including artificial plants as both prey refugia and predator cover, fathead minnows (*Pimephales promelas*) exposed to cadmium were more vulnerable to largemouth bass predation than were unexposed minnows (Sullivan et al., 1978). Concentrations producing a significant increase in vulnerability were lower than those measured for any other sublethal effect. Increased predation was discussed in the context of the abnormal schooling behavior of the exposed minnows. Similar studies examined the influence of fire ant bait (mirex) on pinfish (*Lagodon rhomboides*) predation of grass shrimp (*Palaemonetes vulgaris*) (Tagatz, 1976). More recently, turbellarian predation of isopods as influenced by cadmium (Ham et al., 1995) and *Hydra* predation on *Daphnia* after lindane (γ-hexachlorocyclohexane) exposure (Taylor et al., 1995) were quantified, but in contrast to the above assays, exposure involved both predator and prey.

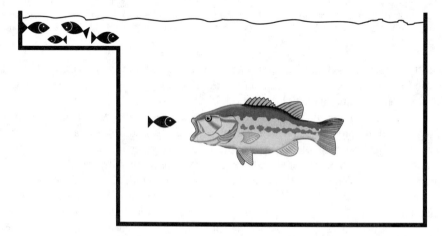

Figure 11.2   Experimental arena used to measure the effect of radiation (Goodyear, 1972) or inorganic mercury (Kania and O'Hara, 1974) on mosquitofish avoidance of predation by largemouth bass. The stressors diminished the ability of the mosquitofish to avoid predation by remaining in a shallow refuge.

Information gleaned from such simple studies can be enriched greatly by applying the principle of allocation (see Chapter 10). Organisms must effectively allocate energy among many activities in order to optimize Darwinian fitness. For example, a predator may change its prey consumption rate as prey densities increase. Such a **functional response** (a change in some predator function, such as prey consumption rate, as a response to changes in prey density) was studied for a largemouth bass–mosquitofish system. Both predator and prey were exposed to ammonia (Woltering et al., 1978). Changes in prey consumption and bass weight were monitored at different ammonia concentrations. Increases in prey consumption rate with an increase in prey density were slowed at high ammonia concentrations, as was the increase in bass weight. In fact, because the mosquitofish were more tolerant of ammonia than bass, the mosquitofish harassed the bass in high ammonia and prey density treatments, resulting in a weight lose of the predator. Clearly, the influence of toxicants on predator-prey interactions is important and complicated.

Atchison and coworkers (Sandheinrich and Atchison, 1990; Henry and Atchison, 1991; Atchison et al., 1996) provide the richest description of predator-prey interactions in the context of energy allocation. They argue from the extensive literature on **optimal foraging theory** (the ideal forager reaches a maximum net rate of energy gain by optimally allocating its time and energy to the various components of foraging) that many important components of foraging are influenced by toxicants. For example, a predator must optimize time and energy spent in prey searching, identification, choice, pursuit and capture, handling, and ingestion. Any or all of these might be altered by the presence of toxicants. Atchison and coworkers cite numerous studies in which predator foraging activities are modified by toxicant exposure.

Kersting (1984) demonstrated that grazing is influenced by toxicants in a study in which he examined the deviation from normal dynamics for a herbivore-plant (*Daphnia magna-Chlorella vulgaris*) micro-ecosystem resulting from pesticide (Dichlobenil) exposure (Figure 11.3). He allowed the *Daphnia-Chlorella* system to come to steady state and plotted grazer density vs. algal density through time. A lag function was incorporated in the plot because there was a delay of

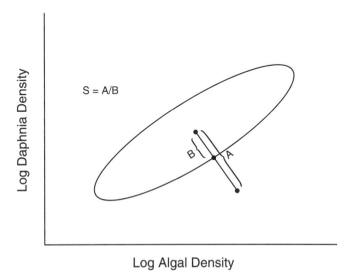

Figure 11.3 Calculation of normalized ecosystem strain as described by Kersting (1984). Strain is equal to A/B in this diagram. Ecosystem strain is measured relative to the normal behavior of the system as reflected by the 95% tolerance ellipse. These particular dimensions can be replaced by variables other than algal and grazing *Daphnia* densities. (Modified from Figure 4 in Kersting, K., *Ecol. Bull.*, 36, 150–153, 1984.)

approximately 7 days before the grazer density[2] could respond fully to any change in algal density. A 95% tolerance ellipse was drawn to define the limits of the system's behavior in the absence of the pesticide. In theory, one would get a point outside of this ellipse only 1 in 20 times as a consequence of random chance alone, i.e., points outside of this ellipse would be judged to be outside of normal dynamics. The pesticide was introduced, and grazer density vs. algal density points were plotted. Because these points fell outside the tolerance ellipse, the grazer-plant system was judged to have changed as a consequence of the pesticide exposure. A normalized ecosystem strain index (S) was used to quantify the degree to which the system had changed relative to its normal state.

II.B Competition

Interspecies competition, the interference with or inhibition of one species by another, can be influenced by toxicants and also contribute to changes in community structure. This could involve **interference competition**, in which one species interferes with another, as might occur with territoriality. A toxicant-induced change in aggressive behavior (e.g., bluegill exposed to cadmium [Henry and Atchison, 1979]) could shift the balance of interference competition. **Exploitation competition**, where species compete for some limiting resource such as food, can also be affected by toxicants. For various freshwater zooplankton under toxicant exposure, Atchison et al. (1996) noted differences in filtration rates and suggested that toxicants can produce shifts in exploitation competition among these potentially competing filter feeders.

Atchison and coworkers (Atchison et al., 1996) lament the paucity of studies in behavioral ecotoxicology and attribute this deficiency to the indirect, and relatively complicated, means by which interspecies competition is measured, i.e., as changes in population dynamics of competing species. In the absence of direct information, indirect information was used as evidence to suggest that exploitation competition could be affected significantly by contaminants. Newman (1995) details methods for quantifying the effects of toxicants on interspecies competition.

To digress for a moment, it is implied in this and the previous discussion that changes in species niches result from toxicant exposure. Here, we define niche with the **Hutchinsonian niche** concept, "… the certain biological activity space in which an organism exists in a particular habitat. This space is influenced by the physiological and behavioral limits of a species and by effects of environmental parameters (physical and biotic, such as temperature and predation) acting on it" (Wetzel, 1982). A species in a particular habitat has a **fundamental niche**, in which it could exist based on its physiological and other limitations, and a **realized niche**, which is that portion of its fundamental niche that it actually occupies. With shifts in competition and foraging behavior, the realized niche of a species is modified by the toxicant. If balanced competition among species in the community is assumed to enhance species packing by fostering optimal niche separation among species, then competitive dysfunction might decrease species diversity. As we will see shortly, shifts associated with species-abundance curves support this speculation. Interestingly, Chattopadhyay (1996) suggests that toxicants can also stabilize fluctuations in competing species populations under some conditions.

III COMMUNITY QUALITIES

III.A General

Several community or species assemblage qualities are measured routinely to assess toxicant effect. The number of species inhabiting a toxicant-impacted site can be compared with the number at a nonimpacted site. The presence or absence of indicator species may also be noted. For example, a species that is extremely sensitive to a pollutant might be used much as the proverbial canary in the coal mine, with the presence of a particularly sensitive species suggesting no toxicant effect.

---

[2] In contrast to a functional response, a change in predator or grazer number through increased reproductive output, decreased mortality, or increased immigration in response to changes in prey or food densities is called a **numerical response**.

The decline in osprey populations on Long Island has already been mentioned as an example of such change for a sensitive species. Another example is the disappearance of pH-sensitive species from lakes undergoing acidification. The mysid shrimp, *Mysis relicta*, disappeared from an experimentally acidified lake when pH was lowered to 6 (Schindler, 1996). Not only did this sensitive species provide an early warning of deteriorating conditions, it was functioning as a keystone species in this Canadian lake. A final example includes mayflies (e.g., *Baetis* sp.) that tend to be sensitive indicators for a variety of pollutants in freshwater systems (Ford, 1989).

Alternatively, a rise to predominance of pollution-tolerant species might suggest a deteriorating community. Benthic communities found below high-BOD (biochemical oxygen demand) outfalls from sewage treatment plants are typically dominated by heterotrophs tolerant of low dissolved-oxygen concentrations. The oligochaete, *Tubifex tubifex*, is a common benthic species at such polluted sites. The *Sphaerotilus* bacterium also forms extensive filamentous mats below sewage discharges and is used as an obvious indicator species. Indeed, based on this concept, Kolkwitz and Marsson (1908) described a **saprobien spectrum**, a characteristic change in community composition at different distances below a discharge of putrescible organic waste into a river or stream. Characteristic species define zones (e.g., polysaprobic, mesosaprobic, and oligosaprobic zones) below a sewage discharge relative to the oxygen concentrations, amounts of putrescible organic material, and stage of stream recovery.

Several qualities apparently influence community (or "ecosystem") **vulnerability** (susceptibility to irreversible damage) to toxicants (Cairns, 1976). Low **elasticity** (the ability to return to a prestressed condition), **inertia** (ability to resist change), and **resilience** (the number of times a community can return to its normal state after perturbation) all contribute to vulnerability. Elasticity is enhanced by the ease with which new individuals can move back into the affected area. Inertia can be influenced by the structural redundancy in the community and previous adaptation of the community to environmental variability. Resilience is influenced by the elasticity of the community and the frequency of perturbation. Too frequent perturbation of a community with low elasticity gradually ratchets the community downward toward a degraded state.

Implicitly, all of this discussion assumes that a community is in some kind of balance and can be expected to return to that balance after a perturbation. Pratt and Cairns (1996) point out that this concept of community steady state is pervasive in ecotoxicology. It extends into regulations such as those setting environmental criteria that seek to protect "balanced biological communities."

In contrast to this concept of a community deviating from and then returning to a steady state condition after the stressor is removed, Matthews et al. (1996) suggest that disturbed communities will not return to their original states. This suggestion is consistent with the increasingly expressed view of ecologists that communities are not steady state systems (Pratt and Cairns, 1996). Matthews et al. (1996) argue that communities retain information about occurrences in their past, dubbing this argument the **community conditioning hypothesis**. Any dynamics back toward some norm will also reflect the history of the community: One cannot assume that the community will return to its original state. The implication here is that pollution effects to communities will be present long after the toxicant is removed and that any assumed return to an original state is presumptuous. It is this author's opinion that both views (steady state and community conditioning hypotheses) are useful if applied appropriately. One might expect a general recovery of some community qualities to a near "normal," but unique, state after community disruption by toxicants.

III.B Structure

III.B.1 Community Indices

The most commonly used indices of community change are species richness, evenness, and diversity (heterogeneity). **Species richness** is the number of species present in a community. Because the tally of species in a community increases as more and more individuals are sampled—and because it is often impractical to sample all individuals—species richness is often expressed relative to that of a sample with a standard number of individuals in it. A **rarefaction estimate of richness** produces a number such as 25 expected species in a standard sample of 250 individuals. **Species evenness** is the extent to which the individuals in the community are evenly or uniformly distributed among species. For example, let three species be present in two communities composed of 500 individuals each. In the first community, 450, 41, and 9 individuals are from species A, B, and C, respectively. The numbers of individuals in species A, B, and C in the second community are 134, 138, and 228, respectively. The individuals are more evenly distributed among the species in the second community.

Both species richness and evenness contribute to **species diversity** (= heterogeneity) and are reflected in species-diversity indices. Species diversity can be quantified with several formulations; the **Shannon diversity index** (Equation 11.1) and the **Brillouin diversity index** (Equation 11.2) are the most common.

$$H' = \sum_{i=1}^{S} p_i \ln p_i \tag{11.1}$$

$$H = \frac{1}{N} \ln \frac{N!}{\prod_{i=1}^{S} n_i!} \tag{11.2}$$

where S = total number of species, and p_i = proportion of all individuals that are species i as estimated by the number of individuals of species i (n_i) divided by the total number of individuals (N) in the S species.

Both give similar estimates, but Equation 11.2 gives estimates lower than Equation 11.1. This difference arises because the Shannon index is a diversity estimate *for the community* from which the sample was taken, but the Brillouin index is a diversity estimate *for the sample*. The diversity in the sample will be lower than the diversity predicted for the entire community from which the sample was taken.

Similarly, species evenness can be estimated for the community (Equation 11.3, **Pielou's J'**) or for the sample (Equation 11.4, **Pielou's J**). The $\ln S$ and H_{MAX} in these equations are the maxima for Shannon (H') and Brillouin (H) indices, respectively. Consequently, these evenness indices are the estimated species diversity divided by the maximum possible species diversity for that community or sample.

$$J' = \frac{H'}{\ln S} \tag{11.3}$$

$$J = \frac{H}{H_{MAX}} \tag{11.4}$$

where

$$H_{MAX} = \frac{1}{N} \ln \left[\frac{N!}{([N/S]!)^{S-r}(([N/S] + 1)!)^{r}} \right]$$

with $[N/S]$ being the integer part of the quotient, N/S, and r being $N - S[N/S]$.

Often, but not always, values for species indices decline as a consequence of pollution. Species diversity dropped in periphyton communities below heavy metal mine discharges (Austin and Deniseger, 1985) in periphyton communities in the presence of high zinc concentrations (Williams and Mount, 1965) and in stream macroinvertebrate communities below coal mine drainage (Chadwick and Canton, 1983). Both diversity and richness dropped for lake algal communities exposed to mining wastes with a few tolerant species becoming very abundant (Austin et al., 1985). Ford (1989) indicates that richness is a good measure of effect to plankton and benthic communities, although richness may increase slightly at low levels of pollution.

Often, a statistically significant difference in species richness or species diversity is used to suggest an adverse impact of toxicants on communities. Aside from the problem of equating statistical and biological significance (i.e., maulstick incongruity), this approach suffers from our lack of knowledge regarding functional redundancy within communities. **Functional redundancy** involves an apparently unaltered maintenance of community functioning despite changes in structure. Species may drop out or be replaced; yet the community will still appear to function normally.

There are two unresolved hypotheses involving functional redundancy that are germane to the impact of toxicants: the rivet-popper and redundancy hypotheses. The **rivet-popper hypothesis** suggests that species in a community are like rivets that hold an airplane together and contribute to its proper functioning (Ehrlich and Ehrlich, 1981). Each loss of a rivet weakens the structure by a small but noticeable amount. The loss of too many rivets eventually leads to a catastrophic failure in function. In contrast, the **redundancy hypothesis** holds that many species are redundant, and the loss of some species will not influence the community function as long as crucial (keystone and dominant) species are maintained (Walker, 1991). There are often guilds[3] of similarly functioning species to provide consistency of function if one or a few member species are lost. Pratt and Cairns (1996) emphasize the importance to ecotoxicology of determining which of these hypotheses best describes real biological communities. The answer is needed to decide how much toxicant-induced change in a community is required to degrade its functioning. Currently, there is some evidence to support the rivet-popper hypothesis (Baskin, 1994). Given our present lack of understanding, it seems prudent to assume that the conservative rivet-popper hypothesis should be the working model.

Species-abundance curves are also used to describe community shifts as a consequence of toxicant exposure (Figure 11.4). These curves are based on the **law of frequencies**, which states that there is a relationship between the numbers of species and the number of individuals in a community (Fisher et al., 1943). The numbers of species falling into different abundance classes are plotted against abundance class. In the classic approach of Preston (1948), abundance classes are defined as doublings in abundances, e.g., 2, 4, 8, 16, 32, etc. individuals present for a species. These $\log_2$ classes (e.g., 1, 2 to 3, 4 to 7, 8 to 15, 16 to 31, etc. individuals) are called **octaves**. A plot similar to that in Figure 11.4 is produced and describes a log-normal distribution. This **log-normal model** for species abundance is thought to reflect a community structure in which several factors influence species interactions and subsequent allocation of resources.

Patrick (1973) noted that this log-normal curve shifts in a predictable way for diatom communities exposed to organic pollution. The mode drops down and the right tail extends out to include more octaves with high numbers of individuals. There is a shift toward more very dominant species and fewer species of intermediate or rare abundances. Herricks and Cairns (1982) suggested that this shift results from a rise to dominance of opportunistic species and a disruption of equilibrium. To May (1976a) and Odum (1985), this shift suggested reversion to an earlier successional stage as a consequence of the disordering effects of pollution. The diverse processes allowing better species packing in a mature community (k-strategy) become less important in shaping community

---

[3] An ecological **guild** is a "group of functionally similar species whose members interact strongly with one another but weakly with the remainder of the community" (Smith, 1986).

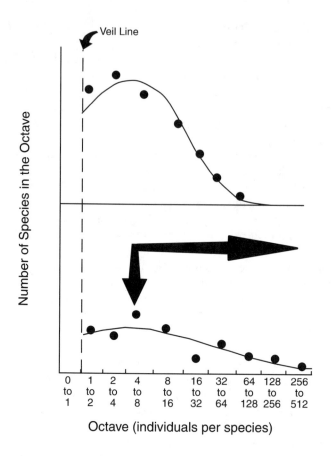

Figure 11.4   Species abundance curves (log-normal model) before (top panel) and after toxicant exposure. Note the transition from a common log-normal distribution to a curve with a lowered mode and extended right tail. May (1976a) suggests that such a transition may reflect a reversion to an earlier successional stage when interspecies interactions were less important in shaping community species structure (i.e., away from the importance of species interactions toward an r-selection strategy).

structure than those associated with earlier successional stages, i.e., the classic r-strategy.[4] Based on this premise, Bongers (1990) proposed a community **maturity index** for pollution based on the proportions of species in a soil nematode community that fell into various categories ranging from colonizers (r-strategists) to persisters (k-strategists). Regardless, this shift in the log normal species abundance curve is useful as an indicator of pollutant effect (Gray, 1979; Gerhart et al., 1977).

Finally, more comprehensive indices can be used to detect changes in the biota inhabiting potentially impacted sites. Currently, one of the most widely applied for aquatic systems is the **index of biological integrity** (IBI). Originally, this index combined 12 qualities of fish assemblages of warm-water, low-gradient streams to determine the degree of stream degradation. These qualities included information on species richness and composition, trophic characteristics, and abundance and condition (Table 11.1). Numerical scores are generated for each quality and summed to produce

---

[4] Although inadequate to fully explain individual success, these two strategies do provide sufficient information here. An **r-strategy** (r = intrinsic rate of increase) or opportunistic strategy may be taken by species coming into an uninhabited/unexploited habitat such as a newly plowed field. Selection favors species that establish themselves quickly, grow quickly to exploit as many resources as possible, and produce many offspring quickly. A **k-strategy** (k = carrying capacity) or equilibrium strategy involves important interactions among species that allow coexistence of many species in the community. Equilibrium species are more effective competitors than opportunistic species. Many factors, including interactions among species, determine the structure of such a mature community, whereas the early successional community structure may be determined more by **niche preemption**, a rapid use and preemption of resources by any species that exploits them before another can. In actuality, the r- and k- strategies are extremes in a spectrum of possible strategies.

Table 11.1   Qualities (Metrics) Included in the Original Index of Biological Integrity (IBI)

Category	Specific Quality
Species richness and composition	Total number of species
	Number of darter species
	Number of sunfish species
	Number of sucker species
	Number of intolerant species
	Proportion of all individuals that were green sunfish, a pollution-tolerant species
Trophic composition	Proportion of all individuals that are omnivores
	Proportion of all individuals that are insectivorous cyprinids
	Proportion of all individuals that are piscivores
Abundance and condition	Total number of individuals in the sample
	Proportion of total that are hybrids
	Total number of individuals with signs of disease or some abnormality

Source: Modified from Newman, M.C., *Quantitative Methods in Aquatic Ecotoxicology,* Lewis Publishers, Boca Raton, FL, 1995.

the IBI for a site. These IBI scores are compared with those expected in the particular area for an undisturbed system. This specialized index has been successfully modified for a variety of aquatic habitats (e.g., Steedman, 1988) and enjoys widespread application today.

Vignette 11.1   Biological Integrity and Ecological Health

James R. Karr
University of Washington, Seattle, WA

Six decades ago, Aldo Leopold discovered that important lessons often come when least expected. "In those days," he wrote, "we had never heard of passing up a chance to kill a wolf" (Leopold, 1949). Fewer wolves, the young Leopold thought, would mean more deer—a hunter's paradise. After pumping lead into a pack of big pups led by a female, Leopold and his friends "reached the old wolf in time to watch a fierce green fire dying in her eyes. I realized then, and have known ever since, that there was something new to me in those eyes. …"

Leopold spent decades afterward protecting and putting words to the fire he saw in the old wolf's eyes—a fire peculiar to living things, including living landscapes. His writings speak often of "land health" as "the capacity of the land for self-renewal"; they also speak of "integrity": "A thing is right when it tends to preserve the integrity, stability, and beauty of the biotic community. It is wrong when it tends otherwise."

Since Leopold's time, the terms *health* and *integrity* have become lightning rods, especially among scientists. Some argue that such value-laden words should not be applied to multispecies assemblages, such as ecosystems or landscapes; others hold that talking about ecological health or biological integrity is beyond the purview of science. Yet the words are particularly useful in policy-making arenas precisely because they are familiar and imply values worth protecting. It seems a natural intuitive leap from "my health" or the nation's "economic health" to "ecological health" or "land health." And, as a goal of policy or law, protecting a place's health and integrity has greater direct appeal than abstractions like "system dynamics" or "ecosystem functions."

Like people, ecosystems or landscapes can be more or less "ill." An unhealthy person may be suffering from a cold or dying of cancer. An unhealthy landscape may be degraded by loss of a few sensitive species or all of its vegetation. An unhealthy river may have game fish populations depleted by overfishing, or after severe chemical pollution it may have no fish at all or only a few of the river's most tolerant invertebrates.

The healthiest places are those that have undergone little or no disturbance at human hands. These areas support the full range of biological parts and processes characteristic of the region; that is, they show a full complement of plants, animals, and microbes, and their genetic diversity, as well as a full array of ecological processes, such as nutrient cycling, births, deaths, competition, and mutualisms. Because such places support a thriving, living system, they retain the capacity to regenerate, reproduce, sustain, adapt, develop, and evolve; that is, they retain the full legacy of wild nature, or in Leopold's words, they still have "all the parts."

Complete, unimpaired living systems possess biological integrity. They constitute one end in the spectrum of biological condition and provide a benchmark against which other sites can be evaluated. A "normal," or benchmark, body temperature of 37°C (98.6°F) provides a similar standard for humans.

Defining biological integrity, however, is only the first step toward using the concept in science, policy making, or law. For credibility in any of these fields, practitioners need tools for translating the subjective concept into something objective. They need tools both to quantify and to describe. Fortunately, the toolbox has been expanded in recent decades, enabling practitioners to evaluate sites and rank them along a gradient of biological condition according to how far they diverge from integrity.

Links between biology and human impacts came to the forefront more than a century ago when pollution, particularly from raw sewage, was found to harm living systems. Through much of the 20th century, efforts to track the health of water bodies have focused on the presence of chemical contaminants. The assumption was that chemically clean water was sufficient to protect river health. This assumption proved wrong.

We now know that human influences on living systems fall into five major classes: changes in energy sources, chemical pollution, modification of seasonal flows, physical habitat alteration, and shifts in biotic interactions (Figure 11.5). Given the choice of measuring all such influences or of measuring the condition of the biota—which includes the prime witnesses, and victims, of environmental change—many agencies and institutions are shifting to direct measurement of biological condition. Biological monitoring and assessment, or biomonitoring, detects and evaluates human-caused biotic changes apart from those occurring naturally; the techniques are gaining widespread acceptance as part of the water and land manager's toolkits.

Two major approaches to river biomonitoring have emerged in the past 20 years: the index of biological integrity (IBI) and the river invertebrate prediction and classification system (RIVPACS). The IBI and other multimetric indexes are composed of biological metrics that count, for example, the number of kinds of organisms present at a site (taxa richness or biodiversity) or the relative abundance of trophic groups such as predators. The RIVPACS analyses compare the number of taxa found at a test site with the taxa richness predicted by multivariate statistical models to be present at an undisturbed site. Common themes in these two approaches include: (1) focus on biological endpoints to define river health; (2) use of natural or "undisturbed sites" as a benchmark; (3) standardized sampling, laboratory, and analytical methods, including statistics; (4) numerical ranking of sites according to their condition; and (5) a scientifically rigorous

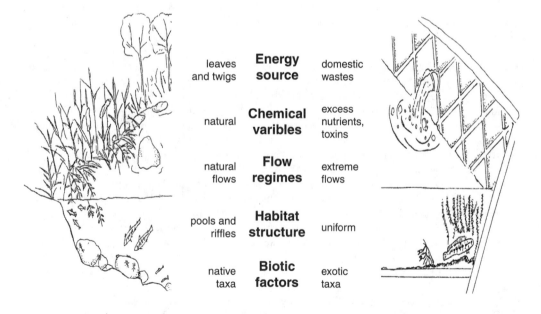

	Energy source	
leaves and twigs		domestic wastes
natural	Chemical varibles	excess nutrients, toxins
natural flows	Flow regimes	extreme flows
pools and riffles	Habitat structure	uniform
native taxa	Biotic factors	exotic taxa

Figure 11.5   Five principal features, with examples, that determine river health. Left, a natural river; right, a modified river. (From Karr, J.R. and E.M. Rossano, *Ecol. Civ. Eng.*, 4, 3–18, 2001. With permission.)

foundation for water policy. Such measures provide better information about the biological dimensions of environmental quality than physical and chemical measures incorrectly assumed to be surrogates of biological condition.

The IBI is an especially powerful biomonitoring tool for several reasons. First, like the index of leading economic indicators, IBI bases its conclusions about river health on an ensemble of biological indicators (metrics), each measuring a different aspect of the biotic community. Second, the metrics in the index reflect tested and predictable responses to human influences; each metric has its own "dose-response" curve associated with human land uses or other impacts (Figure 11.6). Third, metrics are chosen to reflect the effects of diverse human actions, such as logging, urbanization, or agriculture. Adaptations of IBI are now available for various organisms (fishes, insects, birds, vascular plants, and algae) and for diverse environments (rivers, wetlands, coastal areas, and terrestrial areas).

In seven Japanese watersheds, for example, IBI's multiple biological metrics reveal a much more refined picture of river health than the single parameter of biochemical oxygen demand (BOD) (Figure 11.7). Minimally impaired biological condition (high IBI values) occurs only at sites where BOD is low (<1.75 ppm), but not all sites with low BOD had high IBIs. Management or policy decisions based solely on BOD would fail to recognize the wider biotic degradation at a substantial number of sites in the seven

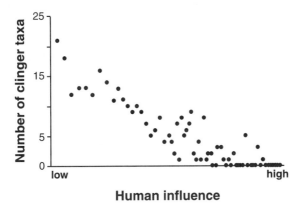

Figure 11.6 Dose-response curve for taxa richness of clingers—benthic invertebrates that cling to rocks, enabling them to live in the interstitial spaces between rocks—in standard samples from 65 Japanese streams ranked according to intensity of human influence. (From Karr, J.R. and E.W. Chu, *Restoring Life in Running Waters: Better Biological Monitoring,* Island Press, Washington, D.C., 1999. With permission.)

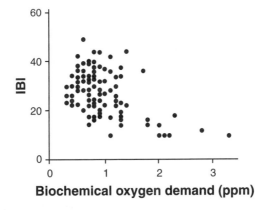

Figure 11.7 Index of biological integrity (IBI) compared with biochemical oxygen demand (BOD) for 100 sites from nine watersheds in Chugoku district, Japan. (From Karr, J.R. and E.M. Rossano, *Ecol. Civ. Eng.,* 4, 3–18, 2001. With permission.)

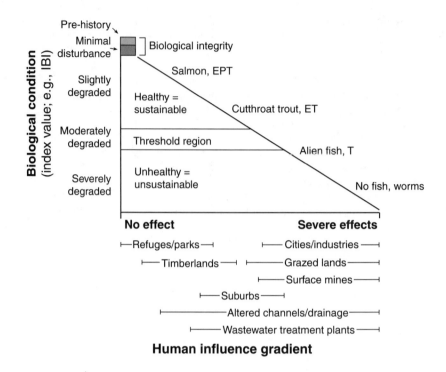

Figure 11.8   Relationship between biological condition and a hypothetical, synthetic measure of human
activity, with examples. Different human activities result in biological changes such as different
dominant organisms along a descending slope of biological condition. EPT stands for mayflies
(Ephemeroptera), stoneflies (Plecoptera), and caddisflies (Trichoptera); see text for details.
(Modified after Karr in Pimentel, D. et al., Eds., *Ecological Integrity: Integrating Environment,
Conservation, and Health,* Island Press, Washington, D.C., 2000. With permission.)

watersheds. In short, IBI describes a river's ecological health with greater relevance, precision, and clarity
than does water chemistry, thus offering a better guide for river protection or restoration.

Built into biomonitoring tools such as IBI is the biological reality that the condition of living systems
varies continuously with human influence (Figure 11.8). Instead of designating water bodies "impaired"
or "unimpaired," scientists and managers can express ecological health with greater precision along a scale
of biological condition (Figure 11.8, y-axis) in relation to human influence (Figure 11.8, x-axis).

In the U.S. Pacific Northwest, for example, benthic IBIs (B-IBI) have been calculated on the basis of
ten metrics, including diversity within three key insect orders: mayflies (Ephemeroptera), stoneflies (Plecoptera), and caddisflies (Trichoptera); index values range up to 50. Healthy streams support a high diversity
of fish and invertebrates. As human influence increases, salmon and most stoneflies disappear if B-IBI
drops below 35. Cutthroat trout largely disappear, and only the most tolerant mayflies and caddisflies are
present when B-IBI drops below 20. These shifts in the biota—and the progressively declining IBI values
that summarize those shifts quantitatively—are associated with a variety of human land uses, also varying
continuously in their impacts, from protected areas in parks and refuges through lightly or heavily logged
forestlands and farms through suburban and urban development.

Watershed managers can also overlay legal and regulatory categories—such as the U.S. Clean Water
Act's "designated uses"—on the sliding scales of biological condition and human influence (Figure 11.8).
Ultimately, society decides whether sites or regions are impaired or unimpaired, acceptable or unacceptable,
or within legal or regulatory limits. Those limits can have different unimpaired/impaired thresholds,
depending on context (e.g., a national park vs. a large city). Managers must use care to avoid declines in
biological condition that drop below a threshold, or "tipping point," where neither important components
of the natural biota nor human activity can persist in a place.

Backed by tools such as the IBI, laws and international agreements that invoke integrity and health
finally have a scorecard for their implementation and success. For example, the goal of the Clean Water

Act is "to restore and maintain the chemical, physical and biological integrity of the Nation's waters." The spirit of that mandate has since been included in the 1987 Great Lakes Water Quality Agreement between the United States and Canada, the 1988 amendment to Canada's National Park Act, the 1998 Water Framework that guides the European Union, and the 1999 Freshwater Strategy for British Columbia. Similar calls are present in legislative and policy initiatives in other regions throughout the world. Such laws and policies lay out goals for protecting Leopold's "green fire," "land health," or "integrity of the biotic community." Biomonitoring can help drive actual practice toward those goals.

III.B.2 Approaches to Measuring Community Structure

Laboratory, microcosm, mesocosm, and field approaches are applied to study change in community structure. These range from straightforward experiments, such as those described above for simple species interactions, to whole ecosystem exposures, such as the one shown in Figure 11.9. Laboratory experiments have many advantages, including the ability to randomly assign treatments, the ability to achieve adequate treatment replication, control of potentially confounding factors, and control over exposure dosing. Laboratory studies also can include two or more species.

Figure 11.9 The Biology Gamma Forest at the Brookhaven National Laboratory (Long Island, NY) as it appeared in 1964. This eastern deciduous forest, which was dominated by white oak, scarlet oak, and pitch pine, was exposed to 9,500 curies of ^{137}Cs for approximately six months beginning in 1961 (Woodwell, 1962; 1963). The radiation source was drawn up remotely from inside an underground pipe to expose the woodland. Exposure was many thousand roentgens at this γ source (center of barren spot) and decreased inversely with distance from the source. Zones composed of species of different tolerances ringed the source. Pitch pine (*Pinus rigida*) was the most sensitive, with death occurring at 20 r per day. At the other extreme, sedge (*Carex pensylvanica*) was the most tolerant, surviving 350 r per day. (Courtesy of Brookhaven National Laboratory. With permission.)

Laboratory systems designed to simulate some component of an ecosystem (such as multiple species assemblages) are called **microcosms**. Pontasch et al. (1989) examined changes in stream macroinvertebrate assemblages in response to an industrial discharge by exposing assemblages in laboratory "stream" microcosms. Niederlehner et al. (1985) examined cadmium's influence on species richness of protozoan communities by exposing naturally colonized substrates to cadmium in laboratory microcosms. In addition to these aquatic microcosms, terrestrial microcosms can involve plant growth chambers or soil columns (Gillett, 1989).

Between field and laboratory studies are those involving **mesocosms**, relatively large experimental systems also designed to simulate some component of an ecosystem. Mesocosms are delimited and enclosed to a lesser extent than are microcosms. They are normally used outdoors or, in some manner, incorporated intimately with the ecosystem that they are designed to reflect. They differ from microcosms by being larger, being located outdoors as a rule, and as having a lower degree of control by the researcher (Gillett, 1989). Although mesocosms vary considerably in their design (e.g., Figure 11.10), mesocosm studies all have the common goal of obtaining more realism than achievable with microcosms and more tractability than afforded by field surveys. Liber et al. (1992) conducted a mesocosm-based study to examine natural zooplankton community response to 2,3,4,6-tetrachlorophenol with *in situ* plastic bags extending upwards from the sediments to the surface of a freshwater body. The bags allowed treatments of different concentrations of toxicant and replication within treatments. Goldsborough and Robinson (1986) used similar *in situ* marsh enclosures to study periphyton assemblage response to the triazine herbicides, simazine and terbutryn. Flowing systems can also be studied with mesocosms, as evidenced by the work of McCormick et al. (1991), who studied diatom and protozoan assemblages in experimental stream channels dosed with the surfactant, dodecyl trimethyl ammonium chloride.

Field studies can also be applied to a wide range of scenarios, from surveys of contaminated systems to whole or partial ecosystem manipulations. Because experimental manipulations of natural systems afford stronger inference than surveys, a variety of studies have attempted such large-scale and expensive manipulations (e.g., Figure 11.11). To examine effects of radiation exposure, terrestrial systems in several geographical regions were irradiated, and the changes in associated communities were studied. The communities included old fields in South Carolina (Monk, 1966), woodlands in Georgia (Schnell, 1964), and forests and old fields in New York (Woodwell, 1962; 1963). The influence of acidification was examined by Schindler and coworkers (Schindler, 1996) by adding acid to an entire experimental lake in Canada. Sensitive species were identified as they disappeared from the lake. Functional redundancy was demonstrated by a shift in lake trout (*Salvelinus namaycush*) predation. As the pH-sensitive fathead minnow (*Pimephales promelas*) population declined, lake trout shifted their predation effort to the more pH-tolerant pearl dace (*Semotilus margarita*). Community shifts were also examined as a consequence of copper spiking of streams in Ohio (Winner et al., 1980) and California (Leland and Carter, 1984). High concentrations of copper shifted the insect assemblage away from caddisflies toward more tolerant midges. It also reduced diatom species richness.

More often used than field manipulations, field surveys provide less structured yet much less expensive observation of the consequences of toxicant introduction to communities. Such **biomonitoring**[5]—the widely applied monitoring (with selected sampling protocols) of a subset of an entire community with the goal of assessing community condition (Herricks and Cairns, 1982)—can involve a simple listing of species or much more complex analysis of data. Herricks and Cairns (1982) suggest three general types of biomonitoring efforts. The first simply describes the biota, perhaps summarizing results as a species-abundance list. The second involves the formulation of

---

[5] Qualifiers are frequently made for the term *biomonitoring*. As an example, Hopkin (1993) defines the monitoring of community changes along a gradient or among sites differing in levels of pollution as **Type 1 biomonitoring**. **Type 2 biomonitoring** involves the measurement of bioaccumulation in organisms among sites notionally varying in the level of contamination. **Type 3 biomonitoring** attempts to define the effects on organisms using tools such as biochemical markers in sentinel species or some measure of diminished fitness of individuals. **Type 4 biomonitoring** involves the detection of genetically based resistance in populations of contaminated areas.

Figure 11.10 Two types of mesocosms used to study fate and effects of contaminants. The top panel shows the indoor mesocosms of the Procter & Gamble Company's experimental streams facility (ESF). This system has the great advantage of more control over conditions (e.g., light and temperature) than normally afforded by outside mesocosms. Eight 12-m-long channels allow replication of treatments and production of exposure-concentration gradients. The top (head) section of each stream is paved with clay tiles for colonization by algae and microorganisms. Trays of gravel and sand are placed downstream of the tiles and afford substrate for invertebrates. (Courtesy of John Bowling of Procter & Gamble Co. With permission.) The bottom panel shows several outdoor pond mesocosms used in similar fashion for examining pollutant effects. (Courtesy of Thomas La Point, North Texas University. With permission.)

Figure 11.11   Whole-lake or split-lake studies such as shown here afford *in situ* but expensive information on responses of aquatic communities to anthropogenic materials. Although not associated with a conventional "toxicant," this particular split-lake study of nutrient enrichment (Canadian experimental Lake 226: P, N, and C added to the bottom section and only N and C added to the top section of the lake) clearly shows the phytoplankton community response. (Courtesy of Ken Mills. With permission.)

a hypothesis that is then tested with field observations. The third type combines both approaches, involving a formal test of conclusions (hypotheses) derived from the descriptive phase of the biomonitoring effort. Clearly, inferential strength is highest for this last type of biomonitoring.

III.C   Function

Ecotoxicologists tend to focus on structural changes rather than changes in community function. The reason for this bias is the general belief that feedback loops and functional redundancies make community functions less sensitive to toxicants than community structure (Odum, 1985; Forbes and Forbes, 1994). Regardless, some important community functions can be modified by toxicants. Certainly, modified functioning is implied by any change in functional groups or guilds, such as the shift in macroinvertebrate shredder and collector groups measured around coal mine drainage (Chadwick and Canton, 1983). Blanck (1985) also suggested that natural periphyton photosynthetic activity, measured as $^{14}CO_2$ incorporation, could be used as an ecotoxicological test. Giesy (1978) measured a significant drop in leaf-litter decomposition rates at elevated cadmium concentrations, suggesting another important function influenced by toxicants. Cairns and coworkers (e.g., Nieder-lehner et al., 1985; Cairns et al., 1986; McCormick et al., 1991) demonstrated a clear concentration-response relationship for colonization by protozoa of artificial substrates. They fit colonization data under various toxicant concentrations using the **MacArthur-Wilson model of island colonization**:

$$S_t = S_{EQ} \ (1 - e^{-Gt}) \qquad\qquad (11.5)$$

where S_t = number of species present at time t, S_{EQ} = equilibrium number of species for the island, and G = rate constant for colonization of the island. A sensitive assay was developed and demon-strated with a series of toxicants (Figure 11.12).

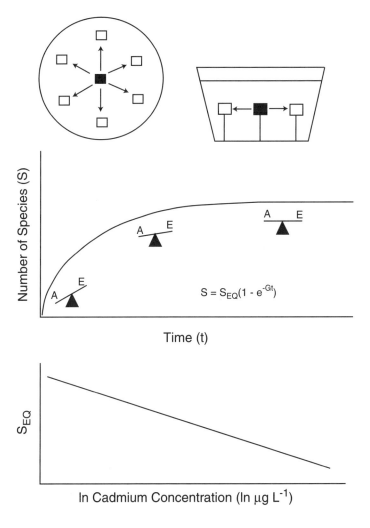

Figure 11.12 The protozoan community colonization assay developed by Cairns and coworkers (e.g., Niederlehner et al., 1985; Cairns et al., 1986). A polyurethane foam substrate (filled square in top diagram) is allowed to accumulate species in a natural stream and then brought to the laboratory to serve as a source or epicenter for colonization of other, uncolonized foam substrates (open squares in top diagram). The dynamics of species colonization on foam substrates is measured using the MacArthur-Wilson model for island colonization (center plot) (A = arrivals and E = extinctions). This is done with substrates submersed in containers filled with different concentrations of toxicant (bottom plot) to determine the effect of toxicants on the process. (Modified from Figures 1, 2, and 4 in Cairns, J., Jr. et al., *Environ. Monit. Assess.*, 6, 207–220, 1986.)

Change in community tolerance has also been proposed as a measure of function change with pollutant exposure. It is measured as **pollution-induced community tolerance** (PICT), an increase in tolerance to pollution resulting from species-composition shifts in the community, acclimation of individuals, and genetic changes in populations in the community. Procedurally, a previously exposed community and an unexposed community might be challenged with a toxicant, with the difference in responses of the two communities used to reflect adaptation. Kaufman (1982) reported that a periphyton community adapted to copper displayed a smaller decrease in ATP and chlorophyll upon repeated exposure to high copper concentrations relative to a periphyton community with no previous exposure. Similar results were obtained for 4,5,6-trichloroguaiacol-adapted periphyton communities exposed for a second time to this toxicant (Molander et al., 1990).

IV ECOSYSTEM QUALITIES

Many of the changes predicted by Odum (1985) to occur in ecosystems impacted by toxicants (Table 11.2) have already been discussed in the context of communities. However, a few remain to be discussed more fully. Particularly relevant are those related directly to nutrient cycling and energy flow.

Energy flow in agricultural ecosystems is changed by pesticide application, as is obvious from the resulting increase in crop biomass. Species dominance and structure of soil arthropod communities can also shift dramatically in ecosystems to which pesticides (e.g., DDT, parathion, or aldrin [Pimental and Edwards, 1982]) are applied. Implied in such changes is a change in energy flow via trophic exchange. Perhaps less obvious is the diminished decomposition rates in soils and recycling of minerals by soil organisms in agricultural systems and adjacent natural ecosystems. As evidence, application of organochlorine pesticides can shift the elemental composition of crops such as corn and beans (Pimental and Edwards, 1982).

Changes in energy flow and material cycling for nonagricultural systems have been demonstrated over a wide range of scales. Acidification of streams in the Great Smoky Mountains diminished leaf decomposition, bacterial production, and microbial respiration rate (Mulholland et al., 1987).

Table 11.2 Odum's Predicted Changes in Ecosystems Experiencing Toxicant Stress

Component/Quality	Predicted Change
Energetics	Increased community respiration
	Imbalance production:respiration, i.e., P/R < 1 or P/R >1
	Increased maintenance:biomass, i.e., increase in production/biomass and respiration/biomass
	Increased importance of energy from outside the ecosystem
	Increased export of primary production
Nutrients	Increased turnover of nutrients
	Decreased cycling of nutrients
	Increased loss of nutrients as a result of the two above changes
Community structure	Increased proportion of species that are r-strategists
	Decreased size of organisms
	Decreased life span of organisms
	Shortened food chains
	Decreased species diversity with increased species dominance
Ecosystem	Decreased internal cycling and increased importance of input and output from outside sources
	Regression to an earlier successional stage
	Functions (e.g., community metabolism) changed less than structural components (e.g., species richness)
	Decreased positive (e.g., mutualism) and increased negative (e.g., disease or parasitism) interactions

Source: Modified from Odum, E.P., *Bioscience,* 35, 419–422, 1985.

Odum (1985) indicates that enhanced losses of calcium from forested watersheds are a good indicator of functional damage to the watershed. At the other extreme of scale, microcosm and mesocosms studies also have been used to examine energy flow and material cycling under the influence of toxicants. For example, copper addition to microcosms reduced primary production and dissolved organic carbon production (Hedtke, 1984). In contrast to the results of Giesy (1978) with cadmium, spiking of experimental streams with triphenyl phosphate (Fairchild et al., 1987) did not lower leaf decomposition rates relative to control streams. Also, rooted flora increased and net nutrient retention increased with treatment. Clearly, these important functions of ecosystems are influenced by toxicants. However, there will be exceptions (e.g., Fairchild et al. [1987]) to general predictions such as Odum's due to complex changes in the community structure.

V SUMMARY

Beginning with a brief discussion of the ecological community, species assemblages, and niche, this chapter outlines general laboratory, mesocosm, and field approaches to determining the influence of toxicants on communities or species assemblages. General methods of estimating community effects include the use of indicator species (sensitive and tolerant species), community-level NOEC estimation, and biomonitoring. Simple species interactions, e.g., predator-prey and interspecies competition, were shown to be susceptible to toxicants. The reduction in fitness of individuals participating in such simple interactions was placed into the context of the principle of allocation. Assuming an equilibrium model for communities, qualities contributing to community vulnerability to toxicant effect were detailed, including community elasticity, inertia, and resilience. The question of whether a community can be rendered accurately to an equilibrium context was brought up and contrasted with the community conditioning hypothesis. Measures of community structure and function were then discussed relative to toxicant effects. Such changes were placed into the context of community successional regression, functional redundancy theory, and the law of frequencies. Changes in ecosystem energy flow and material cycling, although also implied in discussions of community shifts, were then described briefly. Some of these ecosystem changes will be discussed again in a wider geographical context in Chapter 12.

SUGGESTED READINGS

Atchison, G.J., M.B. Sandheinrich, and M.D. Bryan, Effects of environmental stressors on interspecific interactions of aquatic animals, in *Ecotoxicology: A Hierarchical Treatment*, Newman, M.C. and Jagoe, C.H., Eds., CRC Press, Boca Raton, FL, 1996.

Clements, W.H. and M.C. Newman, *Community Ecotoxicology*, John Wiley & Sons, Chichester, U.K., 2002.

Gillett, J.W., The role of terrestrial microcosms and mesocosms in ecotoxicological research, in *Ecotoxicology: Problems and Approaches*, Levin, S.A., M.A. Harwell, J.R. Kelly, and K.D. Kimball, Eds., Springer-Verlag, New York, 1989.

Graney, R.L., J.P. Giesy, and J.R. Clark, Field studies, in *Fundamentals of Aquatic Toxicology*, 2nd ed., Rand, G.M., Ed., Taylor & Francis, Washington, D.C., 1995.

Hopkin, S.P., *In situ* biological monitoring of pollution in terrestrial and aquatic ecosystems, in *Handbook of Ecotoxicology*, Calow, P., Ed., Blackwell Scientific Publications, London, 1993.

Odum, E.P., Trends expected in stressed ecosystems, *Bioscience*, 35, 419–422, 1985.

Woodwell, G.M., The ecological effects of radiation, *Sci. Am.*, 208, 2–11, 1963.

CHAPTER 12

Landscape to Global Effects

Even though the pattern of our relationship to the environment has undergone a profound transformation, most people still do not see the new pattern. ... The sights and sounds of this change are spread over an area too large for us to hold in our field of awareness.

Gore (1992)

I GENERAL

"Is it bigger than a bread box?" This is the conventional opening to a familiar guessing game in which an object is eventually identified from answers to a series of questions. It reflects our tendency to categorize things by size or scale. This tendency even extends to topics traditionally classified as within or outside the purview of ecotoxicology. Customarily, but not always correctly, ecotoxicology focuses on scales up to the traditional ecosystem, e.g., the fate and effects of pollutants in a lake, stream, field, or forest. Some studies do extend beyond this framework, but they are not common.

Divergent answers would result if one asked established ecotoxicologists to decide whether a topic such as global warming, widespread forest decline in central Europe, or global distillation of persistent organic pollutants were within the purview of ecotoxicology. Some would feel that, if the context of the problem were bigger than a traditional ecosystem, it would be better handled in biogeochemistry, landscape ecology, soil sciences, or atmospheric chemistry. There would be a contrastingly uniform affirmation if the question involved PCB (polychlorinated biphenyls) bioaccumulation in trout of a lake or a pollution-induced decrease in arthropod species diversity in forest litter. One obvious reason for this bias is that much of ecotoxicology was derived from the science of ecology. Until a few decades ago, the dominating context of ecology was the ecosystem or lower levels of biological organization.

In Chapters 9 and 10, we suggested that the single-species bias in much of ecotoxicology grew out of the early transplanting of ideas and approaches from mammalian toxicology. Although still present in ecotoxicology, this single-species focus is generally accepted as inadequate to addressing many important topics. Higher level effects can be equally or more important than those at the level of the individual. Similarly, the conventional ecosystem[1] bias is opined here to be inadequate

---

[1] Note that the term *biosphere* was used instead of the usual *ecosystem* in the definition of ecotoxicology (Chapter 1). The intent in doing so was to untether discussion from the conventional ecosystem context and allow free consideration of landscape, regional, continental, and global scales.

Figure 12.1   Landscape modification by smelting and mining activities in Copperhill, TN. Copperhill is situated in the Blue Ridge Mountains at the convergence of northern Georgia, western North Carolina, and southern Tennessee. The Ducktown Mining District began smelting circa 1854 and rapidly developed during the next four decades. Sulfuric acid and sulfur dioxide releases were greatly reduced after 1910. Tree growth, as measured from growth rings, was slowed from 1863 to 1912 in the nearby Great Smoky Mountains National Park (88 km upwind) due to the emissions from smelting (Baes and McLaughlin, 1984). This photograph (1982) was taken more than 70 years after emission reductions occurred and shows a desertlike landscape instead of the typical, forested landscape.

in many cases too. This opinion is reinforced by Cairns, (1993), Catallo (1993), and Holl and Cairns (1995), who argued that a landscape context for ecotoxicology is also needed. This traditional, but now too confining, bias toward the ecosystem or lower levels is designated the **ecosystem incongruity** here.

Supporting examples are easy to find. A landscape example involves copper mining and smelting in Copperhill, TN. By killing vegetation and stripping nutrients from the soil, acidic fumes from smelting transformed a lush forested landscape to the desertlike surroundings shown in Figure 12.1. A larger scale example is pollution from the Kuwait oil fires (Figure 1.2), an event influencing significant land (desert and urban) and marine components of a country. The final and most encompassing example is the TransAlaska Oil Pipeline (Figure 12.2). It extends south from Prudhoe Bay (Arctic Ocean) up the North Slope over the Brooks Range to cross the Arctic Circle, Yukon River, and the Alaska Range to end at the Valdez marine terminal on Prince William Sound (Pacific Ocean). For this one project, risk of damage exists for tundra, taiga, boreal forest, river, mountain, lake, fjord, and intertidal "ecosystems." One accident associated with only one segment of this project, the 1989 Exxon Valdez spill, spread oil out into parts of Cook Inlet and Alaska Sound and covered 30,000 km^2 of Alaskan waters. A large hypothetical spill onto the tundra could conceivably have an impact beyond that "ecosystem" because many bird species spend part of their time there and migrate to Asia (e.g., the wheatear, *Oenanthe oenanthe*, nesting in rocky fields of the tundra),

Figure 12.2 The TransAlaska Oil Pipeline as it passes across the taiga, a transitional community between the tundra and boreal forest communities.

North America (e.g., sandhill crane, *Grus canadensis*, breeding in tundra marshes), and South America (e.g., golden plover, *Pluvialis dominica*, nesting on tundra hillsides). Bird populations on several continents could be impacted by an oil spill in Alaska. Clearly, any preoccupation with an ecosystem, rather than a landscape or larger, context would result in an insufficient description of potential consequences of the TransAlaska Oil Pipeline. Our "stress signature" on the Earth now extends up to a global context (e.g., Figure 12.3).

There is a second and equally important reason why the ecosystem bias is no longer acceptable. The ecosystem focus draws attention away from important qualities of landscapes that are a heterogeneous matrix of "ecosystems." Unique properties emerge in this landscape[2] context. The source-sink framework for population dynamics discussed in Chapter 10 is an obvious example. Maurer and Holt (1996) modeled effects of pesticides on mobile wildlife in a complex landscape, and inclusion of source-sink dynamics emerged as crucial in predicting impact on populations. Another classic example involves **ecotones**, areas of transition between two or more community types (Odum, 1971). Ecotones often have species assemblages with high species richness and high abundance of individuals relative to those of the adjacent communities. There are several reasons underlying this **edge effect**. Species from contiguous habitats are present in the ecotone, increasing species richness. Some species can exploit both habitats in different ways, increasing their abundances at the ecotone. A species may nest in the forest but forage on grains in an adjacent field. Finally, unique species adapted to the ecotone, e.g., estuarine species, add to species richness. Consequently, the application of pesticides to agricultural fields may not have the same predicted ecotoxicological consequences for areas with an extensive network of hedgerows or patches of

---

[2] **Landscape** is an ambiguous term used in many contexts. Here, it is used to denote the sum total aspect of any geographical area (Monkhouse, 1965).

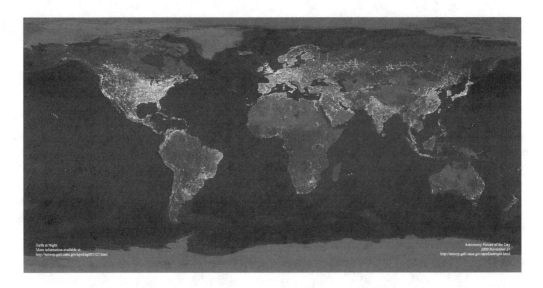

Figure 12.3 The global pattern of night lights as visualized by compositing 200 Defense Meteorological Satellite Program (DMSP) images (http://antwrp.gsfc.nasa.gov/apod/earth.html). (Courtesy of NASA.)

woods among the fields compared with those without. Such differences due to ecotones become important as the trend toward large agroindustrial farming and away from small farms continues in many parts of the world. Another important class of ecotones, estuaries at the mouths of rivers, is extremely vulnerable to contaminants from upriver sources and from port cities along their shores. Any unwarranted preoccupation with the traditional "ecosystem" context tends to draw attention away from the unique qualities of ecotones and other important features.

The third and final reason why we should extend our spatial context for ecotoxicology is simple: We now have the tools and data to do so. Affordable computer costs and increase in computational power allow diverse data sets, including inexpensive high-altitude and satellite data, to be integrated into a coherent and informative form by researchers and managers. Computerized **geographic information systems** (GIS) have emerged to handle these data at a reasonable cost. Most allow one to archive, organize, integrate, statistically analyze, and display many kinds of spatial information using a common coordinate system (Avery and Berlin, 1985). Data of different types such as land use, vegetation, rates of pesticide application, soil type, weather, and air or water quality can be merged and compared statistically to provide invaluable insights for effective stewardship of resources and environmental regulation. Books, such as that by Michener et al. (1994), detail methods for doing so, and a wide range of affordable imagery and maps are available.

Some imagery is produced by remote sensing. **Remote sensing** technologies allow the acquisition and analysis of data without requiring physical contact with the land or water surface being studied. Most determine qualities or characteristics of areas of interest based on measurements of visible light, infrared radiation, or radio energy coming from them (Sabins, 1987). For example, infrared spectral characteristics can be used to define vegetation community types over a wide area. Data from sensitive radiation sensors mounted in an airplane are used to map γ irradiation over large areas of U.S. Department of Energy nuclear facilities where releases occurred. Oil slicks on sea surfaces are detected and tracked by their higher radiance of ultraviolet and blue light (Sabins, 1987).

This type of spatial information is quickly becoming incorporated into environmental regulation and management activities. The U.S. EPA now has placed U.S. vegetation types, a toxic release inventory (TRI), air pollution, areas of air quality nonattainment, and Superfund sites into a GIS format (Reichhardt, 1996).

In a departure from the approach used in previous chapters, this chapter is based largely on examples. Each will be selected to represent an ecotoxicological topic at a particular spatial scale, i.e., landscape, regional, continental, hemispheric, or global scale. From the examples, the general trend will become obvious that a contaminant's potential for dispersal and its spatial scale for concern increases with the degree to which it is associated with the more mobile components of the environment (atmosphere mobility > hydrosphere mobility > pedosphere[3] mobility > lithosphere mobility). For example, contaminants associated primarily with the atmosphere, such as those giving rise to acid precipitation, will have effects over wide expanses. Some exceptions to this trend occur if large amounts of a material (e.g., a pesticide) are applied to a wide region[4] committed predominantly to one human activity (e.g., a large agricultural region of North America), or if the human activity giving rise to the contamination is occurring over extensive areas (e.g., lead contamination of North American soils resulting from widespread use of leaded gasoline). An unfortunate corollary is that cause-and-effect relationships are clearest locally but become increasingly difficult to assign with distance from a source (Cairns and Pratt, 1990). Consequently, some of the widest-spread problems of global concern, such as ozone depletion or global warming, are quite difficult to assign a cause and to enact an effective remedy.

II LANDSCAPES AND REGIONS

Often, landscape studies are based on some physical feature such as a watershed. Richards et al. (1993) used GIS methods to categorize land use in a Michigan catchment and linked land use with macroinvertebrate community composition of associated water bodies. There was a direct linkage between agricultural activity and stream substrate quality. In turn, substrate quality influenced the abundances of Ephemeropteran, Plecopteran, and Trichopteran insect taxa in benthic communities. From this study, recommendations for modifying land use and predictions of change under various restoration scenarios were generated. Richards and Host (1994) successfully applied this method again to Minnesota catchments along the shores of Lake Superior. A similar approach was taken to categorize and then project future problem areas for nonpoint pollution in the St. Johns River Basin in Florida (Adamus and Bergman, 1995). Analysis integrated information on contaminant amounts and concentrations in surface runoff, sites of storm water treatment and efficiency of that treatment, projected changes in land use, soil types, rainfall, hydrology, and current water quality. The inset of Figure 12.4 shows the predicted sites of significant pollutant generation along the St. Johns River.

Also illustrated in Figure 12.4 are features important in predicting contaminant impact at a larger scale. The dominant vegetation changes considerably in the lower half of the state and determines the specific communities at risk and the milieu in which the contaminant effect may or may not be expressed. The scale of an entire state may also be important. Because laws and regulations are applied by states, the arena for dealing with contaminants may be defined by state boundaries. For example, state fish-consumption advisories and bans are determined by concentrations in game species. States establish their own, occasionally divergent, criteria using Food and Drug Administration (FDA) action levels or EPA risk-based methods (Cunningham et al., 1994).

Often transcending state borders are **ecoregions** ("mapped classification[s] of ecosystem regions of the U.S. ... generally considered to be regions of relative homogeneity in ecological systems or

---

[3] The **pedosphere** is that part of the earth made up of soils and where important soil processes are occurring (Ugolini and Spaltenstein, 1992).

[4] As used here, a geographic **region** is an "area of the earth's surface differentiated by its specific characteristics" (Monkhouse, 1965).

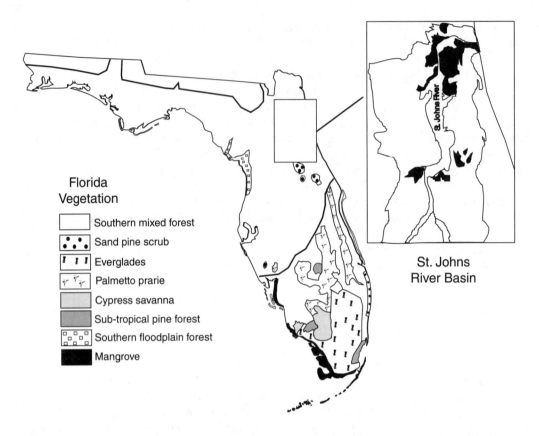

Figure 12.4 Three scales (river basin, vegetation type, and ecoregion) of consideration for Florida. (Modified by combining spatial data from three sources: Figure 5 in Adamus, C.L. and M.J. Bergman, *Water Resour. Bull.,* 31, 647–655, 1995; an ecoregion map from Omernik, J.M., *Ann. Assoc. Amer. Geographers,* 77, 118–125, 1987; and a U.S. Geological Survey vegetation map [Sheet 90].)

in relationships between organisms and their environment"[5] (Omernik, 1987). Inherent in their use is the working principle that contaminant effects vary significantly among ecoregions, e.g., ecoregions with high carbonate soils will be less sensitive to acid precipitation effects than those with mineral chemistries reflecting underlying granite or sandstone mineralogy (Glass et al., 1982). Ecoregions are used to manage aquatic and terrestrial resources of the United States based on land use, land surface features, vegetation, and soil types. In Figure 12.4, the lines crossing central Florida and those dipping down along the northern border of the panhandle define the edges of the three ecoregions in Florida. The Southeastern Coastal Plain ecoregion is the northernmost; the Southern Coastal Plain (approximately the upper half of Florida) and Southern Florida Coastal Plain (approximately the lower half of Florida) ecoregions occupy most of the state.

Hughes and Larsen (1988) applied the ecoregion classification to the formulation of a surface water protection strategy for the contiguous United States. Their aim was to develop a more accurate and appropriate framework for water quality criteria based on ecoregions, rather than the entire United States. Assuming correctly that water bodies within ecoregions were more similar to each other than to water bodies of other ecoregions, such diverse qualities as the index of biological integrity for fish assemblages, phosphorus concentrations, and dissolved oxygen concentrations were successfully classified for various ecoregions.

[5] It is apparent from Figure 12.4 that some ecoregions, such as the Southern Florida Coastal Plain ecoregion, are more heterogeneous than others. It contains the Everglades, palmetto prairie, sub-tropical pine forest, sand pine scrub, cypress savanna, and mangrove swamps.

III CONTINENTS AND HEMISPHERES

Problems such as acid rain[6] fit somewhere between the spatial scales of ecoregions and continents (Figure 12.5). Continental networks of precipitation monitoring have documented the spatial scale of the acid precipitation problem and concordance of precipitation pH with sources of acid-generating contaminants (Barrie and Hales, 1984). Industrialized areas emit sulfur and nitrogen

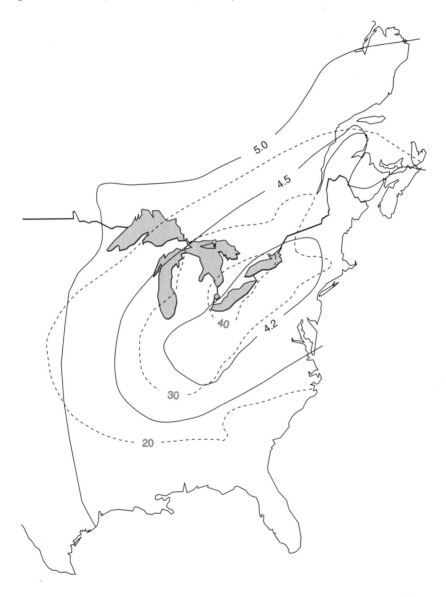

Figure 12.5 The pH (solid lines and associated numbers) and sulfate ion concentration ($\mu m\ l^{-1}$) (dashed lines and associated numbers) in precipitation as measured in 1980. In general, pH and sulfate contours coincide with the spatial pattern of sulfur dioxide emissions reported by a joint U.S./Canadian working group. (Modified from Figures 2 and 3 in Barrie, L.A. and J.M. Hales, *Tellus,* 36B, 333–335, 1984.)

---

[6] In equilibrium with gaseous carbon dioxide (Equation 12.1), liquid water in the atmosphere is predicted to have a pH of 5.7. **Acid precipitation**, including rain, fog (Hileman, 1983), snow, or other forms of precipitation, is defined as that with a pH below 5.7.

oxides that combine with atmospheric water to form H^+, SO_4^{2-}, and NO_3^-. Much of the sulfur dioxide produced by North American sources involves the burning of coal, which can contain 1.5 to 5% S, and the roasting of Ni, Zn, and Pb sulfides to produce metals. Another source is the burning of oil that can contain 2.5. to 3.5% S (Bridgman, 1994). In contrast to sulfur dioxide sources, which tend to be industrial or commercial sources with tall smoke stacks, nitrogen oxides sources are primarily near-ground sources such as automobiles (Bridgman, 1994). This, combined with the fact that the acid-producing reactions for sulfur dioxide are slower than those for nitrogen oxides, gives explanation for the general observation that N-related pH problems are more localized than S-related pH problems.

The equations below summarize the general reactions leading to precipitation with high H^+ concentrations: Equations 12.2 to 12.3 for sulfur dioxide and Equations 12.4 to 12.6 for nitrogen oxides (Bunce, 1991). An oxidation occurs in the first step of Equation 12.3 and in Equation 12.4. Although not explicitly indicated as such, Equation 12.6 is a catalyzed reaction.

$$CO_{2\,(g)} + H_2O_{(l)} \leftrightarrow H_2CO_{3\,(aq)} \leftrightarrow H^+_{(aq)} + HCO^-_{3\,(aq)} \tag{12.1}$$

$$SO_2 + H_2O \rightarrow H_2SO_3 \tag{12.2}$$

$$SO_2 \rightarrow SO_3 \rightarrow H_2SO_4 \tag{12.3}$$

$$NO \rightarrow NO_2 \tag{12.4}$$

$$2\,NO_2 + H_2O \rightarrow HNO_2 + HNO_3 \tag{12.5}$$

$$NO_2 + OH \rightarrow HNO_3 \tag{12.6}$$

These gases can disperse hundreds to thousands of kilometers from their sources (Cowling and Linthurst, 1981), causing widespread problems in parts of North America, northern Europe (Likens, 1976), and China (Bridgman, 1994).[7]

The impact of low pH precipitation is not solely a function of proximity to and magnitude of a source (Ravera, 1986). Different regions are inherently more sensitive than others. Soil type and underlying mineralogy influence the capacity to buffer pH changes and, consequently, influence sensitivity to acid precipitation's effects. An area with an underlying geology of granite, granitic gneisses, or quartz sandstones will have very poor buffering capacity and be sensitive to low-pH precipitation. Those areas with sandstone or shale mineralogies are poorly to moderately buffered, and those with limestone or dolomitic geologies will have high buffering capacity and be insensitive to acid precipitation (Glass et al., 1982). Bedrock geology maps can be combined with maps of the distribution of acid precipitation to predict areas of high or low concern. Glass et al. (1982) related bedrock geology, sources of acid precipitation, and stream alkalinity[8] to define pH-sensitivity classes of surface water bodies for New York State. Schindler (1988) examined the acid-neutralizing capacity of North American lakes and documented a gradual decrease in the northeastern

---

[7] Although the discussion here revolves around wet precipitation, fluxes of both dry and wet material can contribute pollutants to sites of effect. **Wet deposition** includes pollutants formed in the liquid media of the precipitation and that incorporated into the precipitation during rain out. **Dry deposition** is the flux of particles and gases such as SO_2, HNO_3, and NH_3 to surfaces (Stumm et al., 1987).

[8] **Alkalinity** is the capacity of a natural water to neutralize acid and is measured by titration of a water sample with a dilute acid to a specific pH endpoint. Most often, it is a function of carbonate (CO_3^{2-}), bicarbonate (HCO_3^-), and hydroxide (OH^-) concentrations, i.e., the carbon dioxide-bicarbonate-carbonate buffering of the water. However, dissolved organic compounds, borates, phosphates, and silicates can also contribute to alkalinity.

United States (New England and New York), northeastern Canada, and areas of Canada above the Great Lakes.

Effects of acid precipitation on aquatic biota can be sudden or gradual. Releases of pollutants accumulated in snowpack during seasonal thaws can cause high mortality to or diminished spawning success of downstream fish (Cowling and Linthurst, 1981; Bridgman, 1994). Schindler (1988) suggests that, in general, autumn-spawning fishes will be more sensitive than spring-spawning fishes to such releases because their pH-sensitive hatchlings tend to be in shallow, nearshore waters when the spring thaw brings pulses of low pH and high aluminum water. Slow deterioration of aquatic systems involves lowered buffering capacity as acid in precipitation "titrates" the entire system downward toward damagingly low pH conditions. Slow deterioration can involve the shift in equilibria for various biogeochemical processes until a dysfunctional condition emerges. For example, acidic conditions can increase dissolved aluminum flux into overlying waters from solid forms in sediments until toxic concentrations are reached. Low pH conditions can also increase leaching of aluminum and other metals from soils and minerals of the watershed, having toxic consequences to aquatic biota. Aluminum, calcium, and magnesium leaching can increase in a watershed as a consequence of acid precipitation (Smith, 1981; Schindler, 1988). Cronan and Schofield (1979) showed that atmospheric inputs of sulfuric and nitric acid to the pH sensitive aquatic systems of the Adirondack Mountain region of New York resulted in high dissolved Al concentrations in surface waters. The geology of this sensitive area is dominated by granitic gneisses, resulting in poorly buffered waters. They (Cronan and Schofield, 1979) expressed concerns regarding aluminum toxicity there and in other areas of the United States and Europe that have silicate bedrock.

Regardless of the exact mechanism of demise during the decline in aquatic systems, it is clear that aquatic systems, spread over wide areas of continents, are being damaged by acid precipitation. Baker et al. (1991) estimated that the atmospheric input of acid anions represents the dominant anion flux into 75% of 1180 acid sensitive lakes and 47% of 4670 acid sensitive streams surveyed in the United States. In a survey of 5000 lakes of southern Norway, 1750 had lost fish species and another 900 were seriously impacted due to acid precipitation (Bridgman, 1994). In southern Ontario, 56% of surveyed lakes had reduced fish populations, and an extraordinary 24% had no fish at all due to acid precipitation (Bridgman, 1994).

Effects of acid precipitation on terrestrial components of the biosphere are also significant and widespread. At low levels, the nitrogen and sulfur added to a forest in acid precipitation can enhance growth via the **fertilization effect** (Bridgman, 1994). However, acid precipitation can also increase nutrient leaching from foliage and forest soils and can accelerate the weathering of minerals (Smith, 1981). Acid leaching of calcium, magnesium, potassium, and sodium from decomposing forest litter, and calcium, potassium, and magnesium from soils has been measured (Smith, 1981). Leaching of magnesium and potassium from soils can produce a deficiency of these essential plant nutrients. Release of aluminum from solid phases in soils can result in direct toxicity to vegetation (Cowling and Linthurst, 1981; Smith, 1981). Acid precipitation can also cause necrotic lesions on foliage, increased plant susceptibility to disease, increased rate of wax erosion from foliage surfaces, and lower nitrogen fixation by legumes via the inhibition of root nodule formation (Cowling and Linthurst, 1981). The composite of all of these effects of acid precipitation on forests is a major explanation forwarded for the widespread forest damage in large tracts of North America (Nihlgård, 1985) and the **waldsterben**, "the widespread and substantial decline in growth and the change in behavior of many softwood and hardwood forest ecosystems in central Europe" (Schütt and Cowling, 1985).

Atmospheric dispersal of pollutants in aerosols or particulates can also encompass wide areas. Certainly, dry deposition plays an important role in the acid deposition related problems just discussed. Much of the widespread metal deposition associated with the Copperhill smelting shown in Figure 12.1 was associated with particulate transport. Elevated iron concentrations were measured in tree rings 88 km from the Copperhill source (Baes and McLaughlin, 1984), indicating that

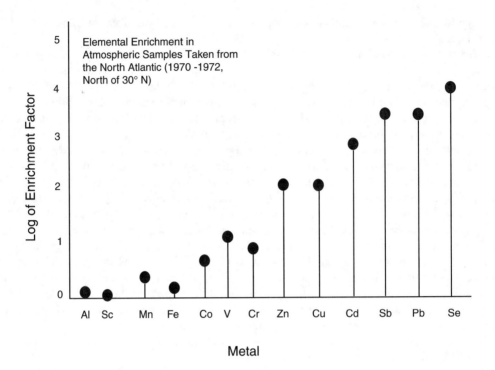

Figure 12.6   Metal particulate dispersal over the remote North Atlantic Ocean as evidenced by increased enrichment factors. The **enrichment factor** (EF_{crust}) for an element is its concentration (X) measured in air samples divided by that expected in the earth's crust: $EF_{crust} = [X/Al]_{air}/[X/Al]_{crust}$. Both air and crustal concentrations are normalized to Al concentrations, since aluminum is an ubiquitous element comprising about 8% of crustal material. Increases in EF_{crust} above 10^0 (= 1) imply enrichment from anthropogenic sources. (Modified from Figure 1 in Duce, R.A. and W.H. Zoller, *Science,* 187, 59–61, 1975.)

particulate-associated iron moved long distances from its source at ore processing. Generally, widespread deposition of particulate-associated cadmium, manganese, lead, and zinc in forests of Tennessee has been documented: One-third or more of the annual flux of cadmium, lead, and zinc to a Tennessee Valley forest was from atmospheric deposition (Lindberg et al., 1982). Hirao and Patterson (1974) found that most of the lead in sedge (*Carex scopulorum*) and voles (*Microtus montanus*) inhabiting a remote High Sierra valley in California came long distances from automotive and industrial sources in Los Angeles and San Francisco. Much of the lead (12 kg versus 1 kg from other sources) entered the valley associated with snow. Still wider transport of Pb has been documented. Analysis of Antarctic ice cores shows an increase in Pb flux to the Antarctic after the worldwide onset of industrialization (Boutron and Patterson, 1983). At the earth's other pole, the flux of particulate-associated contaminants is significant from Eurasia to the Arctic (Pacyna, 1995). In air above remote parts of the northern Atlantic Ocean, levels of many metals are elevated far above background concentrations (Figure 12.6) (Duce and Zoller, 1975). Particulate-associated radionuclides released into the atmosphere during the Chernobyl reactor meltdown were distributed over most of the Northern Hemisphere and are predicted to result in elevated cancer deaths throughout Europe and portions of the former Soviet Union (Barnaby, 1986; Anspaugh et al., 1988).

Another phenomenon with a scale encompassing an entire continent is ozone depletion by chlorofluorocarbons (Figure 12.7). The introduction of chlorofluorocarbons (CFC)[9] such as CFC12

---

[9] The CFC structures can be derived easily from their names. Ninety is added to the number in the CFC's name, e.g., CFC11 produces 90 + 11 or 101. This number codes for 1 carbon ("hundreds" digit), 0 hydrogens ("tens" digit), and 1 fluorine ("units" digit). Because the carbon atom forms four covalent bonds and only one is occupied (by a fluorine atom), the other three bonds must be with three chlorine atoms. So, CFC11 is $CFCl_3$.

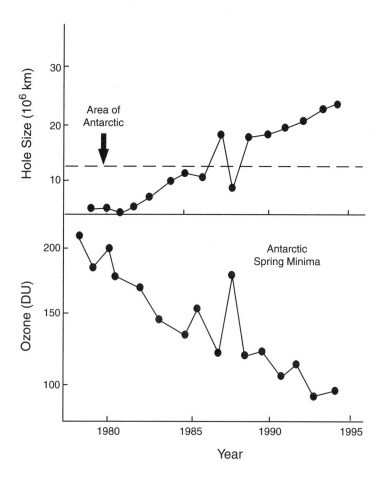

Figure 12.7   Increase in the size of the Antarctic ozone hole (top panel) and decrease in the spring minima ozone concentration (bottom panel). Ozone concentrations are expressed in **Dobson units** (DU). One DU is the equivalent of 0.001-mm thickness of pure ozone at 1 atm. To give some scale to this unit, if all of the ozone in the atmosphere were to be brought down to form a pure layer of ozone at sea level, it would be only 3 mm thick (Bunce, 1991). (Data taken from NASA web site: jwocky.gstc.nasa.gov, files O3holes2.gif and O3holemin.gif.)

(dichlorofluoromethane or CF_2Cl_2) and CFC11 (trichlorofluoromethane or $CFCl_3$) has shifted the balance among reactions taking place in the stratosphere to disfavor the maintenance of normal levels of ozone (O_3).

Ozone is formed by the **Chapman mechanism** (Reactions 12.7 to 12.9) (Bunce, 1991). Energy from sunlight of wavelengths <240 nm and <325 nm is required for Reactions 12.7 and the second step of 12.8, respectively. A catalyst is required for Reaction 12.9.

$$O_2 \rightarrow 2O \qquad\qquad (12.7)$$

$$O + O_2 \rightarrow O_3 \rightarrow O_2 + O \qquad\qquad (12.8)$$

$$O + O_3 \rightarrow 2O_2 \qquad\qquad (12.9)$$

Chloride also becomes involved in the ozone generating and depleting reactions in the stratosphere (Reactions 12.10 to 12.14) (Zurer, 1988). Reactions 12.12 and 12.14 require a catalyst, and 12.13 requires light energy.

$$Cl + O_3 \rightarrow ClO + O_2 \tag{12.10}$$

$$ClO + ClO \rightarrow Cl_2O_2 \tag{12.11}$$

$$ClO + O \rightarrow Cl + O_2 \tag{12.12}$$

$$Cl_2O_2 \rightarrow Cl + ClOO \tag{12.13}$$

$$ClOO \rightarrow Cl + O_2 \tag{12.14}$$

Nitrogen species can shift these reactions such that some of the Cl is bound up in nitrogen compounds and unavailable to react with ozone, e.g., $ClO + NO_2 \rightarrow ClONO_2$ (Zurer, 1987; Kerr, 1988a). However, excess Cl from the breakdown of CFCs can overwhelm this sequestering process, with a net effect of decreasing ozone concentrations. Molecular chlorine (Cl_2) and hypochlorous acid (HOCl) generated by Reactions 12.15 and 12.16 are readily converted to free radicals which destroy ozone (Zurer, 1987).

$$H_2O + ClONO_2 \rightarrow HNO_3 + HOCl \tag{12.15}$$

$$HCl + ClONO_2 \rightarrow HNO_3 + Cl_2 \tag{12.16}$$

Ozone destruction as a consequence of CFC accumulation in the stratosphere and circulation patterns above the Antarctic has produced an alarming **ozone hole** recently. Although meteorological conditions are less favorable for the formation of a similar hole above the Arctic (i.e., the polar vortex does not last as long), elevated chlorine monoxide (ClO) concentrations and ozone thinning have been reported there too (Zurer, 1989; Kerr, 1992).

The concern for the destruction of vast parts of the ozone layer is heightened by the realization that the expected lifetime of CFCs in the atmosphere is extremely long, i.e., 70 years for CFC11 and 110 years for CFC12 (Thompson, 1992). Any remedial action now will take considerable time to reverse the damage caused by CFC releases. Fortunately, the **Montreal Protocol**, an international treaty to limit and eventually eliminate the use of CFCs, was endorsed by 70 countries and then signed into law by 1987 (Crawford, 1987; Bunce, 1991). Although it will take a long time to reduce CFCs in the stratosphere, there are already encouraging indications that chlorine concentrations have leveled off in the stratosphere (Kerr, 1996).

Predictions of ozone depletion effects to humans and ecological systems remain vague. Ozone absorbs UV light with wavelengths of 290 to 330 nm, or roughly the UV-B range (280 to 320 nm) as it enters the earth's atmosphere. This UV-B can cause skin cancer, and speculations are that the incidence of skin cancers could increase as stratospheric ozone levels drop. However, Bunce (1991) provides the tempering comparison that "for people living in the middle latitudes, the increased risk of skin cancer due to each 1% decrease in ozone levels is equivalent to that posed by moving 20 km closer to the equator." Effects to ecological entities also remain equivocal at this time. Some speculation exists linking ozone depletion to the global decline in amphibian species. Increased UV-B penetration into the surface waters of the oceans is also speculated to decrease marine phytoplankton photosynthesis (Baird, 1995).

IV BIOSPHERE

IV.A General

Some human activities can stretch out to involve hemispheres or the entire planet. As depicted in Figure 12.7, the Antarctic ozone hole extended beyond that continent in the late 1980s. Two general phenomena that are even more encompassing will be discussed in this section: global distillation of persistent organic pollutants and global warming.

IV.B Global Movement of Persistent Organic Pollutants

Many persistent organic pollutants (POP)[10] are subject to extensive movement and redistribution on a global scale (Table 12.1). A POP will vaporize and move in the atmosphere until it reaches a temperature at which it condenses. It then becomes associated with a less mobile solid or liquid phase. The extent of such movement of POPs from their sources of release to cooler latitudes depends on each POP's rate of degradation, vapor pressure, and lipophility (Simonich and Hites, 1995).

According to the **cold condensation theory**, POPs in the air will condense onto soil, water, and biota at cool temperatures. Consequently, the ratios for POP concentrations in the air and on condensed phases decreases from warmer to cooler climates (Wania and Mackay, 1995). This leads

Table 12.1 Persistent Organic Pollutant Mobility in a Global Context

Pollutant Classes (Subclassified by Number of Cl Atoms, Rings, or by Pesticide Type)	Relative Mobility Class			
	Rapidly Deposited and Retained Near Source	Preferential Deposition and Accumulation in Mid-Latitudes	Preferential Deposition and Accumulation in Polar Latitudes	Worldwide Dispersion and Deposition
Chlorobenzenes	—	—	5 to 6 Cl	0 to 4 Cl
PCBs[a]	8 to 9 Cl	4 to 8 Cl	1 to 4 Cl	0 to 1 Cl
PCDDs[a] and PCDFs[a]	4 to 8 Cl	2 to 4 Cl	0 to 1 Cl	—
PAHs[a]	>4 rings	4 rings	3 rings	2 rings
Organochlorine pesticides	Mirex	polychlorinated camphenes, DDT, DDE, chlordanes	HCB,[a] HCCHs,[a] dieldrin	—
Pollutant Quality				
Log K_{OA}[b]	>10	8 to 10	6 to 8	<6
P_L[c]	<−4	−4 to −2	−2 to 0	>0
T_C[d]	>+30	−10 to +30	−50 to −10	<−50

[a] PCB = polychlorinated biphenyls; PCDD and PCDF = polychlorinated di-benzo-*p*-dioxins and -furans; PAH = polycyclic aromatic hydrocarbons; HCB = hexachlorobenzene; HCCHs = hexachlorocyclohexanes.

[b] K_{OA} is the partition coefficient between octanol and air. Like the K_{OW}, it is a measure of lipophilicity.

[c] P_L is the **subcooled liquid-vapor pressure** (Pa), a measure of a compound's volatility. Specifically, it is the liquid vapor pressure corrected or adjusted for the heat of fusion, the energy needed to convert a mole of a compound from a solid to a liquid phase. Its use allows the expression of *liquid* vapor pressures at a specific temperature for organic compounds with widely varying melting temperatures.

[d] T_C is the **temperature of condensation**, the temperature (°C) at which the compound condenses or partitions from the gaseous to the nongaseous phase.

Source: Modified from Wania, F. and D. Mackay, *Environ. Sci. Technol.*, 30, 390A–396A, 1996.

[10] **Persistent organic pollutants** (POPs) are those organic pollutants that are long-lived in the environment and tend to increase in concentration as they move through food chains (Wania and Mackay, 1996). According to Wania and Mackay (1996), they are also called **bioaccumulative chemicals of concern** (BCCs) and **persistent toxicants that bioaccumulate** (PTBs).

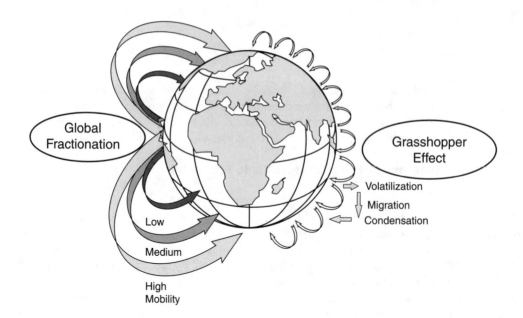

Figure 12.8 The movement of persistent organic pollutants on a global scale. (Modified from Figure 1 in Wania, F. and D. Mackay, *Environ. Sci. Technol.*, 30, 390A–396A, 1996. With permission.)

to the phenomenon of **global distillation** in which POPs migrate from warm regions of release to cold regions of condensation (Figure 12.8). This distillation can involve seasonal cycling of temperatures, such that movement toward the higher latitudes occurs in annual pulses or jumps (the **grasshopper effect**) (Wania and Mackay, 1996). Because POPs differ in their individual rates of degradation, vapor pressures, and lipophilities, a **global fractionation** occurs in which some POPs move more rapidly than others toward the polar regions. Some, because of their temperatures of condensation, may be unable to move beyond a certain point toward cooler latitudes. The net result is a redistribution of the different POPs from the equator or site of origin toward the cold polar regions of the earth. The K_{OW}s of POPs also influence their movement via the **retention effect**. Those POPs with high lipophility tend to be held more firmly than less lipophilic POPs in solid phases such as soil and vegetation. Consequently, they spend less time in the atmosphere and are less available for transport in that medium (Wania and Mackay, 1995). All else being equal, the more lipophilic POPs move more slowly toward higher latitudes than less lipophilic POPs.

Actual global distributions of POPs conform to the global distillation and fractionation scheme outlined above. In a survey by Simonich and Hites (1995) of tree bark from sites around the world, global distillation was apparent for the volatile hexachlorobenzene but not apparent for the less volatile endosulfan and DDT. Clear movement of volatile organic compounds into Arctic systems has also been documented (Chernyak et al., 1995; Muir et al., 1995). Mackay and Wania (1995) produced a model that accurately predicted POP global distributions based on these processes.

IV.C Global Warming

Global warming is thought to result primarily from the increased atmospheric carbon dioxide (CO_2) concentrations produced by fossil fuel burning and the worldwide destruction of forests (Woodwell, 1978; Khalil and Rasmussen, 1984). Much of the concern centers on carbon dioxide, methane, and nitrous oxide, which have rapidly increased in the atmosphere since preindustrial times (Hileman, 1989). These gases, along with water vapor, are relatively transparent to light entering the earth's atmosphere but absorb long wave, infrared radiation radiating back from the earth's surface toward space. Because the net energy balance of sunlight influx and infrared radiation efflux from the earth's surface determines the steady-state temperature of the earth (the **greenhouse effect**), increases in these greenhouse gases[11] are believed to produce an increase in global temperatures. Although there is still some disagreement about whether this is occurring (e.g., Hileman, 1984; White, 1990; Thompson, 1992; Lindzen, 1994; Masood, 1995), global warming could have widespread effects to human and nonhuman species. Abrupt changes in economic and agricultural activities might occur, sea level might change, and tropical diseases could extend into other regions of the world (Kerr, 1988b). Relative to changes occurring between ice ages, species ranges would have to change very rapidly, with some species becoming extinct and global biodiversity decreasing (Roberts, 1988). Species, including important plant species that normally migrate during the ice age interglacial cycles, would be severely challenged by the extremely rapid rate of temperature change. In the current landscape that is highly fragmented by human activity, some would be unable to migrate successfully. Such species in fragmented habitats would be "man-locked" (Roberts, 1988), unable to migrate, and at high risk of extinction.

Vignette 12.1   Biomass Burning: Global Effects

Joel S. Levine

Atmospheric Sciences, NASA Langley Research Center, Hampton, VA

Biomass Burning as a Source of Gases and Particulates to the Atmosphere.

Biomass burning is a significant local, regional, and global source of gases and particulates to the atmosphere. Biomass or vegetation burning is the burning of living and dead vegetation (i.e., trees, grass, agricultural stubble after the harvest) for land clearing and land-use change. It has been estimated that as much as 90% of global burning is human initiated, with only about 10% of all burning occurring naturally due to atmospheric lightning. Most of the world's burning occurs in the tropics. However, recent studies indicate that significant burning occurs in the world's boreal forests (Levine, 1991; 1996a; 1996b).

Biomass burning produces a whole suite of environmentally significant gases, including carbon dioxide (CO_2), carbon monoxide (CO), methane (CH_4), more than a dozen species of hydrocarbons, nitric oxide (NO), ammonia (NH_3), hydrogen cyanide (HCN), acetonitrile (CH_3CN), cyanogens (NCCN), and nitrous oxide (N_2O). Laboratory biomass burning experiments conducted by Lobert et al. (1991) have identified and quantified the various carbon (Table 12.2) and nitrogen (Table 12.3) species released to the atmosphere by burning.

Carbon dioxide, methane, and nitrous oxide are greenhouse gases, which trap Earth-emitted infrared radiation and lead to global warming. Carbon monoxide, methane, and the oxides of nitrogen lead to the photochemical production of ozone (O_3) in the troposphere. In the troposphere, ozone is an irritant and harmful pollutant and, in some cases, is toxic to living systems. Nitric oxide leads to the chemical production of nitric acid (HNO_3) in the troposphere. Nitric acid is the fastest growing component of acidic precipitation. Ammonia is the only basic gaseous species that neutralizes the acidic nature of the troposphere. Nitrous oxide leads to chemical destruction of ozone in the stratosphere, where ozone shields the Earth's surface from solar ultraviolet radiation.

Particulates—small (usually about 10 μm or smaller) solid particles, such as smoke or soot particles—are also produced during the burning process and released into the atmosphere. These solid particulates absorb and scatter incoming sunlight and hence impact the local, regional, and global climate. In addition, these particulates (specifically, particulates 2.5 μm or smaller) can lead to various human respiratory and general health problems when inhaled. The gases and particulates produced during biomass burning lead to the formation of "smog." The word *smog* was coined as a combination of smoke and fog and is now used to describe any smoky or hazy pollution in the atmosphere.

---

[11] **Greenhouse gases** include water vapor, carbon dioxide, methane, nitrous oxide (N_2O), CFCs, methylchloroform, carbon tetrachloride, and the fire retardant, halon®. Ozone *in the troposphere* may also act as a greenhouse gas (Hileman, 1989). All greenhouse gases are not equal relative to infrared radiation absorption. For example, one molecule of CFC11 absorbs 10,000 times more infrared radiation than one molecule of carbon dioxide (Anon, 1985).

Table 12.2   Carbon and Gases Produced during Biomass Burning

Compound	Mean Emission Factor Relative to the Fuel C (%)
Carbon dioxide (CO_2)	82.58
Carbon monoxide (CO)	5.73
Methane (CH_4)	0.424
Ethane (CH_3CH_3)	0.061
Ethene ($CH_2 = CH_2$)	0.123
Ethine (CH = CH)	0.056
Propane (C_3H_8)	0.019
Propene (C_3H_6)	0.066
n-butane (C_4H_{10})	0.005
2-butene (cis) (C_4H_8)	0.004
2-butene (trans) (C_4H_8)	0.005
i-butene, i-butene ($C_4H_8 + C_4H_8$)	0.033
1,3-butadiene (C_4H_6)	0.021
n-pentane (C_3H_{12})	0.007
Isoprene (C_5H_8)	0.008
Benzene (C_6H_6)	0.064
Toluene (C_7H_8)	0.037
m-, p-xylene (C_8H_{10})	0.011
o-xylene (C_8H_{10})	0.006
Methyl chloride (CH_3Cl)	0.010
NMHC (As C) (C_2 to C_8)	1.18
Ash (As C)	5.00
Total Sum C	**94.92** (including ash)

Source: Lobert, J.M. et al., in *Global Biomass Burning: Atmospheric, Climatic, and Biospheric Implications,* Levine, J.S., Ed., MIT Press, Cambridge, MA, 1991.

Table 12.3   Nitrogen Gases Produced during Biomass Burning

Compound	Mean Emission Factor Relative to the Fuel N (%)
Nitrogen oxides (NO_x)	13.55
Ammonia (NH_3)	4.15
Hydrogen cyanide (HCN)	2.64
Acetonitrile (CH_3CN)	1.00
Cyanogen (NCCN) (As N)	0.023
Acrylonitrile (CH_2CHCN)	0.135
Propionitrile (CH_3CH_2CN)	0.071
Nitrous oxide (N_2O)	0.072
Methylamine (CH_3NH_2)	0.047
Dimethylamine (($CH_3)_2NH$)	0.030
Ethylamine ($CH_3CH_2NH_2$)	0.005
Trimethylamine (($CH_3)N$)	0.02
2-methyl-1-butylamine ($C_5H_{11}NH_2$)	0.04
n-pentylamine (n-$C_5H_{11}NH_2$)	0.137
Nitrates (70% HNO_3)	1.10
Ash (As N)	9.94
Total sum N (as N)	**33.66** (Including ash)
Molecular nitrogen (N_2)	21.60
Higher HC and particles	20

Source: Lobert, J.M. et al., in *Global Biomass Burning: Atmospheric, Climatic, and Biospheric Implications,* Levine, J.S., Ed., MIT Press, Cambridge, MA, 1991.

Table 12.4 Global Estimates of Annual Amounts of Biomass Burning and of the Resulting Release of Carbon to the Atmosphere

Source	Biomass Burned (Tg dry material year^{-1})	Carbon Released (Tg C year^{-1})
Savanna	3690	1660
Agricultural waste	2020	910
Fuel wood	1430	640
Tropical forests	1260	570
Temperate/boreal forests	280	130
World totals	8680	3910

Source: Andreae, M.O., in *Global Biomass Burning: Atmospheric, Climatic, and Biospheric Implications,* Levine, J.S., Ed., MIT Press, Cambridge, MA, 1991.

Table 12.5 Comparison of Global Emissions from Biomass Burning with Emissions from All Sources (Including Biomass Burning)

Species	Biomass Burning (Tg element[a] year^{-1})	All Sources (Tg element[a] year^{-1})	Emissions Due to Biomass Burning (%)
CO_2 (gross)	3500	8700[b]	40
CO_2 (net)	1800	7000[c]	26
CO	350	1100	32
CH_4	38	380	10
NMHC[d]	24	100	24
N_2O	0.8	13	6
NO_x	8.5	40	21
NH_3	5.3	44	12
Sulfur	2.8	150	2
COS	0.09	1.4	6
CH_3Cl	0.51	2.3	22
H_2	19	75	25
Tropospheric O_3	420	1100	38
TPM[e]	104	1530	7
POC[f]	69	180	39
EC[g]	19	<22	>86

[a] Tg element year^{-1} where C, N, S, Cl are the elements.
[b] Biomass burning plus fossil fuel burning.
[c] Deforestation plus fossil fuel burning.
[d] Nonmethane hydrocarbons (excluding isoprene and terpenes).
[e] Total particulate matter (Tg year^{-1}).
[f] Particulate organic matter (including elemental carbon).
[g] Elemental (black-soot) carbon.

Source: Andreae, M.O., *Global Biomass Burning: Atmospheric, Climatic, and Biospheric Implications,* Levine, J.S., Ed., MIT Press, Cambridge, MA, 1991.

It has recently been reported that the burning of vegetation results in the complete release of mercury contained in the biomass (Friedli et al., 2001). About 95% of the mercury was emitted as elemental mercury and the remainder emitted as particulate mercury. Friedli et al. (2001) concluded that the mercury released by burning becomes part of the global mercury reservoir and undergoes chemical transformation in clouds and in the free troposphere, eventually returning to the surface via wet and dry deposition.

The world's biomass burning occurs in four ecosystems—tropical savannas, agricultural lands following the harvest, tropical forests, and temperate/boreal forests—and through an activity largely confined to the tropics, the domestic burning of biomass material for home heating and cooking. The amount of global burning varies from year to year due to variations in weather patterns, precipitation, soil moisture, etc. Estimates of the average annual amount of burning have been made for each of the ecosystems and for domestic burning (Andreae, 1991). These estimates are given in Table 12.4 in units of teragrams of dry material per year (Tg d.m. year^{-1}) (1 teragram = 10^{12} grams = 10^6 metric tons) (Andreae, 1991). Biomass contains about 45% carbon by mass. The total carbon liberated to the atmosphere by burning in these ecosystems is also summarized in

Table 12.4 in units of teragrams of carbon per year (Tg C year^{-1}) (Andreae, 1991). Combining the estimates of total global annual biomass burned (summarized in Table 12.4) with measurements of the species and amount of gases and particulates produced by burning biomass, estimates can be made of the global annual production of various species by burning (Andreae, 1991). These estimates are summarized in Table 12.5, which gives the species production due to burning, the production of the species due to all sources (in units of teragrams of the element per year), and the percentage of production due to biomass burning (Andreae, 1991).

Biomass Burning as a Source of Local, Regional, and Global Pollution

There is mounting observational evidence indicating that gases and particulates produced from biomass burning and released into the atmosphere are a significant source of atmospheric pollution on the local and regional scales. High concentrations of carbon dioxide from unknown origins were measured episodically on the southeastern United States during June and July 1995. Wotawa and Trainer (2000) have shown that these greatly enhanced episodes of elevated carbon monoxide were produced by forest fires in Canada. Over a period of two weeks, the fire-produced carbon monoxide emissions increased carbon monoxide concentrations in the southeastern United States as well as along the eastern seaboard, a region with one of the world's highest rates of anthropogenic emissions. Within the forest fire plumes, there were also high concentrations of ozone, volatile organic compounds, and particulates. These findings suggest that the impact of boreal forest fire emissions on air quality in the mid-latitudes of the Northern Hemisphere, where anthropogenic sources of pollution predominate, needs to be carefully reevaluated (Wotawa and Trainer, 2000).

During the 1997–1998 period, there were a series of very extensive and widespread wildfires in Southeast Asia, South America, Africa, Mexico, Russia, and the United States (Florida). The greatly enhanced world fire activity was all related to the very strong El Nino and its global perturbations on atmospheric circulation and precipitation patterns. Throughout Southeast Asia, including Indonesia, the major cause of the extensive fires was the regular seasonal burning for land clearing and land use change, a practice going back over 100 years (Eaton and Radojevic, 2001). However, in 1997–1998, the El Nino-induced drought caused normally well-controlled land clearing fires to quickly become uncontrolled wildfires. To make matters worse, in Indonesia, the aboveground uncontrolled wildfires ignited underground deposits of peat and coal. It is estimated that from August to December 1997, some 46,000 km^2 burned in Indonesia alone, with more than 250,000 separate peat and coal fires producing very large amounts of carbon dioxide, carbon monoxide, methane, oxides of nitrogen, and particulates (Levine, 1999), resulting in an extensive cloud of smoke, gas, and smog over Southeast Asia that lasted for months.

The 1997–1998 fires in Indonesia were called "one of the most broad-ranging environmental disasters of the century" (Hamilton et al., 2000). Some of the consequences of the 1997–1998 Indonesian fires included: (1) More than 200 million people were exposed to very high levels of air pollution and particulates, resulting in more than 20 million smoke-related health problems. (2) Due to very poor atmospheric visibility, on September 26, 1997, a Garuda Airlines Airbus 300-B4 crashed while attempting to land in Sumatra, killing all 234 passengers. (3) Due to very poor atmospheric visibility, on September 27, 1997, two ships collided in the Strait of Malacca, off the coast of Malaysia, killing 29 crewmembers. (4) The cost of fire-related damage in Indonesia has been estimated to be in excess of $4 billion (Levine, 2001). In addition to the problems in the immediate vicinity of the Indonesian fires, gases and particulates produced during the wildfires also had regional and global impacts. Enhanced atmospheric levels of carbon monoxide, ethane, and hydrogen cyanide produced in the Indonesian fires were measured as far away as Hawaii (Rinsland et al., 1999).

The impact of gaseous and particulate emissions from biomass burning on atmospheric composition and chemistry, on climate, and on air quality and human health may become a more important concern in the coming years. Coupled climate-biosphere numerical models indicate the wildfires will be more frequent and more widespread as a consequence of global warming.

V SUMMARY

In this chapter, an argument was made that a context larger than the ecosystem is required to fully grasp all ecotoxicological problems facing us today. The argument is made using examples of important problems at increasingly wider spatial scales. Unfortunately, as the scale of problems becomes wider, the potential for widespread harm increases, while the ability to assign a cause-effect relationship decreases. This makes environmental assessment and management of such problems exceedingly difficult and prone to divergent conclusions.

SUGGESTED READINGS

Anspaugh, L.R., R.J. Catlin, and M. Goldman, The global impact of the Chernobyl reactor accident, *Science*, 242, 1513–1519, 1988.

Cowling, E.B., Acid precipitation in historical perspective, *Environ. Sci. Technol.*, 16, 110A–123A, 1982.

Cowling, E.B. and R.A. Linthurst, The acid precipitation phenomenon and its ecological consequences, *Bioscience*, 31, 649–654, 1981.

Hileman, B., Global Warming, *Chem. Eng. News*, Mar. 13, 1989, pp. 25–44.

Houghton, R.A., The global effects of tropical deforestation, *Environ. Sci. Technol.*, 34, 416–422, 1990.

Omernik, J.M., Ecoregions of the conterminous United States, *Ann. Assoc. Am. Geogr.*, 77, 118–125, 1987.

Sabins, F.F., Jr., *Remote Sensing: Principles and Interpretation*, W.H. Freeman and Co., New York, 1987, p. 449.

Stumm, W., L. Sigg, and J.L. Schnoor, Aquatic chemistry of acid deposition, *Environ. Sci. Technol.*, 21, 8–13, 1987.

Wania, F. and D. Mackay, Tracking the distribution of persistent organic pollutants, *Environ. Sci. Technol.*, 30, 390A–396A, 1996.

Zurer, P.S., Arctic Ozone Loss: Fact-Finding Mission Concludes Outlook Is Bleak, *Chem. Eng. News*, Mar. 6, 1989, pp. 29–31.

Risk from Pollutants

CHAPTER 13

Risk Assessment of Contaminants

Inferences are movements of thought within the sphere of belief.

Josephson and Josephson (1996)

I OVERVIEW

I.A Logic of Risk Assessment

This chapter brings our discussion squarely into the realm of technology as applied to environmental problem solving. Techniques estimating both human and nonhuman risk are described together using the reasonable approach developed recently for the CERCLA (Comprehensive Environmental Response, Compensation and Liability Act) assessment process. However, many germane particulars of the approach have been discussed already. Chief among them are the logic of scientific inquiry (Chapter 1), bioaccumulation (Chapters 3 and 4), trophic transfer of contaminants (Chapter 5), biomarkers (Chapter 6), indicators of effects to individuals (Chapters 6 to 8), the NOEL/LOEL approach to sublethal and chronic lethal effects (Chapter 8), models of toxic response including survival-time models (Chapter 9), life tables (Chapter 10), and Hill's aspects of disease association (Chapter 10).

The logic of scientific inquiry—indeed, the logic of any effort to enhance belief about physical phenomena—is frequently more complicated than presented in Chapter 1. For example, the straightforward "reject or accept" context for testing the mettle of a working hypothesis is a logical luxury not always available to the ecotoxicologist. More often, an ecotoxicologist assessing the risk from contamination must bolster belief using tools such as Hill's aspects of disease association. Information is gathered until a balanced judgement can be made based on a **weight of evidence** (preponderance of evidence) approach. Note that this vague term, *weight of evidence*, might mean that a *reasonable person* reviewing the available information *could* agree that the conclusion was plausible (Apple et al., 1986). At the other extreme, the statistical use of the term implies to Kotz and Johnson (1988) a quantitative or semiquantitative estimate of the degree to which the evidence supports or undermines the conclusion. The first definition seems the most accurate for the majority of risk assessment activities. Ideally, evidence describing clear, consistent, and plausible toxic effects would be judged as having considerable weight. Regardless of its shortcomings, the weight-of-evidence approach allows movement toward belief using incomplete knowledge. Although this approach is often labeled as less "scientific" (i.e., lacking logical rigor) than that described in

Chapter 1, the distinction is one of degree. The more precise the logic and more quantitative its expression, the more the weight-of-evidence approach resembles the "scientific" approach.

The weight-of-evidence approach as applied in risk assessment has major elements of abductive inference and probabilistic (Bayesian) induction. (Both abductive inference and probabilistic induction permeate traditional and modern scientific methods too.) **Abductive inference** is simply inference to the best explanation. It uses information gathered about a phenomenon or situation to produce the hypothesis that best explains the data. Josephson and Josephson (1996) give the following example of abductive inference.

1. D is a collection of data.
2. H explains D or, if true, H would explain D.
3. No other explanation (hypothesis) explains D as well as H
∴ H is *probably* true.

This linkage of the weight-of-evidence approach in risk assessment to abductive logic is not a trivial point. Abductive inference can be formalized in artificial intelligence computer programs (Josephson and Josephson, 1996) that could easily be adopted for risk assessments.

Probabilistic (Bayesian) induction uses probabilities associated with competing theories or explanations to decide which is most probably true. Credibilities are assigned to competing explanations based on their associated probabilities (Howson and Urbach, 1989). Instead of the quantal (accept or reject) falsification of a working hypothesis as described in Chapter 1, Bayesian induction considers a hypothesis falsified if it becomes sufficiently improbable.[1] Like the traditional conclusion to designate an accepted explanation as the best, current approximation of reality, this conclusion can be reconsidered later if conflicting facts emerge. In reality, if not tradition, these approaches are no less valuable in fostering the growth of knowledge than the classic methods described earlier. Indeed, the enhanced status of a hypothesis surviving repeated and rigorous testing is a straightforward permutation of abductive inference. It would also lead to probabilistic induction if done rigorously and with ample statistical power so as to provide probabilities. Both abductive inference and probabilistic induction can contribute to the weight-of-evidence approach as applied to risk assessments.

Vignette 13.1   Why Risk Assessment?

Glenn W. Suter II

U.S. Environmental Protection Agency, Cincinnati, OH

Risk assessment is technical support for decision making under uncertainty. It is based on the realization, dating to the 17th century, that, although the future is unpredictable, one can estimate the likelihood of alternative outcomes of an action. Risk assessment is used when a decision must be made that has uncertain outcomes due to varying conditions or uncertainty concerning the nature of the situation. It implies alternatives with qualitatively or quantitatively different possible outcomes. Hence, risk assessment estimates the absolute or relative probabilities of prescribed negative outcomes of alternative choices. It has been plausibly argued that basing decisions on estimates of risk rather than auguries, prayers, astrology, or intuition is the defining feature of modern culture (Bernstein, 1996). Risk assessment is applied to many activities including insurance, engineering, forest fire management, investment, and, as discussed in this book, the management of chemicals in the environment.

The principal alternative to risk assessment is rule-based decision making. For example, offshore disposal of sewage sludge was banned in the United States, not because of risks it posed, but because of

---

[1] Obviously, probabilistic induction is one important logical extension of conventional statistical analyses used in many risk assessments.

popular concern. No analysis indicated that deep offshore disposal of sludge has significant ecological or health risks, or that the risks from land disposal or incineration of sludge are less. The **precautionary principle**[2] is often cited as an alternative to risk assessment. However, this principle simply requires that an action be demonstrated to be safe with reasonable confidence before an action is approved. It is a management principle that is applied after a technical analysis, not an alternative to analysis.

Ecological risk assessment is concerned with assessing risks to nonhuman organisms, populations, or ecosystems. It consists of a problem formulation, an analysis of exposure and of the relationship of exposure to response, and a characterization of risk. It is connected to the risk management process at the beginning, when the problem to be assessed and the goals of the assessment are defined, and at the end, when the results are communicated. Individuals and organizations, termed stakeholders, that have an interest in the risk-management decision may also be involved in helping to define the problem and goals and in communicating or evaluating the results. The nature of the interactions among risk assessors, risk managers, and stakeholders depends on the cultural and legal context.

The case of Bt corn and monarch butterflies may serve to illustrate the risk assessment process. Genetically modified corn, containing the *Bacillus thuringiensis* (Bt) delta endotoxin was developed to combat corn borers and corn earworms. Because the corn pollen also contains the endotoxin, it constitutes a potential hazard to lepidopteran larvae in or adjacent to a cornfield. This issue was raised by studies that demonstrated that experimental exposures to Bt corn pollen increased mortality and arrested development of monarch butterfly (*Danaus plexippus*) larvae. The extreme precautionary approach, advocated by some environmental groups, would ban the genetically modified corn based on this hazard. However, a risk-based regulatory approach would determine whether there was a significant probability of adverse effects, and a comparative risk approach would determine whether that risk was greater than the risks from alternative methods of controlling corn borers and earworms. The following summary is based on the U.S. EPA's assessment of plant-incorporated Bt endotoxin that considered many risks other than those to monarch butterflies (EPA, 2001).

The goal of the assessment is to determine what restrictions on the use of Bt corn, if any, are required to protect monarch butterflies. The problem formulation could identify the Bt pollen as the stressor of concern, the U.S. corn belt as the environment of concern, and the abundance of monarch butterflies as the assessment endpoint.

The analysis of exposure determined the distribution and abundance of the butterflies and the distribution of the larval food (milkweed — *Asclepias* spp.) with respect to cornfields. It then estimated the density of corn pollen or endotoxin protein on milkweed leaves in cornfields and outside fields as a function of distance from cornfields and the rate of degradation of the endotoxin in pollen. The exposure estimate could be refined by determining the frequency of oviposition in corn fields relative to other habitats and the larval feeding pattern in terms of leaf height, leaf age, leaf surface, etc., relative to the distribution of pollen on milkweed plants. Monarch butterflies were found to oviposit on milkweed in cornfields and there was zero to 75% overlap of their presence with the period of pollen shedding.

The analysis of effects estimated the response of larvae to pollen consumption. Available studies examined the effects of pollen from various modified corn varieties in the laboratory and field as well as the effects of the endotoxin proteins. The responses that were considered included death, growth, maturation, emergence, and wing length.

The risk characterization concluded that significant effects are unlikely. Mean observed pollen densities were 170 per cm^2 in cornfields and 63 per cm^2 at the edge. The maximum observed density was 900 per cm^2, and the estimated worst case was 1400 per cm^2. Experimental concentrations at least as high as 4000 per cm^2 caused no observed significant effects. Conservatively, 1 in 100,000 monarch butterflies were estimated to be exposed to sublethal levels of the pollen. This estimate is based on assumptions that 50% of monarchs use the corn belt, 18% of that habitat is corn fields, 25% of that is Bt corn, there is a 50% overlap of pollen shedding and larval occurrence, and 0.1% of those larvae experience sublethal toxic effects. A field study found that Bt pollen had no detectable effects on monarch larvae, while larvae exposed to conventional pesticide sprays or drift were killed (Stanley-Horn et al., 2001). Hence, the risk of Bt pollen to monarch butterflies is estimated to be low in absolute terms. Although a formal comparative assessment was not performed, the risk appears to be much lower than that associated with the most likely alternative.

---

[2] The precautionary principle is based on the conservative policy that, even in the absence of any clear evidence and in the presence of high scientific uncertainty, action should be taken if there is any reason to think that harm might be caused.

However, because of remaining uncertainties concerning effects of long-term exposures on monarch populations, the U.S. EPA is requiring that the registrants conduct additional studies addressing that issue.

This case demonstrates that risk is a function of the magnitude and likelihood of exposure and of the quantitative relationship of effects to exposure, and that the existence of a hazard does not provide an adequate basis for decision making. The analysis is relatively simple because the data are relatively abundant, so little modeling is required, and because the results are far from suggesting a potentially significant risk. If the assessment had suggested that larval mortality would occur in cornfields, a more complex assessment might be required. Such an assessment might model the demographics of monarch butterflies and the distribution of exposure levels given weather, the behavior of the butterflies, and the abundance of milkweed in cornfields and elsewhere. It might also formally assess risks from conventional pesticides and from Bt sprays for comparison to risks from Bt corn pollen.

I.B Expressions of Risk

In this chapter, the preoccupation will be on the expression of **risk** as a probability of some adverse consequence occurring to an exposed human or to an exposed ecological entity. For example, one might estimate a less than 1 in 100,000 chance of dying due to exposure to a particular toxicant. More precisely, risk is "the product of the probability and frequency of effect [e.g., (probability of an accident) × (the number of expected mortalities)]" (Suter, 1993). The concept, as applied to environmental risk assessment includes the probability of an event occurring that *could* lead to an adverse effect and the probability of an adverse effect given that the event *did occur*. This will be the context for the term in most of this chapter. However, other expressions of risk have already been discussed that are equally valuable. In an epidemiological context (Chapter 10), risk was expressed as a relative risk ratio. In discussions of time-to-event models (Chapter 9), risks were expressed as relative risks and modeled as proportional hazards. Although not used as often as probabilities in environmental risk assessments, relative risks are often used to explain risk to the general public. For example, people living in U.S. states with very low selenium levels in soils and waters have a higher risk (expressed as a percentage above the national average) of dying from heart disease than those living in states with normal levels of selenium (Anon., 1976). Survival functions (Chapter 9) and tabulations of age-specific life expectancies (Chapter 10) are also useful expressions of fatal risk that can easily be incorporated into predictive models.

In communicating with the public, the conventional expression of risk as a probability (e.g., 10^{-5}, or 1 chance in 100,000 of the consequence occurring) can be less intuitive than its expression as a change in life expectancy. Consequently, expression of risk in terms of life expectancy is worth exploring for a moment before proceeding to the more conventional context of risk. Also this expression of risk is very closely tied to and extends the mathematical foundations laid down in Chapters 9 and 10. The **loss of life expectancy** (LLE) is estimated as the simple difference between the life expectancy with (E_x) versus without (E) the risk factor being present (Cohen and Lee, 1979).

$$\Delta E = E_x - E \tag{13.1}$$

Loss of life expectancy is expressed in days, months, or years depending on the magnitude of the risk factor's effect. For example, the LLE for the average American (age 0 to 55 years) due to cancer is 0.34 (males) and 0.32 (females) years (Cohen and Lee, 1979). A uranium miner (1970–1972 statistics) has a mortality rate of 232×10^{-5} year^{-1}, or a LLE of 1160 days relative to that of the average person. Cohen (1981) gives an example of a gain in life expectancy (negative LLE) that is a particularly fascinating statistic to the authors. There is a gain of 500 days if one's occupation were that of a university teacher. There is an intuitive statistic with which we can live!

I.C Risk Assessment

Assessment of contaminant-associated risk is mandated in key U.S. federal laws, including RCRA (Resource Conservation and Recovery Act) and CERCLA as amended by the Superfund Amendments and Reauthorization Act (SARA). It is also implied by use of the term *unreasonable risks* in the Federal Insecticide, Fungicide and Rodenticide Act (FIFRA) and the Toxic Substances Control Act (TSCA) (Suter, 1993). For this reason, the EPA has developed numerous documents and regulations ensuring that the intent of these laws is met relative to protecting human health and the environment. For example, assessment of risk to humans is detailed in guidelines for Superfund sites (EPA, 1989d; 1996) and to ecological entities in several EPA guidance documents (EPA, 1989c; 1996). The remainder of this chapter is a condensed version of these guideline documents.

As defined in such regulations, **risk assessment** is the process by which one estimates the probability of some adverse effect(s) of a present or planned release to either human or ecological entities. Two general categories of risk assessments exist: retroactive and predictive. A **retroactive risk assessment** deals with an existing condition such as a contaminated seepage basin, and the **predictive risk assessment** deals with a planned or proposed condition such as a planned discharge of a waste.[3] In some situations, the predictive risk assessment might also deal with a future consequence of an existing situation. For example, a predictive assessment might estimate the consequences of a contaminated plume of groundwater that will outcrop soon to a stream or reach nearby drinking water wells.

A risk assessment is carried out by a **risk assessor**, a person or group of people "who actually organizes and analyses site data, develops exposure and risk calculations, and prepares a risk assessment report" (EPA, 1989d). The assessment is provided to a **risk manager**, "the individual or group who serves as primary decision-maker for a site" (EPA, 1989d). This distinction between the roles of the risk assessor and manager is important: The assessor does not make any decisions regarding the action to be taken as a consequence of the assessment, although potential remedial actions may be detailed in the assessor's report. A decision is the responsibility of the risk manager, who must also weigh all costs, benefits, and risks. As a human risk example, the extremely low risk of cancer associated with consuming residual pesticide in food can be deemed acceptable relative to the widespread economic and nutritional risk associated with abandoning pesticide use in agriculture. A common ecological example involves the certain risk of damage via the destruction of habitat during removal of contaminated soil from an area that is predicted to have a small, but measurable, toxic risk to an endemic species. Is it wise, and is it in the spirit of federal law, to destroy an invaluable and fragile habitat in order to remove a contaminated soil that presents a very low risk?

Permutations of the **NAS (National Academy of Sciences) paradigm** (National Research Council, 1983) (Figure 13.1) are used for both human and ecological risk assessments. As already mentioned in Vignette 13.1, there are four components to this paradigm: hazard identification, exposure assessment, dose-response assessment, and risk characterization. In the first, relevant data on the situation are gathered and chemicals of potential concern highlighted. **Exposure** (contact with the contaminant) assessment estimates the magnitude of releases, identifies possible pathways of exposure, and estimates potential exposure. Dose-response assessment gathers together relevant toxicological data relating exposure to relevant effects. Risk characterization then integrates this information to assess the potential or existing risk of an adverse effect. Also included in this fourth

---

[3] Formally, a risk assessment is different from a hazard assessment. A **hazard assessment** compares the expected environmental concentration (EEC) to some estimated threshold effect (ETT) with the intent of deciding if (1) a situation is safe, (2) a situation is not safe, or (3) there is not enough information to decide. Often a **hazard quotient** (HQ = EEC/ETT) is used as a crude indicator of hazard. (This use of the EEC divided by some endpoint value for adverse effect is called the **quotient method**.) A risk assessment is like a hazard assessment except that it has as its goal the generation of a quantitative estimate (probability) of some adverse effect occurring.

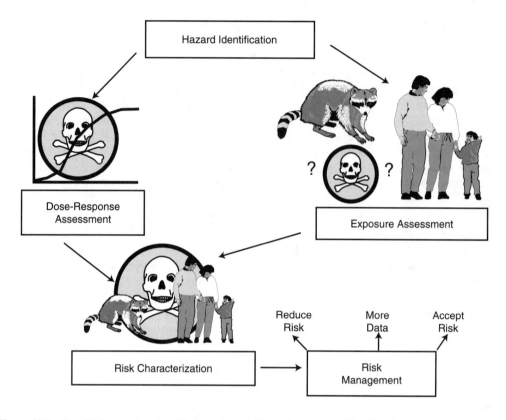

Figure 13.1   The NAS paradigm for risk assessment. The risk assessor identifies the hazard, assesses the potential for exposure (or current exposure), and the dose-response relationship for the relevant entities being protected and the relevant toxicants. This information is integrated to characterize the risk. At that point, the assessment is provided to the risk manager, who then decides to accept or reduce the associated risk or, if insufficient information is available to make a good decision, to seek out more information.

component is a statement of the uncertainty involved in the risk estimate and the general quality of the data used in the assessment. These risk assessment components are described in the following chapter sections as they apply to human and ecological risk assessments.

II HUMAN RISK ASSESSMENT

II.A   General

... scientific activity is not the indiscriminate amassing of truths; science is selective and seeks the truths that count for most, either in point of intrinsic interest or as instruments for coping with the world.

Quine (1982)

Obviously, human risk assessment has as its goal the estimation of the probability or likelihood of an adverse effect to humans occurring as a result of a defined exposure. For Superfund sites, it is one part of a general **remedial investigation and feasibility study** (RI/FS)[4], which has the larger goal of implementing "remedies that reduce, control, or eliminate risks to human health and the environment" or, more specifically, the accumulation of "information sufficient to support an informed risk management decision regarding which remedy appears to be most appropriate for a given site" (EPA, 1989d). The human risk assessment part of the RI/FS has three components: a baseline risk assessment, the refinement of the initial remediation goals, and the evaluation of the risk associated with candidate remedial activities. The first stage involves the application of the NAS risk assessment paradigm to the specific site, with the intention of determining if the contaminants at the waste site present, or could present in the future, a risk if left alone, i.e., the **"no action" alternative** to remediation of the site. In the assessment, consideration should include exposure from one or several routes, and effect to one or several subpopulations of humans. The four sections that follow (Sections II.B to II.E) give the details for the baseline human risk assessment based on the NAS paradigm.

II.B   Hazard Identification (Data Collection and Data Evaluation)

From among the many chemicals present at the site, a tentative list of those of most concern (**chemicals of potential concern**) is made in this phase of the risk assessment. The concentrations of these chemicals are compared with background concentrations. Concentrations are measured in all relevant media, e.g., air, surface water, ground water, soils, sediments, game species, or edible fish. In contrast to ecological risk assessments, contaminant concentrations in biological media most relevant in human risk assessment are edible parts, not the entire organism. Concentrations in all relevant media from the waste site are compared with background levels.

The various contaminant sources are described. Qualities of the environmental setting are also compiled, especially those qualities "that may affect the fate, transport, and persistence of the contaminants" (EPA, 1989d). (As part of the entire RI/FS, a preliminary series of tests might also be done to assess various treatability or remediation alternatives at this stage.) The quality of the data needed for the assessment should also be defined, e.g., types of effects data or required detection limits for the chemicals of potential concern. Also, preliminary exposure scenarios are formulated at this early stage. Usually, risk assessors establish a formal scoping process to identify what data must be collected at this point.

After the data have been collected, they are used to do the following: (1) determine the analytical data quality and adequacy of sampling, (2) compare site concentrations with background concentrations, (3) identify chemicals of potential concern from the initial list of compounds, and (4) determine if the data are adequate to proceed. A rough screening is done to highlight contaminants for further consideration. According to the EPA (1989d), a chemical concentration can be multiplied by a toxicity value to generate an associated **risk factor**.

$$R_{ij} = C_{ij} T_{ij} \qquad (13.2)$$

where R_{ij} = risk factor for the chemical i in association with media j, C_{ij} = concentration of i in media j, and T_{ij} = toxicity value for i in j. It is further suggested that the total risk score for a mixture of contaminants in a medium can be estimated as the sum of the individual R_{ij} values: $R_j = R_{1j} + R_{2j} + R_{3j} + \cdots + R_{ij}$. The size of the risk factor dictates whether the chemical is retained for further consideration. Obviously, the main virtue of this approach is expediency, not accuracy. It is felt that the crude estimate of risk is adequate, assuming effect additivity and a Selyean stress context for mixture effects.

---

[4] The **remedial investigation** (RI) has three parts: characterization of the type and degree of the contamination, human risk assessment, and ecological risk assessment (EPA, 1994). The **feasibility study** (FS) explores the various options for remediation. The chapter focuses on the last two parts of the RI.

Determining a **toxicity value** can involve either one of two methods. One method uses the slope of a published effect-dose relationship: $R_{ij} = \text{Slope} \times C_{ij}$. The other uses a reference dose (RfD) value in a similar manner. Details of these approaches are described below in Section II.D, Dose-Response Assessment.

Note that, although the methods below focus on long exposure periods, short-duration exposures such as those associated with consumption of tainted drinking water can be assessed using one-day, ten-day, or other short-duration health advisories. The U.S. EPA Office of Drinking Water has developed such health advisories based on ingestion of individual chemicals. These **health advisory concentrations** identify concentrations below which no impact to human health is expected for the specified duration of exposure. They are usually derived by a process like that described below for RfD values. Often they are applied by public officials in dealing with spills, short-term exposures, or similar situations, although longer-term advisory concentrations are also available.

II.C Exposure Assessment

The goal of human exposure assessment is to determine or estimate the route, magnitude, frequency, and duration of exposure (EPA, 1989d). In such an assessment, exposures are estimated for specific subpopulations (e.g., hypersensitive individuals, elderly, children, remediation workers) by specific routes (e.g., dust inhalation during remediation, drinking water, game consumption). **Exposure pathways** (the avenues by which an individual is exposed to a contaminant, including the source and route to contact) can include inhalation, ingestion from various media, and dermal contact and absorption.

The **reasonable maximum exposure** (RME) is calculated for chemicals of potential concern. This conservative estimate of exposure is computed differently, depending on the route of exposure. Exposure concentrations are often taken from among the highest measured, e.g., the concentration at the 95% upper confidence limit instead of the mean concentration. The intake is then estimated using an equation such as the general Equation 13.3 (EPA, 1989d).

$$I = C \, \frac{(CR)(EFD)}{BW} \, \frac{1}{AT} \tag{13.3}$$

where I = intake [e.g., (mg of contaminant) (kg of body mass)$^{-1}$ (day)$^{-1}$], C = contaminant concentration (e.g., mg l^{-1}), CR = contact rate (e.g., l day^{-1}), EFD = an estimate of frequency and duration of exposure composed of EF (exposure frequency, e.g., days year^{-1}) and ED (exposure duration, e.g., years), BW = body weight or mass (kg), and AT = time over which exposure is averaged. The AT can be estimated as ED $\times$ 365 days per year for a noncarcinogen or 70 years $\times$ 365 days per year for a carcinogen. The exact formulation of this equation changes, depending on the exposure route, but the general approach remains the same, allowing estimation of the appropriate RMEs. Appendix 6 is a listing of those formulae as supplied in EPA guidelines (EPA, 1989d). EPA documents and other sources are drawn on for specific variable values used in such calculations.

II.D Dose-Response Assessment

The goal of the dose-response assessment is to gather all information useful in establishing a relationship between the extent of contamination and the likelihood or magnitude of an adverse effect. A wide range of information is drawn upon for human risk assessments, including human epidemiological data, data derived from study of nonhuman animals, and predictive models such as QSARs. General mechanistic information is also sought in making judgments. The most valuable data are those from long-term studies involving humans, e.g., Japanese atomic bomb studies discussed in Chapter 14. But such studies are often less structured than nonhuman animal studies.

There is an obvious reason for this. Most human information comes from effects observed after accidents, often occupational accidents with very high exposures. From these unstructured experiences, it is difficult to isolate dose effects from covariates (e.g., age, sex, or other risk factors) and to extrapolate downward to the lower, chronic exposure scenarios normally associated with environmental contaminants. The use of nonhuman animal studies carries the uncertainty of extrapolation to humans. Most animal-to-human extrapolations involve allometric modification of effects, e.g., adjusting effect by weight$^{2/3}$ (or weight$^{3/4}$) if mg day^{-1} intakes are used. If intake is expressed as mg kg^{-1} day^{-1}, the allometric adjustment for the difference in weights between the study animal and humans is weight$^{1-2/3}$, or weight$^{1/3}$. (See Chapter 9 for further explanation.) Sometimes, the allometric PBPK (physiologically based pharmacokinetics) models discussed in Chapters 3 and 4 are applied to reduce inaccuracies in extrapolation. In general, results with high consistency among animal species provide increased confidence in extrapolating to humans (EPA, 1989d). As you will see, various uncertainty factors are employed to compensate for the uncertainty associated with such problems. For carcinogens, a weight-of-evidence classification has also been established to aid the risk manager in understanding the degree of uncertainty involved in each risk calculation.

For noncarcinogenic effects, the **reference dose** (RfD) is applied to risk estimation. It is the best estimate of the daily exposure for humans, including the most sensitive subpopulation, that will result in no significant risk of an adverse health effect if not exceeded. It is assumed that it is accurate to only within an order of magnitude, i.e., 10-fold of the true value. There are different types of RfDs. In most assessments, a **chronic RfD** is assumed unless indicated otherwise. It is the RfD associated with exposures spanning an individual's lifetime. There are also **subchronic RfDs** (RfD$_s$) that are derived from short-term exposure data and **developmental RfDs** (RfD$_{dt}$) that focus on developmental consequences of a single, maternal exposure during development. Which RfD is most appropriate depends on the exposure scenario of concern. Calculations can be done for several scenarios if warranted. For example, a chronic scenario might assess the risk of leaving the site as it exists, and an acute scenario might assess risk for workers during remediation.

RfDs incorporate toxicity data (i.e., an NOAEL) and qualitative uncertainty factors (UF) associated with these data. First, all data sets for the toxicant in the appropriate media are compiled. In the absence of data for the appropriate media or exposure route, data from other forms or routes of exposure might be used after adjustment. Unless there is sound evidence indicating otherwise, the conservative assumption is made that humans are as sensitive as the most sensitive species tested. Under this assumption, the human or nonhuman animal study with the lowest observed adverse effect level (LOAEL) is taken to be the **critical study**. The associated effect is called the **critical toxic effect**. Notice that the LOAEL comes from a suite of LOAELs derived by methods discussed in Chapter 8, and that there are compromises associated with the application of LO(A)ELs and NO(A)ELs. Once the LOAEL is found, the no observed adverse effect level (NOAEL) or highest toxicant level tested that had no adverse effect is identified for the critical study/effect. This NOAEL is the measure of toxicity used in the derivation of the RfD. The LOAEL can be used after adjustment if an NOAEL is not available.

Uncertainty factors (UF) and a modifying factor (MF) are now applied to the NOAEL to compensate in a conservative direction for uncertainty or unaccounted factors. **Uncertainty factors** are often, but not always, factors of ten that lower the NOAEL to compensate for various sources of uncertainty, including variation in sensitivity within the human population (UF$_H$), extrapolation from other species to humans (UF$_A$), use of subchronic rather than chronic or lifetime exposure data (UF$_S$), and use of a LOAEL instead of the NOAEL (UF$_L$). Uncertainty factors can take other values, e.g., EPA's database (IRIS) lists those for hexavalent chromium as UF$_H$ = 10, UF$_A$ = 10 (data from rats), and UF$_S$ = 5 (data involved a chronic, but less than lifetime, exposure). Expert opinion can be applied to modify a NOAEL even further. Expert opinion is incorporated through a **modifying factor** that ranges from >1 to 10. The RfD is then generated using the NOAEL (or LOAEL), UFs, and MF. Obviously, the default values for the UFs and MF are 1 if the associated uncertainties or modifying circumstances are insignificant. Commonly, the UF$_H$ is set at 10 to ensure that the most sensitive humans are protected. The UF_L is 10 if LOAEL is used instead of NOAEL in Equation 13.4; otherwise, it is 1.

$$RfD = \frac{NOAEL}{(UF_H)\,(UF_A)\,(UF_S)\,(UF_L)\,(MF)} \tag{13.4}$$

Clearly, the approach just described for noncarcinogenic effects is based on a threshold model of effect. Below a certain level, the human individual is protected from any adverse effect. That protective level is assumed to be above the RfD but below the LOAEL.

In contrast, a nonthreshold model is assumed in dealing with the risk of carcinogenic effects. A **slope factor** (SF) (risk or probability of occurrence per unit of dose or intake) for the risk-dose model is then used to estimate the probability of a cancer under a particular exposure scenario. Although based primarily on intake, models can be expressed in terms of concentration, dose, or intake. The slope factor is defined by EPA as "a plausible upper-bound estimate of the probability of a response per unit intake of a chemical over a lifetime" (EPA, 1989d). Usually, the upper 95% confidence limit for the estimated slope of the risk-dose or risk-intake curve is used as the slope factor. The exposure dose or intake is multiplied by the slope factor to estimate risk.

$$Risk = (CDI)(SF) \tag{13.5}$$

where Risk = probability of developing cancer, and CDI = chronic daily intake averaged over a lifetime (70 years) ($mg\ kg^{-1}\ day^{-1}$). Estimation of daily intakes is illustrated generically in Equation 13.3 and specifically for various sources in Appendix 6 at the end of this book. All risk estimations are accompanied by a qualitative **EPA weight-of-evidence classification** because the strength of evidence for specific chemicals being human carcinogens varies widely. This classification informs the risk manager about the strength of the evidence supporting the risk calculation.

The EPA has compiled RfDs, slope factors, drinking water health advisories (one day, ten day, longer term, and lifetime advisories), and important associated information into a large database, **integrated risk information system** (IRIS). At this time, IRIS is accessible through the U.S. government's right-to-know web site (http://www.rtk.net/T866). As an example of how easily these data can be found, the first author spent less than 10 minutes logging onto IRIS to retrieve the following information for cadmium. The cadmium RfD is 0.0005 $mg\ kg^{-1}\ day^{-1}$ based on proteinuria[5] (the critical effect) in humans after imbibing cadmium in drinking water. The NOAELs were 0.005 $mg\ kg^{-1}\ day^{-1}$ for water and 0.01 $mg\ kg^{-1}\ day^{-1}$ for food. The UF_H is 10 and MF is 1. Confidence in these data was judged to be high because of the extensive human and animal data sets and because there were sound PBPK models available to the assessor.[6]

II.E   Risk Characterization

The data collected in the previous steps are now combined in this last step to generate a statement of risk. To this end, the exposure information, including intake rates and dose-response data, are used in the calculations below. The final statement of risk can be qualitative or quantitative. Regardless, it must include an explanation, specific details, and qualifiers for the final expression of risk.

---

[5] **Proteinuria** is the presence of protein in the urine. The suggestion is kidney damage caused by cadmium that has accumulated in the renal cortex.

[6] Here, again, is another opportunity for confusion if the distinct goals of scientific, technical, and practical ecotoxicologists are not kept in mind and respected. Data from very sound scientific studies may be judged of "low" confidence relative to its use in estimating an RfD. On the other hand, an RfD is meaningless in a scientific sense. What might be good for one purpose is inadequate for another.

A hazard quotient (Equation 13.6) is estimated for each noncarcinogenic effect using the exposure level or intake (E) and associated RfD. If the quotient does not exceed 1, the human population is assumed to be safe. Hazard quotients for chronic, subchronic, and shorter term exposure may need to be estimated also, depending on the exposure scenario of concern.

$$\text{Hazard Quotient} = \frac{E}{RfD} \tag{13.6}$$

However, the quotient estimated above is not useful if there is more than one chemical of potential concern.

A hazard quotient can be calculated for situations involving several (x) chemicals of potential concern:

$$\text{Hazard Index}_{\text{Total}} = \sum_{i=1}^{x} \frac{E_i}{RfD_i} \tag{13.7}$$

where E_i = exposure levels (or intakes), and RfD_i = reference doses for the i chemicals expressed in similar units and covering the same exposure durations as E_i. With temperance, summations could be done for contaminants in different media. Again, a quotient less than 1 suggests protection of the human population. For chronic exposures, a chronic hazard index can be estimated by Equation 13.7 if chronic RfD_is are used and CDI_i is used as E_i. Similarly, a subchronic hazard index is generated with a subchronic EfD_i and subchronic daily intake rates (SDI_is).

The risk of carcinogenic effects is estimated with Equation 13.5, assuming a multistage model of carcinogenicity. This linearized, multistage model is appropriate only at low doses. Consequently, it should be replaced by the first-order or **one-hit risk model** if estimated risk is 0.01 or higher:

$$\text{Risk} = 1 - e^{-(CDI)(SF)} \tag{13.8}$$

Of course, other plausible models such as a gamma multiple-hit or Weibull models exist, as discussed regarding the dynamics of carcinogenesis in Chapters 7 and 9 and as expressed mathematically in Chapter 9.

If several carcinogens are considered together, a total cancer risk is approximated as the simple sum of the individual risks, assuming independence of effects. This might not be valid, as suggested from our previous discussions of carcinogenesis. Again, risk can also be summed across different media, with an understanding of the limits of such a calculation.

II.F Summary

I prefer the errors of enthusiasm to the indifference of wisdom.

Anatole France (Quoted in Casti [1989])

The techniques described above for assessing risk from noncarcinogenic and carcinogenic contaminants (Figure 13.2) can be criticized easily for inconsistency with scientific knowledge. However, as detailed in Chapter 1, the goal of this crucial process is not scientific and should not be judged from that context alone. The goal is to protect human health. Proponents are acutely aware of the many approximations and compromises in the approach. They conditionally accept

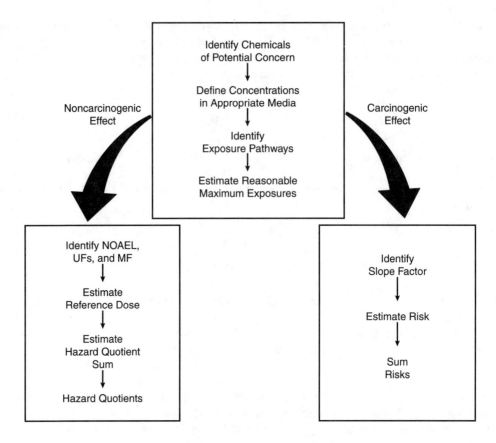

Figure 13.2 The general sequence of steps for assessing the hazard/risk associated with noncarcinogenic and carcinogenic effects of contaminants. The final step of summing hazard quotients or risks for the individual chemicals or media may not be required or may not be appropriate. (See text for more details.)

those errors in order to fulfill their immediate goal. Indeed, they incorporate uncertainty factors and conservative slope factors so as to ensure that the results will be biased toward excessive caution and away from any possible harm to humans. These modifications are not made to more accurately define the threshold levels for the contaminant effects. Unfortunately, this dictates that more time and money than necessary be spent in remediation. It also assures that more valuable habitat than necessary will be damaged or destroyed during unnecessary or excessive remediation activities.

III ECOLOGICAL RISK ASSESSMENT

"If seven maids with seven mops
Swept it for half a year,
Do you suppose," the Walrus said,
"That they could get it clear?"
"I doubt it," said the Carpenter,
And shed a bitter tear.

Carroll (1872)

III.A General

The first two parts of a remedial investigation (characterization of the contamination and human risk assessment) have been described briefly to this point. Now let us consider the third part of a remedial investigation, the ecological risk assessment. Like human risk assessment, the goal of ecological risk assessment is the estimation of the likelihood[7] of a specified adverse effect or ecological event due to a defined exposure to a stressor.[8] Relevant effects can range from the suborganismal to the landscape scale. Unlike human risk assessment, ecological risk assessments must consider many species with diverse niches and phylogenies. It might even consider ecological entities, e.g., communities composed of many species occupying a heterogeneous landscape. Also, in contrast to human risk assessment in which extrapolation to one species (humans) is often done from many species (e.g., mouse, rat, or dog toxicity data), ecological risk assessment extrapolates from one or a few species to many.

Ecological assessments can be retroactive or predictive. The predictive assessment generally adheres to the NAS paradigm (Figure 13.1), but retroactive assessment relies less on this paradigm and more on surveys of contamination and ecological impact, models of fate and effects, and epidemiological data.

The ecological risk assessment process is organized slightly differently from the NAS paradigm, although the overall logic remains the same (Figure 13.3). The first step is problem formulation, a process involving both the risk assessor and risk manager. Next is the analysis step. The analysis step has two components similar to the NAS paradigm's exposure and dose-response assessments. It also has parts of the hazard-identification component of the NAS paradigm. In the analysis and risk characterization stages, there might be reexamination of various actions or decisions as new information arises. In the last step, risk characterization, the information generated in the analysis step, and the context developed in the problem formulation step come together. After the risk characterization step, the risk assessors and managers review the results relative to the original needs set out during problem formulation and any needs that might have emerged during the process.

III.B Problem Formulation

Problem formulation includes the initial planning and scoping that establishes the framework around which the assessment is done (Norton et al., 1992). It includes the selection of assessment endpoints, a conceptual model, and a plan of analysis. The **assessment endpoint** is the valued ecological entity to be protected (e.g., bald eagles nesting by a contaminated lake) and the precise quality to be measured for this entity (e.g., adult survival and nesting success of bald eagles).[9] Including the measured quality in the definition ensures that the assessor specifies clearly an effect or improvement after remediation that can be quantified. For example, if one incorrectly decided that the assessment endpoint was the vague "functional integrity of the stream ecosystem," it would be very difficult to know what exactly should be measured to establish whether an adverse effect is present or to document any improvement after remediation.

---

[7] Likelihood is interpreted even more loosely in ecological assessments than in human assessments (Norton et al., 1992). "Descriptions of risk may range from qualitative judgments to quantitative probabilities. While risk assessments may include quantitative risk estimates, the present state of the science often may not support such quantitation" (EPA, 1996). The term may not always imply the generation of a probability with an associated statistical statement of confidence.

[8] **Stressor** is defined in ecological risk assessments as "any chemical, physical, or biological entity that can induce adverse effects on *ecological components*, that is, individuals, populations, communities, or ecosystems" (Norton et al., 1992). Obviously, there is extreme latitude in this definition and, as discussed before, many effects at higher levels will more often be ambiguous than clearly adverse.

[9] In many publications (e.g., EPA, 1989c; Norton et al., 1992), the distinction is made that the assessment endpoint or receptor is the ecological entity or value to be protected (e.g., a population of the endangered bald eagle nesting by a contaminated lake), and the **measurement endpoint** is a measurable response to the stressor (e.g., number of fledglings produced per nest each year) that is related to the valued qualities of the assessment endpoint (e.g., reproductive success of the bald eagles). Some logical or quantitative model must link the two endpoints.

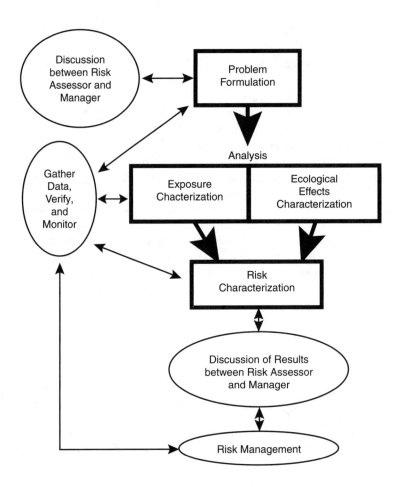

Figure 13.3 The general framework of an ecological risk assessment. The boxes and thick arrows reflect the risk assessment proper, similar to that of the NAS paradigm (Figure 13.1). The narrow, two-way arrows reflect interactions between the risk assessor and risk manager during the process. They also reflect the continual accrual and evaluation of information during the entire process. (Modified from Figure 2 in EPA, Ecological Risk Assessment Guidance for Superfund: Process for Designing and Conducting Ecological Risk Assessment, review draft, U.S. EPA, Washington, D.C., Sept. 26, 1994.)

Sometimes, the specific quality of the assessment endpoint cannot be measured in the valued entity, and it is measured in a surrogate instead. For example, risk assessment for an endangered species might require unacceptable destructive sampling to determine egg viability or body burden of a contaminant. Instead, these qualities may be measured for a closely related or ecologically similar species, and the associated results applied to predicting risk to the endangered species.

The EPA (1996) suggests three qualities for a good assessment endpoint: (1) It should be ecologically relevant to the ecosystem being assessed. (2) It should be susceptible to the stressor. (3) It is desirable, but not necessary, that it be valued by society. The last quality increases the value of the assessment to the risk manager. Suter (EPA, 1989c) further suggests that an ideal assessment endpoint should have an unambiguous operational definition and that it should be readily measurable or predictable from measurements. He gives the example of "balanced indigenous populations" as an ambiguous, and therefore compromised, assessment endpoint. Assessment endpoints are often, but not always, legally protected entities (e.g., survival and reproduction of an endangered or threatened species) or economic entities (e.g., successful reproduction of a salmon species). They can also be important ecological qualities, e.g., species diversity or some reflection of biodiversity.

Table 13.1 Examples of Endpoints for Ecological Risk Assessments

Level of Ecological Organization	Assessment Endpoint	Measurement Endpoint
Population	Extinction	Occurrence
	Abundance	Abundance
	Yield or production	Reproductive performance
	Age or size class structure	Age or size class structure
	Mass mortality	Frequency of mass mortality
Community	Market sport value	Number of species
	Recreational value	Species diversity and richness
	Change to a less useful or appealing state	IBI (index of biological integrity)
Ecosystem	Productive capacity	Biomass
		Productivity
		Nutrient dynamics

Source: Taken from EPA, Ecological Assessment of Hazardous Waste Sites, EPA 600/3-89/013, National Technical Information Service, Springfield, VA, 1989, p. 260.

Suter (EPA, 1989c) tabulated (Table 13.1) some examples of assessment and measurement endpoints that can be applied to various levels of ecological organization.

The **conceptual model** links the assessment endpoint and the stressor of concern. It evaluates possible exposure pathways, effects, and ecological receptors. Conceptual models include hypotheses of risk and a diagram of the conceptual model. The **risk hypotheses** are clear statements of postulated or predicted effects of the stressor on the assessment endpoint. The **conceptual model diagram** (Figure 13.4) shows the pathways of exposure and illustrates areas of uncertainty or concern. It is a visual aid for communicating to the risk manager the model from which the risk hypotheses emerge.

In the final step of problem formulation, the risk hypotheses are examined carefully, and a plan of analysis is produced. "Here, risk hypotheses are evaluated to determine how they will be assessed using available and new data" (EPA, 1996). An **analysis plan** defines the format and design of the assessment, explicitly states the required data, and describes the methods and design for data analysis. It describes what will or will not be analyzed. Measurement endpoints, those qualities that will be measured to assess effect to the assessment endpoint, are also stated in the analysis plan. A measurement endpoint may involve measurements derived directly from the valued ecological entity or from its surrogate.

III.C Analysis

Again, the analysis step has two components (exposure characterization and ecological effects characterization) that are very similar to the exposure and dose-response assessments of the NAS paradigm. The exposure and ecological effects characterizations are done in tandem, with considerable exchange of information occurring between the two components.

Some aspects of hazard assessment of the NAS paradigm are inserted as part of the analysis step of an ecological risk assessment. The gathering of relevant data and identification of chemicals of potential concern done in the hazard-assessment step of the NAS paradigm are also done in the analysis step of the ecological risk assessment. However, some of the initial data gathering also occurred during the problem formulation step of the ecological risk assessment.

III.C.1 Exposure Characterization

Exposure characterization describes the characteristics of any contact between the contaminant and the ecological entity of concern. It summarizes this information in an exposure profile. Temporal and spatial patterns in contaminant distribution are defined in addition to the amount of contaminant present. The source of the contaminant, any potential costressors, transport pathways

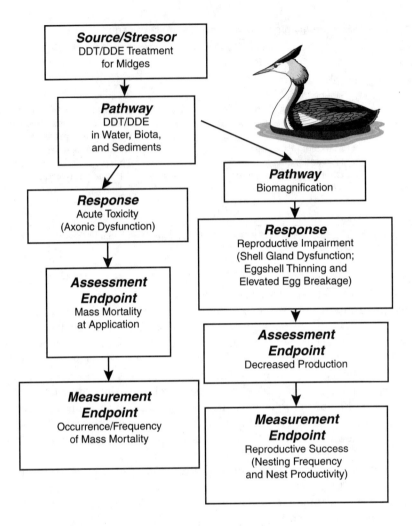

Figure 13.4 A conceptual model diagram for the pesticide spraying of nonbiting midges as described in Chapter 1 and depicted in Figure 1.1. Here, the assessment and measurement endpoints are separated for clarity. They could have been combined as the assessment endpoint. The valued ecological entity is the western grebe population.

(e.g., outcropping of contaminated groundwater to the stream, or ingestion of prey after biomagnification), and type of contact are defined. In this characterization of exposure, quantitative methods described earlier for human exposure are applied. Concentration, duration, and frequency of exposure must be considered, including consideration of factors such as seasonal cycles and home ranges of species. For example, the contaminated region might be used only 10% of the time by a free-ranging species. Or a contaminated food might be ingested only during one season. Estimates of exposure duration and frequency should take such factors into consideration. The final **exposure profile** "quantifies the magnitude, and spatial and temporal pattern of exposure for the scenarios developed during problem formulation" (Norton et al., 1992).

III.C.2 Ecological Effects Characterization

Ecological effects characterization "describes the effects that are elicited by a stressor, links these effects with the assessment endpoints, and evaluates how effects change with varying stressor levels" (EPA, 1996). It also specifies the strength of evidence associated with the effects characterization and the level of confidence in the causal linkage between the contaminant and the effect (Norton et al., 1992). Information generated with many of the methods described previously are brought together to develop a stressor-response profile for the valued ecological entity.[10] Which methods are applied would depend on the nature of the exposure and the ecological entity of concern. Unfortunately, most dose-response information is generated for effects to individuals, yet those of most use here are dose-response information for populations, communities, ecosystems, and landscapes. The unfortunate consequence of this imbalance between available information and need is compromised ecological risk assessments. (See quote at the beginning of Chapter 10.)

III.D Risk Characterization

Risk characterization draws together the information from previous steps to produce a statement of the likelihood of an adverse effect to the assessment endpoint (EPA, 1996). Risk can be expressed in several ways, including a simple qualitative judgment or hazard quotient. It could involve a richer interpretation, including description of the influences of concentration and temporal variations on estimates of effect. Or it could employ complex models that also generate some estimate of confidence in the risk predictions.

The final statement of risk must include details about the adequacy of the data going into the judgment, about the uncertainty involved in the conceptual mode or calculations, and about the weight of evidence for each causal relationship. If a surrogate were used in place of the assessment endpoint, confidence in the associated extrapolation should be addressed. For example, the confidence in the extrapolation would be high if there were very little variation among raptors in the dose-effect relationship and a surrogate raptor with a very similar niche were used for bald eagles. Some statement about the significance of the adverse effect should also be made so that the risk manager can better judge the seriousness of risk consequences.

III.E Summary

The expedient compromises and conservative inaccuracies already discussed for human risk assessments are also present in ecological risk assessments. More are added because the manifestation of effects could involve several species or levels of ecological organization. There is a tendency to address the level at which most information exists, i.e., the individual level. Understandably, implications are most often to the level of population viability.

Vignette 13.2   A Nationwide Survey and Risk Assessment of Organochlorines in New Zealand

Paul D. Jones and John P. Giesy Jr.
Michigan State University, East Lansing, MI

Simon J. Buckland and Howard K. Ellis
Ministry for the Environment, Wellington, New Zealand

In 1995, the New Zealand government undertook a country-wide environmental survey of organochlorine contaminants (OC), including dioxins, polychlorinated biphenyls (PCB), persistent organochlorine pesticides, and chlorophenols, as part of a national Organochlorines Program. Concentrations of OC were measured in a range of terrestrial and aquatic media, including air, soils, river waters, estuarine sediments, and biota, in a sampling program covering the entire country. This survey of over 250 samples represented

---

[10] Databases like IRIS are being developed for ecological effects. For example, Oak Ridge National Laboratory has databases on its web site (http://www.hsrd.ornl.gov/ecorisk/benchome.html) for aquatic biota, wildlife, terrestrial plants, sediments, and soil invertebrates/microbial processes. Details are provided in extensive documentation obtained from this same web site. Also, EPA has an ECOTOX database for ecological effects.

the single largest nationally based environmental monitoring program ever undertaken in New Zealand. An assessment of the ecological risk posed by organochlorine chemicals was carried out using the concentrations of these chemicals found in terrestrial and aquatic ecosystems.

Although New Zealand is a relatively small country, there was still a vast area of land and aquatic ecosystems that needed to be investigated to obtain a truly representative picture of the extent of OC contamination. In addition, the survey wished to focus on assessing OC concentrations based on a variety of land-use categories. Therefore a sampling program was required that would provide a representative sampling of environmental media based on region and land-use. While such a sampling plan would be relatively easy to design if a very large number of samples were collected, the cost of some of the analyses (most notably ultra-trace analysis for dioxins) placed a limitation on the size of the study that could be undertaken.

The Organochlorines Program

The organochlorines program as a whole comprised a study of environmental and human levels of OC, development of an inventory of ongoing dioxin emissions and reservoir sources, and an estimation of the possible health and ecological risks posed by these substances. The desired outcomes from the program were: (1) the development of national environmental standards for dioxins and other OC, (2) the identification of technologies to safely and effectively destroy OC, (3) the development of an integrated management strategy for OC, and (4) a mechanism for public input to decisions on the management of OC.

Sampling

The design of the sampling plan for the organochlorines program had to balance the needs for representative sampling with the resources available for the project. The design of this plan was critical to the eventual success of the environmental survey and, consequentially, the findings of the risk assessment. The design focused on the collection of environmental media to reflect known reservoirs and sources of OC. Therefore, sampling focused on air, soils, sediments, and aquatic biota. The sample collection methods were designed to cover as wide an area as practicable to produce a representative sample of the media from the selected site.

Air

Air was sampled using GPS-1 high-volume air samplers from ten sites over 1 year: two reference, two rural, five urban, and one industrial site (Buckland et al., 1999). At all sites except Baring Head (a reference site), sampling was undertaken continuously over a period of typically 20 days at a flow rate of approximately 150 l per minute. Air sample volumes of approximately 4000 m³ of air were collected. At the Baring Head site, which was located on the southern coastline of the North Island, the sampler was operated during a 20-day period only during periods of southerly winds. The samples from this site therefore reflect southern ocean maritime conditions. Typically four to six samples were collected from each site every second or third month. Quality assurance data showed no evidence of analyte breakthrough and loss from the application of these extended sampling periods.

Soil

Soils were collected from a range of climates and landforms (Buckland et al., 1998). Samples were taken from indigenous forests, grasslands, agricultural soils, and parks and reserves in urban (provincial and metropolitan) centers. All soil samples were collected as 25 mm × 100 mm depth cores.

Indigenous. Indigenous forest and indigenous grassland soils were taken from National Parks and Department of Conservation estate land in remote areas. For the purpose of this study, they were defined as reference sites. Four forested and three grassland sites were sampled in the North Island, and three forested and two grassland sites were sampled in the South Island. Typically, for each sample, soil cores were taken at 150 m equidistant points along the sides of an equilateral triangle, with sides measuring 1.5 km in length (i.e., a total of 27 soil cores per sample). Particular care was taken to ensure that samples were not collected from areas impacted by human activity. Sampling criteria included: no sample must be collected within 1 km of any road or track carrying motor vehicles, or within 1 km of any recreational area; any sampling point accessed from a walking track must be no closer than 40 m from the track.

Agricultural. Samples were taken from hill country pastoral and flatland pastoral sites. Soil cores were collected from two randomly selected sampling stations (27 cores per sampling station). Sampling stations were based on an equilateral triangle with sides measuring 1.5 km in length. Sampling criteria applied were: each sampling station was restricted to a single farm property; no samples to be collected within 1 km of a major state highway, 500 m of a secondary sealed road, or 200 m of an unsealed road; no soil cores to be collected within 5 m of any wooden building or structure (such as fences or telegraph poles) to avoid any possible contamination from treated timbers.

Urban. Soil samples were collected from eight provincial centers (one sample per center), and two major metropolitan centers, Auckland (nine samples) and Christchurch (six samples). For each provincial center, the sample was a composite of soil cores collected from four parks and reserves. Nine cores were collected from each park or reserve, preferably using a 3 × 3 grid, with the spacing between the individual cores dependent upon the size of the park. In some cases, a random sampling pattern was employed. Ideally, a 50 m spacing was set between each core, but for smaller parks and reserves where this was not possible, no cores were closer than 25 m to each other. No cores were taken closer than 10 m from any park boundary, fence line, or building. A similar approach was adopted for metropolitan centers, with the exception that 12 cores were collected from each park using either a 4 × 3 grid or a random sampling pattern for parks where stratified sampling could not be undertaken.

Rivers

Thirteen rivers of regional and/or national significance were studied (Buckland et al., 1997). Three were selected as reference or background sites, no wastes having been discharged at or above the sampling sites. Other rivers were selected because they received a variety of effluent discharges or ran through agricultural areas with the potential for pastoral runoff. River water and eel (*Anguilla* spp.) were captured from all 13 rivers, and trout (*Salmo trutta* or *Oncorhynchus mykiss*) were captured from 9 of the 13 rivers. The OC analyses were undertaken on composite samples.

Estuaries

Sediment and shellfish were collected from 12 estuaries from remote, agricultural, and urbanized catchments (Buckland et al., 1997). For all estuaries, up to three composite sediment samples were collected, depending upon the size and nature of each particular estuary. Each composite consisted of samples taken from five sampling stations randomly selected within a defined (1 to 2 km^2) region of the estuary. Each sampling station encompassed an area of approximately 25 m^2 based on a 5-m × 5-m grid. Within each station, five sediment cores were randomly collected and composited together. Each sample therefore consisted of 25 individual sediment cores taken to a depth of 5 cm. Cores were placed in widemouthed glass sample jars, and any excess water was decanted off from the top as the sediment settled.

Shellfish were collected to provide some indication of the extent of bioaccumulation of OC contaminants within the estuarine ecosystem. The species chosen was the bottom-dwelling filter feeder, New Zealand cockle (*Austrovenus stutchburyi*), or in the absence of cockles, Pacific oysters (*Crassostrea gigas*).

Stressors

The stressors or chemicals of interest for the study were the dioxins (polychlorinated dibenzo-p-dioxins and polychlorinated dibenzofurans), the PCBs (25 congeners, including the dioxinlike PCB congeners), persistent OC pesticides (including ΣDDTs, dieldrin, and HCB), and chlorophenols, including pentachlorophenol.

Survey Results

The results of the environmental survey can be summarized as follows:

Air

1. Dioxins: Concentrations were measured at 0.77 to 7.48 fg I-TEQ (toxic equivalent units) m^{-3} (reference); 0.94 to 31.7 fg I-TEQ m^{-3} (rural); 6.15 to 262 fg I-TEQ m^{-3} (urban); 40.3 to 1170 fg I-TEQ m^{-3} (industrial). Generally, most dioxin congeners were not measured at the reference

sites, with the TEQ level arising from inclusion of half limit of detection (LOD) values. The major source of dioxins at urban sites was domestic emissions from the burning of wood for space heating.

2. PCBs: Concentrations measured were 5.72 to 18.2 pg m^{-3} (reference sites); 4.99 to 30.0 pg m^{-3} (rural); 29.9 to 129 pg m^{-3} (urban); 210 to 471 pg m^{-3} (industrial). Variations in concentrations were observed, broadly corresponding to seasonal changes (minimum levels in winter, maximum levels in summer), suggesting that volatilization and environmental recycling from reservoir sources play a dominant role in PCB air concentrations.

3. OC pesticides were detected at all sites. The most abundant and frequently detected were lindane (1.28 to 154 pg m^{-3}); hexachlorobenzene (HCB) (8.14 to 99.4 pg m^{-3}); dieldrin (1.17 to 288 pg m^{-3}); ΣDDTs (1.04 to 57.4 pg m^{-3}).

Soil

1. Dioxins concentrations (excluding outliers) were 0.17 to 1.99 ng I-TEQ kg^{-1} dry weight (d.w.) (forest/grassland); 0.17 to 0.90 ng I-TEQ kg^{-1} d.w. (agricultural); 0.52 to 6.67 ng I-TEQ kg^{-1} d.w. (urban). For forest, grassland, and agricultural soils, the primary contributor (typically >80%) to the determined TEQ arose from including half LODs for nondetected congeners.

2. One hill-country pastoral site had an atypical dioxin concentration of 9.14 ng I-TEQ kg^{-1} d.w., and one urban site had an atypical concentration of 33.0 ng I-TEQ kg^{-1} d.w. Concentrations at these sites were elevated compared with other similar sites due to higher than usual levels of 2,3,7,8-tetrachlorodibenzo-p-dioxin, associated with the manufacture and use of the phenoxy herbicide, 2,4,5-T.

3. PCBs were only infrequently detected in forest/grassland and agricultural soils. Including half the LOD for nondetected congeners, maximum concentrations for the sum of the 25 PCBs analyzed were 1.20 µg kg^{-1} d.w. (forest/grassland) and 3.37 µg kg^{-1} d.w. (agricultural).

4. PCBs were more commonly detected in urban soils to a maximum concentration of 9.74 µg kg^{-1} d.w.

5. In indigenous forest and grassland soils, only HCB (0.065 to 0.28 µg kg^{-1} d.w.), dieldrin (<0.03 to 0.83 µg kg^{-1} d.w.), and ΣDDTs (0.082 to 5.33 µg kg^{-1} d.w.) were found.

6. HCB, dieldrin, and DDTs were also the pesticides most frequently detected in urban soils, with dieldrin (0.15–42.1 µg kg^{-1} d.w.) and ΣDDTs (1.5–853 µg kg^{-1} d.w.) being measured in all samples. Agricultural soils were not analyzed for any of the persistent pesticides. However, concentration data for DDTs were available from other monitoring studies, and these data were used in the risk assessment.

Rivers

1. No OC were measured in any river water sample despite the analysis of large sample volumes (up to 4.8 l) and the application of sensitive analytical methods.

2. Dioxins were measured in less than 50% of the finfish samples collected, at concentrations between 0.016 to 0.39 ng I-TEQ kg^{-1} wet fillet weight (w.f.w.) for eel and 0.016 to 0.20 ng I-TEQ kg^{-1} w.f.w. for trout.

3. All but one fish sample contained PCB, to a maximum concentration of 18.5 µg kg^{-1} w.f.w. (eel) and 8.80 µg kg^{-1} w.f.w. (trout).

4. OC pesticides, most commonly HCB, dieldrin, and DDTs, were measured in both eel and trout, including those samples collected from the reference sites. Concentrations for HCB were 0.030 to 0.52 µg kg^{-1} w.f.w. (eel) and <0.01 to 0.17 µg kg^{-1} w.f.w. (trout); for dieldrin 0.24 to 11.4 µg kg^{-1} w.f.w. (eel) and 0.021 to 1.12 µg kg^{-1} w.f.w. (trout); and for ΣDDTs 0.80 to 214 µg kg^{-1} w.f.w. (eel) and 2.05 to 76.6 µg kg^{-1} w.f.w. (trout).

Estuaries

1. Dioxins were typically detected in sediment samples, although concentrations were low, in the range of 0.081 to 2.71 ng I-TEQ kg^{-1} d.w. (13.2 to 700 ng I-TEQ kg^{-1} organic carbon). In shellfish, concentrations were also low, in the range 0.015 to 0.26 ng I-TEQ kg^{-1} wet weight (w.w.) (1.77 to 28.9 ng I-TEQ kg^{-1} lipid).

2. PCBs were less frequently detected in estuarine samples than the dioxins, being quantified in fewer than 35% of sediment samples (0.12 to 8.80 µg kg^{-1} d.w.; 9.72 to 926 µg kg^{-1} organic carbon) and in 70% of shellfish samples (0.11 to 12.9 µg kg^{-1} w.w.; 13.3 to 1430 µg kg^{-1} lipid).
3. The most common pesticides found were dieldrin and DDT. Sediment concentrations were in the range <0.05 to 0.38 µg kg^{-1} d.w. (dieldrin) and <0.01 to 5.22 µg kg^{-1} d.w. (ΣDDTs), and shellfish concentrations <0.02 to 0.56 µg kg^{-1} w.w. (dieldrin) and <0.01 to 4.36 µg kg^{-1} w.w. (ΣDDTs).

Risk Assessment

An assessment of the ecological risk posed by OC in New Zealand was undertaken based on concentrations present in terrestrial and aquatic ecosystems as determined in the environmental survey, together with other available exposure data. The specific objectives of the assessment were to:

1. Determine whether current background concentrations of OC in the New Zealand environment result in their accumulation in biota to levels that can cause adverse effects at a population level.
2. Provide a basis for setting national concentration-based limit values for OC in biotic and abiotic media that will protect the health of ecosystems.
3. Provide the means to determine toxicity reference values for use in the assessment of OC-contaminated sites.

Several risk methodologies were applied because of the varying quantity and quality of data for specific OC in specific media, in particular with reference to values that were below the method detection limit. These ranged from a full probabilistic assessment to less-rigorous screening-level assessments for specific chemicals in certain ecosystems. The methods aimed to determine risks to invertebrates, fish, and invertebrate-eating wildlife, as well as to top-of-the-food-chain predators such as piscivorous (fish-eating) birds and marine mammals.

Freshwater Fish. Risks to freshwater fish were estimated based on literature-derived toxic reference values and measured tissue concentrations. Chronic exposure risks were derived from acute estimates using an acute to chronic ratio from the U.S. EPA water quality criteria. Minimal risk from OC was identified.

Fish-Eating Birds. Risks to fish-eating birds were modeled based on predicted egg concentrations using freshwater finfish (trout or eel) as the food source. Accumulated concentrations were predicted to pose minimal risk.

Marine Mammals. Although not specifically collected as part of the organochlorines program environmental survey, a considerable amount of data have previously been reported on OC concentrations in New Zealand marine mammals (Jones et al., 1999). Risks to marine mammals were calculated using both dietary intake data and tissue residue concentrations. Risks, especially from the dioxinlike compounds (i.e., dioxins + dioxinlike PCBs) were greatest for inshore feeding species such as the protected Hector's dolphin (*Cephalorhynchus hectori*).

Terrestrial Birds. Risks to insectivorous terrestrial birds were estimated by modeling soil OC concentrations through soil invertebrates and into bird eggs using conservative bioconcentration factors. This procedure indicated possible risks, but the approach was so conservative that the risks would appear to be small.

The assessment did not consider effects at localized areas, where higher exposures may occur (e.g., contaminated sites). This assessment aimed to estimate background levels of risk to which impacted sites could be compared.

Conclusions

Overall, the environmental survey found generally low concentrations of OC in the New Zealand environment, particularly of the dioxins and PCBs. Concentrations measured were typically lower than levels reported in industrialized countries in the Northern Hemisphere. The risk assessment found that the concentrations of OC measured are less than thresholds for effects for most wildlife species. However, there are indications that some pesticides and dioxinlike compounds require further assessment and management. Specifically, the assessment reached the following conclusions:

1. Based on method detection limits, concentrations of persistent OC pesticides in river water are less than those known to be acutely or chronically toxic to aquatic species.
2. Dioxins and PCBs in freshwater fish are unlikely to be causing adverse effects.
3. OCs are unlikely to be causing adverse effects on soil invertebrates.
4. OC pesticides in nonagricultural soils are unlikely to pose risks to birds or mammals, either acutely or due to biomagnification.
5. For some OCs, the highest soil concentrations measured are close to effect levels. In these situations, no increases in these OCs should be allowed.
6. The historical application of DDT to agricultural soils has resulted in concentrations greater than background. Concentrations of ΣDDTs in predatory birds indicates that these OCs are entering and bioaccumulating in the food chain.
7. Modeling of the accumulation of dioxins from soil into bird eggs predicts concentrations close to effect levels. Therefore this conservative assessment cannot discount the possibility of risk.
8. The risk of effects on shore birds due to consumption of dioxins and other OCs in shellfish appears to be minimal.
9. Dioxinlike compounds that have accumulated in some marine mammal species are approaching or above effects thresholds for other marine species. Risks to open ocean marine mammals appear small.

Research Reports Available Online

Four reports and a summary document generated from the New Zealand organochlorines program environmental survey are available online at:

http://www.mfe.govt.nz/issues/waste/ocreports.htm

The following reports are available: Reporting on Persistent Organochlorines in New Zealand (a public summary document), Ambient Concentrations of Selected Organochlorines in Rivers, Ambient Concentrations of Selected Organochlorines in Estuaries, Ambient Concentrations of Selected Organochlorines in Air, and Ambient Concentrations of Selected Organochlorines in Soil.

Each of these reports provides an introduction to the study along with detailed information on the sampling design, the sampling and analytical programs, and the concentrations of organochlorines measured, and an assessment of this exposure data. In addition, to enable other researchers to have access to this data set, which is thought to be one of the most comprehensive of its type ever collected on a national basis, the appendices within each report provide fully tabulated concentration data for every sample collected from each site. The data set is also available electronically, in the form of an Access database, from the above web site.

Other research components of the organochlorines program have included a dioxin emissions inventory, the measurement of human exposure to organochlorines, and an appraisal of the risks to the New Zealand population from dioxinlike compounds. The reports on these phases of the program are also available on the organochlorine program Web site.

IV CONCLUSION

Risk-assessment technology is evolving quickly. Indeed, terminology and emphasis have changed from 1989 to 1996, as evidenced in the EPA documents used to frame this chapter. It is anticipated that this treatment will be passé within just a few years. Regardless, the basic risk-assessment paradigm is intelligent and insightful: It will probably remain intact for the near future.

What is painfully needed at this time is a sound data set for effects at all levels of organization. Also, basic ecotoxicological testing approaches produce data that are inadequate to the task, e.g., NOAEL approaches for estimating toxic thresholds. As an important example, temporal dynamics are given minimal consideration in most ecotoxicological tests. As we discussed in Chapter 9, only tradition in the field inhibits the construction of more complete models incorporating time effectively.

Neglected methods for expressing many toxicant-related effects in terms of risk probabilities are also discussed in Chapter 9. Finally, unjustified conceptual shortcuts are made in ecological risk assessments, such as assuming that the most sensitive life stage is the critical life stage for a population or that effects are (concentration) additive. As we already discussed in earlier chapters, these are dubious assumptions. Hopefully, these and other shortcomings in risk assessments will be resolved soon.

SUGGESTED READINGS

EPA, Risk Assessment Guidance for Superfund, Volume I: Human Health Evaluation Manual (Part A), Interim Final, EPA 540/1–89/002, NTIS, Springfield, VA, 1989.

EPA, Ecological Assessment of Hazardous Waste Sites: A Field and Laboratory Reference Document, EPA 600/3–89/013, NTIS, Springfield, VA, 1989.

EPA, Summary Report on Issues in Ecological Risk Assessment, EPA 625/3–91/018, NTIS, Springfield, VA, 1991, p. 46.

EPA, Proposed Guidelines for Ecological Risk Assessment, EPA/630/R-95/002B, U.S. Environmental Protection Agency, Washington, D.C., 1996, p. 247.

Neely, W.B., *Introduction to Chemical Exposure and Risk Assessment*, CRC Press, Boca Raton, FL, 1994, 190 pp.

Page, N.P., Human health risk assessment, in *Basic Environmental Toxicology*, Cockerham, L.G. and Shane, B.S., Eds., CRC Press, Boca Raton, FL, 1994.

Suter, G.W., II, *Ecological Risk Assessment*, Lewis Publishers, Chelsea, MI, 1993, p. 538.

CHAPTER 14

Risks from Exposure to Radiation

Thomas G. Hinton

Savannah River Ecology Laboratory, University of Georgia

It is still an unending source of surprise for me to see how a few scribbles on a blackboard or on a sheet of paper could change the course of human affairs.

Stanislaw Ulam (Quoted in Rhodes [1986])

I INTRODUCTION

This chapter introduces the reader to the fundamentals of radioactive contamination and its associated risks. Emphasis is placed on topics that illustrate differences between radioactive and nonradioactive contaminants. Three key pathways by which organisms are exposed to radioactive contamination and the concepts used to calculate dose are highlighted. The effects of exposure to radiation and the data from which risk factors have been derived are presented. Both human and ecological risks are considered. The chapter ends by addressing the widely held view among radiation protection organizations that if man is adequately protected, then so are the aquatic and terrestrial biota.

II FUNDAMENTALS OF RADIOACTIVITY

During much of the 15th to 17th centuries, a respectable branch of chemistry was engaged in trying to "transmutate" ordinary base metals into gold: to change the ordinary into the extraordinary. Fortunes were spent, careers wasted, and heads rolled with the finality of the guillotine, but no one could get one element to change into another. Human attempts at alchemy failed.

Radiation, however, is Nature's alchemist. The process of radioactive decay transforms one element into another. Indeed, there are entire chains of transformations that occur naturally within our ecosystems. For example, uranium-238 changes into thorium-234, thorium-234 into protactinium, and eventually—approximately 10^{10} years later and having undergone 14 different transformations—the original U atom changes from radioactive bismuth into stable lead. At each step, the resulting product loses all the characteristics of the parent element and acquires the characteristics of the newly formed daughter element. Characteristics such as color, melting point, hardness, even physical state, are changed; for example, radium, a solid, transforms into radon, a gas.

These transformations occur because radionuclides are excited atoms. They have an excess amount of energy and, therefore, are unstable. Radionuclides gain stability by releasing the excess energy in the form of electromagnetic photons (x- or gamma [γ] rays) and/or particles (alpha [α] or beta [β]). In doing so, changes in the atomic structure occur, resulting in a gain or loss of a proton. The number of protons largely determines the characteristics of an atom. All elements above bismuth (z = 83) in the periodic table are naturally unstable (radioactive), and a few of the lighter elements have one or more radioactive **isotopes**.

If only the 17th century alchemists could have duplicated what was occurring naturally all around them. It was not until 1942, when Enrico Fermi and his colleagues engineered the first sustained nuclear chain reaction, that man's attempts at alchemy proved successful. Their goal, however, was not to produce gold; they, instead, sought the tremendous energy released from the fission of U atoms.[1] Their "scribbles on a blackboard" led to the development of nuclear energy and nuclear weapons, thus profoundly affecting the 20th century (Rhodes, 1986).

II.A   Types of Radiation

Three principal types of radiation are emitted from radionuclides during radioactive decay: electromagnetic photons, beta particles (β), and alpha particles (α). Electromagnetic photons consist of gamma rays (γ) and x-rays that differ in their source. **Gamma rays** are emitted from the nucleus and **x-rays** are emitted from the shells of electrons that surround the nucleus. Once emitted, differences are not discernible. The essentially massless γ rays are photons similar to those of visible light and radio waves, all with the same velocity; but γ rays have shorter wavelengths and much greater energies.[2] All photons travel at the velocity of light and have energies inversely proportional to their wavelengths. The most energetic x-rays have wavelengths less than 10^{-12} cm, while the wavelengths of weak x-rays approach 10^{-6} cm. The energy of a photon affects its ability to penetrate matter. Energetic photons have sufficient energy to pass entirely through a human and require a meter thickness of dense material, such as concrete, to be fully absorbed (i.e., to have all of their energy dissipated into the surrounding media).

Gamma rays are emitted at monoenergetic energies. Thus the γ ray emitted by cesium-137 (^{137}Cs) always has an energy of 662 keV; cobalt-60 (^{60}Co) has two gamma rays that always have energies of 1170 and 1330 keV; and naturally occurring potassium-40 (^{40}K) emits a γ photon at 1440 keV. This unfaltering reliability of the emitted photon energies facilitates identification of unknown gamma-emitting radionuclides.

Beta (β) particles are electrons or positrons ejected from the atom during the radioactive decay process. They have the same mass as an electron (5.49×10^{-4} amu)[3] and can be either positively (**positron**) or negatively charged (electron). Beta particles have a continuous spectrum of energies,

---

[1] **Nuclear fission** is the splitting of atomic nuclei with neutrons, resulting in the release of energy, other neutrons, and radioactive fragments called **fission products**. Naturally occurring ^{235}U and man-made ^{233}U and ^{239}Pu have high probabilities of undergoing fission when bombarded by neutrons. The fissile nucleus is split into two or more fragments. The most probable mass partitioning results in fragments having mass numbers roughly in the ranges of 90 to 106 and 134 to 144. Fission products contain excess energy and are thus radioactive, and most of these products decay by the emission of beta particles. Approximately 200 MeV of energy is released for each fission, as well as two or three neutrons. The neutrons can cause additional fission, and thus a chain reaction can take place, governed by the density and geometry of fissile nuclei and the presence of material that slows or captures the neutrons. A nuclear reactor controls the fission process at a specific rate, thereby producing heat to boil water, turn turbines, and generate electricity. In a fission weapon, the geometry of the constituents is such that the chain reaction proceeds in an explosive manner. A nuclear explosion is not physically possible in a reactor because of fuel density, geometry, and other considerations (Lapp and Andrews, 1972; Whicker and Schultz, 1982; Rhodes, 1986).

[2] Energy can be expressed in terms of electron volts (eV). Gamma rays have energies of 10^2 to 10^7 eV compared with 20 to 80 eV for visible light and 10^{-4} to 10^{-8} eV for radio waves. An electron volt is the energy acquired by an electron when it passes through a potential difference of 1 V (1 eV = 1.602×10^{-12} erg). Kiloelectron volts (keV, 10^3 eV) and megaelectron volts (MeV, 10^6 eV) are commonly used extensions of the base unit.

[3] Masses of atomic constituents are measured in atomic mass units (amu), defined as 1/12th the mass of an atom of ^{12}C. There are 1.6605×10^{-24} g per amu.

with a maximum characteristic for that radioisotope. The average β energy of a radioisotope is equal to approximately one-third the characteristic maximum energy. Beta particles are less penetrating than γ photons. A sheet of aluminum a few millimeters thick can stop β radiation, as can 1 to 2 cm of flesh. Tritium, one of the commonly emitted radioisotopes from nuclear reactors, emits low energy β particles.

Alpha (α) particles are a chunk of the nucleus ejected from a radioactive atom to reduce excess energy and gain stability. They are relatively massive (7345 times larger than a β particle), consist of 2 neutrons and 2 protons, and carry a +2 charge. Their emissions are monoenergetic. The large mass and double charge of α particles cause them to react strongly with matter. Their energy is quickly deposited within a very short distance of the material they interact with, greatly reducing their penetrating abilities. High-energy α particles are stopped by a sheet of paper, or outer skin surfaces. Thus, their hazard is greatest if they are inhaled or ingested. The α emissions from plutonium, for example, are of concern when Pu is ingested and translocated to bone surfaces, or when Pu is inhaled and the sensitive lining of the lung exposed.

More detailed information on the types of nuclear decay and the physics of radiation can be found in Wang et al. (1975), Johns and Cunningham (1980), and Knoll (1989).

II.B Concentrations, Decay Constants, and Half-Life

Concentrations of nonradioactive contaminants are generally expressed on a mass basis (e.g., mg lead kg^{-1} tissue). In contrast, concentrations of radioactive contaminants are expressed on an activity basis (Bq ^{137}Cs kg^{-1} tissue). The base unit of activity is the becquerel (Bq), which is one disintegration per second.[4] A **disintegration** is the event in which a radioactive element releases photons or particles to gain stability, i.e., radioactive decay. Thus, **activity** is a measure of the rate at which a given quantity of radioactive material is emitting radiation.

Each radionuclide has a characteristic half-life ($T_{1/2}$) and decay constant (λ). **Half-life** is the amount of time required for one-half of the number of radioactive atoms to decay. At the end of one half-life, 50% of the original activity will remain; after a second half-life, 25% of the original activity will still be present, and so on. Half-life is an indication of a radioactive element's instability. Those with very short half-lives are particularly unstable and decay quickly.

The decay constant (λ) indicates the fraction of radioactive atoms (N) that will decay per unit time (t):

$$\frac{dN}{dt} = -\lambda N \qquad (14.1)$$

and is related to half-life by:

$$T_{1/2} = \frac{0.693}{\lambda} \qquad (14.2)$$

The activity of a radioactive sample can be calculated at any time (A_t)—if the original activity (A_0), decay constant, and elapsed time (t) are known—using the formula:

$$A_t = A_0 e^{-\lambda t} \qquad (14.3)$$

---

[4] The traditional unit of activity (Curie) has been replaced in the International System of units with the Bq; 1 Ci = 3.7×10^{10} Bq.

II.C   Radionuclide Detection

Samples containing radionuclides have two unique traits that facilitate determining their contaminant levels. If the sample contains γ-emitting radionuclides, some of the γ rays emitted from the contaminant within the tissue of the living plant or animal actually pass through the tissues and can be detected externally to the organism. Very sensitive instrumentation is used to detect the emitted γ rays[5] without having to destructively process the sample or take an aliquot from it. This provides a powerful way of resampling a living organism and thus determining the accumulation or elimination of the radioactive contaminant with time. Such techniques are not applicable to α and β emitters because most of the radiation is absorbed within the organism and is not externally emitted. Detection of α- and β-emitting contaminants generally requires sample preparation that precludes repetitive sampling on living organisms.

The second interesting aspect of radioactively contaminated samples is that the level at which instruments detect the radiation is improved merely by analyzing the sample for a longer period of time. A longer assay allows activity within a sample to be better distinguished from background activity and electronic noise of the measurement instruments. The longer a radioactive sample is assayed, the more radiation is emitted and thus measured. Nonradioactive samples either have an adequate level of contaminant to be detected or they do not. Assaying a sample containing stable potassium for a longer period of time will do nothing to improve the probability of detecting it. In contrast, if radioactive ^{40}K is not distinguishable in a sample after a 10-min assay, analyzing the same sample for 10 hours might increase the signal-to-noise ratio sufficiently to detect it.

These traits, coupled with the extreme sensitivity of today's radioanalytical instruments, make it possible to detect minute quantities of radiation. For example, radioactive contaminant levels on PAR Pond, a 10 km^2 lake on the Department of Energy's Savannah River Site, are such that detection of ^{137}Cs is easily accomplished within all components of the ecosystem. The lake contains a total ^{137}Cs inventory of 1.6×10^{12} Bq (44 Ci). When converted to mass, the entire 10^{12} Bq amount to a total of only 0.5 g of ^{137}Cs, which is distributed throughout the entire lake. Samples from the lake components containing as little as 0.3 Bq g^{-1} of ^{137}Cs can easily be assayed in one hour, an activity equivalent to a mass of 1.0×10^{-13} g of ^{137}Cs within the sample.

II.D   Effects

Damage from exposure to radioactive contaminants is initiated by **ionization**[6] caused by the energetic rays and particles released during radioactive decay. Ionization occurs in biological material exposed to radiation, resulting in some probability of molecular or genetic damage, either directly or through a multistep process. Part of the process often involves the formation of free radicals.[7] Free radicals are a principal cause of damage from exposure to radiation because they can easily break chemical bonds within DNA molecules. Among the most common is the OH[•] free radical, which is formed when cellular water is ionized.

Ionization results in biological damage. For humans, cancer is the biological consequence upon which most irradiation risk calculations are based. Four steps are thought to occur in cancerous tumor formation (Hall, 1978). The first step, *initiation*, involves damage to the DNA and, most

---

[5] See Knoll (1989) and Wang et al. (1975) for information on radioanalytical instruments.

[6] If the radiation has sufficient energy to eject one or more orbital electrons from the atom or molecule with which it interacts, then the process is referred to as ionization. Ionizing radiation is characterized by a large release of energy (approximately 33 eV per event), which is more than enough to break strong chemical bonds; for example, only 4.9 eV are required to break a C=C bond (Hall, 1978).

[7] Free radicals are atoms or molecules with an unpaired or odd orbital electron. In addition to orbiting around the nucleus, electrons also spin, either clockwise or counterclockwise, about their own axis. In atoms or molecules with an even number of electrons, spins are matched; for every electron spinning clockwise, another is spinning counterclockwise. This leads to a high degree of chemical stability. Molecules or atoms with an odd number of electrons, free radicals, have an electron that is left with an unmatched spin, creating instability.

likely, damage to both strands of the DNA double helix. Although a portion of the double-strand damage is repaired, completely error free repair is not expected. *Promotional events* in the intra- and extracellular environment, brought about by dietary constituents, hormones, or other environmental agents, cause the initiated cells to abnormally proliferate. Further, gene mutations within the rapidly dividing cell population cause a *conversion* to full malignancy. The *progression* stage of the disease allows invasion of adjacent normal tissues, eventually resulting in fatality if not successfully treated. It is important to realize that the time between the breakage of chemical bonds and the expression of a biological effect, the latency period, might take from days to decades, depending on the circumstances involved.

Cancer and genetic disorders are classified as **stochastic health effects**, meaning that the initiation of effects is probabilistic and that the risk of incurring cancer or genetic effects is proportional, without threshold, to the dose in the relevant tissue. The severity of a stochastic health effect is independent of the dose. In contrast, **nonstochastic health effects** (acute radiation syndrome, opacification of the eye lens, erythema of the skin, and temporary impairment of fertility) are dependent on the magnitude of the dose in excess of a threshold (ICRP, 1977).

III DOSE

Risk can be defined as the probability of a deleterious effect from a specific exposure to a contaminant. The derivation of risk factors, or slope factors if the EPA terminology is used, requires detailed knowledge about dose-response relationships. For nonradioactive contaminants, dose is often expressed as a concentration (e.g., mg lead kg^{-1} tissue). For radioactive contaminants, however, dose goes beyond concentration and refers to the energy deposited within biological tissues as a result of radioactive decay (i.e., 1 J energy kg^{-1} tissue = 1 gray, or Gy).[8]

If it is assumed that all of the released energy is deposited within the tissue, **absorbed dose rate** (Gy day^{-1}) can be calculated from a given radionuclide concentration using the general equation:

$$\text{Dose rate(Gy d}^{-1}) = (C)(SEE)(1.602 \times 10^{-13} \text{ Joules MeV}^{-1})\left(\frac{\text{dis}}{\text{s Bq}}\right)\left(\frac{86,400}{\text{d}}\right) \quad (14.4)$$

where C = concentration (Bq kg^{-1}), SEE = specific effective energy (MeV dis^{-1}), and dis = disintegration. The radionuclide specific effective energy is found in reference tables (ICRP, 1983).

Dose, however, is not entirely adequate to relate the amount of radiation absorbed to its effects. The effectiveness of absorbed energy at causing biological damage depends on the physics of radioactive decay, characteristics of the exposed tissues (some organs and tissues are more sensitive to radiation than others), and characteristics of the radiation. For example, a 1 MeV α particle creates approximately 20 times more damage over a given distance than a 1 MeV γ ray because the α particle has a higher **linear energy transfer** (LET): More of its energy is transferred into surrounding tissues per micron distance traveled. The relative biological effectiveness of various radiation types has been incorporated into quality factors that, if multiplied by absorbed dose, estimates *dose equivalent*. Dose equivalent was expressed traditionally in rem, but it is now expressed in the International System of Units (SI) of Sievert, where 1 Sv = 100 rem (ICRP, 1977; 1991). Dose equivalent normalizes the different types of radiation to the same propensity for biological damage.

A further improvement in radiation dosimetry was made in 1977 (ICRP), when the concept of **effective dose equivalent** was introduced. Recognizing that biological effects from a uniform

---

[8] The traditional unit for absorbed dose was the rad, where 100 rad = 1 Gy.

Table 14.1   Comparison of Parameters Used to Describe Various Aspects of Radioactive and Nonradioactive Contamination

Parameter	Contamination Type	Measures	Typical Unit
Activity	Radioactive	Rate at which radiation is emitted	Disintegration per second (Bq)
Concentration	Nonradioactive	Mass of contaminant per mass of tissue	mg lead per kg tissue
Concentration	Radioactive	Activity of contaminant per mass of tissue	Bq ^{137}Cs per kg tissue
Dose	Nonradioactive	Concentration	mg lead per kg tissue
Dose	Radioactive	Energy absorbed in biological tissue due to radiation emitted from contaminant	Energy absorbed per kg tissue (1 joule per kg = 1 Gray; traditional unit was rad)
Dose equivalent[a]	Radioactive	Normalizes the different types of radiation to the same propensity for biological damage	Absorbed dose × radiation-specific quality factors, expressed as Sieverts (Sv); traditionally expressed as rem
Effective dose equivalent[a]	Radioactive	Normalizes for different biological effects when a dose is concentrated in specific organs compared with the same dose distributed uniformly over the whole body	Sv
Committed effective dose[a]	Radioactive	Integrates effective dose equivalent over a 50 year period to account for physical and biological decay of ingested radioactivity	Sv

[a] No analog for chemical carcinogen.

irradiation of the whole body are different than effects from a similar dose concentrated in specific tissues (as happens if some radionuclides are ingested, e.g., ^{131}I goes directly to the thyroid gland), effective dose equivalent weights the dose to different organs or tissues. Thus, the fractional contribution of organs and tissues to the total risk of stochastic health effects is normalized to when the entire body is uniformly irradiated (ICRP, 1991). Exposures with equal effective dose equivalents are assumed to result in equal risks, regardless of the distribution of the deposited energy among different body tissues.

The last consideration made when calculating dose involves the time period of exposure from internally deposited radioactivity. Dose rates are integrated over a 50 year period for adults and 70 years for children because a one time intake of radionuclides commits an individual to a future dose, due to the time required for the contaminant to be removed from the body by biological elimination and radioactive decay. The final product of all quality factors, organ weighting factors, and integration of dose over the individual's working lifetime results in the **committed effective dose** (CED). CED is the fundamental unit to which risk factors are multiplied in order to estimate the probability of an individual human acquiring a fatal cancer from exposure to radiation (NCRP, 1993).

The CED is a rigorous attempt to specify not only the amount of energy absorbed by tissues, but also to account for numerous factors that influence the biological effectiveness of that energy at causing damage. From this perspective, radiation dosimetry is better developed than current methods used to determine dose from nonradioactive contaminants. Table 14.1 compares various parameters used for radioactive and nonradioactive contaminants.

The unit of CED is intended for use in setting human exposure limits or in assessing risk in general terms (e.g., for hypothetical exposure situations). The basic framework of radiological protection is designed to provide an appropriate standard of protection against ionizing radiation without unduly limiting the beneficial uses of radiation (Clark, 1995). The 1990 dose-limit recommendations (ICRP, 1991) keep doses below the relevant threshold for deterministic effects and demand that all reasonable steps are taken to reduce the incidence of stochastic effects to acceptable levels (Clark, 1995).

IV ENVIRONMENTAL TRANSPORT

In order to estimate risks from exposure to radioactive contaminants, knowledge of contaminant transport and fate is required. All processes governing the transport of stable contaminants described in previous chapters are applicable to radioactive contaminants as well. Any process by which an organism can acquire a nutrient, or stable contaminant, is also a plausible route for radioactive contaminant uptake. Thus, processes such as ingestion, inhalation, root uptake, and surface absorption are often necessary considerations. It is important to recognize that the movement of an element through the environment is not affected by whether or not it is radioactive. Radioactive cesium (^{137}Cs) moves through the soil and is taken up by plants in the same manner and rates as stable cesium (^{133}Cs). There are, however, characteristics of radioactive contaminants that can cause their transport kinetics to differ from stable analogs; particularly important is the contaminant's half-life.

If a radioactive isotope has a short half-life, it may decay into another element before it has time to participate in all of the environmental pathways that its stable isotope does. For example, ^{131}I has a half-life of 8 days. With such a short life, it is not possible for aerially deposited ^{131}I, released from a nuclear reactor, for example, to migrate through the soil and into the rooting zone of plants. Such a transport process takes time, and the ^{131}I will have decayed long before it reaches the plant roots. Ubiquitous stable iodine, however, is in contact with plant roots and can be taken up by the plant. Thus, in this case, the behavior of the radioactive element does not entirely duplicate the stable one.[9]

The other situation where a radioactive contaminant may not behave like its stable counterpart occurs when a recent introduction of a radionuclide is not in equilibrium with the environment and, therefore, behaves differently than its equilibrated stable isotope. Cesium uptake by plants represents a good example. Cesium's bioavailability is partially dependent on the soil clay content. Cesium has a strong tendency to bind to three different exchange sites on clay: surface sites, frayed-edged sites, and nonexchangeable interlayer sites (Cremers, 1988). Each site has different kinetics associated with the attachment of cesium atoms. Cesium adsorbs to the surface sites most readily, but it is also least tightly bound there and can be dislodged by other ions with a greater affinity for those sites (such as K^+ or NH_4^+). If a cesium atom gets into the interlayer sites, then the clay tends to collapse, tightly trapping the cesium atom. Cesium is not easily displaced from these inner sites by other ions, and the exchange rates are very slow. The exchange dynamics of cesium on the frayed-edged sites are intermediate in tenacity and exchange rates.

The interaction of cesium with the various exchange sites in the soil is time dependent. Different dynamics of these exchange sites cause newly deposited ^{137}Cs to behave differently than stable cesium that has been in the soil long enough to attain equilibrium. The ^{137}Cs deposited from the Chernobyl accident was taken up by plants at a faster rate than stable cesium in the same soils (Salbu et al., 1994; Bunzl et al., 1995). Stable cesium (^{133}Cs) is a natural, ubiquitous component of soils that, in essence, has been in the soil since its formation. It is in equilibrium with all of the exchange sites within the clay matrix. In contrast, newly deposited ^{137}Cs has a higher probability of dominating the surface and frayed-edged sites, locations where other ions can more readily displace the ^{137}Cs into the soil solution and thus increase its availability for root uptake. Once ^{137}Cs has remained in the soil long enough, it will reach a similar equilibrium to that of stable cesium, and their kinetics will be similar. This concept of radionuclides changing bioavailability over time

---

[9] Iodine-131 is a particularly important dose-contributing contaminant following a nuclear accident such as at Chernobyl. It was not necessary for the iodine to migrate to the rooting zone for plants to become contaminated. The I-131 was deposited on the surface of the grasses and taken up by the leaves. Cattle forged on the contaminated grass, ingested the contaminant, and a portion was transferred to their milk. Iodine was deposited in the thyroid of humans who drank the contaminated milk, thus increasing their risk of thyroid cancer. The Chernobyl accident occurred on April 26, 1986. Such cancers have a latency period of about 10 years, which is why we are just now observing an increased rate of thyroid cancers in children exposed to high levels of ^{131}I following the Chernobyl accident.

has been termed **aging** and has been observed numerous times in field and laboratory studies (Schimmack et al., 1989; Sanzharova et al., 1994; Velasko et al., 1993).

Newly deposited stable cesium would also undergo a similar aging process, but it would be difficult to distinguish its dynamics from the stable cesium already in the soil. The ^{137}Cs isotope serves as a tag, or **tracer**, that provides information about its behavior as a contaminant. By using a radioactive tracer, we also learn about clay particles and the behavior of stable cesium within the environment, illustrating that radioactive tracers can be powerful research tools to increase our knowledge of chemical, physical, and biological processes.

Excellent information on the environmental transport and fate of radionuclides can be found in Eisenbud (1973), Till and Meyer (1983), and Whicker and Schultz (1982). Table 14.2 provides general environmental transport properties of some common radionuclides.

IV.A Models Using Rate Constants

Predicting the impact of radioactive contamination on humans and the environment generally requires some form of mathematical model to simulate the contaminant's transport through the environment. Models have a wide range of complexities, depending on their intended use. Radioactive contaminant transport has been most commonly predicted with computer simulation models that use first order differential equations to describe the rate at which contaminants move among environmental components. Details of such models are beyond the scope of this introductory chapter, but it is important to understand that the *rate* at which a contaminant moves among all the components is required. It is not too difficult to measure the current location and concentration of an environmental contaminant. One can sample the water, soil, and various species of plants and animals and determine the concentration (Bq kg^{-1}) currently in each component. The difficult task is being able to predict what concentrations will be in those same components in 10 weeks or 10 years. The challenge is to identify the ecological processes by which contaminants move from one environmental component to another, and then to determine the corresponding **rate constants**. Simulation models that utilize rate constants to predict the movement and future concentrations of radioactive contaminants in the environment are exemplified by PATHWAY (Whicker and Kirchner, 1987) and ECOSYS-87 (Müller and Pröhl, 1993).

IV.B Screening Level Models

Many decisions regarding radionuclide releases and their potential impacts are based on results from much less sophisticated calculations referred to as screening level models. The complex intricacies of radionuclide movement are replaced by a handful of dominant processes and then further compensated for by inserting very conservative numbers for those processes. For example, among the parameters governing the uptake of a radionuclide by a plant are (1) characteristics of the radionuclide (chemical form, physical form, particle size, solubility, valence state), (2) local climatic conditions (humidity, temperature, solar irradiation, barometric pressure), (3) soil conditions (pH, moisture content, texture, clay type and abundance, percent organic matter, abundance of competing ions), and (4) plant characteristics (species, growth stage, area of leaves, area of roots, depth of roots, type of leaf surface). Rather than describe each of these individual parameters by separate mathematical formulas, a screening model might aggregate them all into a single parameter, a soil-to-plant transfer factor, and then choose a number for the transfer factor that would conservatively maximize the estimate of contaminant uptake by the plant.

Similar aggregation and insertion of conservative factors are done for other processes. Calculations are conducted, and the model prediction is then compared with some predetermined action level, such as a maximum allowable concentration or dose limit. If the simplified, conservatively obtained model prediction is less than the action level, then compliance with the legal limit has been demonstrated (NCRP, 1989). The assumption is that the overly conservative aggregation of

Table 14.2 General Ecological Properties of Selected Radionuclides

Radionuclide ($T_{1/2}$; emission)	Sources[a]	Nutrient Analog[b]	Important Exposure Modes	Degree of Food-Chain Transport[c]	Successive Trophic-Level Concentration[d]	Critical Organs (vertebrates)	Gastrointestinal Assimilation	Biological Retention
^{3}H (12 years; β)	Cosmic, fission, activation	H	Ingestion, uptake, absorption, inhalation	High	Approaches unity	Total body	Complete	Low (days)
^{131}I (8 days; β, γ)	Fission	I	Ingestion, absorption, inhalation	High	Up to ten times	Thyroid	High	Moderate (weeks–months)
^{40}K (1.3×10^9 years; β, γ)	Primordial	K	Ingestion, absorption, uptake, external γ	High	Approaches unity	Total body	High	Moderate (weeks)
^{137}Cs (30 years; β, γ)	Fission	K	Ingestion, absorption, external γ	High	Approaches 3.0	Total body	High	Moderate (weeks–months)
^{90}Sr (28 years; β)	Fission	Ca	Ingestion, absorption, uptake	High	<1.0	Bone	Moderate	High (years)
^{222}Rn (3.8 days; α, γ)	^{238}U decay	None	Inhalation of daughters	Negligible	Negligible	Lung (from daughters)	Negligible	Negligible
^{60}Co (5.2 years; β, γ)	Activation	Co	Ingestion, adsorption, inhalation, external γ	Moderate–high	1.0–10^2	GI, total body, lung	Moderate	Low (days)
^{144}Ce (285 days; β, γ)	Fission	None	Ingestion, inhalation, adsorption, external γ	Low–moderate	<0.1	GI, bone, lung, liver	Very low–negligible	Moderate (1–5 years)
^{232}Th (1.4×10^{10} years; α, γ)	Primordial	None	Ingestion, inhalation	Very low	$<10^{-2}$	Bone, lung	Very low–negligible	High (years)
^{238}U (4.5×10^9 years; α, γ)	Primordial	S, Se?	Ingestion, inhalation, uptake, external γ	Low–moderate	<1.0	GI, kidney, lung	Very low	Moderate (months)
^{239}Pu (2.4×10^4 years; α, γ)	Activation	None	Ingestion, inhalation, adsorption	Very low	$<10^{-2}$	Bone, lung	Very low–negligible	High (years)

a Activation: produced in a nuclear reactor or during a nuclear explosion due to interactions with neutrons; cosmic: produced in the atmosphere by interactions of cosmic rays with matter; fission: produced as a by-product from the fission of U or transuranics in nuclear reactors or nuclear explosions; primordial: radionuclides that appeared at the Earth's formation and are still present today due to their extremely long half-lives; ^{238}U decay series: a daughter product in the primordial ^{238}U decay scheme.

b The presence of a nutrient analog is important, as it often increases the ecological mobility of the radionuclide.

c This is a relative ranking of the radionuclide's bioavailability.

d Some contaminants tend to bioaccumulate up successive tropic levels; radionuclides generally do not, although some exceptions are noted.

Source: Adapted from Whicker, F.W. and V. Schultz, Radioecology: Nuclear Energy and the Environment, CRC Press, Boca Raton, FL, 1982.

parameters has caused the model prediction of contaminant concentration to be much larger than reality. Because the exaggerated model prediction is less than the regulatory limit, the risk manager is confident that the true concentration is also less than the limit.

When applying mathematical models to assess radionuclide contamination, it is often recommended that the simplest model that adequately addresses the problem be used first (NCRP, 1984). Recent trends in risk analyses, however, are moving away from screening level models toward models that produce realistic predictions with associated estimates of uncertainties.

IV.C Models Using Equilibrium Conditions and Dose-Conversion Factors

At a level of complexity slightly greater than screening models, the next three subsections introduce the reader to a general approach of deriving the committed effective dose (CED) from radioactive contaminants by assuming equilibrium conditions and using dose-conversion factors. The three subsections cover the principal pathways by which humans come in contact with radioactive contaminants. Recall that before we can assign a risk, it is first necessary to determine the dose.

IV.C.1 Inhalation Pathway

If the source term of the contaminant is atmospheric, then the committed effective dose from inhaling airborne material can be calculated from Equation 14.5:

$$CED(Sv) = (C)(IR)(ET)(EF)(EED)(IDF) \qquad (14.5)$$

where C = concentration in air (Bq m^{-3}), IR = inhalation rate (m^3 h^{-1}), ET = exposure time (h day^{-1}), EF = exposure frequency (day $year^{-1}$), EED = effective exposure duration (year), and IDF = inhalation dose factor (Sv Bq^{-1}). The effective exposure duration accounts for radioactive decay by:

$$\int_0^t e^{-\lambda t} dt \qquad (14.6)$$

where t is the residence time and λ is the physical or radioactive decay constant. Some parameters in Equation 14.5 depend on the particular scenario being modeled, such as the inhalation rate, exposure time, and frequency. In any risk calculation, assumptions about the contaminant and exposure scenario must be made. For example, calculations are often made for a future hypothetical resident of a contaminated site. Standardized data for numerous assumptions related to dietary consumption patterns, breathing rates, skin surface area available for contact, and exposure frequency and duration for various levels of occupancy are located in guidance documents for Superfund sites (EPA, 1991c). Other standardized parameters can be found in NCRP (1984; 1989), Pao et al. (1982), and Yang and Nelson (1984). A brief listing of some useful parameters is provided in Table 14.3.

The inhalation dose conversion factor (Equation 14.5) converts the inhaled activity to a collective dose equivalent per unit intake. Dose conversion factors for each radionuclide are found in Eckerman et al. (1988) and Eckerman and Ryman (1993) for inhalation, ingestion, submersion in a contaminated plume, submersion in contaminated water, and external irradiation from contaminated soils. The authors have taken the primary guides[10] for radiation protection, issued by the International Commission on Radiological Protection (ICRP, 1977) and EPA (1987c), and derived dose conversion factors and **annual limits on intake** (ALI).

---

[10] Primary guides are the current recommendations for radiological protection (e.g., the committed effective dose for a radiation worker in a given year should not exceed 50 mSv).

Table 14.3   Usage Factors for Calculating Dose to Humans

Parameter	EPA[a]	NCRP[b]
Drinking water	2 l day^{-1} (adult, 90th percentile) 1.4 l day^{-1} (adult, average)	800 l year^{-1}
Ingestion of surface water while swimming	50 ml h^{-1}	—
Swimming frequency	2.6 h day^{-1} (national average) 7 days year^{-1} (national average)	300 h year^{-1}
Soil ingestion	200 mg day^{-1} (1–6 years of age) 100 mg day^{-1} (>6 years of age)	—
Inhalation	30 m^3 day^{-1} (adult, upper limit) 20 m^3 day^{-1} (adult, average)	8000 h year^{-1}
Ingestion rate		
Fish	0.284 kg meal^{-1} (95th percentile)[c] 0.113 kg meal^{-1} (50th percentile)[c] 6.5 g day^{-1} (averaged over a year) 48 days year^{-1} (average per capita)	Freshwater: 10 kg year^{-1} Marine: 20 kg year^{-1}
Milk[d]	—	300 l year^{-1}
Meat[d]	—	100 kg year^{-1}
Vegetables[d]	—	200 kg year^{-1}
Home grown (fraction)[e]		
Fruit	0.2 average; 0.3 worst case	—
Vegetables	0.25 average; 0.4 worst case	—
Beef	0.44 average; 0.75 worst case	—
Dairy products	0.40 average; 0.75 worst case	—
Fruit, vegetables, and grain		200 kg year^{-1}
External exposure		
Contaminated surface from air deposition	—	8000 h year^{-1}
Shoreline	—	2000 h year^{-1}
Submersion in water	—	300 h year^{-1}
Submersion in air	—	8000 h year^{-1}

[a] EPA, 1989d.
[b] NCRP (National Council on Radiation Protection and Measurements), Effects of Ionizing Radiation on Aquatic Organisms, NCRP Report No. 109, Bethesda, MD, 1991, p. 115.
[c] Pao, E.M. et al., Food Commonly Eaten by Individuals: Amount Per Day and Per Eating Occasion, U.S. Department of Agriculture, Springville, VA, 1982.
[d] NCRP-recommended screening values (NCRP [National Council on Radiation Protection and Measurements], Screening Techniques for Determining Compliance with Environmental Standards, NCRP Commentary No. 3, Bethesda, MD, 1989, p. 134); derived from: NCRP (National Council on Radiation Protection and Measurements), Radiological Assessment: Predicting the Transport, Bioaccumulation, and Uptake by Man of Radionuclides Released to the Environment, NCRP Report No. 76, Bethesda, MD, 1984, p. 304; Yang, Y. and C.B. Nelson, An Estimation of the Daily Average Food Intake by Age and Sex for Use in Assessing the Radionuclide Intake of Individuals in the General Population, EPA Report No. 520/1-84-021, Office of Radiation Programs, U.S. Environmental Protection Agency, Washington, D.C., 1984; Rupp, E.M., Health Phys., 39, 141–146, 1980; and Rupp, E.M., F.L. Miller, and C.F. Baes III, Health Phys., 39, 1965–1970, 1980.
[e] For contaminated backyard gardens, the fraction of ingested food that is contaminated and consumed daily.

IV.C.2   Ingestion Pathway

Dose from consuming contaminated food can be calculated as:

$$CED(Sv) = (C)(INR)(EF)(EED)(INDF) \qquad (14.7)$$

where C = concentration in food (Bq kg^{-1}), INR = ingestion rate (kg day^{-1}), EF = exposure frequency (day year^{-1}), EED = effective exposure duration (year), and INDF = ingestion dose factor (Sv Bq^{-1}). Ingestion dose conversion factors are listed in Eckerman et al. (1988) and account for the fractional uptake from the small intestine to blood, and then to specific organs, of the common chemical forms of the radionuclides. They also account for losses from tissues via excretion and radioactive decay.

IV.C.3 External Irradiation

In addition to the pathways of contaminant transfer already discussed above and in Chapters 3 to 5, external irradiation has to be considered with radioactive contamination. This pathway is generally not needed for nonradioactive contamination. For example, if a lake is contaminated with stable lead, bound largely to the sediments, swimming over the contaminated sediments poses no health risk to fish. A fish would receive a chemical dose from the lead contamination in its food, in water consumed, and in lead passing across the gills. However, if the same system were contaminated with radioactive ^{214}Pb, the fish would receive an additional exposure merely by swimming above the contaminated sediments. Some of the emitted γ rays from ^{214}Pb would be projected out from the sediments, upon radioactive decay, and irradiate organisms in the overlying water column.

Once emitted from the sediments, however, the radiation is not capable of traveling indefinitely. Attenuation occurs exponentially and can be expressed as:

$$I = I_0 e^{-\mu d} \tag{14.8}$$

where I_0 is the initial γ incident rate (photons cm^{-2} s^{-1}) dependent on the activity of the source, μ is the attenuation coefficient (cm^{-1}), and d is the thickness of the attenuating material (cm; the water column in our example). Attenuation coefficients for most materials can be found in Johns and Cunningham (1980). Attenuation is dependent on the energy and type of radiation (i.e., α, β, or γ) as well as the density of the absorbing medium (e.g., air versus lead shielding). In our example, a 1 MeV γ ray would have an attenuation coefficient of 0.0706 in water. Calculations using Equation 14.8 reveal that 50% of the ^{214}Pb γ rays would be attenuated by water within 9.8 cm of the sediment surface. Thus, bottom-dwelling catfish would receive a greater external dose from the contaminated sediments than pelagic sunfish because the emitted γ rays are quickly attenuated by the water.

Calculating the external dose to humans exposed to radioactively contaminated soil, water column, or plume of air is possible using external dose coefficients provided in Eckerman and Ryman (1993). Equation 14.9 can be used in situations involving contaminated soil:

$$CED(Sv) = (C)(EF)(EED)(ISF)(SDC) \tag{14.9}$$

where C = the concentration in soil (Bq m^{-3}), EF = exposure frequency (s $year^{-1}$), EED = effective exposure duration (y), ISF = indoor shielding factor (unitless), and SDC = soil dose coefficient ($[Sv$ $m^3][Bq$ $s]^{-1}$). An indoor shielding factor accounts for time spent indoors, where the irradiation from the contaminated soil would be reduced. Shielding factors are discussed and provided in Eckerman and Ryman (1993). The soil dose coefficient is dependent on the distribution of contaminant within the soil as well as the energy of radiation emitted.

External dose from a plume of contaminated air can be calculated as:

$$CED(Sv) = (C)(EF)(EED)(ADC) \tag{14.10}$$

where C = concentration in air (Bq m^{-3}), EF = exposure frequency (s $year^{-1}$), EED = effective exposure duration (years), and ADC = air dose coefficient ($[Sv$ $m^3][Bq$ $s]^{-1}$).

A similar calculation can be performed for immersion in water by replacing the air concentration with water concentration and using the appropriate dose coefficient (Eckerman and Ryman, 1993).

Each of these calculations (Equations 14.5, 14.7, 14.9, and 14.10) results in an estimate of the committed effective dose for a particular pathway of contaminant exposure. Effective dose equivalents for all pathways are summed to yield a total CED. The total CED can then be multiplied by an appropriate risk factor to obtain the probability of harmful effects. An example of this approach related to a DOE Superfund site is provided by Whicker et al. (1993). Derivation of the risk factors is the subject of the next section.

V DERIVATION OF RISK FACTORS

V.A Epidemiological Studies

Human epidemiological data form the basis for determining the probability of deleterious effects from exposure to radionuclides. In contrast, risks from exposure to most chemical carcinogens are largely extrapolated from laboratory experiments using nonhuman subjects (EPA, 1989e). The latter involves a greater degree of uncertainty. The major epidemiological studies used to determine risks from radionuclide exposures are presented below.

In the late 1890s, the diagnostic potential of newly discovered x-rays was realized. These x-rays provided a unique tool for physicians to peer inside the body of a living human. The harmful effects of overexposure to radiation, however, were immediately observed, often in parallel with the discovery of radiation's beneficial medical uses. Early radiologists placed their hands in the x-ray beam to gauge its intensity from the red discoloration to the skin. For a short period, skin erythema was a commonly used metric, but the developed side effects of infections, tumors, and, in extreme exposures, a required amputation of fingers quickly alerted practitioners to the damage of radiation.

The delayed effects of radiation became apparent when some individuals hired to paint luminous dials on clocks and watches subsequently developed bone cancer. The painters would often lick their radium-ladened brushes to sharpen the point, thereby, ingesting some radioactive material. Other groups of exposed individuals for whom the effects of radiation have been observed include uranium miners exposed to radon gas and individuals medically treated with acute partial body x-rays to counter the effects of various illnesses. Some of the miners developed lung cancers, and certain medical patients developed an assortment of solid cancers.

The single most important source of information to understand radiation effects comes from the Japanese atomic bomb survivors. A very meticulous reconstruction of each individual's physical position relative to ground zero at the time of the blast has been performed, taking into consideration the presence of buildings and other structures that shielded the individuals from the radiation. This has allowed careful dose estimates to be made for individuals that are now being followed for lifetime health consequences.

The average dose to the Japanese cohort of 75,991 individuals was 0.12 Gy.[11] The ongoing study of the Japanese population indicates that 344 of the 5936 cancer deaths are attributable to radiation. No dose-related increase in genetic effects has been observed (NCRP, 1993). Such epidemiological studies have resulted in today's consensus that induction of cancer is the predominant risk from exposure to radiation rather than genetic effects (NCRP, 1993). EPA (1989d) concurs that the risk of cancer can be used as the sole basis for assessing the radiation-related human health risks for a site contaminated with radionuclides.

V.B Dose-Response Relationships

When a conflation of the Japanese data occurs with other sources from which radiation effects to humans have been observed (Table 14.4), it becomes evident that the doses at which a significant

---

[11] Epidemiological studies use the base unit of absorbed dose (Gy) rather than Sv because of the additional uncertainties associated with the quality and weighting factors.

Table 14.4 Features of Some Epidemiological Studies from Which Radiation Risk Factors Were Derived

Parameter	Life Span Study (LSS) of Japanese Atomic Bomb Survivors	Ankylosing Spondylitis Study (ASS)	Canadian Tuberculosis Patients Given Chest Fluoroscopies	Children in Israel Irradiated for Ringworm of the Scalp	U.K. National Registry for Radiation Workers
Population size	75,991	14,106	31,701	10,834	95,217
Period of follow-up	5 to 40 years after exposure	Up to 38 years (mean of 13 years)	Up to 30 years (mean of 27 years)	Up to 32 years (mean of 26 years)	Up to 34 years (mean of 12.7 years)
Ranges of:					
Ages at exposure	All	Virtually all ≥15 years	At least 10 years	0–15 years	18 to 64 years
Sexes	Similar numbers of males and females	83% males	Female	Similar numbers of males and females	92% male
Ethnic groups	Japanese	Western (U.K.)	Western (North America)	African, Asian	Western (U.K.)
Setting in which exposure was received	War	Medical: therapy for nonmalignant disease	Medical: diagnostic	Medical: therapy for nonmalignant disease	Occupational
Range of organs irradiated	All	All, but mainly those in proximity to spine	Mainly breast and lung	Mainly brain, bone marrow, thyroid, skin, breast	All
Range of doses	Mainly 0–4 Gy	Mainly 0–20 Gy	Mainly 0–10 Gy	Brain: 0–6 Gy (mean: 1.5 Gy) Thyroid: 0 to 0.5 Gy (mean: 0.09 Gy)	Mainly 0 to 0.5 Sv (mean: 0.034 Sv)
Dose rate	High	High	High, but highly fractionated	High	Low

Source: Adapted from Clark, R.H., Managing radiation risks, presented at Pathway Analysis and Risk Assessment for Environmental Compliance and Dose Reconstruction Workshop, Kiawah Island, SC, Nov. 6–10, 1995.

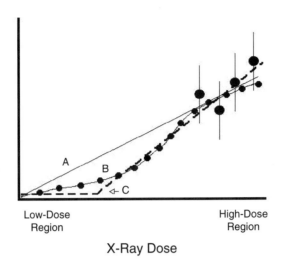

X-Ray Dose

Figure 14.1   A hypothetical illustration of how various models can be used to extrapolate the incidence of cancer from data generated at high doses to the low-dose region. Curve A represents a linear extrapo- lation; curve B depicts a linear-quadratic relationship; and curve C illustrates a threshold-type response. All three curves pass through the data, yet each gives quite different results when extrapolated to the low-dose region. (Adapted from Hall, E.J., *Radiobiology for the Radiologist*, Harper & Row Publishers, San Francisco, CA, 1978.)

increase in carcinogenesis has been documented are relatively high (0.1 to 0.2 Gy). Such doses are 100 to 200 times greater than the maximum allowable exposures received by today's public. Most importantly, effects have not been demonstrated at the lower doses.

A problem arises as to how to extrapolate effects observed at high doses to potential effects at low doses. There are three fundamental dose-response options: a linear extrapolation, a linear- quadratic relationship, and a threshold-type response (Figure 14.1). Much debate centers on which is the appropriate dose-response relationship. Choosing the appropriate extrapolation model is critical because the risk factors used to estimate the probability of harmful effects are derived from the dose-response relationships. Some method is required, however, to project the risks beyond those observed to have occurred at high doses to those that might occur at the lower doses encountered in today's radiation environment.

V.B.1   Threshold Option

Threshold models suggest that there is a dose below which there is no risk. Proponents for a threshold response (Curve C, Figure 14.1) cite as supporting evidence that an increased cancer incidence has not been observed in populations exposed to unusually high background radiation,[12] three times the nominal rate (NAS/NRC, 1990).

The international committees responsible for establishing radiation protection criteria, however, have assumed that a threshold does *not* exist, partially because of the method by which cancers are formed. It is generally accepted that tumors initiate from damage to single cells, although as described earlier, a complex series of multistage promotional events are required for a neoplasia to progress to full malignancy. A single mutational event in a critical gene in a single target cell can create the potential for neoplastic development (Clark, 1995). This means that there is some probability, albeit very low, that a single radiation track (the lowest dose and dose rate possible)

[12] **Natural radiation background** is composed of cosmic radiation emitted from stars and long-lived terrestrial radionuclides that are ubiquitously present in Earth's soils. An NCRP (1987) monograph thoroughly describes background radiation and cites that the estimated average CED from background radiation is 3.0 mSv year^{-1} in the United States.

hitting the nucleus of an appropriate target cell could cause damage to the DNA and initiate a tumor (Clark, 1995). Thus, at the DNA level, there is no basis for assuming that a dose exists below which the risk of tumor induction is zero.[13]

V.B.2 Linear versus Linear-Quadratic

A linear model suggests that there is no dose-rate effect. That is, a dose given over a short period of time (e.g., 50 mSv in 1 day) is just as effective at causing damage as the same dose protracted over a longer period of time (e.g., 50 mSv over 50 years). This model is contrary to numerous animal experiments where significant dose-rate effects have been observed (Ullrich et al., 1987). Animal experiments over a wide range of doses suggest that the linear-quadratic model (Curve B, Figure 14.1) is more appropriate. Human data, however, are inconclusive. For all cancers, other than leukemia, the Japanese data fit a linear dose-response model (Curve A, Figure 14.1). The same data also fit a linear-quadratic model (Curve B). Neither model fits the data significantly better than the other. Interestingly, the Japanese data also indicate that the dose-response curve of solid cancers differs from that of leukemia. The data for leukemia suggest a linear-quadratic model is significantly better than a linear dose-response relationship (NCRP, 1993).

The United Nations Scientific Committee on Effects of Atomic Radiation (UNSCEAR, 1986) concluded that linear extrapolation from high dose data to low doses (less than 0.2 Gy) could result in an overestimation of risk by a factor as high as five. The Committee on Biological Effects of Ionizing Radiation (NAS/NRC, 1990) also considered the animal experimental data and found that the linear model overestimated risk from two to ten times for the endpoints of specific locus mutation, reciprocal translocations, tumor formation, and longevity.

Currently, a linear dose-response model is used to obtain risk factors. The linear model produces the largest risk factors, and the agencies concerned with health protection have elected to use the most conservative numbers until sufficient data are accumulated to warrant otherwise. However, the ICRP (1991) has recommended that estimates of cancer risks associated with exposures to low doses or low dose rates be reduced by a factor of two when results are based on exposure to high doses at high dose rates.

V.C Currently Accepted Risk Factors

Recommended risk factors, furnished by the International Commission on Radiological Protection, are derived from the dose-response relationships obtained in the epidemiological studies presented above. The **risk factors** give the probability of a deleterious effect for each mSv of dose received. The ICRP (1991) has quantified risks into four categories, as seen in Table 14.5.

Table 14.5 Risk Factors for Each mSv of Dose Received

Deleterious Effect	Risk Factor per mSv Dose
Fatal cancer	5.0×10^{-5}
Severe genetic effects	1.3×10^{-5}
Nonfatal cancer	1.0×10^{-5}
Total Detriment	7.3×10^{-5}

Source: ICRP (International Commission on Radiological Protection), *1990 Recommendations of the International Commission on Radiological Protection,* ICRP Publication 60, Annals of the ICRP 21, Pergamon Press, New York, 1991, p. 211.

---

[13] The Health Physics Society, a professional organization responsible for protecting humans from overexposure to radiation, recently questioned current radiation protection criteria. They issued a position statement stating that doses below 0.1 Sv are either too small to be observed or are nonexistent and, therefore, should not warrant expenditures of conducting full-scale risk analyses. (Mossman et al., 1996).

EPA slope factors differ slightly from the ICRP recommendations. The incidence of cancer is the endpoint of interest in the EPA slope factors, rather than the probability of fatality from a cancer as used by the ICRP.

VI RISKS TO HUMANS FROM EXPOSURE TO RADIATION

Risk, in the context of this section, is the lifetime probability of a human experiencing a deleterious effect due to radiation exposure. Risk from exposure to radionuclides is calculated as the committed effective dose (presented earlier in the chapter) times a risk factor:

$$\text{Risk} = \text{CED (mSv)} \times \text{Risk Factor (mSv}^{-1}) \qquad (14.11)$$

Using the risk factors presented above, individuals with a CED of 2.0 mSv, an amount twice that recommended by the ICRP for public exposure, would have a probability of acquiring a fatal cancer during their lifetimes of 1.0×10^{-4} (a probability of 1 in 10,000; Equation 14.11).

Declaration of acceptable risk is purely a societal decision. Figure 14.2 compares the EPA (1989d) guideline of acceptable risks (10^{-4} to 10^{-6}) to other risks to which we are routinely subjected. Risks associated with radiation, and currently deemed acceptable, are of a similar magnitude to the probability of an individual being struck by lightning (4×10^{-5}). Such risks are orders of magnitude less than the natural cancer incidence (unrelated to radiation exposure) of 2×10^{-1}.

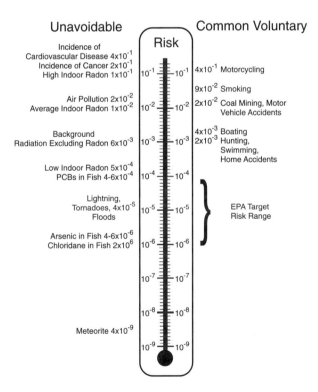

Figure 14.2 A comparison of risks from various voluntary and unavoidable sources. Numbers indicate the probability of death from the disease, natural phenomenon, contaminant, or activity. (Adapted from Hoffman, F.O., paper presented at Pathway Analysis and Risk Assessment for Environmental Compliance and Reconstruction Workshop, Kiawah Island, SC, 1991.)

VII ECOLOGICAL EFFECTS FROM RADIOACTIVE CONTAMINATION

The propensity for nonhuman species to take up radioactive contaminants is demonstrated by an increased ^{131}I burden in the thyroids of herbivores following radioactive releases from atmospheric testing of nuclear weapons (Figure 14.3). Radioactive releases generated in China were quickly transferred to Colorado deer and elk (Whicker and Schultz, 1982), illustrating the mobility and potential impact of some contaminants.

Notable reviews of radiation effects have been prepared on specific groups of organisms: protozoa (Wichterman, 1972), brine shrimp (Metalli and Ballardin, 1972), insects (O'Brien and Wolfe, 1964), amphibians (Brunst, 1965), reptiles (Cosgrove, 1971), birds (Mellinger and Schultz, 1975), plants (Sparrow et al., 1958), terrestrial and aquatic animal populations (Turner, 1975; Blaylock and Trabalka, 1978), and plant communities (Whicker and Fraley, 1974). The National Council on Radiation Protection and Measurements has examined the effects of ionizing radiation on aquatic organisms (NCRP, 1991); the International Atomic Energy Agency has considered whether or not nonhuman species are adequately protected by radiation standards designed for humans (IAEA, 1992); and the lower limits of radiosensitivity in nonhuman species have been reviewed (Rose, 1992).

Research on effects to plants and animals documented over the last 70 years have led to the formulation of some basic paradigms. The effects of radiation on reproduction have been most extensively studied in mammals, and the majority of results suggest that natality is a more radiosensitive parameter than mortality (Carlson and Gassner, 1964). Among the vertebrates, mammals are generally more radiosensitive than birds, fish, amphibians, or reptiles (Casarett, 1968). Rose (1992) reviewed the literature for lower limits of radiosensitivity in nonhumans and found that the lowest dose from an acute exposure with measurable effects was 10 mGy. This dose was delivered to pregnant rats and ultimately impaired the reflexes of their offspring (Semagin, 1986). Mice are among the organisms whose ovaries are the most sensitive to irradiation, and Gowen and

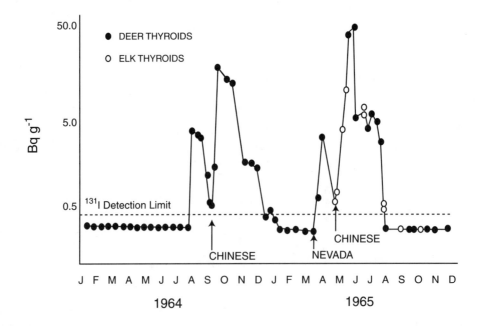

Figure 14.3 ^{131}I concentrations in Colorado mule deer and elk thyroids following nuclear testing activities in China and the United States during 1964 and 1965. (Adapted from Whicker, F.W. and V. Schultz, *Radioecology: Nuclear Energy and the Environment,* CRC Press, Boca Raton, FL, 1982, Vol. I, p. 212; Vol. II, p. 228.)

Stadler (1964) found that reproduction was impaired in females at doses of 200 mGy, with permanent sterility occurring at 1000 mGy. Males were less sensitive and had impaired reproduction at 3200 mGy (Rugh and Wolff, 1957).

Studies on radiation effects to nonmammalian organisms have indicated higher radioresistance and, as with mammals, that the early life-cycle stages are the most radiosensitive (NCRP, 1991). The lowest dose reported to have an impact on amphibians was 20 mGy, a dose that was lethal to newt eggs (*Triturus alpestris*; Peters, 1960). Anderson and Harrison (1986) found that radiation doses in excess of 10 mGy were necessary to damage the most sensitive stages of fish development.

The National Council on Radiation Protection and Measurements (NCRP, 1991) recently established a maximum dose rate of 10 mGy day^{-1} from chronic exposure for the protection of populations of aquatic organisms. The NCRP recognized that other environmental stresses might act in combination with radiation and cause an impact at the maximum reference level of 10 mGy day^{-1}. Therefore, they conservatively recommended that a comprehensive ecological evaluation of the radiation exposure and environmental stressors be conducted when populations are exposed to 2.4 mGy day^{-1}. The International Atomic Energy Agency (IAEA, 1992) has also addressed the issue of effects of ionizing radiation on plants and animals, concluding that "there is no convincing evidence from the scientific literature that chronic radiation dose rates below 1 mGy day^{-1} will harm animal or plant populations."

Sufficient data exist on acute lethality to rank groups of organisms according to their sensitivities to radiation (Figure 14.4). Mammals seem to be the most sensitive group, and humans are among

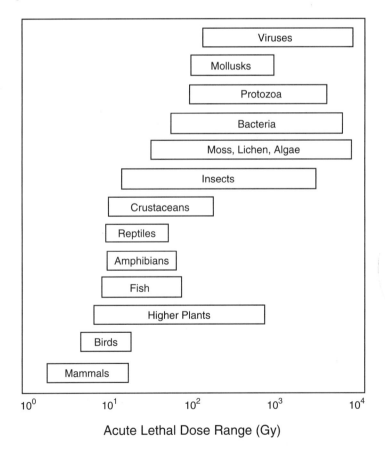

Figure 14.4 A comparison of the radiation sensitivity among groups of organisms. (Adapted from Whicker, F.W. and V. Schultz, *Radioecology: Nuclear Energy and the Environment,* CRC Press, Boca Raton, FL, 1982, Vol. I, p. 212; Vol. II, p. 228.)

the most sensitive mammals. This has caused some organizations to conclude that if man is adequately protected from radiation, then other organisms are also likely to be sufficiently protected (ICRP, 1977). This assumption has been generally accepted and adopted by those establishing radiation protection standards, even though "sufficient protection" has never been quantified.

The International Atomic Energy Agency (IAEA, 1992) set about to determine if the statements made by the ICRP were consistent with current knowledge and whether radiation protection standards were needed for aquatic and terrestrial biota. The IAEA approached the problem by developing hypothetical contamination scenarios that resulted in humans acquiring a maximum permissible dose (1 mSv year^{-1}). They then calculated the dose rates aquatic and terrestrial biota would receive if living in the same hypothetical environment. Doses received by the biota were then compared with doses known to have caused observable effects to nonhumans. They concluded that limiting the dose to the most-exposed group of humans to 1 mSv year^{-1} would lead to dose rates to plants and animals in the same area of less than 1 mGy day^{-1} (notice the different dose rates: year^{-1} for humans and day^{-1} for nonhumans). Because convincing evidence does not exist within the scientific literature that chronic radiation doses below 1 mGy day^{-1} will harm plant or animal populations, the IAEA (1992) concluded that specific radiation protection standards for nonhuman biota are not needed. However, they stated that, under some situations, site-specific analyses would be required, such as analyses involving endangered species.

VIII CONFIDENCE IN RISK ANALYSES

The conclusion of the IAEA should not imply that ecological risk analyses are not needed or that our understanding of risk to humans and the environment is complete. Numerous uncertainties exist in risk analyses that result in our having a reduced confidence in the risk estimate. A thorough risk analysis should include uncertainty estimates that indicate the level of confidence in the risk prediction (NCRP, 1996). Estimating the uncertainty associated with risk analyses is critical to a meaningful use of the risk estimate. A cumulative uncertainty of a factor of ten would not be unreasonable in a risk analysis conducted for human exposure to radiation. Some of the dominant sources of uncertainty (Sinclair, 1993) are the following: (1) estimating the dose received by the Japanese cohort; (2) using data from a population that received high dose rates and extrapolating to the low-dose range; and (3) using epidemiological data derived from a population with strikingly different cultures and life styles, some of which seem to influence cancer incidence. For example, the natural stomach cancer incidence among male Japanese (unrelated to the atomic weapons explosion) is about six times greater than among males from North America, and incidence of breast cancer is six times greater in North American females compared with Japanese (NCRP, 1993). If natural cancer incidences differ among populations, then the question arises as to how pertinent are radiation risk factors to populations other than those from which the factors were derived.

As complex as human risk analyses are, determining risks to other environmental components is even more so, and with much greater uncertainties. Problems exist for the following reasons:

1. Measuring the dose received by most nonhuman species is not a trivial task. Dose measurement instruments can be placed on some species, but recapturing the same animal and recovering the instrument is problematic.
2. Mathematical models that predict dose to free-ranging organisms have not been developed and validated as well as human dosimetry models.
3. Tables of dose conversion factors that transform a unit concentration of contaminant to a dose received by the organism exist for only a few nonhuman species. A recent compilation of dose conversion factors for nonhumans (Amiro, 1997) has many shortcomings when compared with dose conversion factors derived for humans. Amiro (1997) states that the factors for nonhumans are generic and useful only as conservative indicators of radiological dose. They are not intended to give exact dose estimates. Precise doses are not possible because the nonhuman dose conversion

factors: (1) do not account for the relative biological effectiveness of different radiation types (e.g., γ versus α emissions); (2) conservatively assume that all energies emitted by radionuclides from within the organism are also absorbed by the organism, resulting in an overestimation of dose to smaller organisms; and (3) do not consider the shape of different organisms. Thus, the same dose conversion factors are used regardless of an organism's geometry, which "could range from an ant to an elephant, or from a lichen to a tree" (Amiro, 1997). The dose conversion factors presented by Amiro (1997) are useful, but it is apparent that they are far less rigorous than those calculated for humans.

4. There are no risk factors established for nonhumans. Dose-response data pertinent to nonhumans are inadequate to quantify the probability of deleterious effects for the common exposure conditions (i.e., chronic, low-dose conditions).

5. Agreement has not been reached as to what the proper endpoint should be for an ecological risk analysis. The endpoint for humans is health of the individual. However, risk to individuals is generally not considered for nonhuman species. Seldom are we concerned with an individual fish or tree but, rather, that their populations remain viable.[14] The critical endpoint for humans (cancer incidence) is not likely to be relevant for most populations of plants and animals. Contaminants can exert effects on levels of biological organization ranging from molecules to ecosystems. Often, because of the complexity of documenting effects at the levels of population, community, and ecosystem, most ecotoxicological research concentrates on effects at lower levels of organization (cellular or molecular). Also, many of the mechanisms that connect effects between individuals and populations are poorly understood. However, it is the population-level effects that are of primary concern for nonhuman species. Cellular damage generally represents a sublethal endpoint that may provide good early-warning signs of potential contaminant impact, but specific relationships between cellular damage and consequences at upper levels of biological organization have not been made (Clements and Kiffney, 1994; Underwood and Peterson, 1988). Current environmental risk analyses that use cellular/molecular effects as endpoints are of questionable value, primarily because of large uncertainties about the significance of cellular/molecular effects to the population. Many cellular and molecular effects are repaired within the organism or are removed from the population through natural selection.

6. Most of the dose-effect data for nonhumans are not pertinent to conditions under which the majority of organisms are now exposed (chronic, low-level exposures). Most of the data in the literature were generated under conditions of high dose rates and short-term laboratory experiments.

The findings of a recent committee on risk assessment methodology, convened by the National Research Council, are particularly pertinent. They identified four major areas where scientific consensus was lacking in ecological risk analysis (Barnthouse, 1994):

1. Extrapolation across scales of time, space, and ecological organization
2. Quantification of uncertainty
3. Validation of predictive tools
4. Economic valuation of ecological resources

Until these areas are adequately addressed, as well as numerous others brought out in this chapter, ecological risk analyses will have greater uncertainties than those currently associated with estimating risk to humans.

IX SUMMARY

Risk, when rigorously defined, should include a probability statement about the incidence of a deleterious event. The science supporting the determination of hazards to humans from radiation exposures is sufficiently developed to meet that rigorous definition. Documentation of effects from human exposures to radiation were initiated with the discovery of x-rays in the late 1890s and have

---

[14] Concern for individual nonhuman species does occur when the species is endangered.

since developed into a science that crosses disciplines of nuclear physics, human physiology, chemistry, biology, and the medical professions. The uncertainties associated with risks to humans from exposure to radiation are probably less than for many nonradioactive contaminants. The concept of dose seems to be much better developed in the radiological sciences, and human epidemiological data are available to derive the risk factors for radioactive contaminants, whereas most risk factors for nonradioactive contaminants must rely on extrapolation from animal laboratory data.

Quantifying ecological risks from radioactive contamination is more difficult; indeed, the term *risk* may not be appropriate, as the science is not sufficiently developed to attach a probability statement to the analysis. Thus, ecological risk analyses are more qualitative than quantitative. Risk factors for radiological or nonradiological contaminants do not even exist for nonhuman organisms.

Risk factors are a powerful tool for making decisions about contaminated sites. They provide a common metric with which comparisons can be made. Most importantly, they are generated from scientific input. Many decisions regarding the management of contaminated sites will be influenced by sociopolitical motivations and the public's perception of the hazard. Risk analysis is the primary mechanism by which science can enter into the decision-making process. It is, therefore, critical for risk analyses to be conducted with the utmost professionalism and scientific rigor. Anything less will result in the risk-analysis process losing credibility and, thereby, the loss of scientific input into issues concerning public health.

X ACKNOWLEDGMENTS

This work was supported in cooperation by the University of Georgia and the Department of Energy Award Number DE-FC09–96SR18546. The critical reviews by C. Bell, J. Graves, J. Joyner, M. Malak, L. Marsh, P. Mulvey, S. Rich, C. Strojan, and F. W. Whicker were much appreciated.

SELECTED READINGS

Eisenbud, M., *Environmental Radioactivity*, Academic Press, New York, 1973.

Hall, E.J., *Radiobiology for the Radiologist*, Harper & Row, San Francisco, CA, 1978.

IAEA, *Effects of Ionizing Radiation on Plants and Animals at Levels Implied by Current Radiation Protection Standards*, International Atomic Energy Agency, Vienna, Austria, 1992.

Lapp, R.E. and H.L. Andrews, *Nuclear Radiation Physics*, Prentice-Hall, Englewood Cliffs, NJ, 1972.

NAS/NRC (National Academy of Sciences/National Research Council, Committee on the Biological Effects of Ionizing Radiations), *Health Risks of Exposure to Low Levels of Ionizing Radiation*, BEIR V, National Academy Press, Washington, D.C., 1990.

UNSCEAR (United Nations Scientific Committee on the Effects of Atomic Radiation), *Sources, Effects and Risks of Radiation*, Publication E.88.IX.7, United Nations, New York, 1988.

Whicker, F.W. and V. Schultz, *Radioecology: Nuclear Energy and the Environment*, CRC Press, Boca Raton, FL, 1982.

SECTION FIVE

Summary

CHAPTER 15

Conclusions

There will come soft rains and the smell of the ground,
And swallows circling with their shimmering sound;
…
Not one would mind, neither bird nor tree,
If mankind perished utterly.

Teasdale (1937)

I OVERVIEW

So far, we have discussed the diverse goals of ecotoxicology and the many disciplines contributing to our understanding of contaminant fate and effects. We discussed contaminant sources, cycling, and accumulation in components of the biosphere, including transfers to individuals and trophic levels. Then effects from the suborganismal to the biosphere levels were discussed in a series of chapters (Chapters 6 to 12) that took up most of the book. Finally, the practical issue of estimating chemical risk to humans and the environment was explored. One of the intentions of detailing risk assessment methods was to point out the presently unavoidable shortcomings in the associated technologies and science, and then to emphasize the need for more effort in this extremely important aspect of ecotoxicology. Now all that is left to do is remind the reader of the importance and context of ecotoxicology. That is the goal of this short chapter.

II PRACTICAL IMPORTANCE OF ECOTOXICOLOGY

In looking back at the enormous human suffering of World War I, Sarah Teasdale wrote the poem from which the above excerpt was taken. She correctly observed that, although the suffering was of such enormity that it will stand out in history for centuries, it was trivial in the larger context of Nature. The biosphere can do just fine without our help. Indeed, Stephen Jay Gould (quoted in Wheeler, 1996) states: "Humans are simply a dot on one side of the curve of biological complexity … a dot that could easily disappear."

Does this mean that we are free to contaminate the biosphere with abandon? This is certainly the case in the context of ensuring that life continues on the earth. Our influence on the permanence of life on earth is minuscule. One meteorite striking the earth at the end of the Cretaceous period had more influence on the trajectory of life on earth than we will ever have. In a few billion years, life will end when the Sun expands to engulf our planet: Our efforts cannot change the physics of star aging. However, this is certainly not true in the context of ensuring that the biosphere remains compatible with and appealing to human life. It would be a catastrophic mistake to continue to contaminate the biosphere. Presently, we are having enormous influence on the biosphere relative to its ability to provide an appealing home for us and for valued species. For this reason, ecotoxicology has now become an essential component of human knowledge and activities—not because life on earth needs us to protect it, i.e., the Lorax incongruity. It is extremely important to understand that ecotoxicological knowledge and activities are no more or less important than our industrial knowledge and activities. Few people would willingly abandon the right to clean water and an aesthetically pleasing environment. Similarly, few people would deny themselves or their children the fruits of our industry—an industry that produces waste. The key is striking an informed and insightful balance between these two facets of our lives. "The ultimate goal of development and of our concern for the environment is to increase the quality of life" (Gallopin and Öberg, 1992).

III SCIENTIFIC IMPORTANCE OF ECOTOXICOLOGY

There is intrinsic value to the material described so far, although that may not be readily apparent if you are reading this chapter as part of a college course that, at long last, is finally drawing to an end. There is enormous value in simply learning more about your world. With ecotoxicology, one has the added value of the knowledge having obvious and immediate utility.

Our world is filled with fascinating concepts, and more appear daily. In the two years that have passed since writing the first edition of this book, many new discoveries were made about the world. The conclusion from the Viking lander missions that life did not exist on Mars was thrown open to question based on evidence from a Martian meteorite found on the Antarctic ice. As this second edition was being finished, NASA had detected large amounts of water on Mars. While the first edition was being written, the first mammalian clone, a sheep named Molly, was produced from a mature cell of another individual. As this second edition goes to press, the field of cloning has shot forward at such a rate that all countries of the world are intently focused on the legal and moral issues surrounding human cloning. What were once the props of science fiction writers are now the realities of our world. New facts enhance our appreciation of the world and facilitate wise decisions. Although we admit to more than a small bias here, those facts associated with ecotoxicology are certainly among the most interesting and useful.

Study Questions

CHAPTER 1 (AND APPENDICES 3 AND 4)

1. Define the terms highlighted in the text and check your definitions against those provided in the glossary.
2. Briefly describe the transition and events leading to the shift from the Dilution Paradigm to the Boomerang Paradigm.
3. How does one maintain conceptual coherency in dealing with a hierarchical field of study such as ecotoxicology?
4. Define the scientific, technical, and practical goals of ecotoxicologists.
5. Explain the evolution of scientific inquiry from the ruling theory to the working hypothesis to the multiple working hypotheses stages. Is the multiple working hypothesis prevalent in ecotoxicology today?
6. Compare and contrast normal and innovative science. What are their relative values in young versus mature sciences?
7. What are the abnormal practices called precipitate explanation, the tyranny of the particular, and *idola quantitatus*? Are they a problem in ecotoxicology today?
8. What qualities are valued in scientific, technological, and practical ecotoxicology?
9. What are the roles of the Federal Register and Code of Federal Regulations in defining and establishing environmental regulations?
10. Briefly give the purpose and major qualities of the following legislation:
National Environmental Policy Act
Clean Air Act
Federal Insecticides, Fungicide and Rodenticide Act
Marine Protection, Research and Sanctuaries Act
Safe Drinking Water Act
Toxic Substances Control Act
Resource Conservation and Recovery Act
Clean Water Act
Comprehensive Environmental Response, Compensation and Liability Act
11. What is the role of the Environmental Protection Agency relative to the environmental legislation listed above?
12. Relative to environmental regulation, how does the European Union (EU) differ from the United States?
13. What are the three general categories of environmental legislation adopted by the EU?
14. Briefly describe the following EU directives:
Council Directive 67/548
Plant Protection Products Directive
Council Regulation 793/93 on Existing Chemicals
Council Directive 76/464 on Dangerous Substances
Water Framework Directive

CHAPTER 2

1. Define the terms highlighted in the chapter and check your definitions against those provided in the glossary.
2. If metals are naturally occurring elements, how do they become environmental contaminants and pollutants?
3. The total tributyltin (TBT) concentration in a 2 liter water sample is 100 ng l^{-1}. Assume the sorption coefficient for TBT is $K_p = 10^3$. If the suspended sediment load is 20 mg l^{-1}, how much of the TBT in the sample is in the dissolved state? What assumptions are you making when conducting this calculation?
4. Explain how dissolved humic material could enhance the degradation of an organic contaminant in the surface waters of an estuary.
5. What are the commercial uses for dibenzodioxin and dibenzofuran compounds?
6. What are some of the management strategies that are being employed to fight eutrophication?
7. Both organochlorine and pyrethroid insecticides are synthetic organic molecules. Why are organochlorines less prone to microbial degradation?
8. You are told that soil samples near an abandoned creosote facility contain high levels of pyrene. No other organic contaminants are reported in significant concentrations by the analytical lab. As a risk assessor, you must evaluate the potential for ecological harm at this site. Why should you be suspect of these data?
9. EPA standard methods for priority pollutants are often used to generate contaminant data for risk assessments. We now know that there are potential problems with this approach. What are the problems, and what are some of ways to prevent them in the future?

CHAPTER 3 (AND APPENDIX 5)

1. Define the terms highlighted in the chapter and check your definitions against those provided in the glossary.
2. Distinguish between steady state and chemical equilibrium. Which term is most accurate in describing bioaccumulation? Why?
3. What contaminants are most likely to pass into an organism via the lipid route? Which are least likely?
4. What parameter would you use to quantify the difference in adsorption affinity of a series of contaminants to a surface?
5. Would ionized or nonionized ammonia pass through the membrane fastest?
6. How could you determine if passage of a contaminant across a biological membrane takes place by active or passive diffusion?
7. What is the reaction order of the following equation? $dC/dt = kC^1$? Is this reaction order common or uncommon for elimination kinetics?
8. Describe the role of biomineralization in the bioaccumulation of metals and metalloids.
9. Describe Phase I and II reactions in the metabolism of organic contaminants.
10. What is enterohepatic circulation, and why might it be important relative to realized effect of toxicant exposure?
11. Distinguish among elimination, depuration, and clearance.
12. What are the three phases of renal elimination?
13. Describe the differences and similarities of rate constant-based, clearance volume-based, and fugacity models.
14. The rate constant for a single component, rate constant-based elimination model is 0.1 h^{-1}. What are the biological half-life and mean residence time predicted for this model?
15. Describe the backstripping method. Does it have weaknesses?
16. Explain the clearance volume concept.
17. Describe the models given in Equations 3.20 to 3.22. What are the meanings of the associated variables and constants?
18. The units of k_u are $ml(g^{-1} h^{-1})$, but the units of k_e are h^{-1}. Why?

CHAPTER 4

1. Define the terms highlighted in the chapter and check your definitions against those provided in the glossary.
2. Discuss the definition of bioavailability relative to ecotoxicology and pharmacology.
3. Describe the various approaches to estimating bioavailability.
4. How could you estimate the relative bioavailability of a compound in imbibed water versus ingested plant matter?
5. Discuss the free ion activity model (FIAM) and exceptions to this model.
6. What factors influence bioavailability of contaminants in food?
7. What measures can be used to reflect the bioavailability of metals from oxic and anoxic sediments?
8. Describe the QSAR (quantitative structure-activity relationship) approach. What is the basis for the extensive use of K_{OW} in these models?
9. Explain the pH-partition hypothesis and relate the Henderson-Hasselbalch relationship to the bioavailability of a monobasic acid using this hypothesis.
10. How would you normalize for animal size effects on bioaccumulation?
11. A series of clam populations are sampled and analyzed for tissue cadmium concentrations. Cadmium concentrations are found to be directly dependent on ambient cadmium concentrations. In this case, is cadmium acting as a biologically determinant or indeterminant element? Explain your answer.
12. What are the major biological factors influencing bioaccumulation in bivalve mollusks?

CHAPTER 5

1. Define the terms highlighted in the chapter and check your definitions against those provided in the glossary.
2. Describe how isotopic ratios of nitrogen, sulfur, and carbon might be used to define the trophic structure of a lake community. What are the advantages and disadvantages of each of these elements for such use?
3. Describe how you would estimate biomagnification using body mass-weighted mean concentrations.
4. What elements might biomagnify? Why would these specific elements biomagnify?
5. What is "liquid" digestion by zooplankton, and how does it relate to metal bioavailability from algae?
6. Would fish-eating sea birds accumulate more mercury than seabirds consuming other prey? Why?
7. How do birds eliminate mercury taken up in their diets?
8. How do calcium sinks influence transfer of calcium analogs?
9. What qualities of organic compounds would foster biomagnification?

CHAPTER 6

1. Define the terms highlighted in the chapter and check your definitions against those provided in the glossary.
2. What are the qualities of the ideal molecular biomarker?
3. Describe in detail the components of the mixed function oxidase system.
4. What biomarkers are associated with Phase II reactions?
5. List the qualities of metallothioneins.
6. What are the functions of metallothioneins?
7. What is the spillover hypothesis? Are there any situations that you can identify where this concept would not be valid, e.g., mixtures of metals, etc.? Why?
8. Describe proteotoxicity and the roles that stress proteins play in lessening the effects of proteotoxic agents.

9. Describe the types of molecular damage that can be produced by oxyradicals. How does the cell lessen the damage of oxyradicals?
10. Hydrogen peroxide is not an oxyradical, yet it contributes to oxidative stress. How?
11. How might you measure lipid peroxidation in a field population of a species thought to be experiencing oxidative stress?
12. How would you measure DNA damage?
13. A group of subsistence fishermen are concerned that they are subject to high mercury exposure from their diet. They hire you to determine if this is a legitimate concern. Design a biomarker study to assess the potential exposure.

CHAPTER 7

1. Define the terms highlighted in the chapter and check your definitions against those provided in the glossary.
2. Define and contrast the holistic and reductionist approaches to hierarchical subjects such as ecotoxicology. Which is the correct or best approach to ecotoxicology?
3. Why are cellular, tissue, and organ biomarkers valuable in ecotoxicology?
4. Describe some of the characteristics of necrosis.
5. Describe the process of inflammation, including its four cardinal signs. Are all relevant to all animal species?
6. If you suspected that a genotoxic agent was affecting mice inhabiting an area, what biomarkers would you use to test your suspicions? In your answer, explain why you would use each biomarker.
7. An individual was exposed to high concentrations of a chemical with the properties of an initiator. Will that individual develop cancer? Explain your answer.
8. You are asked to determine if trout in a lake modified by acid precipitation (low pH and high aluminum concentrations) have been affected at the cellular, tissue, or organ level. What would you examine to provide the answer to this question? What changes might you expect? Link these biomarkers to consequences to individuals and to the trout population.

CHAPTER 8

1. Define the terms highlighted in the chapter and check your definitions against those provided in the glossary.
2. What sublethal effects are most often studied by ecotoxicologists?
3. What is Selyean stress? Which of the following responses is not associated with Selyean stress: increased adrenal size, induction of EROD (ethoxyresorufin O-deethylase) activity, kidney damage by Cd, elevated blood pressure due to an emotional shock, or depletion of secretory granules in cells of the adrenal cortex?
4. A data set for growth retardation of an endemic, endangered fish suggests that concentrations of zinc in the range measured in your discharge consistently stimulate growth, but local officials responsible for regulating your discharge insist on a linear-no-threshold extrapolation of effect down to no more than a 5% reduction in growth (i.e., extend the dose-effect line downward from high-effect concentrations to a concentration that is predicted to have only a 5% reduction in growth based on a linear-no-threshold model). Extrapolation predicts 5% inhibition, but your data set suggests 20% enhancement of growth. Give your reasoning in deciding to conform to or contest the regulators' decision.
5. You are told that voles near a contaminated site have necrotic lesions in their livers. Discuss one reason why this observation may suggest that abnormal development will also be occurring in the young of this rodent.
6. Hatchlings of an endangered and long-lived sea turtle are closely monitored by state fish and wildlife technicians. Data from 20 years ago describe a sex ratio of 50:50 male:female of hatchlings for many populations of these turtles. Now, almost all turtles are females. Describe what you would

do to determine the cause of this apparent shift and what you might do to counterbalance the shift. Should you intercede?

7. Periodic episodes of meadow lark mortality are noted near an agricultural field. The field was used for many years to grow cotton and, consequently, has a history of DDT and lead arsenate application. At present, carbamate pesticides are periodically applied to the fields. How would you determine which, if any, of these potential toxicants is producing the mortality?

8. Why are behavioral abnormalities used less often than other sublethal effects in the assessment of contaminant effects to individuals?

9. Tabulate the advantages and disadvantages of the post-ANOVA methods described in this chapter.

10. Compare and contrast the hypothesis testing and regression methods for analyzing sublethal effects data.

11. Given the following table of NOEC and LOEC values for toxicant X, estimate the concentration ("safe concentration") that will protect an endangered siren (amphibian) living below a continuous discharge containing toxicant X.

| | | Concentration (μg l^{-1}) | | |
| | | NOEC[a] | LOEC[b] | Exposure Duration (days) |
Species	Effect			
Fathead minnow	Growth	1	10	14
Bluegill sunfish	Egg hatch success	5	46	7
Leopard frog	Tadpole growth	0.09	0.60	7
FETAX assay	Heart development	0.17	1.7	4
Daphnia magna	Fecundity	0.001	0.008	7

[a] No observed effect concentration.
[b] Lowest observed effect concentration.

CHAPTER 9

1. Define the terms highlighted in the chapter and check your definitions against those provided in the glossary.

2. The most critical (sensitive) life stage of a particular invertebrate is its first instar, which has an incipient LC50 (median lethal concentration) of 10 μg l^{-1} and a NOEC of 5 μg l^{-1} for growth for toxicant X. Can a viable population of this invertebrate be maintained below an effluent with a maximum concentration of 5 μg l^{-1} for this toxicant?

3. Compare the advantages and disadvantages of static, static-renewal, and flow-through (continuous and intermittent) tests.

4. What are the relative advantages of the dose- or concentration-response versus the survival time approach to measuring toxicity?

5. A channel in a heavily industrialized estuary must be deepened. How would you assess the potential for acute and chronic toxicity to endemic species resulting from the dredging activities?

6. Defend the concept of individual effective dose.

7. You are required to accurately estimate the LC50, LC10, and LC5 for a toxicant. Describe the methods you would use to calculate these numbers. Give the reasons for selecting one approach over another.

8. Would normit and probit analyses (MLE [maximum likelihood estimation] method) of a data set produce the same LC50 estimate? Would they produce the same LC5 estimate? Would the probit and logit analyses produce the same LC50 and LC5 estimates? Give reasons for each of your responses.

9. Describe how you would estimate toxic incipiency.

10. What quantitative models would you use for the joint action of two toxicants with identical modes of lethal action? Which model would you use for the joint action of two toxicants with completely different modes of lethal action? How would you determine if their joint effects were additive or non-additive?

11. What is the difference between concentration additivity and effect additivity?
12. Could a pair of chemicals that are functional antagonists also be receptor antagonists? Could they also be dispositional antagonists?
13. What is the difference between the proportional hazard model and the accelerated failure time model? Are there any differences in the data collected to fit these two models?
14. Describe the Bliss method of accounting for the effect of individual size on toxic effect.
15. What is the difference between photo-induced toxicity and photosensitivity?
16. What is a split probit? What factors would result in a split probit plot?

CHAPTER 10

1. Define the terms highlighted in the chapter and check your definitions against those provided in the glossary.
2. Review the previous chapters and identify ten specific risk factors. Five should be qualities of individuals and five should be etiological factors.
3. Liver tumors in flounder are surveyed in a contaminated bay. Calculate the incidence rate as cases per 1000 flounder-years of exposure. Assume the tumors kill each diseased individual within the sampling year. What was the mean prevalence rate over the five years of sampling?

Year	Number of Flounder Sampled	Number of Diseased Flounder
1	1583	112
2	2032	100
3	1911	216
4	2118	96
5	1625	97

4. The population in the above table is compared with a reference population ("control") that has the qualities tabulated below. Use the mean incidence rate for both populations to calculate the rate ratio. Assuming that the contamination in the bay was the "treatment," calculate the odds ratio for each of the five years of the survey.

Year	Number of Flounder Sampled	Number of Diseased Flounder
1	3008	3
2	5025	11
3	4091	7
4	3557	6
5	4253	17

5. Describe the nine aspects of disease association. Include an example of each.
6. Explain the predisposition of ecotoxicologists to using laboratory sublethal and lethal test results to predict the viability or distribution of populations about a contaminated field site.
7. In one year, a population increased from 100 to 150 individuals. Calculate λ and r for this population for that time. Assuming no change in growth dynamics, calculate the population size in 20 years of growth beginning with 100 individuals.
8. How could knowledge of V_A (reproductive value) for the age classes of a population aid in estimating the viability of a population exposed to a toxicant over long periods of time?
9. Explain how metapopulation processes might influence the outcome of chronic exposure of a metapopulation to an environmental toxicant.
10. State with reasons whether individual- or population-based effect metrics would be more sensitive to the effects of an environmental toxicant.

11. Using the principle of allocation and concept of strategy, relate how life histories of individuals in a population can shift in response to a toxicant.
12. Contrast how a stressor might influence longevity of an individual under the disposable-soma and stress theories of aging.
13. How might a toxicant influence the genetics of a population?
14. What factors influence the rate of tolerance acquisition in a population?
15. Describe the concept of selection components.

CHAPTER 11

1. Define the terms highlighted in the chapter and check your definitions against those provided in the glossary.
2. What are the differences among the terms *community, species assemblage*, and *guild*?
3. Describe the community-level NOEC concept, including its advantages and disadvantages.
4. How might optimal foraging theory help us to understand the effects of toxicants on populations in a community?
5. Describe the factors contributing to the ability of a community to successfully avoid irreversible damage by a pollutant.
6. What is the major difference between the Shannon and Brillouin indices of species diversity? Which one will have the largest value for any sample?
7. Describe the rivet popper hypothesis versus redundancy hypothesis debate. What is the significance of this debate relative to assessing an adverse effect of pollutants to ecological communities?
8. In general, how would the log-normal, species abundance curve change with the introduction of a toxicant to an ecosystem? What is (are) the possible mechanism(s) for this shift in the curve?
9. What are the differences between microcosms and mesocosms? What are the relative advantages and disadvantages of these tools?
10. Assuming that the MacArthur-Wilson model accurately reflects colonization, predict the number of species inhabiting a beach devastated by an oil spill if the number of species originally on the beach was 23 ($S_{EQ} = 23$?), $G = 0.10$ (units of years), and 10 years has passed since the devastation occurred. How long would it take for 95% of the species to reestablish themselves? Would the community (species assemblage) be back to normal at that point?

CHAPTER 12

1. Define the terms highlighted in the chapter and check your definitions against those provided in the glossary.
2. Give three reasons why the traditional ecosystem context for ecotoxicology should be expanded to a larger scale in order to address ecotoxicological problems facing us today.
3. What factors would influence the sensitivity of a watershed to acid precipitation? What effects might you expect for aquatic and terrestrial biota in a watershed with a granite bedrock geology?
4. How does the increased use of CFCs (chlorofluorocarbons) result in decreased ozone levels in the stratosphere above the Antarctic? Why is the problem less noticeable above the Arctic?
5. Describe the processes by which POPs (persistent organic pollutants) are transported and differentially distributed from the equator to the earth's poles.

CHAPTER 13

1. Define the terms highlighted in the chapter and check your definitions against those provided in the glossary.
2. What is the major difference between the risk assessment principle and the precautionary principle for making management decisions?

3. What are the four components of the NAS (National Academy of Sciences) paradigm for risk assessment? What are their specific goals?

4. What is the major distinction between a hazard assessment and a risk assessment? As described in this chapter, is the assessment done for noncarcinogenic effects hazard or risk assessment? Explain your answer.

5. Outline the steps in a human risk assessment. Check your outline against that shown in Figure 13.2.

6. Groundwater below a closed landfill contains high concentrations of trichloroethylene (TCE), and the contaminated water is predicted to outcrop to a stream within the next two years. You are asked to do an assessment of the situation. Are you doing a predictive or retroactive risk assessment? Give the reasons for your answer.

7. What is the function of uncertainty factors? Is it the same as the slope factor? What is the function of the modifying factor?

8. Get onto IRIS (EPA's Integrated Risk Information System, [http://www.rtk.net/T866]) to find the RfD (reference dose) and associated information for hexavalent chromium, i.e., chromium (IV). If the concentration in drinking water taken from a representative well near a hazardous waste site was 6 mg l^{-1}, what would the hazard quotient be? Provide the qualifiers needed by the risk manager in order to use this calculated value intelligently.

9. Contrast and give examples of assessment and measurement endpoints. What are the qualities of good assessment and measurement endpoints? Give some good and bad examples of both endpoints.

10. Describe the steps and products of an ecological risk assessment.

CHAPTER 14

1. Define the terms highlighted in the chapter and check your definitions against those provided in the glossary.

2. Why are alpha particles of little concern from an external irradiation standpoint but of major concern if inhaled?

3. Assume that one batch of milk is contaminated with 1000 Bq of ^{131}I and another batch with 1000 Bq of ^{134}Cs. The half lives of ^{131}I and ^{134}Cs are 8 days and 2.4 years, respectively. Recognizing that the milk is contaminated, we opt to make cheese from it and store the product until it is safe for human consumption (which we will arbitrarily set at a level of 100 Bq). How long would you have to store each batch of cheese before it has a safe activity? Hint: Convert Equation 14.3 into natural logarithms and solve for t.

4. Explain the concept of a screening-level model.

5. A hypothetical individual is exposed to a ^{137}Cs plume of 900 kBq m^{-3}. The individual is outside, directly in the plume for 30 h, and indoors for an additional 100 h. The residence structure provides a shielding factor of 0.1. The air-dose coefficient for ^{137}Cs is 7.74 10^{-18} Sv-m^3 /Bq-s. Calculate the probability that this individual will acquire a fatal cancer from the external exposure.

6. How do risk analyses for human exposures to radioactive contaminants differ from exposure to nonradioactive contaminants? Consider the pathways of contaminant transport and derivation of the risk factors.

7. Which provides the more rigorous quantitative answer—human or ecological risk analyses? Explain why.

8. Why do some radiation protection agencies think that if levels of radiation are kept low enough to protect humans, then the natural biota living in the same contaminated environment should be adequately protected as well?

International System (SI) of Units Prefixes

International System (SI) of Units Prefixes

Factor	Unit Prefix (Symbol)	Factor	Unit Prefix (Symbol)
10^{12}	tera- (T)	10^{-1}	deci- (d)
10^{9}	giga- (G)	10^{-2}	centi- (c)
10^{6}	mega- (M)	10^{-3}	milli- (m)
10^{3}	kilo- (k)	10^{-6}	micro- (μ)
10^{2}	hecto- (h)	10^{-9}	nano- (n)
10^{1}	deka- (da)	10^{-12}	pico- (p)
		10^{-15}	femto- (f)

Miscellaneous Conversion Factors

Conversion Factors for Miscellaneous Units

acre	4.046873×10^3 meter^2 (m^2)
atmosphere (atm)	1.013250×10^5 pascal (Pa)
bar	10^5 pascal (Pa)
barrel of oil (bbl)	4.2×10^1 gallons (gal) $= 1.589873 \times 10^{-1}$ meter^3 (m^3)
becquerel (Bq)	1 disintegration per minute (dpm) $= 27.027 \times 10^{-12}$ Ci
British thermal unit (Btu)	1.055×10^3 joule (J)
calorie (Cal)	4.186800 joule (J)
cubic foot (ft^3)	2.831685×10^{-2} cubic meter (m^3)
curie (Ci)	2.2×10^6 disintegrations per minute (dpm) $= 3.700 \times 10^{10}$ Bq
Dalton (Da)	an arbitrary atomic mass unit that is 1/12 the mass of a carbon atom with mass number of 12
degrees Celsius (°C)	degrees Kelvin (K), -273.16°C
degrees Celsius (°C)	$(5/9) \times$ (degrees Fahrenheit $- 32$)
electron volt (eV)	1.602×10^{-12} erg $= 1.602 \times 10^{-19}$ Joule
foot (ft)	3.048000×10^{-1} meter (m)
gallon (gal)	3.785412×10^{-3} cubic meter (m^3)
Gray (Gy)	1 joule (J) of energy deposited per kg of tissue
liter (l) of crude oil	≈ 868.63 gram (g) based on the density of Kuwait crude oil at 20°C
micron (µm)	1×10^{-6} meter (m)
mile (mi)	1.6093×10^3 meter (m)
millibar (mbar)	100 pascal (Pa)
millimeter of mercury (mm Hg)	1.33322×10^2 pascal (Pa)
pound (lb)	4.535924×10^{-1} kilogram (kg)
rad, absorbed radiation dose (rd)	1×10^{-2} Gray (Gy)
rem, Radiation dose equivalent	1×10^{-2} sievert (Sv)
Roentgen (R)	2.58×10^{-4} coulomb (C) per kg of air
Sievert (Sv)	100 mrem
short ton (ton)	2×10^3 pound (lb) $= 9.071847 \times 10^2$ kilogram (kg) $= 0.907$ tonne (t)
torr (mm Hg)	1.33322×10^2 pascal (Pa)
yard (yd)	9.144×10^{-1} meter (m)

Source: Extracted from Lide, D.R., *CRC Handbook of Chemistry and Physics*, 73rd ed., CRC Press, Boca Raton, FL, 1992; and miscellaneous sources.

Summary of U.S. Laws and Regulations

Laws and regulations are important aspects of practical ecotoxicology that deserve discussion. Greater detail can be acquired using the suggested readings at the end of Chapter 1 and the references provided here for the specific laws. Discussion is limited to U.S. federal law, with European regulations being covered in Appendix 4. Laws pertinent to Canada and the **European Community** (EC) are also outlined in Foster (1985) for water, Elsom (1987) for air, and Frankel (1995) for marine resources. European community activities and the **Organization for Economic Cooperation and Development** (OECD) policies are described in Blok and Balk (1995), Grandy (1995), and Crane et al. (2002b).

Once the U.S. law is established, a federal agency such as the EPA is charged with developing the associated regulations that are intended to enforce the particulars of the law. When EPA is so charged, its proposed rules are published initially in the *Federal Register (FR)* and remain open for public comment for a specified period of time. After that time has elapsed, EPA repeats the internal process of rule development with consideration of any public comment and other input. Final rules are then established and published in the *FR*. These regulations may be open to interpretation by affected parties, but redress occurs through the courts after this point in the process. During the next revision of the *Code of Federal Regulations (CFR)*, the regulations are incorporated into this compilation of federal code.

The acts and laws established by the U.S. Congress are identified using references such as 42 USC §§6901 to 6992k or PL 91-190 for the National Environmental Policy Act (NEPA). USC and PL are abbreviations for U.S. code and public law, respectively. As mentioned, a **law**[1] requires a federal agency such as the EPA to develop **regulations** that provide specific details ensuring that the intent of the law is met. According to Mackenthun and Bregman (1992), regulations "specify conditions and requirements to be met, ... provide a schedule for compliance, and ... record any exceptions to the regulated community." Within the text of these materials, legal definitions are also provided for terms such as contaminant (40 CFR Sections 230.3, 300.6, 310.11, etc.), point source (40 CFR Sections 233.3 and 260.10), or pollutant (40 CFR Section 122.2) that may vary among laws and deviate from scientific definitions. Therefore, it is important to understand the legal context of such terms. Legal dictionaries of environment-related terms (e.g., King, 1993) should be consulted to avoid confusion when working under a specific set of regulations. Regulations from EPA are published in the *CFR* under Title 40 (Protection of Environment) and updated as needed in the *FR*. The daily issuances of the *FR* augment the *CFR*, and consequently, these

[1] Although we are discussing only federal law here, **environmental law** is much more inclusive. McGregor (1994) defines environmental law as "a body of federal, state, and local legislation, in the form of statutes, bylaws, ordinances, and regulations, plus court-made principles known as common law."

sources together define current federal regulations. These specific codifications of environmental regulation are referenced with numbers such as 40 CFR Parts 1500 to 1517. The numbers 1500 to 1517 identify specific parts of Title 40 of *CFR*, in this case, a section providing details on the National Environmental Policy Act (NEPA).

With this background established, specific laws and associated regulations can now be outlined. Each is presented with relevant references noted from the *CFR* and *FR*.

National Environmental Policy Act or NEPA (1969, PL 91-190; 42 USC §§6901–6992k; see 40 CFR 1500–1517) was signed into law on January 1, 1970, by President Richard Nixon. Combined with the Environmental Quality Improvement Act (EQIA, 1970), NEPA formed the cornerstone for U.S. environmental policy (Foster, 1985). NEPA established a federal commitment to judge the government's actions relative to environmental impacts and to take action fostering "harmony" between human activities and nature. It established the **Council on Environmental Quality** to aid and advise the president during preparation of the annual **Environmental Quality Report**. This council assesses federal programs relative to NEPA and researches various topics on environmental quality.

NEPA requires federal agencies to prepare an **environmental impact statement** (EIS) for any major federal action that could have an adverse environmental impact. In addition to direct actions, federal actions include issuance of federal permits, leases, or licenses and the granting, contracting, or loaning of federal monies (McGregor, 1994). NEPA does not cover nonfederal actions, and there are no defined rights of individuals to live without environmental damage included in NEPA (Freedman, 1987). The EIS outlines possible adverse impacts; details alternatives to the action, including the "no-action" alternative; and describes irreversible impacts to resources (McGregor, 1994). Prior to an EIS, a shorter and preliminary **environmental assessment** (EA) can be used to determine if a full EIS is necessary. A statement of no significant effect ("**finding of no significant impact**" or FONSI) is made if no EIS is thought to be needed. Otherwise, a full EIS is developed. This process of incorporating environmental impact into the decision-making of the federal government is open to public comment and involves relevant state, local, and federal agencies, with the EPA overseeing much of the process.

The **Clean Air Act** or CAA (1977, PL 9595, 42 USC §§7401–7671q; see 40 CFR50–99; see also 1990 Clean Air Act Amendments [CAAA] PL 101-549, 104 Stat. 2399) is designed to regulate air pollution so as to protect human health and the environment. It is overseen and administered by the EPA, but responsibility for control and prevention of air pollution is passed to state and local governments. This act sets maximum allowable levels of pollution and dictates emission levels designed to achieve these limits. It establishes national emission standards for pollutants. Stationary sources are controlled by a state permit system. As detailed in the CAA, mobile sources are regulated through processes such as motor vehicle emissions standards or vehicle inspection programs. States are required to develop **state implementation plans** (SIP) outlining steps toward meeting and maintaining **national ambient air quality standards** (NAAQS). Section 109 of this act requires the EPA to establish and periodically revise NAAQS. The act also includes details for **prevention of significant deterioration** (PSD) in geographical areas that have attained the NAAQS.

The **Federal Insecticide, Fungicide and Rodenticide Act** or FIFRA (1972, 7 USC 136, PL 95396; see also 40 CFR 150–189) controls the production, labeling, shipping, sale, and use of **pesticides** ("substances used to prevent, destroy, repel, or mitigate any pest") (Mackenthun and Bregman, 1992). Pesticides may be individual substances or mixtures including plant regulators, insecticides, fungicides, rodenticides, desiccants, and defoliants (Freedman, 1987). FIFRA defines certification of restricted-use pesticide applicators, defines the conditions for canceling the registration for a pesticide, and establishes pesticide disposal requirements (Mackenthun and Bregman, 1992). FIFRA requires submittal of materials for pesticide registration, which includes the specific labeling and directions for use and the results of product tests. Specifics of ecological effects testing and environmental fate are given in 40 CFR 158.145 and 40 CFR 158.130, respectively (Touart, 1995). Submitted information must demonstrate that the product will work as intended without unacceptable human or ecological risk. Registration is renewed periodically and, if a pesticide is found after registration to cause unreasonable risk, the EPA can cancel its registration under Section 6 of FIFRA (see 40 CFR 154).

The **Marine Protection, Research and Sanctuaries Act** or MPRSA (1972, 33 USC §1401; see 40 CFR 220–233) recognizes that the practice of ocean dumping is unsound; states the U.S. policy regarding dumping; and establishes limitations and prohibitions on dumping. It prohibits or limits dumping of various materials from the United States, or by U.S. vehicles (Freedman, 1987). The EPA limits dumping of wastes except dredge spoils through a permit system. The Army Corp of Engineers uses EPA criteria and regulates dredge-spoils dumping (Mackenthun and Bregman, 1992). The Ocean Dumping Ban Act (1988) amended the MPRSA and put a time limit on dumping of materials such as sewage sludge, industrial waste, and medical waste (Mackenthun and Bregman, 1992). Also prohibited is the dumping of radioactive waste and radioactive, chemical, or biological weapons material (Freedman, 1987).

The **Clean Water Act** or CWA (1977, PL 95-217, 33 USC §§1251–1387; see 40 CFR 100–140, 400–699; see also PL 100-4, Feb. 4, 1987 amendments) limits the discharge of pollutants (primarily industrial and municipal discharges) into navigable waters. This is accomplished by imposing limits on discharges using either a state or EPA permitting system (**National Pollutant Discharge Elimination System** or NPDES). State limits are developed under the guidance of the EPA. The immediate goal of this legislation was to improve water quality to insure that waters are swimmable and fishable (McGregor, 1994). The overall goal is to "restore and maintain the chemical, physical, and biological integrity" of U.S. water bodies.

The **Safe Drinking Water Act** or SDWA (1974, 42 USC §§300f–300j-26; see 40 CFR 141–149; amended in 1986) protects the public water supply (systems supplying more than 25 people or possessing more than 15 service outlets) by setting quality standards, and it protects source aquifers from contamination by establishing permitting requirements for underground disposal of waste. The EPA has oversight, but states are assigned responsibility for enforcement. The EPA establishes primary and secondary "at the tap" standards for public water, which the states can either use without modification or use to establish even more stringent standards (Mackenthun and Bregman, 1992). Primary standards ("**maximum contaminant levels**" or MCL) are those for a list of specific contaminants that adversely affect health; secondary standards are EPA recommendations for qualities that protect the public welfare (e.g., water taste, color, smell, etc.) (McGregor, 1994). The present MCL list includes qualities under the categories of volatile organic compounds, organic compounds, and inorganic chemicals, including radionuclides, microbiology, and turbidity (Mackenthun and Bregman, 1992). This list of substances of concern is reviewed and revised every three years by the EPA. In the 1986 amendment, a ban that specified limits was added to restrict lead use in public water system components such as solder and fixtures.

Disposal of wastes into potential underground drinking-water sources is also closely controlled under the SDWA through the use of a permit system or designation of aquifers as "sole-source aquifers" to be protected. Federal activities are restricted in areas with sole-source aquifers. The EPA also sets strict regulations for underground waste-injection activities under the SDWA.

The **Toxic Substances Control Act** or TSCA (1976, 15 USC §2601; see 40 CFR 700–799) regulates, through the EPA, the manufacture, processing, transport, use, import, and disposal of chemicals or chemical mixtures that may pose an unreasonable risk to human health or the environment. Consideration is given to the quality and quantity of substances being produced. It does not cover items already regulated by other laws such as cosmetics, drugs, firearms and ammunition, food additives, pesticides, and nuclear materials (Mackenthun and Bregman, 1992). However, it does include products of biotechnology such as genetically altered microbes intended for release into the environment. It specifically regulates some existing chemicals not already regulated by other laws, e.g., PCBs. The EPA requires producers to give 90-day notification of intent to manufacture or import, or to put a chemical product to new use. It requires testing to clarify the potential for unreasonable risk and a statement of the intended use of the material.

The EPA may then permit, deny permission within 90 days of notice, or delay permit issuance until further study has been completed. The EPA may inspect facilities or investigate any actions that may violate this law. The law provides for criminal liability under specific conditions (i.e., knowing and willful violation) (McGregor, 1994). There are also record-keeping requirements and an obligation to notify the EPA if new information arises relative to the risk associated with the chemical or mixture (Freedman, 1987).

The **Resource Conservation and Recovery Act** or RCRA (1976, 42 USC §§6901–6987; see 40 CFR 240–299 and 1984 Hazardous and Solid Waste Amendments, PL 98-616) regulates production, storage, treatment, and final disposal of hazardous wastes (McGregor, 1994), i.e., "cradle-to-grave" regulation. The EPA is required to develop standards for generation, transport, storage, and treatment of hazardous waste and to establish a permit system for storage, treatment, and disposal (Mackenthun and Bregman, 1992). Disposal requirements for small-quantity generators, such as research laboratories, are also defined by this law. RCRA requires accurate record-keeping. With the 1984 amendment, clear legal penalties were specified for knowing violation of a permit, illegal transport or treatment of waste, or falsifying or altering of records. Disposal of waste into underground formations was also restricted or prohibited by the 1984 amendment.

The **Comprehensive Environmental Response, Compensation and Liability Act** or CERCLA or Superfund Act (1980, 42 USC §§9601–9615, 9631–9633, 9651–9657, 26 USC §§4611–4612, 4661–4662, 4681–4682; see 40 CFR 300–374, amended in 1986 with the Superfund Amendments and Reauthorization Act or SARA) gives EPA the authority to respond to hazardous releases, cleanup, or require cleanup of hazardous-waste sites and to assign liability for released hazardous waste (Mackenthun and Bregman, 1992). Although costs are first sought from responsible parties, CERCLA also established two funds (Hazardous Substance Response Trust Fund and Post-Closure Liability Trust Fund) with various tax sources (oil, chemical, and hazardous waste activities) to cover some cleanup costs (Freedman, 1987). CERCLA also establishes a **National Contingency Plan** to identify sites for cleanup and to define the required cleanup activities (Freedman, 1987). At this time, the EPA has compiled a listing (National Priority List or NPL) of sites for cleanup under CERCLA.

Summary of European Union Laws and Regulations

Mark Crane

Crane Consultants, Chancel Cottage, 23 London Street, Faringdon, Oxfordshire, U.K.

The European Union (EU), established in 1957 by the Treaty of Rome, differs from the U.S. federal government in that EU member states are sovereign nations that have surrendered only some law-making and enforcing powers. The Treaty of Rome was a tool to establish a common market, expand economic activity, promote living standards, and perhaps most importantly, encourage political stability in Western Europe. The European Community formed by the Treaty of Rome unified the European Economic Community, the European Atomic Energy Community, and the European Coal and Steel Community. This has expanded in both membership and aims over the decades since its formation. The postwar organization, set up primarily for economic and security reasons, has now evolved into the EU with 15 member states: Belgium, Luxembourg, the Netherlands, France, Germany, Italy, the United Kingdom, Ireland, Denmark, Greece, Spain, Portugal, Finland, Sweden, and Austria. More countries, principally from the old Soviet bloc, are likely to join the EU in the future.

There are four major EU institutions responsible for legislation under the Treaty of Rome: the European Commission, European Parliament, Council of Ministers, and European Court. The Commission is the supreme EU executive, comprising independent members appointed by individual member states. Members of the commission are charged with operating in the interests of the community as a whole, not as national representatives. Each commissioner has responsibility for an area of community policy, which includes a commissioner for the environment, and the commissioner's main function is to propose EU legislation. The commission civil servants are divided among different directorates that report to the council of ministers.

Legislation generally emerges from the commission in the form of directives. These are usually enforceable across all member states and are the main basis for statutory controls in EU environmental legislation. They empower the commission to define objectives, standards, and procedures while allowing member states some flexibility in implementation so that they can use their own national legislative processes and develop their own regulations. These national legislative processes can differ fundamentally among the different EU states. For example, in the United Kingdom, case law forms the legal base, while in much of continental Europe legal principles derive from the Napoleonic Code.

The European Parliament is a consultative and advisory body, and its function is to assess proposals for legislation by commenting on the commission's proposals. The council of ministers comprises government ministers from each member state and is the primary decision-making body in the EU, with a main function to consider proposals from the commission. The main function of the European Court is to interpret and apply all community law. A European Environment Agency exists, but this is currently a small body responsible largely for gathering data on environmental indicators from member states so that general trends in Europe can be reported. The practical aspects of environmental policy implementation, monitoring, and enforcement fall largely to individual agencies within the member states.

Legislation adopted by the EU over the past two decades can be divided into three broad categories:

1. Directives that try to limit or prohibit discharges of dangerous substances into the environment by industrial plants
2. Directives and regulations setting environmental quality objectives for various uses of the environment (e.g., use of potable water, protection of shellfisheries, or protection of the wider aquatic environment)
3. Geographically specific regulations on pollution to help protect areas such as the North, Baltic, and Mediterranean Seas

A brief review of some important EU environmental directives is presented below.

The marketing of new chemicals was brought within European legislation under 1967 Council Directive 67/548 on the Classification, Packaging and Labelling of Dangerous Substances. The purpose of this was to ensure that new chemicals coming onto the European market in sufficient volumes would be given a classification and hazard label when being transported into or within countries. However, it was only with the sixth amendment to Council Directive 67/548 in 1979 that a series of biological, physical, and chemical tests became a requirement before new chemicals could be marketed. The amount and type of information required for each chemical depends on its production volume. Testing within this scheme comprises three levels: Level 0 (or base level) is for production volumes of up to 10 t/year and requires acute-toxicity data for *Daphnia magna* and fish (e.g., trout) or algal-growth-inhibition tests. These are known as base-level tests in the EU. For higher production volumes of up to 1000 t/year (Level 1) and greater than 1000 t/year (Level 2), more thorough toxicological testing is required, such as prolonged toxicity to *Daphnia* and freshwater fish.

The Plant Protection Products Directive (Directive 91/414/EEC) came into force in 1991. The purpose of this directive is to harmonize arrangements for the marketing of plant pesticides within the community. Many of the data requirements for this directive are similar to those required for new substances, e.g., toxicity profiles for fish and algae, but higher-tier testing, including the use of microcosms, is more likely because pesticides are poisons that are deliberately spread in the environment at lethal concentrations. A risk assessment of product impacts on humans and wildlife must be carried out for both new and existing pesticides. There are also directives covering biocides and veterinary medicines, which contain many similar provisions for environmental protection to the Plant Protection Products Directive, including risk assessments.

Council Regulation 793/93 on existing chemicals was developed to harmonize the different national systems for risk assessment within the member states and to bring about the control of substances that are not new to the market. The regulation came into effect in 1993 and covers about 100,000 chemical substances thought to be used in the EU. Under the regulation, information must be provided to the commission by industry for substances produced or imported into the EU in quantities over 1000 t/year. The information collected by the commission is then used for setting priorities on the basis of a preliminary risk assessment. The priority substances are assessed by distributing them among "reporting" member states. The reporters in the reporting member states can ask for additional information from manufacturers and importers if the substance is suspected to be dangerous. On the basis of these data and more refined risk assessments, a substance can be deemed as dangerous by the commission and subsequently banned or restricted. The risk assessment methodology that should be used is described in the technical guidance document (TGD) (European Union, 1996). This largely consists of estimating a reasonable worst-case predicted environmental concentration (PEC) for each environmental compartment and potential exposure scenario and then comparing this with a predicted no-effect concentration (PNEC). The PNEC is usually estimated by finding the most-sensitive and reliable toxicity value that is relevant for the environmental compartment of concern and then applying a safety factor to this value to account for various uncertainties, such as interspecies differences in sensitivity, acute-to-chronic ratios, and laboratory-to-field extrapolations. The use of species sensitivity distributions (Posthuma et al., 2002) to establish a PNEC is also allowed under the TGD. The EU existing substances regulations are coordinated with the Organization for Economic Cooperation and Development (OECD) program for developing test guidelines. The OECD is the main coordinating body for the development of new ecotoxicity testing strategies and specifies tests that then become mandatory data requirements in EU directives.

The 1976 Council Directive 76/464 on dangerous substances was the first to provide a framework for the elimination or reduction of historical water pollution in both freshwater and marine systems by particularly dangerous substances. Member states were required to take appropriate steps to eliminate pollution by 129 toxic and bioaccumulable List I substances (otherwise known as blacklist substances) and to reduce pollution by List II substances (otherwise known as gray-list substances). List I contains organohalogen and organophosphorus compounds, organotin compounds, carcinogenic substances, and mercury and cadmium compounds. List II includes biocides not included in List I, metalloids/metals and their compounds, toxic or organic compounds of silicon, inorganic compounds of phosphorus, ammonia and nitrites, cyanides and fluorides, nonresistant mineral oils, and hydrocarbons of petroleum origin. List II compounds are regarded as less dangerous to the environment than List I compounds and may be contained within a given area, depending on the characteristics and location of that area.

The Dangerous Substances Directive requires member states to draw up authorization limits for emissions of substances on both lists. In the case of List I, the limit values are to be at least equivalent to those adopted by the commission. So far, several daughter directives have established emission limits for specific substances on List I, including mercury, cadmium, hexachlorocyclohexane, carbon tetrachloride, DDT, and pentachlorophenol. For List II substances, member states are required to set water quality objectives and prior-authorization requirements on industry to reduce pollution from these substances. The groundwater directive (80/68/EEC) extended the List I and List II approach to protect groundwater resources, although action to implement this directive by member states has generally been poor.

Several directives have focused on protection of specific aquatic systems. For example, the Shellfish Water Directive (79/923/EEC) is intended to protect and improve the quality of coastal and brackish waters designated by member states for shellfish growth. Council Decision 75/437/EEC approved the Paris Convention. The aim of this is to prevent marine pollution from land-based sources (i.e., that emanating from watercourses, underwater pipelines, and ports) to the northeast Atlantic and Arctic Oceans, the North and Baltic Seas, and parts of the Mediterranean. This was extended in 1986 to cover marine pollution by emissions into the atmosphere. In the mid 1990s, a proposed Ecological Quality Directive (COM (93) 0680) defined European standards for surface water by criteria other than those already subject to existing directives. It required member states to define quality objectives for all surface water, to set up water-monitoring and control systems together with an inventory of pollution sources, and to prepare and implement a series of programs to improve water quality. However, this proposed directive was ultimately rejected because it did not appear to add coherence to overall water policy in the EU.

The fragmentary nature of these and other efforts to prevent or reduce aquatic contamination and pollution has been the driving force behind a recent and important piece of legislation that seeks to draw all of these strands together, the Water Framework Directive (2000/60/EC). This directive contains both environmental quality standards and emissions limit values from point sources and will repeal much of the earlier water quality legislation described in this appendix. The main principles of the Water Framework Directive are:

1. Expand the scope of water protection. All of Europe's waters will be subject to protection under the Water Framework Directive. Unlike previous water legislation, the directive covers surface water, groundwater, estuaries, and marine waters.
2. Achieve good status for all waters, except for limited derogations. Good status for surface waters is measured in terms of ecological and chemical quality. Member states will need to establish programs for monitoring these criteria.
3. Base water management on river basins rather than administrative or political boundaries. For each river basin district, a river basin management plan must be established and updated regularly. Coastal waters are assigned to the nearest or most appropriate river basin district.
4. Establish a program of emission limit values, water quality standards, and any other necessary measures. Within the framework of river basin management plans, member states will need to establish a program of measures to ensure that all waters within a river basin achieve good water status. The first step in achieving good water status is the full implementation of relevant national or local legislation as well as EU legislation. If this basic set of measures is not sufficient to reach a good status, the program must be supplemented with whatever further measures are necessary. This can include stricter controls on polluting emissions from industry, agriculture, and urban wastewater sources. The directive takes a combined approach to pollution control in two main ways: firstly by limiting pollution at the source by setting emission limit values, and secondly by establishing water quality objectives for water bodies. In each case, the more stringent approach will apply.
5. Get the citizen involved. The establishment of river basin management plans will require more involvement of, and consultation with, EU citizens, interested parties, and nongovernmental organizations.

The historical use of aquatic systems as repositories for human wastes goes some way toward explaining why so much emphasis has been placed on aquatic pollution prevention in EU law. However, air and land pollution have not been entirely ignored. Framework directives exist for waste disposal (84/360/EEC) and large combustion sites (88/609/EEC), or waste incinerators. Limits for air pollution have been set for lead, oxides of nitrogen, sulfur dioxide, and particulates. These guidelines are intended to protect both human health and to serve as long-term safeguards for the environment. There are also draft atmospheric limit values for volatile organic compounds (as total organic carbon), dioxins and furans, and for inorganic elements and compounds such as chlorine, fluorine, ammonia, and metals.

Finally, several directives on nature conservation have implications for national agencies in member states with responsibilities for regulating the environmental impacts of chemicals in either aquatic or terrestrial environments. The Habitats Directive requires member states to take measures to maintain or restore natural habitats and wild species that are of nature conservation value. Special areas of conservation in each member state have been agreed with the commission, along with necessary measures to protect them. The Birds Directive has a similar nature conservation objective, but with an obvious focus on bird species, their eggs, nests, and wider habitats. The Directive on the Assessment of the Effects of Certain Public and Private Projects on the Environment requires an environmental assessment to be carried out before a decision is taken on whether development consent should be granted for certain types of projects that are likely to have significant environmental effects. Clearly, consideration of the potential effects of toxic chemicals falls under each of these nature conservation directives.

In summary, the EU has developed many directives that seek to protect the environment. These piecemeal legal instruments are gradually being drawn together to provide a coherent framework for controlling the marketing and use of chemicals in the EU. This has not prevented individual member states from implementing their own antipollution legislation to meet local environmental or political concerns, but the general move is away from individual member state initiatives and toward environmental regulation through harmonized procedures throughout the EU.

Derivation of Units for Simple Bioaccumulation Models

Peter Landrum
NOAA Great Lakes Research Laboratories, Ann Arbor, MI

Assume a closed system containing water and a fish in which both the amounts in the fish and water are changing over time. The total amount in the system (A) is $Q_w + Q_f$, where the subscripts w and f refer to water and fish, respectively, and Q denotes the amount (mass) in the compartment. There is no direct reference to the size of the fish or water compartments. The amount in the fish can be described by Equation A5.1:

$$\frac{dQ_f}{dt} = k_1 Q_w - k_2 Q_f \tag{A5.1}$$

where k_1 = rate constant for the fractional reduction in the amount in the water (fraction of mass t^{-1}) and k_2 is the rate constant for the fractional reduction in the amount in the fish (fraction of mass t^{-1}). These are conditional rate constants because they depend on the specific experimental conditions, i.e., the sizes of the compartments. For example, assume two systems differing only in the relative sizes of the associated fish and water compartments. If one system had fish and water compartments of identical size, the rate constants would be different from another system in which the fish compartment was much smaller than the water compartment. Although the flux of material into the fish is the same for the two systems, the fractional reduction in the amount in the waters for the two systems (i.e., the rate constants) would be different. A larger fraction of the amount in the water of the system with equal compartment sizes for the fish and water would be removed per unit time than from the system with the much larger water compartment.

The integrated form of Equation A5.1 is Equation A5.2:

$$Q_a = \frac{k_1 A(1 - e^{-(k_1 + k_2)t})}{k_1 + k_2} \tag{A5.2}$$

The units analysis for this is presented in Equation A5.3:

$$Q_a(ng) = \frac{k_1(h^{-1})A(ng)(1 - e^{-(k_1(h^{-1}) + k_2(h^{-1}))t(h)})}{k_1(h^{-1}) + k_2(h^{-1})} \tag{A5.3}$$

which reduces to ng = ng. Now, the information is referenced to the size of the fish and the concentration in the water. The formulation of the equation for the change of concentrations is given in Equation A5.4:

$$C_f = \frac{k_u C_w}{k_e}(1 - e^{-(k_e t)})$$

(A5.4)

Solving for the units,

$$C_f\left(\frac{ng}{g}\right) = \frac{k_u\left(\frac{mL}{gh}\right) C_w\left(\frac{ng}{mL}\right)}{k_e(h^{-1})}(1 - e^{-(k_e(h^{-1})t(h))})$$

(A5.5)

which reduces to ng/g = ng/g. It is clear from this analysis of units that the units for k_u are ml $(g^{-1}\ h^{-1})$ and those of k_e are h^{-1}. The k_u is the clearance rate of the water by the fish of a specific size.

Demonstration of the associated units for k_u can also shown using Equation A5.6, assuming steady state.

$$\frac{dQ_f}{dt} = k_1 Q_w - k_2 Q_f$$

(A5.6)

At steady state, $k_1 Q_w = k_2 Q_f$, or $Q_f/Q_w = k_1/k_2$ for this relationship describing changes in mass. To get to a bioconcentration factor (BCF = concentration in fish expressed as mass per unit mass of fish divided by concentration in water, expressed as mass per unit of volume of water), the relationship just derived between masses and constants is multiplied by volume/mass, as seen in Equation A5.7.

$$\left(\frac{volume}{mass}\right)\left(\frac{Q_f}{Q_w}\right) = \left(\frac{volume}{mass}\right)\left(\frac{k_1}{k_2}\right) = \frac{\dfrac{Q_f}{mass}}{\dfrac{Q_w}{volume}} = \frac{C_f}{C_w}$$

(A5.7)

Rearranging Equation A5.7 and linking the rearranged equation to the relationship between constants for the concentration-based formulation (i.e., k_u/k_e = BCF; see Equation 3.23 for more detail) yields Equation A5.8:

$$\frac{k_1\left(\frac{volume}{mass}\right)}{k_2} = \frac{C_f}{C_w} = \frac{k_u}{k_e}$$

(A5.8)

Because both k_2 and k_e describe the fractional change in chemical in the fish per unit time, they are equivalent. But the relationship between k_1 and k_u is k_1 (volume/mass) = k_u. The units of k_u for the model of change in concentration over time are ml $(g^{-1}\ h^{-1})$, not h^{-1}.

Equations for the Estimation
of Contaminant Exposure

Equations and example values for variables were extracted directly from EPA guidelines (EPA, 1989d). Each is a derivation of the general equation provided in Chapter 13 (Equation 13.3). Note that units may change for variables among the equations.

6.1 IMBIBED VIA DRINKING WATER OR BEVERAGES
MADE WITH DRINKING WATER

$$\text{Intake (mg per kg-day)} = \frac{(CW)(IR)(EF)(ED)}{(BW)(AT)} \qquad (A6.1)$$

where
- CW = concentration in the drinking water (mg l^{-1})
- IR = imbibing or ingestion rate (l day^{-1}), e.g., 2 l day^{-1} for an adult
- EF = exposure frequency (days year^{-1}), e.g., 365 days year^{-1} for a resident
- ED = exposure duration (years), e.g., 70 years for a lifetime
- BW = body weight or mass (kg), e.g., 70 kg for an adult
- AT = averaging time (days), e.g., ED × 365 days for a noncarcinogen or 70 years lifetime^{-1} × 365 days year^{-1} for a carcinogen

6.2 IMBIBED FROM SURFACE WATER WHILE SWIMMING

$$\text{Intake (mg per kg-day)} = \frac{(CW)(CR)(ET)(EF)(ED)}{(BW)(AT)} \qquad (A6.2)$$

where
- CW = concentration in the drinking water (mg l^{-1})
- CR = contact rate (l day^{-1}), e.g., 0.050 l day^{-1}
- ET = exposure time (h event^{-1})
- EF = exposure frequency (days year^{-1}), e.g., 7 days year^{-1} as an average
- ED = exposure duration (years), e.g., 70 years for a lifetime
- BW = body weight or mass (kg), e.g., 70 kg for an adult
- AT = averaging time (days), e.g., ED × 365 days for a noncarcinogen or 70 years lifetime^{-1} × 365 days year^{-1} for a carcinogen

6.3 ABSORBED DURING DERMAL CONTACT WITH WATER

$$\text{Absorbed Dose (mg per kg-day)} = \frac{(CW)(SA)(PC)(ET)(EF)(ED)(CF)}{(BW)(AT)} \qquad \text{(A6.3)}$$

where

CW = concentration in the drinking water (mg l^{-1})
SA = surface area of skin in contact (cm^{-1}), which is dependent on sex, age, and contact type
PC = dermal permeability constant (cm h^{-1}), which is highly dependent on the specific chemical
ET = exposure time (h day^{-1})
EF = exposure frequency (days year^{-1}), e.g., 7 days year^{-1} as an average
ED = exposure duration (years), e.g., 70 years for a lifetime
CF = volumetric conversion factor for water (1 l (1000 cm^3)$^{-1}$)
BW = body weight or mass (kg), e.g., 70 kg for an adult
AT = averaging time (days), e.g., ED × 365 days for a noncarcinogen or 70 years lifetime^{-1} × 365 days year^{-1} for a carcinogen

6.4 INHALED AS A VAPOR IN THE AIR

$$\text{Intake (mg per kg-day)} = \frac{(CA)(IR)(ET)(EF)(ED)}{(BW)(AT)} \qquad \text{(A6.4)}$$

where

CA = concentration in the air (mg m^{-3})
IR = inhalation rate (m^3 h^{-1}), e.g., 30 m^3 h^{-1} for an adult
ET = exposure time (h day^{-1})
EF = exposure frequency (days year^{-1})
ED = exposure duration (years), e.g., 70 years for a lifetime
BW = body weight or mass (kg), e.g., 70 kg for an adult
AT = averaging time (days), e.g., ED × 365 days for a noncarcinogen or 70 years lifetime^{-1} × 365 days year^{-1} for a carcinogen

6.5 INGESTED IN FISH OR SHELLFISH FROM THE CONTAMINATED AREA

$$\text{Intake (mg per kg-day)} = \frac{(CF)(IR)(FI)(EF)(ED)}{(BW)(AT)} \qquad \text{(A6.5)}$$

where

CF = concentration in the food (mg kg^{-1})
IR = ingestion rate (kg meal^{-1}), e.g., 0.284 kg meal^{-1} (upper 95% confidence level for fish) consumption
FI = fraction ingested from the contaminated source (no units to this fraction)
EF = exposure frequency (meals year^{-1})

ED = exposure duration (years), e.g., 70 years for a lifetime

BW = body weight or mass (kg), e.g., 70 kg for an adult

AT = averaging time (days), e.g., ED × 365 days for a noncarcinogen or 70 years lifetime^{-1} × 365 days year^{-1} for a carcinogen

6.6 INGESTED IN VEGETABLE MATTER FROM THE CONTAMINATED AREA

$$\text{Intake (mg per kg-day)} = \frac{(CF)(IR)(FI)(EF)(ED)}{(BW)(AT)} \qquad \text{(A6.6)}$$

where

CF = concentration in the food (mg kg^{-1})

IR = ingestion rate (kg meal^{-1})

FI = fraction ingested from the contaminated source (no units to this fraction)

EF = exposure frequency (meals year^{-1})

ED = exposure duration (years), e.g., 70 years for a lifetime

BW = body weight or mass (kg), e.g., 70 kg for an adult

AT = averaging time (days), e.g., ED × 365 days for a noncarcinogen or 70 years lifetime^{-1} × 365 days year^{-1} for a carcinogen

Glossary

abductive inference. Inference to the best explanation. It uses information gathered about a phenomenon or situation to produce a hypothesis that best explains the data. (Chapter 13)

absolute bioavailability. The bioavailability of a dose (D) estimated from the area under the curve (AUC) for any route or formulation of the compound divided by the AUC after direct injection of the same dose (D) into the bloodstream. (Chapter 4)

absorbed dose rate (radiation). The released energy that is deposited within the tissue. (Chapter 14)

absorption rate constant (k_a). A first order rate constant for absorption calculated as $MAT = k_a^{-1}$, where MAT is the mean absorption time. (Chapter 4)

accelerated failure time model. A survival time model in which the time to death (ln TTD) of a particular type/class of individual (e.g., smoker) is changed (accelerated) as some function of a covariate (e.g., classification relative to smoking habits). (Chapter 9)

acclimation. The modification of biological functions, especially those physiological, or structures to maintain or minimize deviations from homeostasis despite change in some environmental quality such as temperature, salinity, light, radiation, or toxicant concentration. It is an expression of phenotypic plasticity of individuals in response to a sublethal change in some environmental factor. (Chapter 9)

acclimatization. Like acclimation, acclimatization is the modification of biological functions, especially those physiological, or structures to maintain or minimize deviations from homeostasis despite change in some environmental quality. However, these shifts are taking place under natural conditions, not controlled laboratory conditions as is often the case with studies of acclimation. (Chapter 9)

accumulation factor (AF). The ratio of nonpolar organic compound concentration in the organism to that of sediments ([organism]/[sediment]) with the organism's concentration normalized to grams of lipid and sediment concentration normalized to grams of organic carbon. (Chapter 3)

acetylcholinesterase inhibitors. Compounds such as many organophosphate and carbamate insecticides that inhibit the normal functioning of the enzyme, acetylcholinesterase, which hydrolyzes the neurotransmitter, acetylcholine. (Chapter 8)

acidophilic component. Component of a cell, such as the general cytoplasm, that is readily stained by an acidic dye. (Chapter 7)

acid precipitation. Precipitation including rain, fog, snow, or other forms of precipitation, with a pH below 5.7. (Chapter 12)

acid volatile sulfides (AVS). Sediment-associated sulfides extracted with cold HCl that are assumed to be primarily iron and manganese sulfides. (Chapter 4)

activation. One result of biotransformation in which the effect of an active compound is worsened or an inactive compound is converted to one with an adverse bioactivity. (Chapter 3)

active transport. Movement of a substance up an electrochemical gradient that requires a carrier molecule and energy. (Chapter 3)

activity (radio-). A measure of the rate at which a given quantity of radioactive material is emitting radiation. (Chapter 14)

acute lethality. Death following a short and often intense exposure. The duration of an acute exposure in toxicity testing is generally 96 h (4 days) or fewer of exposure. (Chapter 9)

additive index. An index for quantifying the joint action of toxicants in mixture. (Chapter 9)

additivity. The condition of additivity exists for two or more toxicants in mixture if the mixture effect level was simply that expected by summing the expected individual toxicant effects. (Chapter 9)

adduct. Modification of the DNA molecule produced when a xenobiotic or a metabolite binds covalently to a base (most often) or to another portion of the DNA molecule. (Chapter 6)

adenylate energy charge (AEC). An index reflecting the balance of energy transfer between catabolic and anabolic processes. AEC = (ATP + $\frac{1}{2}$ADP)/(ATP + ADP + AMP), where ATP, ADP, and AMP = concentrations of adenosine tri-, di-, and monophosphate, respectively. (Chapter 8)

adsorption. The accumulation of a substance at the common boundary of two phases, most often adsorption from a solution onto the solid surface. Adsorption of contaminants can result from processes of ion exchange, weak bonding to surfaces such as that associated with van der Waal forces, or even molecular orientation of large, dipolar organic compounds relative to solid (less polar) and water (more polar) phases. (Chapters 2, 3)

age pigment. *See* lipofuscin. (Chapter 7)

age-specific birth rate. The mean number of females born to a female of an age class x. (Chapter 10)

age-specific death rate. The probability of dying as tabulated for a life table interval or age class (x). (Chapter 10)

age-specific number of individuals dying. The number of individuals dying in a life table interval (x). It is estimated as a simple difference, $d_x = l_x - l_{x+1}$. (Chapter 10)

aging (of a radionuclide). A decrease in bioavailability of a radioactive contaminant with time. Generally this is due to the radionuclide's increased adsorption to soil particles. (Chapter 14)

AHH (aryl hydrocarbon hydroxylase). Activity units of benzo[a]pyrene hydroxylation by aryl hydrocarbon hydroxylase used to reflect cytochrome P-450 monooxygenase activity. (Chapter 6)

alkalinity. The capacity of a natural water to neutralize acid as measured by titration of a water sample with a dilute acid to a specific pH endpoint. Most often, it is a function of carbonate (CO_3^{2-}), bicarbonate (HCO_3^-), and hydroxide (OH^-) concentrations, i.e., the carbonate-bicarbonate buffering of the water. However, dissolved organic compounds, borates, phosphates, and silicates can also contribute to alkalinity. (Chapter 12)

alkoxyradicals. Oxyradicals of organic compounds (R) of the form $RO^\bullet$. (Chapter 6)

allometry. The study of size and its consequences. (Chapter 4)

allozyme. An allelic variant of an enzyme coded for by a particular locus. (Chapter 10)

alpha (α) particles. Pieces of the nucleus ejected from a radioactive atom to reduce excess energy and gain stability. They are relatively huge in mass (7345 times larger than a β particle), consisting of 2 neutrons and 2 protons, and carrying a +2 charge. (Chapter 14)

amelia. A developmental abnormality in which the individual is born without limbs. (Chapter 8)

δ-aminolevulinic acid dehydratase (δ-ALAD or ALAD). An enzyme catalyzing the conversion of δ-aminolevulinic acid to porphobilnogen during heme synthesis. (Chapter 6)

analog (elemental). An element that behaves like, but not necessarily identical to, another element in biological processes; e.g., cesium is an analog of potassium, and strontium is an analog of calcium. (Chapter 5)

analysis of variance (ANOVA). One-way ANOVA breaks down the total variance (total sum of squares) in a data set into the variance among and within treatments, e.g., among the different concentration treatments and within replicates for each concentration treatment. The variance within treatments (mean sum of squares$_{within}$) is assumed to reflect the sampling or error variance, and that among treatments (sum of squares$_{among}$) is thought to estimate the error variance plus any additional variance associated with the treatment. ANOVA is used often to test the null hypothesis of equal means among treatments in sublethal or chronic lethal assays. (Chapter 8)

analysis plan (in ecological risk assessment). A plan that defines the exact format and design of the assessment, explicitly states the data needed, and describes the methods and design for analyzing these data. (Chapter 13)

androgen receptor antagonist. A xenobiotic that acts by blocking androgen receptor-mediated processes. (Chapter 8)

aneuploidy. The deviation by loss or addition of chromosomes from the usual (e.g., 2N) number of chromosomes. (Chapter 6)

annual limit on intake (ALI). The activity of a radionuclide that, if inhaled or ingested, would result in a dose equal to the most limiting primary guide (Eckerman et al., 1988). Primary guides are the current recommendations for radiological protection. (Chapter 14)

antagonism. Toxicant antagonism occurs if the actual effect level of the mixture was lower than the sum of the predicted effects for the individual toxicants in the mixture. (Chapter 9)

antisense oligonucleotides. Oligonucleotides that are complementary to messenger RNA and bind to them. (Chapter 6)

antisymmetry. A population quality of bilaterally symmetrical individuals in which the differences (d) between a trait measured from the right and left side of individuals from that population describe a bimodal distribution. (*See also* fluctuating asymmetry.) (Chapter 8)

apparent volume of distribution (V_d). A mathematical volume used in clearance volume-based models. It is expressed in units of volume of a reference compartment, often the blood or plasma compartment. (Chapter 3)

arcsine square root transformation. A common and useful transformation of effects data that often allows one to meet the assumption of homogeneous variances for proportions of exposed individuals responding:

$$\text{Transform} = \arcsin\sqrt{P},$$

where P = the measured effect, e.g., proportion of the exposed organisms. (Chapter 8)

assessment endpoint (in ecological risk assessment). The valued ecological entity that is to be protected and the precise quality to be measured for this entity. Earlier, only the first part of this definition (the valued entity) was considered the assessment endpoint. (*See also* measurement endpoint.) (Chapter 13)

backstripping or backprojection procedure. A method used to extract parameter estimates for elimination involving a multiexponential process. The procedure can be implemented graphically or mathematically with a computer program. (Chapter 3)

Bartlett's test. A statistical test used to test the assumption of homogeneity of variances. In this book, it was used to assess this assumption prior to one-way analysis of variance of sublethal and chronic lethal effects data. (Chapter 8)

basophilic component. Component of a cell such as the nucleus that is readily stained by a basic dye. (Chapter 7)

becquerel (Bq). The official unit of radioactivity now replacing the curie. One curie is 3.7×10^{10} Bq. (Chapter 1)

behavioral teratology. The study of behavioral abnormalities in otherwise normal appearing individuals after exposure as an embryo to an agent. (Chapter 8)

behavioral toxicology. The science of abnormal behaviors produced by exposure to a chemical or physical agent. (Chapter 8)

beta (β) particles. Electrons or positrons (positive electrons) ejected from the atom during radioactive decay. (Chapter 14)

binomial method. A method of estimating LC50, EC50, or LD50 if there are no partial kills. (*See also* median lethal concentration, median effective concentration, median lethal dose.) (Chapter 9)

bioaccumulation. The net accumulation of a contaminant in and on an organism from all sources including water, air, and solid phases of the environment. Solid phases include food sources. (Chapter 3)

bioaccumulation factor (BAF). The ratio of the contaminant concentration in the organism to that in the sediment or some other source, i.e., [organism]/[sediment] or [organism]/[food]. Often, this term is used in a broader sense if the exact source is not known or poorly defined, as in a field survey [*See* bioaccumulation factor (BF)]. For the purposes of avoiding confusion, BSAF (*See* biota-sediment accumulation factor) is used in this textbook to define sediment-associated bioaccumulation, and BAF is used in a more general context. (Chapter 3)

bioaccumulation factor (BF). Like BAF, the BF is the ratio of contaminant concentration in the organism to that in potential sources. The BF is based on the assumption that water and food (including sediments in some cases) may contribute to differences in concentrations measured for individuals at different trophic levels. (*See also* bioaccumulation factor, biota-sediment accumulation factor.) (Chapter 5)

bioaccumulative chemicals of concern (BCC). *See* persistent organic pollutants. (Chapter 12)

bioamplification. *See* biomagnification. (Chapter 5)

bioavailability. The extent to which a contaminant in a source is free for uptake. In many definitions, especially those associated with pharmacology or mammalian toxicology, bioavailability implies the degree to which the contaminant is free to be taken up by the organism *and to cause an effect at the site of action.* (Chapter 4)

bioconcentration. The net accumulation in and on an organism of a contaminant from water only. (Chapter 3)

bioconcentration factor (BCF). The ratio of concentrations of contaminant in the organism and dissolved in water (the presumed or explicit source): ([organism]/[water]). (Chapter 3)

biological half-life ($t_{1/2}$). The time required for the amount or concentration of a contaminant in a compartment to decrease by 50%. (Chapter 3)

biologically determinant. A quality of an element such that its concentration in organisms remains relatively constant over a wide range of environmental concentrations. Many essential elements are biologically determinant due to their metabolic regulation. (Chapter 4)

biologically indeterminant. A quality of an element such that its concentration in organisms is directly proportional to environmental concentrations. (Chapter 4)

biomagnification. An increase in concentration from one trophic level (e.g., prey) to the next (e.g., predator) due to accumulation of contaminant from food. (Chapter 5)

biomagnification factor (B). The contaminant concentration at trophic level n (C_n) divided by that at the next lowest trophic level (C_{n-1}) (e.g., Bruggeman et al., 1981; Laskowski, 1991). This factor can be estimated with individual organisms of known or assumed trophic status. (Chapter 5)

biomagnification power (b). The exponent of the exponential relationship: concentration in an organism = $ae^{b(\delta^{15}N)}$, where a and b are estimated parameters. It quantifies the proportional increase or decrease in concentration along a trophic food web. The $\delta^{15}N$ reflects the trophic status of the individual. (Chapter 5)

biomarker. A cellular, tissue, body fluid, physiological, or biochemical change in extant individuals that is used quantitatively during biomonitoring to either imply presence of significant pollutant or as an early warning system for imminent effects. A biomarker is also called a bioindicator by some authors. (Chapter 1)

biomineralization. Biologically mediated deposition of minerals. (Preface)

biominification. *See* trophic dilution. (Chapter 5)

biomonitor. To use organisms to monitor contamination and to imply possible effects to biota or routes of toxicant exposure to humans. (Chapter 1)

biomonitoring. A widely applied practice of monitoring of a subset of an entire community with the goal of assessing community condition. (Chapter 11)

biomonitoring (Type 1). The monitoring of community changes along a gradient or among sites differing in levels of pollution. (Chapter 11)

biomonitoring (Type 2). The measurement of bioaccumulation in organisms among sites notionally varying in the level of contamination. (Chapter 11)

biomonitoring (Type 3). The measurement of effects on organisms using tools such as biochemical markers in sentinel species or some measure of diminished fitness of individuals. (Chapter 11)

biomonitoring (Type 4). The measurement of genetically based resistance in populations of contaminated areas. (Chapter 11)

biota-sediment accumulation factor (BSAF). The specific term used for a bioaccumulation factor (BAF) if it is clear that the factor relates accumulated contaminant to that in sediments. (*See also* bioaccumulation factor.) (Chapter 5)

biotransformation. The biologically mediated transformation of one chemical compound to another. (Chapter 3)

biphasic dose-effect model. A model of dose effect that, due to hormesis, is shaped like the threshold model in Figure 7.5, but the curve actually dips down from the control level before increasing with dose. The individuals at these low, subinhibitory concentrations are performing better than individuals exposed to lower or higher concentrations. (Chapter 8)

body burden. The total mass or amount of contaminant in (and occasionally on) an individual. (Chapter 3)

boomerang paradigm. What you throw away can come back to hurt you. (Chapter 1)

borderline metal cations. Metal intermediate between Class A and B metals. These metals have one to nine outer orbital electrons. (Chapter 3)

Brillouin diversity index. A measure of the species diversity of a sample taken from a community. (Chapter 11)

calcium sinks. Physical sinks such as arthropod cuticles or bone that render calcium or its analogs less bioavailable during trophic interactions, thus providing a mechanism for trophic dilution. (Chapter 5)

cancer progression. The change in the biological attributes of neoplastic cells over time that leads to malignancy. (Chapter 7)

carcinogenic. Capable of causing cancer. (Chapter 6)

cardinal signs of inflammation. Heat, redness, swelling, and pain, although heat is not relevant in the case of poikilotherms. (Chapter 7)

carrier proteins. Cell membrane-associated proteins that act as carriers to transfer hydrophilic contaminants across the membrane. (Chapter 3)

carrying capacity (K). The maximum population size expressed as total number of individuals, biomass, or density that a particular environment is capable of sustaining. (Chapter 10)

caseous necrosis. Necrosis (cell death) in which cells disintegrate and form a mass of fat and protein. (Chapter 7)

catagenesis. A long-term geochemical alteration to organic matter, involving high temperatures and pressures deep below the surface of the earth. (Chapter 2)

catalase (CAT). An enzyme catalyzing the reaction: $2H_2O_2 \rightarrow 2H_2O + O_2$. It is involved in reducing oxidative stress. (Chapter 6)

catalyzed Haber-Weiss reaction. A greatly accelerated Haber-Weiss reaction ($O_2^- \rightarrow H_2O_2 \rightarrow {}^\bullet OH + OH$) catalyzed by metal chelates. (*See also* Haber-Weiss reaction.) (Chapter 6)

cellular stress response. An "orchestrated induction of key proteins that form the basis for the cell's protein repair and recycling system" (Sanders and Dyer, 1994). (Chapter 6)

channel proteins. Cell membrane-associated transport proteins that form channels to allow solute passage through the membrane. (Chapter 3)

chaperon. Stress proteins that associate with and direct the proper folding and coming together of proteins. They also protect proteins from denaturing and aggregating, and they enhance refolding to a functional conformation. (Chapter 6)

Chapman mechanism. The series of reactions by which ozone is formed in the stratosphere (Reactions 12.7 to 12.9). (Chapter 12)

chelate. A multidentate ligand. (Chapter 4)

chemical antagonism. Antagonism resulting because two toxicants react with one another to produce a less toxic product. (Chapter 9)

chemicals of potential concern. During human risk assessments, a list of those chemicals of most concern (chemicals of potential concern) is made from a list of all chemicals present at the site. (Chapter 13)

"chi" square (χ^2) ratio. Here, the ratio of χ^2 values for two candidate models used to select the model that best fits the dose- or concentration-effect data set. (Chapter 9)

chloride cells. Specialized cells on the gill found predominately on the primary lamellae but also on the secondary lamellae, which function in ion regulation. (Chapter 7)

chlorofluorocarbons (CFC). Chemicals used as propellants and coolants that have been linked to ozone depletion in the stratosphere. (*See also* Montreal Protocol.) (Chapter 1)

chlorosis. The blanching of green color due to the lack of production or the destruction of chlorophyll. (Chapter 8)

chromatid. Prior to cell division, the DNA in each chromosome is duplicated to produce two chromatids. As the chromosome condenses, the chromosome appears as a pair of chromatids connected to a common centromere. (Chapter 7)

chromatid aberration. An aberration that occurs if only one strand (chromatid) is broken in the chromosome. (Chapter 7)

chromosomal aberration. Damage to chromosomes, including breakage and loss of segments of DNA, addition of segments of DNA, or chromosomal rearrangements. Chromosomal breaks involve double-strand breaks. (Chapter 7)

chronic lethality. Death resulting from prolonged exposure. By recent convention, a chronic test should be at least 10% of the duration of the species life span. (Chapter 9)

chronic reference dose (chronic RfD). The RfD for chronic exposure. (Chapter 13)

Class A metal cation. Metal that has an inert gas electron configuration, high electronegativity, and a hard (difficult to polarize) outer sphere. (Chapter 3)

Class B metal cation. Metal with filled d orbital of 10 to 12 electrons and low electronegativity. It has a soft (easily polarizable) sphere readily deformed by adjacent ions, i.e., it easily forms covalent bonds with donor atoms such as sulfur. (Chapter 3)

clastogenic. Capable of causing chromosome damage in living cells. (Chapter 6)

Clean Air Act (CAA). A U.S. federal act designed to regulate air pollution with the goal of protecting human health and the environment. (Appendix 3)

Clean Water Act (CWA). A U.S. federal act designed to improve water quality, i.e., to restore and maintain the chemical, physical and biological integrity of U.S. waters. (Appendix 3)

clearance. As used in modeling, the rate of substance movement among compartments normalized to concentration. Clearance has units of flow, volume time^{-1}. (Chapter 3)

clearance volume-based model. A model of substance uptake, elimination, or bioaccumulation based on the distribution of the substance in, and clearance from and among, compartments of different volumes. (*See also* clearance.) (Chapter 3)

coagulation necrosis. Necrosis (cell death) characterized by extensive cytoplasmic protein coagulation, which makes the cell appear opaque. The cell outline and arrangement in the tissue remain for some time after cell death. (Chapter 7)

cold condensation theory. Persistent organic pollutants (POP) in the air will condense onto soil, water, and biota at cool temperatures. Consequently, the ratios for POPs concentrations in the air and on condensed phases decrease as one moves from warmer to cooler climates. (*See also* persistent organic pollutants.) (Chapter 12)

committed effective dose (CED). The dose unit to which risk factors are multiplied in order to estimate the probability of an individual experiencing a deleterious effect from exposure to radiation. CED accounts for the relative biological effectiveness of different types of radiation, the characteristics of the radiation such as energy and half-life of the radioactivity, and human physiology; CED also integrates dose from ingested radioactivity over a 50-year period. (Chapter 14)

community. "An assemblage of populations living in a prescribed area or physical habitat: it is an organized unit to the extent that it has characteristics additional to its individual and population components ... [it is] the living part of the ecosystem" (Odum, 1971). The community is made up of species that interact to form an organized unit (Magurran, 1988), although some species may interact only loosely. (Chapter 11)

community conditioning hypothesis. Communities retain information about occurrences in their past and will not return to their original state after perturbation. (Chapter 11)

compensatory hyperplasia. An excessive amount of hyperplasia occurring in response to injury or irritation. (Chapter 7)

Comprehensive Environmental Response, Compensation and Liability Act (CER CLA). A U.S. federal act that gives EPA the ability to respond to hazardous releases, clean up or require clean up of waste sites, and identify liability for released hazardous waste. (Appendix 3)

concentration additivity. Concentration additivity occurs if concentrations of toxicants can be added together to predict effects of toxicant mixtures under the assumption of additivity. (Chapter 9)

concentration factor (CF). The quantitative expression of the change in concentration at different trophic levels relative to the concentration in the ultimate or lowest defined source, e.g., relative to the water concentration, $CF = C_n/C_{water}$ The change in concentration is expressed as a multiple of the source concentration. (Chapter 5)

concept of strategy. *See* principle of allocation. (Chapter 10)

conceptual model (in ecological risk assessment). The model that links and interrelates assessment endpoint(s) and stressor. It includes evaluation of potential exposure pathways, effects, and ecological receptors. Conceptual models include hypotheses of risk and a diagram of the conceptual model. (Chapter 13)

conceptual model diagram (in ecological risk assessment). Part of the conceptual model. A diagram showing the pathways of exposure and illustrating areas of uncertainty or concern. They are visual aids for communicating the model from which the risk hypotheses emerged to the risk manager. (Chapter 13)

congeners. "[A term used to] point up the relationship among members of a chemical family such as the PCBs" (Bunce, 1991). For example, the individual PCB congeners in a mixture share a common form but vary by the numbers and positions of Cl atoms. (Chapters 2, 5)

conjugation. In Phase II reactions, "the addition to foreign compounds of endogenous groups which are generally polar and readily available *in vivo*" (Timbrell, 2000). (Chapter 6).

contaminant. "A substance released by man's activities" (Moriarty, 1983). (Chapters 1, 2)

co-tolerance. *See* cross resistance. (Chapter 10)

cough (gill purge). An abrupt, periodic reversal of water flow over the gills that dislodges and eliminates excess mucus from the gills' surfaces. (Chapter 8)

Council on Environmental Quality. A council established by NEPA to aid and advise the president during preparation of the annual Environmental Quality Report. (Appendix 3)

Cox proportional hazard model. A semiparametric method allowing the examination of proportional hazards without taking on the assumption of any specific model for the underlying baseline hazard. (Chapter 9)

criteria. Estimated concentrations of toxicants based on current scientific information that, if not exceeded, are believed to protect organisms or a defined use of a water body. (Chapter 1)

critical life stage testing. Toxicity testing focused on the life stage of a species thought to be most sensitive to the toxicant, such as newly hatched individuals. (Chapter 9)

critical study. During human risk assessment, the human or nonhuman animal study with the lowest-observed adverse effect level (LOAEL). (*See also* lowest observed effect level.) (Chapter 13)

critical toxic effect. In human risk assessment, the effect associated with the critical study. (Chapter 13)

cross resistance. The condition in which enhanced tolerance to one toxicant also enhances tolerance to another. (Chapter 10)

curie (Ci). A measure of radioactivity equivalent to 2.2×10^6 dpm (disintegrations per minute). (Chapter 1)

cysteine-rich intestinal protein (CRIP). A protein that serves in the uptake of zinc by cells in the intestine wall. (Chapter 4)

cytochrome P-450 monooxygenase. A 45 to 60–kDa hemoprotein associated with membranes, especially those of the endoplasmic recticulum, and active in Phase I metabolism of organic compounds, and metabolism of fatty acids, cholesterol, and steroid hormones. (*See also* monooxygenase.) (Chapter 6)

cytolytic necrosis. *See* liquefactive necrosis. (Chapter 7)

cytotoxicity. Toxicity causing cell death. (Chapter 7)

depuration. The loss of contaminant from an organism that is measured after the organism has been placed into a clean environment and allowed to eliminate contaminant. (Chapter 3)

developmental reference dose (RfD$_{dt}$). A RfD determined for developmental consequences of a single, maternal exposure during development. (Chapter 13)

developmental stability. The capacity of an organism to develop into a consistent phenotype in an environment. (Chapter 8)

developmental toxicity. A broad area of toxic effect that considers altered growth and functional deficiencies in addition to classic teratogenic effects. (Chapter 8)

diffusion. The movement of a contaminant down an electrochemical gradient that requires no energy. (Chapter 3)

dilution paradigm. The solution to pollution is dilution. (Chapter 1)

directional asymmetry. The deviation for a population from a mean of zero for the difference between a trait measured from the right and left sides of bilaterally symmetrical individuals from that population. For example, measurement of the difference in weights of left and right arms of humans would display directional asymmetry because most humans are right-handed and have larger right arms. (*See also* antisymmetry, fluctuating asymmetry.) (Chapter 8)

direct photolysis. Degradation of a contaminant through the energetic input of absorbed light. (Chapter 2)

discrimination ratio (or factor). A ratio measuring the degree of isotopic discrimination, with a ratio of 1 indicating no discrimination. In the context of discrimination between elemental analogs such as cesium and potassium in a trophic exchange, a discrimination factor or ratio is expressed as $[Cs]_{food}/[K]_{food}$ divided by $[Cs]_{body}/[K]_{body}$. (*See also* isotopic discrimination.) (Chapter 5)

disintegration. The event in which a radioactive element releases photons or particles to gain stability, i.e., radioactive decay. (Chapter 14)

disposable soma theory of aging. Aging is a consequence of the gradual accumulation of cellular damage via random molecular defects. (Chapter 10)

dispositional antagonism. Antagonism involving toxicant mixture effects on the uptake, movement within the organism, deposition at specific sites, and elimination of the toxicants. The presence of the two toxicants together shifts one or more of these processes to lower the impact of the toxicants at the site(s) of action or target organ(s). (Chapter 9)

distribution coefficient. *See* partition coefficient. (Chapter 2)

Dobson units (DU). A measure of atmospheric ozone levels that is the equivalent of 0.001-mm thickness of pure ozone at 1 atmosphere. (Chapter 12)

doubling time. The doubling time (t_d) of a population is the estimated time required for the population to double its present size. It (t_d) is estimated from the intrinsic rate of increase (r) as $(\ln 2)/r$. (Chapter 10)

dry deposition. The flux of particles and gases such as SO_2, HNO_3, and NH_3 to surfaces. (Chapter 12)

Dunnett's test. A parametric, post-ANOVA test used often in the analysis of sublethal and chronic lethal effects data. (Chapter 8)

dynamic energy budget (DEB) approach. A theory-rich approach utilizing energy budgeting for individuals as the central theme around which survival, growth, and reproduction under the influence of toxicants are modeled. Standard toxicity test data can be incorporated directly into the DEB approach. (Chapter 10)

early life stage (ELS) test. A critical life stage test using early life stages such as embryos or larvae based on the observation or assumption that the early life stage is the most sensitive in the species life cycle. (Chapter 9)

ecological epidemiology. The name given to epidemiological methods applied to determining the cause, incidence, prevalence, and distribution of adverse effects to nonhuman species inhabiting contaminated sites. It is frequently associated with retrospective ecological risk assessment. (*See also* retroactive risk assessment.) (Chapter 10)

ecological mortality (or death). The toxicant-related diminution of fitness of an individual functioning within an ecosystem context that is of a magnitude sufficient to be equivalent to somatic death. (Chapter 8)

ecoregion. A relatively homogeneous region in an ecosystem or association between organisms and their environment. Ecoregions are usually defined on maps. (Chapters 1, 12)

ecosystem. The functional unit of ecology, including the biotic community and its abiotic environment functioning together as a unit to direct the flow of energy and cycling of materials. (Chapter 11)

ecosystem incongruity. The traditional, but now too confining, bias toward the ecosystem or lower levels in the science of ecotoxicology. (*See also* ecotoxicology.) (Chapter 12)

ecotone. Area of transition between two or more community types. (*See also* edge effect.) (Chapter 12)

ecotoxicology. The science of contaminants in the biosphere and their effects on constituents of the biosphere, including humans. (Chapter 1)

edge effect. Ecotones often have species assemblages with high species richness and high abundance of individuals relative to those of the adjacent communities. (Chapter 12)

effect at a distance hypothesis. In metapopulations, the possibility of an individual being exposed in one patch but, after migration to another, having the effect of toxicant exposure expressed in another patch. (Chapter 10)

effective dose. A term used in pharmacology to define the amount of drug entering the blood and available to have a pharmacological effect. It is used in the context of drug bioavailability from different routes of administration. (Chapter 4)

effective dose equivalent (radiation). Recognizing that biological effects from a uniform irradiation of the whole body are different than effects from a similar dose concentrated in specific tissues, effective dose equivalent weights the radiation dose to different organs or tissues. Thus, the fractional contribution of organs and tissues to the total risk is normalized to when the entire body is uniformly irradiated. (Chapter 14)

effective half-life (k_{eff}). An estimated half-life in a compartment model that has numerous elimination mechanisms, each with an associated k_i. It is equal to $(\ln 2)/\Sigma k_i$. (Chapter 3)

elasticity (community). The ability of a community to return to its prestressed condition. (Chapter 11)

elimination. The loss or metabolism of a contaminant, resulting in a decrease in the amount of contaminant within an organism. (Chapter 3)

elutriate test. A test in which a nonbenthic species such as *Daphnia magna* is exposed to an elutriate produced by mixing the test sediment with water and then centrifuging the mixture. (Chapter 9)

emergent properties. Properties emerging in hierarchical systems, such as ecological communities or ecosystems, that cannot be predicted solely from our limited understanding of the system's parts or components. (Chapter 11)

endocrine disruptor. Environmental pollutants that interfere with the normal functioning of human and animal endocrine systems. (Chapter 8)

endocrine system. A system composed of several tissues that broadly includes any tissue or cells that release a chemical messenger (hormone) to signal or induce a physiological response in some target tissue. (Chapter 8)

endocytosis. Uptake of solids (phagocytosis) or liquids (pinocytosis) by cells through a process of engulfing the material and enclosure in a cellular vacuole. (Chapter 3)

enrichment factor (EF_{crust}). A measure of anthropogenic enrichment of an element above natural levels. EF_{crust} is an element's concentration (X) measured in air samples divided by that expected in the earth's crust: $EF_{crust} = [X/Al]_{air}/[X/Al]_{crust}$. Both air and crustal concentrations are normalized to Al concentrations. (Chapter 12)

enterohepatic circulation. Recirculation of toxicant back to the liver after passage into the intestine via the bile and reabsorption in the intestine. (Chapter 3)

environmental assessment (EA). A short, preliminary assessment of potential environmental damage used to determine if a full environmental impact statement (EIS) is required. (Appendix 3)

environmental epidemiology. A subdiscipline of human epidemiology concerned with diseases caused by chemical or physical agents. (Chapter 10)

environmental impact statement (EIS). A document required by NEPA that outlines possible impacts, describes impacts to resources, and details alternatives for any major federal action. (Appendix 3)

environmental law. "A body of federal, state, and local legislation, in the form of statutes, bylaws, ordinances, and regulations, plus court-made principles known as common law" (McGregor, 1994). (Appendix 3)

Environmental Quality Report. An annual report from the U.S. president's office discussing current environmental conditions and trends. Under NEPA, the Council on Environmental Quality aids and advises the president in the preparation of this report. (Appendix 3)

EPA weight of evidence classification. A classification used in human risk assessment involving carcinogenic effects that suggests to the risk manager the strength of the evidence supporting the risk calculation. (Chapter 13)

epidemiology. The science concerned with the cause, incidence, prevalence, and distribution of infectious and noninfectious diseases in populations. (Chapter 10)

epizootic. Outbreak of disease in a population or in a large number of individuals of a species. (Chapter 10)

EROD (ethoxyresorufin O-deethylase). Units of activity for O-deethylation of ethoxyresorufin by ethoxyresorufin O-deethylase. Used to reflect cytochrome P-450 monooxygenase activity. (Chapter 6)

essential element. An element essential for the normal functioning of a living organism. Mertz (1981) lists the following as the essential elements: H, Na, K, Mg, Ca, V, Cr, Mo, Mn, Fe, Co, Ni, Cu, Zn, Cd(?), C, Si, Sn(?), N, P, As, O, S, Se, F, Cl, and I. (Chapter 4)

estrogenic chemicals. Contaminants possessing biological activities like estrogen that cause changes in the sexual characteristics of individuals. (Chapter 8)

etiological agent. An agent responsible for causing, initiating, or promoting a disease. (Chapter 10)

Euler-Lotka equation. An equation used to estimate the intrinsic rate of increase from life table data. (Chapter 10)

European Community (EC). An organization formed in 1957 to foster a stronger union among European countries and, presently, comprising Austria, Belgium, Denmark, France, Finland, Germany, Greece, Ireland, Italy, Luxembourg, Netherlands, Portugal, Spain, Sweden, and the United Kingdom. The EC has its own budget and the power to make and enforce laws, including environmental laws. With the Maastricht agreement (December 11, 1991), the EC became the European Union (EU). (Appendix 3)

exchange diffusion. Diffusion across a membrane by means of a carrier molecule that requires no energy and involves the exchange of two ions across the membrane. (Chapter 3)

exocytosis. Fusion of intracellular vesicles with the cell membrane followed by the emptying of the vesicle contents to the cell exterior. (Chapter 3)

expected life span. The calculated life span for individuals of age class x of a life table. It can be estimated as T_x/l_x. (Chapter 10)

exploitation competition. Interspecies competition in which species compete for some limiting resource such as food. (Chapter 11)

exponential relationship. A mathematical relationship in which the Y variable is related to some constant raised to the X variable, i.e., $Y = a10^{bX}$. (Compare to power relationship.) (Chapter 4)

exposure. In risk assessment, contact with the contaminant or stressor. (Chapter 13)

exposure characterization (in ecological risk assessment). A description of the presence and characteristics of contact between the contaminant and the ecological entity of concern, and a summary of this information in an exposure profile. (Chapter 13)

exposure pathways. The avenues by which an individual is exposed to a contaminant, including the source and route to contact. (Chapter 13)

exposure profile (in ecological risk assessment). A profile that "quantifies the magnitude and spatial and temporal pattern of exposure for the scenarios developed during problem formulation" (Norton et al., 1992). (Chapter 13)

facilitated diffusion. Diffusion down a gradient not requiring energy, but occurring at a rate faster than expected by simple diffusion alone. (Chapter 3)

fat necrosis. Necrosis (cell death) that involves deposits of saponified fats in dead fat cells. (Chapter 7)

feasibility study (FS). Part of a remedial investigation and feasibility study that explores the various options for remediation. (Chapter 13)

fecundity selection. A component of the life cycle of an individual in which natural selection can occur, involving the production of more offspring by matings of certain genotype pairs than produced by other genotype pairs. (Chapter 10)

Federal Insecticide, Fungicide and Rodenticide Act (FIFRA). A U.S. federal law designed to control the production, distribution, labeling, sale, and use of pesticides. (Appendix 3)

fertilization effect. At low levels, the nitrogen and sulfur added to a forest in acid precipitation can enhance growth. (Chapter 12)

FETAX (frog embryo teratogenesis assay—*Xenopus*). A teratogenesis assay using embryos of the frog, *Xenopus laevis*. (Chapter 8)

filaments. *See* primary lamellae. (Chapter 7)

finding of no significant impact (FONSI). A statement of no significant impact of a major federal action concluded after an environmental assessment (EA). (Appendix 3)

finite rate of increase. The rate of increase of population size measured over set intervals, such as between age classes of a life table or generations of a population with nonoverlapping generations, e.g., an annual plant. (Chapter 10)

fission products. Radioactive fragments produced by nuclear fission. (Chapter 14)

Flory-Huggins theory. A quantitative theory relating solubility (partition coefficient) of compounds in dilute solutions to solvent molecular size (volume). In environmental toxicology, it has been used to explain the nonideal behavior of the K_{OW} in reflecting partitioning between water and lipids for very lipophilic compounds. (Chapter 4)

flow-through test. An aquatic toxicity test that has a constant (continuous flow-through test) or nearly constant (intermittent flow-through test) flow of the toxicant solutions through the exposure tanks. (Chapter 9)

fluctuating asymmetry. Deviation from perfect bilateral symmetry for a population that is thought to reflect developmental instability. A quality is measured from the right and left sides of a bilaterally symmetrical species, and the difference (d = Right − Left) is calculated. The variance in d for the population is a measure of fluctuating asymmetry. (*See also* antisymmetry, directional asymmetry.) (Chapter 8)

free ion activity model (FIAM). "The universal importance of free metal ion activities in determining the uptake, nutrition and toxicity of all cationic trace metals" (Campbell and Tessier (1996). (Chapter 4)

free radical. A molecule having an unshared electron. (The electron is usually designated by a dot, •.) Free radicals are extremely reactive. (Chapter 6)

Freundlich isotherm equation. An empirical relationship quantifying adsorption. (Chapter 3)

functional antagonism. Antagonism resulting from two chemicals eliciting opposite physiological effects and, as a consequence, counterbalancing each other. (Chapter 9)

functional redundancy. An apparently unaltered maintenance of community functioning despite changes in structure. (Chapter 11)

functional response. A change in some predator function, such as prey consumption rate, in response to changes in prey density. (Chapter 11)

fundamental niche. A species has a certain (Hutchinsonian) niche in which it could exist and function based on its physiological and other limits. In contrast to the realized niche, this fundamental niche includes all of the possible niche volume. (*See also* Hutchinsonian niche.) (Chapter 11)

Gaia hypothesis. An hypothesis forwarded by James Lovelock that the Earth's temperature, albedo, and surface chemistry are homeostatically regulated by the sum of all the biota of the Earth. (Preface)

gametic selection. A component in the life cycle of an individual during which natural selection can occur, involving differential success of gametes produced by heterozygotes. (Chapter 10)

gamma (γ) rays. Electromagnetic photons emitted from the nucleus. (Chapter 14)

gangrenous necrosis. Combination of coagulation and liquefactive necrosis, often resulting from puncture and subsequent infection. (*See also* coagulation necrosis, liquefactive necrosis.) (Chapter 7)

gastric emptying rate. The rate at which the contents of the stomach are emptied into the small intestine. (Chapter 4)

gastrointestinal excretion. Excretion through the intestinal mucosa by active or passive processes. This may involve loss by normal cell sloughing of the intestine wall. Metals such as cadmium and mercury can experience significant levels of gastrointestinal excretion. (Chapter 3)

general adaptation syndrome (GAS). The specific syndrome associated with Selyean stress composed of three phases: the alarm reaction, adaptation or resistance, and exhaustion phases. The goal in all phases of the GAS is to regain or resist deviation from homeostasis. (*See also* Selyean stress.) (Chapter 8)

genetic hitchhiking. Used in this book to describe the situation in which a scored locus is acting only as a marker for a closely linked gene that is actually responsible for the difference in tolerance among genotypes. More generally, it is the condition "in which a given allele changes in frequency as a result of linkage or gametic phase disequilibrium with another selected locus" (Endler, 1986). (Chapter 10)

genetic risk. The risk to the progeny of the exposed individual of an adverse effect associated with heritable genetic damage, e.g., damage to germ cells leading to a nonviable fetus or an offspring with a birth defect. (Chapter 7)

genotoxicity. Damage by a physical or chemical agent to genetic materials, e.g., chromosomes or DNA. (Chapter 6)

geographic information systems (GIS). Computerized systems to handle spatial data at a reasonable cost. Most allow one to archive, organize, integrate, statistically analyze, and display many kinds of spatial information using a common coordinate system. (Chapter 12)

global distillation. A process by which persistent and relatively volatile organochlorine compounds are distilled from warmer regions of use to cooler regions of the globe. (Preface, Chapter 12)

global fractionation. Because persistent organic pollutants (POP) differ in their individual rates of degradation, vapor pressures, and lipophilities, a fractionation occurs in which some POPs move more rapidly than others toward the polar regions. The net result is a redistribution of the different POPs from the equator or site of origin toward the cold polar regions of the earth. (Chapter 12)

global warming. A general warming of the Earth thought to result from the increased atmospheric carbon dioxide (CO_2) concentrations from fossil fuel burning, release of other greenhouse gases, and the worldwide destruction of forests. (Chapter 12)

glucose regulated proteins (grp). Proteins that, under low glucose or oxygen conditions, are part of the cellular stress response. The grps are structurally similar to heat shock proteins, are present at basal levels in unstressed cells, and are induced in glucose- or oxygen-deficient cells exposed to toxicants that modify calcium metabolism, e.g., lead (Sanders, 1990). (Chapter 6)

glucuronic acid. A carbohydrate that can be conjugated to xenobiotics by UDP-glucuronosyltransferase. (*See also* uridinediphospho glucuronosyltransferase.) (Chapter 6)

glutathione (GSH). A tripeptide composed of cysteine, glutamate, and glycine; specifically, γ-glutamyl-L-cysteinyl-glycine. (Chapter 6)

glutathione peroxidase. An enzyme catalyzing the reaction, 2 reduced glutathione (GSH) + H_2O_2 → oxidized glutathione (GSSG) + H_2O. It is involved in reducing oxidative stress. (Chapter 6)

glutathione s-transferase (GST). A Phase II enzyme that conjugates glutathione with a xenobiotic or its metabolite. (Chapter 6)

granulation tissue. During the repair stage of the inflammation process, small blood vessels begin to form, and connective tissue begins to grow in a mass called the granulation tissue. (Chapter 7)

grasshopper effect. Global distillation of persistent organic pollutants can involve seasonal cycling of temperatures such that movement toward the higher latitudes occurs in annual pulses. This is called the grasshopper effect. (Chapter 12)

greenhouse effect. Greenhouse gases are relatively transparent to light but absorb long-wave, infrared radiation radiating back from the earth's surface. The net balance for sunlight influx, infrared radiation absorption by greenhouse gases, and infrared eflux from the earth's surface determines the steady state temperature of the earth. The net warming of the earth is called the greenhouse effect. (Chapter 12)

greenhouse gases. Atmospheric gases that are relatively transparent to sunlight entering the atmosphere but absorb infrared radiation being generated at the earth's surface. They include water vapor, carbon dioxide, methane, nitrous oxide, CFCs, methylchloroform, carbon tetrachloride, and the fire retardant, halon. Ozone in the troposphere can also act as a greenhouse gas. (Chapter 12)

growth dilution. The decrease in contaminant concentration in a growing organism because the amount of tissue in which the contaminant is distributed is increasing. (Chapter 3)

guild (ecological). A "group of functionally similar species whose members interact strongly with one another but weakly with the remainder of the community" (Smith, 1986). (Chapter 11)

Haber-Weiss reaction. The reaction: $O_2^- \rightarrow H_2O_2 \rightarrow {}^\bullet OH + OH^-$. (Chapter 6)

half-life (radionuclide). The amount of time required for one-half of the number of radioactive atoms to decay. (Chapter 14)

hardness (of water). The sum of the concentrations of dissolved calcium and magnesium. (Chapter 9)

Hardy-Weinberg equilibrium. The frequency of genotypes will remain constant through time if the following conditions are met: (1) the population is large ("infinite") and composed of randomly mating, diploid organisms with overlapping generations, (2) no selection is occurring, and (3) mutation and migration are neglible. (Chapter 10)

hazard assessment. An assessment that compares the expected environmental concentration (EEC) to some estimated threshold effect (ETT) with the intent of deciding if (1) a situation is safe, (2) a situation is not safe, or (3) there is not enough information to decide. (Chapter 13)

hazard quotient. A crude indicator of hazard calculated as the expected environmental concentration (EEC) divided by some estimated threshold effect concentration (ETT): HQ = EEC/ETT. (Chapter 13)

health advisory concentrations. Concentrations below which no impact on human health is expected for the specified duration of exposure. They often are applied by public officials in dealing with spills, short-term exposures, or similar situations, although longer term advisory concentrations are also available. (Chapter 13)

heat shock proteins (hsp). Stress proteins induced by an abrupt shift in temperature that function to reduce associated protein damage in cells. (Chapter 6)

heavy metal. A term that originated from early studies of the harmful effects of metallic elements such as mercury, lead, and cadmium, which all had very high specific gravity. Although this term is properly applied to metals with specific gravities of five or higher, it is sometimes applied to other metallic elements of environmental concern, regardless of their density. (Chapter 2)

Henderson-Hasselbalch relationship. The relationship between pH and the ratio of conjugate base (B^-) to acid (BH), pH = pK_a + log ([B^-]/[BH]) (Piszkiewicz, 1977). (Chapter 4)

Henry's law coefficient. The quantitative expression that describes a contaminant's relative tendency to partition to the vapor phase or the dissolved phase: Henry's law coefficient (H) = vapor pressure (P)/ water solubility (C). As H increases, so does the compound's tendency to partition into the vapor phase. (Chapter 2)

heterosis. The superior performance of heterozygotes relative to homozygotes. (Chapter 10)

histopathology. The change in cells and tissues associated with a communicable or noncommunicable disease. (Chapter 7)

homeopathic medicine. A branch of medicine, founded by Samuel Hahnemann, based on the law of similars, which states that a drug that induces symptoms similar to those of the disease will aid the body in defending itself by stimulating its natural responses. (Chapter 8)

homolog. A pair of homologous chromosomes. (Chapter 7)

hormesis. A stimulatory effect exhibited with exposure to low, subinhibitory levels of some toxicants or physical agents. Hormesis is not normally a toxicant-specific response. (Chapter 8)

hormonal oncogenesis. Tumor production resulting from high levels of hormones (a promoter) with associated hyperplasia. (Chapter 7)

Hutchinsonian niche. A niche is "the certain biological activity space in which an organism exists in a particular habitat. This space is influenced by the physiological and behavioral limits of a species and by effects of environmental parameters (physical and biotic, such as temperature and predation) acting on it" (Wetzel, 1982). (Chapter 11)

hyperplasia. The capacity of cells to multiply and increase in tissues and organs. (Chapter 7)

hypertrophy. An increase in cell size (and function) resulting from an increase in the mass of cellular structural components, often as a compensatory response. (Chapter 7)

imposex. The development (imposition) of male characteristics such as a penis or vas deferens in females. (Chapter 8)

incidence rate. Incidence rate of a disease for a nonfatal condition is calculated as the number of individuals with the disease divided by the total time that the population had been exposed. It is expressed in units of individuals or cases per unit of exposure time, e.g., 10 new cases per year. (Chapter 10)

incipient median lethal concentration. The concentration below which 50% of individuals will live indefinitely relative to the lethal effects of the toxicant. (*See also* median lethal concentration.) (Chapter 9)

independent joint action. Independent action of toxicants occurs if each toxicant produces an effect independent of the other and by a different mode of action. (Chapter 9)

index of biological integrity (IBI). A composite index combining 12 qualities of fish communities of warm-water, low-gradient streams to determine the level of stream degradation. This index has been modified and widely used in the United States. (Chapter 11)

indirect photolysis. Degradation of a contaminant through interaction with other molecules in solution that have absorbed light energy. This can occur through energy transfer or by chemical reaction with short-lived reactive species. Dissolved humic and fulvic acids are good examples of photoactive compounds that can increase the degradation of contaminants through indirect photolysis. (Chapter 2)

individual effective dose (IED) concept. A concept forming the basis for most dose-response models, which holds that there exists a smallest dose needed to kill any particular individual. The IED is a characteristic of an individual. (Chapter 9)

individual tolerance concept. *See* individual effective dose concept. (Chapter 9)

industrial ecology. The study of the flows of materials and energy in the industrial environment and the effects of these flows on natural systems. (Chapter 1)

industrial melanism. The gradual increase to predominance of melanic forms in populations from industrialized regions. (Chapter 10)

inertia (community). A community's ability to resist change. (Chapter 11)

inflammation. A response to cell injury or death that attempts to isolate and destroy the offending agent and any damaged cells. (Chapter 7)

initiator. An agent producing cancer by converting normal cells to latent tumor cells. (Chapter 7)

innovative science. An activity within any scientific discipline that questions existing paradigms and formulates new paradigms. (Chapter 1)

integrated risk information system (IRIS). A large database containing reference doses, slope factors, drinking-water health advisories (one-day, ten-day, longer term, and life time advisories), and associated information compiled by the EPA. (Chapter 13)

interference competition. Interspecies competition in which one species interferes with another, as might occur with territoriality or aggressive behavior. (Chapter 11)

interspecies competition. The interference with or inhibition of one species by another. (Chapter 11)

intrinsic (or Malthusian) rate of increase. Rate of increase in the size of a population growing under no constraints. (Chapter 10)

ionization (radiation-induced). Ion formation in materials such as tissue caused by energetic rays and particles released during radioactive decay. Ionization occurs in biological material exposed to radiation, resulting in some probability of molecular or genetic damage. Part of the process often involves the formation of free radicals. (Chapter 14)

ischemia. Localized inadequacy of blood supply to or anemia of tissue resulting from an obstruction of blood flow, such as that associated with a wound. (Chapter 7)

isobole approach. An approach used to visualize or quantify joint action of chemical mixtures. (Chapter 9)

isoenzymes. Different forms of the same enzyme that are coded by different gene loci. (Chapter 6)

isotopes. Nuclides with the same number of protons but different numbers of neutrons are called isotopes. The number of protons determines the chemical identity of an atom. For example, all atoms with 82 protons are lead; however, lead can have 122, 124, 125, or 126 neutrons. Thus ^{204}Pb (122 neutrons + 82 protons), ^{206}Pb, ^{207}Pb, and ^{208}Pb are all isotopes of lead. (Chapter 14)

isotopic discrimination. The differential behavior of isotopes occurring if the rate or extent of participation in some biological or chemical process depends significantly on the mass of the isotope. Also called the isotope effect. (Chapter 5)

Itai-Itai disease. An epidemic of cadmium poisoning (1940–1960) linked to water contaminated with mine wastes used to irrigate rice fields. Itai-itai literally means "ouch-ouch" and reflects the extreme joint pain of victims. (Chapter 1)

iteroparous species. A species that reproduces more than once. (Chapter 10)

Kaplan-Meier method. *See* product-limit method. (Chapter 9)

Karnofsky's law. Any agent will be teratogenic if it is present at concentrations or intensities producing cell toxicity. (Chapter 8)

karyolysis. The disintegration of the cell nucleus with necrosis (cell death). (Chapter 7)

keystone habitat. A high quality habitat patch essential to maintaining the vitality of the metapopulation. (Chapter 10)

keystone species. A species that influences the ecological community by its activity or role, not its numerical dominance. (Chapter 11)

K_{OA}. The partition coefficient for a compound between n-octanol and air. Like the K_{OW}, it is a measure of lipophilicity. (Chapter 12)

K_{OW}. The partition coefficient for a compound between n-octanol and water, i.e., concentration in octanol/concentration in water at equilibrium. It or its log-transformed value is used to reflect lipophilicity of compounds. (Chapter 3)

k-strategy. An equilibrium strategy for species involving effective interactions with each other in the community, allowing coexistence of many species. Equilibrium species are more effective competitors than opportunistic species. (Chapter 11)

landscape. The sum total aspect of any geographical area. (Chapter 12)

Langmuir isotherm equation. Theoretically derived relationship quantifying adsorption. (Chapter 3)

latent (latency) period. The time or lag between exposure to the carcinogenic agent and the appearance of cancer. (Chapter 7)

law (U.S. environmental). "A body of federal, state, and local legislation, in the form of statutes, bylaws, ordinances, and regulations, plus court-made principles known as common law" (McGregor, 1994). (Appendix 3)

law of frequencies. Refers to the relationship between the numbers of species and the number of individuals in a community. (Chapter 11)

law of similars. The foundation premise of homeopathic medicine: A drug that induces symptoms similar to those of the disease will aid the body in defending itself by stimulating the body's natural responses. (Chapter 8)

lesion. Alterations in cells, tissues, or organs, indicating exposure or damage. (Chapter 7)

Leslie matrix. A matrix used to analyze demographic data that is constructed with the probability (P_x) of a female alive in period x_i to x_{i+1} being alive in period x_{i+1} to x_{i+2} in the matrix subdiagonal, with the number of daughters (F_x) born in the interval t to t + 1 per female of age x to x + 1 in the top row of the matrix. (Chapter 10)

Liebig's law of the minimum. A population's size (number of individuals or biomass) is limited by some essential factor in the environment that is scarce relative to the amount of other essential factors, e.g., phosphorus-limited algal growth in a lake. (Chapter 10)

life cycle studies. Comprehensive studies to determine the impact of a substance or mixture on the survival, growth, reproduction, development, or other important qualities at all stages of a species life cycle. (Chapter 9)

ligand. An anion or molecule that forms a coordination compound or complex with metals. (Chapter 4)

limited life span paradigm. An inherent quality of an individual is its genetically defined maximum life span. (Chapter 10)

linear energy transfer (LET). The average energy released by ionizing radiation per unit path length through a medium, usually expressed in thousands of electron volts per micron of path length. For example, the LET (keV μ^{-1}) in water for a 1.2-MeV gamma ray emitted from ^{60}Co, a 0.6-keV beta particle from tritium, and a 5.3-MeV alpha particle from polonium are 0.3, 5.5, and 110.0, respectively. The more energy released, the greater is the probability of damage. (Chapter 14)

linear no-threshold theory. This theory, relating incidence or risk of cancer to dose, is based on several radiation-induced cancer studies suggesting no threshold dose. It assumes that any lack of cancers below a certain dose reflects our inability to measure low incidences at these exposure levels and does not reflect a threshold of effect. The dose-response curve is a straight line. (Chapter 7)

linear solvation energy relationship (LSER). A class of quantitative structure-activity relationships based on molecular volume, ability to form hydrogen bonds, and polarity or ability to become polarized. (Chapter 4)

lipid peroxidation. The oxidation of polyunsaturated lipids in membranes, resulting in cell damage during xenobiotic exposures. (Chapter 6)

lipofuscin (age pigment). A degradation product of lipid oxidation that accumulates in cell vacuoles with age or exposure to some toxicants, such as copper. (Chapter 7)

liquefactive (cytolytic) necrosis. Necrosis characterized by a rapid breakdown of the cell as a consequence of the release of cellular enzymes. (Chapter 7)

Litchfield method. A simple, semigraphical method for analyzing survival time data and estimating LT50 values. (Chapter 9)

Litchfield-Wilcoxon method. A semigraphical method for estimating an LC50, EC50, or LD50. Although very easy to perform, it is the most subjective method for such estimations because it involves fitting a line to data by eye. (*See also* median effective concentration, median lethal concentration, median lethal dose.) (Chapter 9)

logit. A metameter used for dose- or concentration-response data under the assumption of a log-logistic model. Although it has the form logit(P) = ln [P/(1 − P)], its transform ([logit as just calculated]/2 + 5) is often used because the associated values are very close to those of the probit metameter. (Chapter 9)

log-normal model. A model fit to species-abundance curves that is thought to reflect a community structure in which several factors influence species interactions and subsequent allocation of resources. (Chapter 11)

Lorax incongruity. The delusion of selfless motivation in environmental stewardship or advocacy. The Lorax is a character in a popular children's book by Dr. Seuss who "speaks for the trees, for the trees have no tongues." (Preface)

lordosis. The extreme and abnormal forward curvature of the spine. (Chapter 8)

loss of life expectancy (LLE). A calculated estimate of loss in life time associated with a risk factor. It is estimated as the simple difference between life expectancy without the risk factor and life expectancy with the risk factor. (Chapter 13)

lowest observed effect concentration (or level) (LOEC or LOEL). The lowest concentration in a test with a statistically significant difference in response from the control response. (Chapter 8)

l_x. From a life table, the number of individuals in a cohort alive at age or stage class, x. (Chapter 10)

l_x life table or schedule. A life table that summarizes mortality data for populations. (Chapter 10)

$l_x m_x$ life table. A life table that summarizes both mortality and natality data for populations. (Chapter 10)

MacArthur-Wilson model of island colonization. The model: $S_t = S_{EQ} (1 - e^{-Gt})$, where $S_t =$ the number of species present at time t, $S_{EQ} =$ the equilibrium number of species for the island, and $G =$ the rate constant for colonization of the island. (Chapter 11)

male-mediated toxicity. Disease or birth defects produced by a father's exposure to a physical or chemical agent. (Chapter 8)

malondialdehyde. A breakdown product of lipid peroxidation; used as an indicator of oxidative damage. (Chapter 6)

Malthusian theory. A series of assumptions and observations regarding limitations on human populations developed by Thomas R. Malthus (1766–1834). (Chapter 10)

Marine Protection, Research and Sanctuaries Act (MPRSA). A U.S. federal act that recognizes the unsoundness of ocean dumping, states the U.S. policy regarding dumping, and establishes limitations and prohibition on dumping. (Appendix 3)

maturity index. An index for pollution based on the proportions of species in a soil nematode community that fell into various categories ranging from colonizers (r-strategists) to persisters (k-strategists). (*See also* k-strategy, r-strategy.) (Chapter 11)

maulstick incongruity. The incongruous assignment of ecological or biological significance of a contaminant's effect based primarily on statements of statistical significance. (Chapter 8)

maximum acceptable toxicant concentration (MATC). "An undetermined concentration within the interval bounded by the NOEC and LOEC that is presumed safe by virtue of the fact that no statistically significant adverse effect was observed" (Weber et al., 1989). (*See also* lowest observed effect concentration, no observed effect concentration.) (Chapter 8)

maximum contaminant levels (MCL). Primary drinking water standards established under the SDWA for a specified list of contaminants that can adversely affect health. (*See also* Safe Drinking Water Act.) (Appendix 3)

maximum likelihood estimation (MLE). A parametric method used to fit dose- or concentration-effect data to the log-normal, log-logistic, or other models. Probit and logit approaches are most often applied with MLE methods. (Chapter 9)

mean absorption time (MAT). The mean time required for absorption of a drug or contaminant calculated as the difference in mean residence time (MRT) of the material introduced by the (noninstantaneous) route of interest and the MRT for the same material injected intravenously. (Chapter 4)

mean generation time (T_c). The predicted generation time for a population estimated from life tables as the sum of the $xl_x m_x$ column divided by net reproductive rate, R_0. (Chapter 10)

mean residence time (τ or MRT). An estimated mean time that a particle (molecule or atom) remains in a compartment. (Chapter 3)

measurement endpoint (in ecological risk assessment). A measurable response to the stressor (e.g., fledglings produced per nest each year) that is related to the valued qualities of the assessment endpoint (e.g., reproductive success of bald eagles). (Chapter 13)

median effective concentration (EC50). For sublethal or ambiguously lethal effects, the concentration affecting 50% of exposed individuals within a predetermined time, e.g., 96 h. (Chapter 9)

median effective time (ET50). For sublethal or ambiguously lethal effects, the time until 50% of the exposed individuals respond. (Chapter 9)

median lethal concentration (LC50). The concentration resulting in death for 50% of exposed individuals within a predetermined time, e.g., 96 h. (*See also* incipient median lethal concentration.) (Chapter 9)

median lethal dose (LD50). The dose resulting in death for 50% of the exposed individuals within a predetermined time, e.g., 96 h. (Chapter 9)

median lethal time (LT50). The time resulting in death for 50% of the exposed individuals. (*See also* median time to death.) (Chapter 9)

median teratogenic concentration (TC50). The concentration resulting in developmental malformations for 50% of the exposed individuals within a predetermined time, e.g., 96 h. (Chapter 8)

median time to death (MTTD). Like the LT50, the time resulting in death for 50% of the exposed individuals. (*See also* median lethal time.) (Chapter 9)

meiotic drive. A component of the life cycle of an individual during which natural selection can occur, involving the differential production of gametes by different heterozygous genotypes. (Chapter 10)

membrane transport proteins. Cell membrane-associated proteins involved in transport of solutes. (Chapter 3)

mesocosms. Relatively large experimental systems designed to simulate some component of an ecosystem. Mesocosms are delimited and enclosed to a lesser extent than are microcosms. They are normally used outdoors or, in some manner, incorporated intimately with the ecosystem that they are designed to reflect. (*See also* microcosms.) (Chapter 11)

metal. Metals are elements known for their lustrous appearance, malleability, ductility, and conductivity. With the exception of hydrogen, they make up the left two-thirds of the periodic table. (Chapter 2)

metalloid. Metalloids are intermediate in properties between the metallic and nonmetallic elements and line up between them in the periodic table. They have a less lustrous appearance than metals, are semiconductors, and include elements such as silicon, arsenic, antimony, and selenium. (Chapter 2)

metallothionein. A relatively small (circa 7000 Da) protein with approximately 25 to 30% of its amino acids being cysteine, having no aromatic amino acids or histidine, and having the capacity to bind six to seven metal atoms per molecule. (Chapter 3)

metallothionein-like proteins. Poorly characterized, metal-binding proteins or proteins not conforming precisely to the classic properties of metallothioneins. (Chapter 6)

metameter. A measurement or a transformation of a measurement used in the analysis of biological tests, e.g., the probit metameter. (Chapter 9)

metapopulation. "A set of local populations which interact via dispersing individuals among local populations; though not all local populations in a metapopulations interact, directly with every other local population" (Hanski 1996). (Chapter 10)

metastasis. The process in which pieces of a cancerous growth dislodge and move to other tissues via the circulatory or lymphatic system to establish other loci of cancerous growth. This process leads to the spread of a cancer from the site of origin to other sites in the body. (Chapter 7)

methemoglobinemia. Referred to as the blue-baby syndrome because of the initial skin color of afflicted babies, it is caused by the reaction of nitrite (and some drugs) to hemoglobin (oxidation of ferrous iron to ferric iron) to produce methemoglobin that is incapable of the normal transport of molecular oxygen in the blood of the newborn. It can be caused directly by nitrite in drinking water or by the conversion of nitrate to nitrite in the baby's anaerobic stomach. It can also be caused in ruminants by the consumption of plants with high nitrate content. (Chapter 2)

method of multiple working hypotheses. A method proposed by Chamberlin (1897) to reduce precipitate explanation by considering all plausible hypotheses simultaneously in testing so that equal amounts of effort and attention are provided to each. (Chapter 1)

microcosms. Laboratory systems designed to simulate some component of an ecosystem, such as multiple species assemblages. (*See also* mesocosms.) (Chapter 11)

micronuclei. Membrane-bound masses of chromatin separate from the nucleus proper. McBee (Vignette 7.1) describes them as "cytoplasmic nuclear bodies that are formed when whole or fragmented chromosomes are not incorporated into the nuclei of daughter cells or when small fragments of chromatin are retained from polychromatic erythrocytes after expulsion of the nucleus in the process of erythrocyte maturation in mammals." (Chapters 6, 7)

Microtox® assay. A rapid, bacterial assay in which a decrease in bioluminescence is thought to reflect toxic action. (Chapter 9)

Minamata disease. An epidemic in Minamata and then Niigata, Japan, resulting from organic mercury release from industrial sources (acetaldehyde and vinyl chloride production) and consequent contamination of seafood. The first case was reported in 1953, and almost 1000 victims were identified by 1975. (Chapter 1)

mineralization. Complete degradation of an organic molecule to inorganic components (i.e., CO_2, H_2O, NO_3^-). (Chapter 2)

minimal time to response. For any toxicant effect, there can be a minimum time required to get an effect. Regardless of the toxicant concentration, the effect cannot occur any faster than this minimum time. (Chapter 9)

mixed function oxidase (MFO). The P-450 complex composed of cytochrome P-450, NADPH-cytochrome P-450 reductase, NADPH, and O_2. (*See also* monooxygenase.) (Chapter 6)

modifying factor. A factor based on expert opinion used to decrease the NOAEL (or LOAEL) during risk assessment. (*See also* lowest-observed-effect level, no-observed-effect level.) (Chapter 13)

monodentate ligand. A ligand sharing one pair of electrons with a cation. (Chapter 4)

monogenic control. Control of some quality by a single gene. (Chapter 10)

monooxygenase. One of a general class of enzymes involved in Phase I reactions with xenobiotics. Their action involves the addition of an oxygen atom (from O_2) to the xenobiotic and reduction of the remaining O atom to produce water. Also called mixed-function oxidases (MFO). (*See also* mixed-function oxidase.) (Chapter 3)

Montreal Protocol. An international treaty to limit and eventually eliminate the use of CFCs that was signed into law in 1988. (Chapter 12)

most sensitive species approach. An ecotoxicological approach in which results for the most sensitive of all tested species are used as an indicator of the toxicant concentration below which the entire community is protected from adverse effects. (Chapter 11)

moving average method. A method of estimating LC50, EC50, or LD50. It can be implemented with straightforward equations if the toxicant concentrations are set in a geometric series and there are equal numbers of individuals exposed in each treatment. (*See also* median effective concentration, median lethal concentration, median lethal dose.) (Chapter 9)

MTF-1. Metal transcription factor-1, a zinc-responsive transcription factor responsible for regulating the expression of major metallothionein genes in varied species. (Chapter 6)

multidentate ligand. A ligand sharing more than one pair of electrons with the cation. They are also called chelates. (Chapter 4)

multiple heterosis. A generally higher fitness of an individual as a composite or summed effect of heterozygote superiority (heterosis) at each of a series of loci. (Chapter 10)

multiplicative growth factor per generation. *See* finite rate of increase.

mutagen. A physical or chemical entity capable of producing mutations. (Chapter 7)

mutagenic. Capable of causing mutations. (Chapter 6)

NAS (National Academy of Sciences) paradigm. A paradigm used for both human and ecological risk assessments. There are four components to this paradigm: hazard identification, exposure assessment, dose-response assessment, and risk characterization. (Chapter 13)

national ambient air quality standards (NAAQS). Standards of air quality that Section 109 of the Clean Air Act requires the EPA to establish and periodically revise. (Appendix 3)

National Contingency Plan. A plan to identify sites for cleanup and to identify the required cleanup activities as mandated by CERCLA. (Appendix 3)

National Environmental Policy Act (NEPA). A U.S. federal law that, combined with the Environmental Quality Improvement Act, forms the cornerstone for U.S. environmental improvement policy. NEPA established the federal government's commitment to judge their actions relative to environmental impacts and to action that fostered harmony between human activities and nature. It established the Council on Environmental Quality and the requirement of an environmental impact statement. (Appendix 3)

National Pollutant Discharge Elimination System (NPDES). A state or EPA permitting system mandated by the Clean Water Act that imposes limits on discharges. (Appendix 3)

natural radiation background (radiation). Cosmic radiation emitted from stars and long-lived terrestrial radionuclides that are ubiquitously present in the earth's soils. (Chapter 14)

necrosis. Cell death resulting from disease or injury. (Chapter 7)

neoplasia. "Hyperplasia which is caused, at least in part, by an intrinsic heritable abnormality in the involved cells" (La Via and Hill, 1971). (Chapter 7)

neoplastic hyperplasia. Hyperplasia resulting from a hereditary change in the cell such that it no longer responds properly to chemical signals that normally control cell growth. Such cells can result in cancerous growth. (Chapter 7)

net reproductive rate (R_0). The expected number of females to be produced during the lifetime of a newborn female as estimated with a life table. (Chapter 10)

niche preemption. A rapid use and preemption of resources by a species that exploits them to the exclusion or severe disadvantage of another species. (Chapter 11)

nine aspects of disease association. Hill (1965) defined nine aspects of evidence fostering the accuracy of linkage between a risk factor and disease: strength of association, consistency of association, specificity of association, temporal association, biological gradient (dose-response) in the association, biological plausibility, coherence of the association, experimental support of association, and analogy. (Chapter 10)

"no action" alternative (to remediation of the site). A scenario in which one assesses if the contaminants at the waste site pose, or will pose in the future, a risk if left alone. (Chapter 13)

nonstochastic health effects. In contrast to stochastic health effects of radiation, nonstochastic health effects are those dependent on the magnitude of the dose in excess of a threshold. Some nonstochastic health effects of radiation include acute radiation syndrome, opacification of the eye lens, erythema of the skin, and temporary impairment of fertility. (Chapter 14)

no observed effect concentration (or level) (NOEC or NOEL). The highest concentration in a test for which there was no statistically significant difference in response from that of the control. (Chapter 8)

normal equivalent deviation (NED). As used in this book, the proportion dying in a toxicity test expressed in terms of standard deviations from the mean of a normal curve. (Chapter 9)

normal science. A major activity of any scientific discipline that works within the framework of established paradigms, increasing the amount and accuracy of knowledge within that framework. (Chapter 1)

normit. The metameter equal to the normal equivalent deviation (NED). The resulting analysis of dose- or concentration-effect data with the normit metameter is often called normit analysis and is essentially equivalent to probit analysis. (Chapter 9)

nuclear fission. The splitting of atomic nuclei with neutrons, resulting in the release of energy. (Chapter 14)

numerical response. A change in predator or grazer number through increased reproductive output, decreased mortality, or increased immigration in response to changes in prey or food densities. (Chapter 11)

octaves. Log_2 classes (e.g., 1–2, 2–4, 4–8, 8–16, 16–32, ... individuals) used in species abundance curves and representing doublings of the numbers of individuals in a species. (Chapter 11)

odds ratio. A measure of relative risk in case-control studies in epidemiology. The number of disease cases that (a) were or (b) were not exposed, and the number of controls that (c) were and (d) were not exposed to the risk factor, such as an etiological agent, are used to estimate the odds ratio: odds ratio = (a/b)/(c/d) or (ad)/(bc). (Chapter 10)

Oklo natural reactors. Naturally occurring nuclear reactors arising through biogeochemical processes approximately 1.8 billion years ago in Oklo (Gabon, Africa). (Preface)

oncogene. A gene involved in cancer. Cancer results from the mutation of this gene that was involved in the normal growth and differentiation of cells. (Chapter 7)

one-hit risk model (for carcinogenic effect). In risk assessment, this model (Equation 13.8) is used to predict risk if the estimated risk is 0.01 or higher. (Chapter 13)

optimal-foraging theory. The theory that the ideal forager will obtain a maximum net rate of energy gain by optimally allocating its time and energy to the various components of foraging. (Chapter 11)

optimal stress response. The optimal stress response involves a shift in the balance in energy allocation between somatic growth rate and longevity (survival) to optimize Darwinian fitness under stressful conditions. (Chapter 10)

Organization for Economic Cooperation and Development (OECD). An organization founded in 1960 with the purpose of enhancing economic growth, living standards, and financial stability and developing and holding forums on associated policies. Member countries are Australia, Austria, Belgium, Canada, Denmark, Finland, France, Germany, Greece, Iceland, Ireland, Italy, Japan, Luxembourg, Mexico, Netherlands, New Zealand, Norway, Portugal, Spain, Sweden, Switzerland, Turkey, United Kingdom, and United States. (Appendix 3)

oxidative stress. The damage to biomolecules from free oxyradicals. (Chapter 6)

oxyradical. A free radical involving an unshared electron of oxygen, e.g., $RO^{\bullet}$. (Chapter 6)

ozone hole. A hole or extreme thinning of ozone above the Antarctic due to the combined effects of circulation patterns above the Antarctic and ozone destruction as a consequence of CFC accumulation in the stratosphere. (Chapter 12)

paradigms. Generally accepted concepts that, in a healthy science, have withstood rigorous testing and are given enhanced status as explanations of fact and observation. (Chapter 1)

partial kill. A treatment in a toxicity test in which some, but not all, exposed individuals are killed. (Chapter 9)

partition coefficient. The quantitative expression for the concentration of a contaminant in one phase relative to the concentration in another phase, e.g., K_d or $K_p = [X_{(phase\ b)}]/[X_{(phase\ a)}]$. (Chapter 2)

pedosphere. That part of the earth made up of soils and where important soil processes are occurring. (Chapter 12)

peroxyradicals. Oxyradicals of organic compounds (R) of the form, $ROO^{\bullet}$. (Chapter 6)

persistent organic pollutants (POP). Those organic pollutants that are long-lived in the environment and tend to increase in concentration as they move through food chains. (Chapter 12)

persistent toxicants that bioaccumulate (PTB). *See* persistent organic pollutants. (Chapter 12)

pesticide. A substance used to prevent, destroy, repel, or mitigate any pest. Under FIFRA, insecticides, fungicides, rodenticides, dessicants, defoliants are included. (*See also* Federal Insecticide, Fungicide and Rodenticide Act.) (Appendix 3)

pharmacokinetics. The study and predictive modeling of the internal kinetics of drugs. (Chapter 3)

Phase I reactions. Reactions in the metabolism of organic contaminants in which reactive groups are added or made available. Although oxidation reactions are the most important Phase I reactions, hydrolysis and reduction reactions are also significant. (Chapter 3)

Phase II reactions. Reactions in the metabolism of organic contaminants in which conjugates are formed that inactivate the compound and foster elimination. (Chapter 3)

phocomelia. A developmental abnormality in which the individual is born with extremely short limbs because the long bones have failed to develop properly. The term is derived from the Greek words *phoke* and *melos* that mean seal and fin, respectively (Taussig, 1962), which refer to the appearance of the extremely short limbs of afflicted individuals. (Chapter 8)

photo-induced toxicity. Toxicity of a chemical in the presence of light due to the production of toxic, photolysis products. (Chapter 9)

photolysis. A photochemical process that leads to the degradation of an organic contaminant. (*See also* direct photolysis, indirect photolysis.) (Chapter 2)

photosensitivity. Sensitivity of cutaneous tissues to the effects of light evoked by a chemical. (Chapter 9)

pH-partition hypothesis. Bioavailability is determined by the diffusion of the unionized form through the gastrointestinal lumen, as determined by pK_a and pH. (Chapter 4)

physiologically based pharmacokinetics (PBPK) model. A pharmacokinetics model that includes physiological and anatomical features in describing internal kinetics. (Chapter 3)

physiologic hyperplasia. Nonpathological hyperplasia in response to a variety of usual stimuli such as that involved in the tissue repair process. (Chapter 7)

phytochelatin. A class of peptides in plants that are induced by and bind to metals. They can function in the regulation and detoxification of metals by plants. (Chapter 3)

Pielou's J. A measure of species evenness for a sample from a community. (Chapter 11)

Pielou's J′. A measure of species evenness for a community. (Chapter 11)

pollutant. "A substance that occurs in the environment at least in part as a result of man's activities, and which has a deleterious effect on living organisms" (Moriarty, 1983). (Chapters 1, 2)

pollution-induced community tolerance (PICT). An increase in tolerance to pollution resulting from species composition shifts in the community, acclimation of individuals, and genetic changes in populations in the community. (Chapter 11)

polygenic control. Control of some quality by several genes. (Chapter 10)

population. A group of individuals occupying a defined space at a particular time. (Chapter 10)

porins. Pores in the cell membrane that are nonspecific among ions. (Chapter 3)

porphyrins. Molecules having a tetrapyrrole ring (four simple, heterocyclic nitrogen ring compounds bound together to form a ring) that are produced as intermediates during heme synthesis in animals. (Chapter 6)

positron. A particle with the same mass of an electron but a positive charge. Also called a positive electron. (Chapter 14)

potentiation. Enhanced toxicity of a chemical in the presence of a second chemical that is not itself toxic at its concentration in the mixture. (Chapter 9)

power relationship. A mathematical relationship in which the Y variable is related to the X variable raised to some power, e.g., $Y = aX^b$. (Chapter 4)

precautionary principle. The conservative policy that, even in the absence of any clear evidence and in the presence of high scientific uncertainty, action should be taken if there is any reason to think that harm might be caused. (Chapter 13)

precipitate explanation. The obsolete and unreliable scientific practice of uncritical or untested acceptance of an explanation based on some ruling theory. (Chapter 1)

predictive risk assessment. A risk assessment dealing with a planned or proposed condition. (Chapter 13)

prevalence. The incidence rate of a disease multiplied by the amount of time that individuals were at risk. (Chapter 10)

prevention of significant deterioration (PSD). A program in the Clean Air Act with the goal of no significant deterioration of air quality in areas attaining NAAQS. (*See also* national ambient air quality standards.) (Appendix 3)

primary lamellae (filaments). Gill structures extending outward at right angles from the branchial arches. (Chapter 7)

principle of allocation. There exists a cost or trade-off to every allocation of energy resources. Energy spent by an individual organism on one function, process, or structure cannot be spent on another. Optimal allocation of resources enhances Darwinian fitness. (Chapter 10)

probabilistic (Bayesian) induction. A type of induction that uses probabilities associated with competing theories or explanations to decide which is the most probable. Certainties or credibilities are then assigned to competing explanations based on their associated probabilities. Instead of the quantal (accept or reject) falsification of a working hypothesis, Bayesian induction considers a hypothesis falsified if it were sufficiently improbable. (Chapter 13)

probit. A metameter produced by adding five to the normal equivalent deviation (NED). Forming the basis of probit analysis, the probit was first proposed in the 1930s to avoid negative numbers. (Chapter 9)

problem formulation (in ecological risk assessment). The planning and scoping phase that establishes the framework around which the risk assessment is done. (Chapter 13)

procarcinogen. A compound that is converted to a carcinogen. (Chapter 6)

product-limit (Kaplan-Meier) method. A nonparametric method for analyzing time-to-death or survival-time data that does not require a specific model for the survival curve. (Chapter 9)

promoter. An agent producing cancer by enhancing the growth of mutated cells. (Chapter 7)

promoter elements. Regulatory sequences upstream of the coding region of a gene. (Chapter 6)

propagule rain. Relative to metapopulation dynamics, the presence of a seed bank or dormant stage for a species that continually introduces individuals to the patch regardless of the density of occupancy in the surrounding patches. This propagule rain increases the likelihood of population reappearance and decreases the likelihood of patch extinction. (Chapter 10)

proportional diluter. A special apparatus used in flow-through toxicity tests to mix and deliver a series of dilutions of a toxic solution to exposure tanks. (Chapter 9)

proportional hazard model. A survival or time-to-death model that relates the hazard (proneness to die or risk of dying at any time, t) of one group (e.g., smokers) quantitatively to that of a reference group (e.g., nonsmokers). (Chapter 9)

proteinuria. The presence of protein in the urine. (Chapter 13)

proteotoxicity. A toxic or adverse effect of a chemical or physical agent with an underlying mechanism of protein damage. (Chapter 6)

proto-oncogene. A gene involved in some way with the normal growth (enhancement) and differentiation of cells and which, upon mutation, becomes an oncogene. (Chapter 7)

Ptolemaic incongruity. The false paradigm that any particular level of biological organization holds a more central or important role than another in the science of ecotoxicology. (Chapter 7)

pyrogenic. Literally, born of fire, the term refers to the generation of organic compounds during the high-temperature combustion of complex organic matter. The polycyclic aromatic hydrocarbons (PAH) are examples of pyrogenic compounds. (Chapter 2)

quantitative structure-activity relationship (QSAR). A quantitative, often statistical, relationship between a molecular quality or molecular qualities and some activity, i.e., bioavailability or toxicity. (Chapter 4)

quotient method. The use of the hazard quotient as a crude indicator of hazard. (Chapter 13)

radiosensitizers. Chemicals, such as derivatives of nitro imidazoles, used during radiation treatment to enhance the production of free radicals, which then kill cancer cells. (Chapter 6)

rarefaction estimate of richness. An estimate of species richness expressed relative to that of a sample with a standard number of individuals in it. (Chapter 11)

rate constant based model. A compartment model that employs rate constants to quantify the rate of change in concentration or amount of toxicant. (Chapter 3)

rate constants. Constants used in mathematical models that have units of $time^{-1}$. They describe the rate at which a contaminant transfers from one compartment to another. Use of a rate constant implies a first order kinetic process where the rate of transport is assumed proportional to the amount of contaminant in the compartment. (Chapter 14)

rate-of-living theory of aging. The total metabolic expenditure of a genotype is generally fixed, and longevity depends on the rate of energy expenditure. (Chapter 10)

rate ratio. The ratio of disease incidence rates for two populations. Rate ratio $= I_A/I_0$, where I_A = incidence rate in population A, and I_0 = incidence rate in the reference or control population. (Chapter 10)

realized niche. That portion of a species' fundamental niche that it actually occupies. (Chapter 11)

reasonable maximum exposure (RME). Exposure calculated for a chemical of potential concern during a risk assessment. It is a conservative estimate of exposure that is computed differently depending on the route of exposure. (Chapter 13)

receptor antagonism. Antagonism that involves the binding of the toxicants to the same receptor and one toxicant blocking the other from fully expressing its toxicity. (Chapter 9)

redox cycling. In the context of contaminant (quinones, aromatic nitro compounds, aromatic hydroxylamines, bipyridyls, and some chelated metals) involvement in the generation of oxyradicals, redox cycling occurs if contaminants are reduced to radicals and then participate in redox reactions to produce the superoxide radical from molecular oxygen. The contaminant exists in its original form at the end of the redox reactions and is available to recycle many times through this process and produce more oxyradicals. (Chapter 6)

redundancy hypothesis. Many species are redundant, and their lose will not influence the community function as long as crucial (e.g., keystone and dominant) species populations are maintained. (Chapter 11)

reference dose (RfD) (for noncarcinogenic effects). The best estimate of the daily exposure for humans, including the most sensitive subpopulation, that will result in no significant risk of an adverse health effect if it is not exceeded. (*See also* chronic reference dose, developmental reference dose, subchronic reference dose.) (Chapter 13)

region (geographical). An "area of the earth's surface differentiated by its specific characteristics" (Monkhouse, 1965). (Chapter 12)

regulation (legal). Specific regulations derived by U.S. federal agencies for ensuring that the intent of a law is met. Environmental regulations "specify conditions and requirements to be met, ... provide a schedule for compliance, and ... record any exceptions to the regulated community" (Mackenthun and Bregman, 1992). (Appendix 3)

relative bioavailability. The bioavailability estimated for a dose administered by any route or formulation relative to a dose administered in a reference (or alternate) route or formulation. (Chapter 4)

relative risk. In survival time analysis, the risk of one group expressed as a multiple of that of another. Relative risk is usually estimated with a hazard model. In epidemiology, it is the ratio of occurrences of the disease in two populations. (Chapters 9, 10)

remedial investigation (RI). Part of a remedial investigation and feasibility study that has three parts: characterization or the type and degree of the contamination, human risk assessment, and ecological risk assessment (EPA, 1994). (Chapter 13)

remedial investigation and feasibility study (RI/FS). For an EPA Superfund site, a study that has as its goal the implementation of "remedies that reduce, control, or eliminate risks to human health and the environment" or, more specifically, the accumulation of "information sufficient to support an informed risk management decision regarding which remedy appears to be most appropriate for a given site" (EPA, 1989d). (Chapter 13)

remote sensing. Technologies that allow the acquisition and analysis of data without requiring physical contact with the land or water surface being studied. Most determine qualities or characteristics of areas of interest based on measurements of visible light, infrared radiation, or radio energy coming from them. (Chapter 12)

repair fidelity (of DNA). The accuracy in repairing and returning the DNA to its original state after damage. (Chapter 7)

reproductive value (V_A). The expected contribution of offspring during the life of an individual of an age class x in a life table. (Chapter 10)

rescue effect. The increased probability of a vacated-patch reoccupation in a metapopulation as the number of nearby, occupied patches increases. (Chapter 10)

residual body. A cell vacuole containing lipofuscin, a degradation product of lipid oxidation. (Chapter 7)

resilience (community). The number of times a community can return to its normal state after perturbation. (Chapter 11)

resistance. The term *resistance* is often reserved for the enhanced ability to cope with a factor due to genetic adaptation. The term *tolerance* is often reserved for enhanced abilities associated with physiological acclimation. *Tolerance* is used in this book for both acclimation and genetic adaptation. (Chapter 10)

Resource Conservation and Recovery Act (RCRA). A U.S. federal law that regulates production, treatment, storage and disposal of hazardous wastes. (Appendix 3)

respiratory lamellae. *See* secondary lamellae. (Chapter 7)

retention effect. The K_{OW}s of persistent organic pollutants (POP) influence their global movement toward higher latitudes. Those POPs with high lipophility tend to be held more firmly in solid phases such as soil and vegetation than less lipophilic POPs. Consequently, they spend less time in the atmosphere and are less available for transport in that medium. (Chapter 12)

retroactive risk assessment. A risk assessment dealing with an existing condition. (Chapter 13)

Ricker model. A difference equation model (Eq. 10.10) for growth of populations with nonoverlapping generations or for experimental designs with discrete intervals of population growth. (Chapter 10)

risk. As used in risk assessments, the probability (or likelihood) of some adverse consequence occurring to an exposed human or to an exposed ecological entity. (Chapter 13)

risk assessment. The process by which one estimates the probability or likelihood of some adverse effect(s) of a present or planned release to either human or ecological entities. (Chapter 13)

risk assessor. A person or group of people "who actually organizes and analyzes site data, develops exposure and risk calculations, and prepares a risk assessment report" (EPA, 1989d). (Chapter 13)

risk characterization (in ecological risk assessment). The last step of the ecological risk assessment that draws together the information generated from previous steps to produce a statement of the likelihood of an adverse effect to the assessment endpoint. (Chapter 13)

risk factor. Any quality of an individual (e.g., age) or an etiological factor (e.g., chronically exposed to high levels of the toxicant) that modifies an individual's risk of developing the disease in question (Chapter 10). Specific to chemical risk assessment (Chapter 13), a chemical concentration multiplied by a toxicity value estimates the risk factor. Relative to radiation's effects (Chapter 14), a risk factor gives the probability of a deleterious effect for each millisievert of dose received.

risk hypotheses (in ecological risk assessment). As part of the conceptual model, they are clear statements of postulated or predicted effects of the contaminant on the assessment endpoint. (Chapter 13)

risk manager. "The individual or group who serves as primary decision-maker for a [waste] site" (EPA, 1989d). (Chapter 13)

rivet popper hypothesis. Species in a community are like rivets that hold an airplane together and contribute to its proper functioning. The loss of each rivet weakens the structure. (Chapter 11)

roentgen (R). A measure of the amount of energy deposited in some material by a certain amount of radiation. It is expressed relative to energy dissipation in 1 cm^3 of dry air. Use of R to express dose allows one to normalize for the different amounts of energy that are deposited in materials such as tissue by different types of radiation. (Chapter 10)

roentgen equivalent man (rem). A measure of radiation that takes into account the differences in potential biological effects of different types of radiation. It relates the dose received to potential damage. (Chapter 1)

r-strategy. An opportunistic strategy favoring species that establish themselves quickly, grow quickly to exploit as many resources as possible, and produce many offspring. (Chapter 11)

rules of practical causal inference. Fox's (Fox, 1991) rules of practical causal inference are used in ecotoxicology to infer causality for toxicant exposure/effect scenarios. (Chapter 10)

safe concentration. "The highest concentration of toxicant that will permit normal propagation of fish and other aquatic life in receiving waters. The concept of a 'safe concentration' is a biological concept, whereas the 'no observed effect concentration' is a statistically defined concentration" (Weber et al., 1989). (Chapter 8)

Safe Drinking Water Act (SDWA). A U.S. federal law that protects the water supply of the public (systems supplying more than 25 people or with more than 15 service outlets) by setting water quality standards. The law also protects source aquifers from contamination by establishing permitting requirements for underground disposal of waste. (Appendix 3)

saprobien spectrum. The characteristic change in community composition at different distances below the discharge of putrescible organic waste to a river or stream. (Chapter 11)

scaling. The handling or transformation of allometric data to produce a quantitative relationship between organism (or species) size and some characteristic such as metabolic rate, gill surface area, lung ventilation rate, or biochemical activity. (Chapter 4)

scoliosis. Lateral curvature of the spine. (Chapter 8)

scope of activity. The difference between the rates of oxygen consumption under maximal and minimal activity levels. It reflects the respiratory capacity available for the diverse demands on and activities of an organism. (Chapter 8)

scope of growth. An index (P = production) calculated as the amount of energy taken into the organism in its food (A) minus the energy used for respiration (R) and excretion (U): P = A − R − U. It is the amount of energy available for growth or production of young. (Chapter 8)

secondary lamellae (respiratory lamellae). These gill structures are parallel rows of projections on the dorsal and ventral sides of each primary lamella. They are the primary sites of gas exchange of the gills. (Chapter 7)

selection components. Components of the life cycle of an individual upon which natural selection can act. They are viability selection, sexual selection, meiotic drive, gametic selection, and fecundity selection. (Chapter 10)

Selyean stress. A nonspecific response of the body when extraordinary demands are made of it. "The state manifested by a specific syndrome which consists of all the nonspecifically induced changes within a biological system" (Selye, 1956). (Chapter 8)

semelparous species. A species that reproduces once. (Chapter 10)

sentinel species. A feral, caged, or endemic species used in measuring and indicating the level of contamination or effect during a biomonitoring exercise. The proverbial canary in the coal mine is an example of a sentinel species. (Chapter 6)

sexual selection. A component of the life cycle of an individual in which natural selection can occur, involving differential mating success of individuals. (Chapter 10)

Shannon diversity index. A measure of species diversity of a community. (Chapter 11)

Shapiro-Wilk's test. A statistical test of the null hypothesis that data are normally distributed. Used in this book to test this assumption prior to performing one-way analysis of variance on sublethal or chronic lethal effects data. (Chapter 8)

Shelford's law of tolerance. A species' tolerance(s) along an environmental gradient (or series of environmental gradients) will determine its population distribution and size in the environment. (Chapter 10)

similar joint action. Toxicants in mixture act by the same mode of action and "one component can be substituted at a constant proportion for the other. ... [T]oxicity of a mixture is predictable directly from that of the constituents if their relative proportions are known" (Finney 1947). (Chapter 9)

sister chromatid. At the metaphase plate, chromosomes are composed of two chromatids called sister chromatids. (Chapter 7)

sister chromatid exchange (SCE). The exchange of DNA between sister chromatids as a consequence of DNA breakage followed by reunion and crossing over of DNA segments of the chromatids. (Chapter 7)

slope factor (SF). In human risk estimation, SF is the slope (risk or probability of occurrence per unit of dose or intake) for the risk-dose model used to estimate the probability of a cancer at a specified exposure. (Chapter 13)

solvent drag. The movement of a solute (contaminant) along with the bulk movement of the solution. (Chapter 3)

somatic death. Death of an individual organism. (Chapter 7)

somatic risk. The risk of an adverse effect to the exposed individual associated with genetic damage to somatic cells, e.g., damage leading to cancer. (Chapter 7)

sorption. The term used instead of *adsorption* in this book when the specific mechanism by which a compound in solution becomes associated with a solid surface is unknown or undefined. (Chapter 3)

sorption coefficient. The quantitative expression for the concentration of a contaminant associated with the sorbed (i.e., sediment or particulate) phase relative to the concentration in the dissolved phase, e.g., $K_p = [X_{(sediment)}]/[X_{(water)}]$. (Chapter 2)

Spearman-Karber method. A nonparametric method to estimate the LC50, EC50, or LD50 when it is difficult or unnecessary to assume a specific model for the dose- or concentration-effect data. (*See also* median effective concentration, median lethal concentration, median lethal dose.) (Chapter 9)

species assemblage. An operationally defined subset of the entire community. (Chapter 11)

species diversity (= heterozygosity). The heterogeneity or diversity of the community, considering both species richness and evenness. (Chapter 11)

species evenness. The degree to which the individuals in the community are evenly or uniformly distributed among species. (Chapter 11)

species richness. The number of species present in the community. (Chapter 11)

specific activity concept (for radiotracer use). The radionuclide used to trace or quantify the movement of a stable nuclide (e.g., ^{14}C for stable C) is assumed to behave identically in chemical and biological processes as its nonradioactive analog (e.g., stable C). (Chapter 5)

spiked bioassay approach (SB). A sediment toxicity test method to generate a concentration-response model for or test hypotheses regarding effects to individuals placed in sediments spiked with different amounts of toxicant. (Chapter 9)

spillover hypothesis. Based on the assumption that binding by metallothionein sequesters toxic metals away from sites of action, this hypothesis states that toxic effects will begin to be seen after exceeding the capacity of the metallothionein present at any time to bind metals. The unbound metals then "spill over" to interact at sites of adverse action. (Chapter 6)

stable population. If conditions do not change with time, a population with a particular r will eventually establish a stable distribution of individuals among the various age classes. Such a population is called a stable population. (Chapter 10)

standard. Legal limits (concentration or intensity) permitted for a specific water body, based on criteria and the specified use of a water body. (Chapter 1)

state implementation plan (SIP). State plans required by the Clean Air Act that outline steps toward meeting and maintaining national air quality standards. (Appendix 3)

static-renewal test. A modified static toxicity test in which solutions are completely or partially replaced with new solutions at set periods during exposures or in which organisms are periodically transferred to new solutions. (Chapter 9)

static toxicity test. A type of aquatic toxicity test in which the exposure water is not changed during the test. (Chapter 9)

Steel's many-one rank test. A nonparametric, post-ANOVA test often employed in the analysis of sublethal and chronic lethal effects data. (Chapter 8).

stochastic health effects. Cancer and genetic disorders for which initiation of effects by radiation is probabilistic and for which the risk of incurring cancer or genetic effects is proportional, without a threshold, to the dose in the relevant tissue. (Chapter 14)

stress. "At any level of ecological organization, a response to or effect of a recent, disorganizing or detrimental factor" (Newman, 1995). (*See also* Selyean stress.) (Chapter 2)

stressor. That which produces stress (Chapter 1); in ecological risk assessment, "any chemical, physical, or biological entity that can induce adverse effects on *ecological components*, that is, individuals, populations, communities, or ecosystems" (Norton et al., 1992). (Chapter 13)

stress protein fingerprinting. The proposed use of the patterns of stress-protein induction seen in the field to suggest the particular toxicant inducing the response. Patterns from organisms sampled from the field can be compared with those obtained with single-candidate toxicants in the laboratory. (Chapter 6)

stress proteins. A class of proteins involved in lessening the damage to proteins associated with a variety of stressors, including heat, anoxia, UV radiation, arsenate, metals, and some xenobiotics. This term is also used by several ecotoxicologists (e.g., Sanders and Dyer, 1994) in a more generic context to mean a protein induced in response to a stressor. Such a definition would include proteins such as metallothioneins. (*See also* cellular stress response, heat shock proteins.) (Chapter 6)

stress theory of aging. Stress shortens longevity by accelerating energy expenditure. Selection takes place for resistance to stress, and as an epiphenomenon, individuals resistant to stress will predominate in extreme age classes of a population. The diminution of homeostasis under stress with age should be slowest in individuals with highest longevity. (*See also* rate-of-living theory of aging.) (Chapter 10)

structure-activity relationship (SAR). A relationship between molecular qualities and some activity, such as bioavailability or toxicity. (Chapter 4)

subchronic reference dose (RfD$_s$). A RfD derived from short-term exposure data. (Chapter 13)

subcooled liquid vapor pressure (P$_L$). The liquid-vapor pressure corrected or adjusted for the heat of fusion, the energy needed to convert a mole of a compound from a solid to a liquid phase. Its use allows the expression of *liquid* vapor pressures at a specific temperature for organic compounds with widely varying melting temperatures. (Chapter 12)

sublethal effects. Effects seen at concentrations below those producing direct somatic death, e.g., slowed growth of an individual or diminished reproduction. (Chapter 8)

sulfotransferase. A Phase II enzyme that conjugates sulfates to xenobiotics or their metabolites. (Chapter 6)

summation rule. Concentrations of toxicants can be added together to predict effects under the assumption of additivity. Although applied frequently in different ecotoxicological methods, this rule is not generally valid. (Chapter 9)

superoxide dismutase (SOD). An enzyme catalyzing the reaction: $2O_2^- + 2H^+ \rightarrow H_2O_2 + O_2$. It functions to reduce oxidative stress. (Chapter 6)

suppressor gene. A gene that functions normally to suppress cell growth and can inhibit abnormal growth. (Chapter 7)

synergism. Toxicant synergism occurs if the joint effect level of a mixture was higher than that predicted by summing the predicted, separate effects for the individual toxicants in the mixture. (Chapter 9)

target organ. The specific or characteristic organ in which lesions from a toxicant occur or are expected to occur based on toxicant transport to, accumulation in, or activation by that organ. (Chapter 7).

taxocene. A taxonomically defined subset of the entire community. (Chapter 11)

temperature of condensation (T_C). The temperature (°C) at which a compound condenses or partitions from the gaseous to the nongaseous phase. (Chapter 12)

teratogen. A chemical or physical agent capable of causing a developmental malformation. (Chapter 8)

teratogenic. Capable of causing developmental malformations. (Chapter 6)

teratogenic index (TI). The mortality of eggs expressed as an LC50 divided by the TC50 (EC50 for production of abnormal embryos). The TI is thought to reflect the developmental hazard of a contaminant. (*See also* median effective concentration, median lethal concentration, median teratogenic concentration.) (Chapter 8)

teratology. The science of fetal and embryonic abnormal development of anatomical structures. (Chapter 8)

tetrad. Two homologous chromosomes composed of two chromatids, each coming together at the metaphase plate to form a tetrad. (Chapter 7)

threshold theory. A theory that assumes no response (dose-related incidence or risk of cancer) below a certain low dose. Above the threshold, the slope of the response-vs.-dose curve increases rapidly. The dose-response curve takes on the appearance of a hockey stick. (Chapter 7)

tolerance. This term is often reserved for enhanced ability to cope with a factor due to physiological acclimation. The term *resistance* is used if the enhanced abilities are associated with genetic adaptation; the term *tolerance* is used in this book for both acclimation and genetic adaptation. (Chapter 10)

toxic equivalence (TEQ). The combined toxicity of dioxins, dibenzofurans, or dioxinlike PCBs (i.e., compounds that have very similar toxic modes of action involving binding to the aryl hydrocarbon receptor) expressed in units of toxicity of 2,3,7,8-tetrachlorodibenzo-p-dioxin (often abbreviated TCDD). This summing of compound effects to generate a TEQ is done using TEF factors. (*See also* toxic equivalency factor.) (Chapter 9)

toxic equivalency factor (TEF). An empirically derived factor that scales the toxicity of a dioxin, dibenzofuran, or dioxinlike PCB (i.e., compounds that have toxic modes of action involving binding to the aryl hydrocarbon receptor) to that of 2,3,7,8-tetrachlorodibenzo-p-dioxin (often abbreviated TCDD). (Chapter 9)

toxicity value. In risk assessment, the toxicity value is a factor used to estimate a risk factor. Its estimation involves one of two methods: (1) Use the slope of a published effect-dose relationship to estimate the risk factor: R = Slope * C, where C = toxicant concentration; (2) use a reference dose (RfD) value in a similar manner. (*See also* reference dose.) (Chapter 13)

toxicokinetics. The study and predictive modeling of the internal kinetics of poisons. (Chapter 3)

Toxic Substances Control Act (TSCA). A U.S. law that regulates, through the EPA, the manufacture, processing, transport, use, import, and disposal of chemicals or chemical mixtures that may pose an unreasonable risk to health or the environment. (Appendix 3)

toxic unit (TU). Amount or concentration of a toxicant expressed in units of lethality such as units of LD50 or LC50. For example, if toxic units are based on the LC50, then a chemical with an LC50 of 20 mg l^{-1} would be present at 0.5 TU in a 10-mg l^{-1} solution. (*See also* median lethal concentration, median lethal dose.) (Chapter 9)

trace metal. A term implying that the concentration of the metallic element measured is very low (≤ppm). (Chapter 2)

tracer. Radioactive contaminants are often present in such low concentrations — compared with the concentrations of similar nonradioactive elements — that the behavior of the contaminant is governed by that of the similar element rather than its own mass characteristics. Thus the radionuclide traces or mimics the normal behavior of the similar element. (Chapter 14)

transcription factor. Protein involved in regulation of gene expression, binding with specific promoter elements of a gene. (Chapter 6)

transgenic organism. An organism modified by genetic engineering. (Chapter 6)

trophic dilution. The decrease in contaminant concentration as trophic level increases. Trophic dilution results from a net balance of ingestion rate, uptake from food, internal transformation, and elimination processes favoring loss of contaminant that enters the organism via food. (Chapter 5)

trophic enrichment. *See* biomagnification. (Chapter 5)

t-test with a Bonferroni adjustment. A parametric, post-ANOVA test used often in the analysis of sublethal and chronic lethal effects data. (Chapter 8)

t-test with a Dunn-Šidák adjustment. A parametric, post-ANOVA test used rarely in the analysis of sublethal and chronic lethal effects data that has slightly better statistical power than the t-test with a Bonferroni adjustment. (Chapter 8)

twin-tracer technique. An experimental technique that simultaneously introduces a radiotracer of the substance being assimilated and an inert tracer that will not be assimilated, thus providing a basis for evaluating the assimilation. (Chapter 5)

Type A organism. According to the scheme of Campbell et al. (1988), an organism in contact with sediments but unable to ingest particulates. The implication is bioavailability from interstitial water but not sediment-associated particulates. Some examples include rooted macrophytes and benthic algae. (*See also* Type B organism.) (Chapter 4)

Type B organism. According to the scheme of Campbell et al. (1988), an organism in contact with sediments and capable of ingesting particulates. The implication is bioavailability from interstitial water and sediment-associated particulates. Some examples include detritivores and suspension feeders. (*See also* Type A organism.) (Chapter 4)

uncertainty factors. In risk assessment, factors to decrease the NOAEL (or LOAEL) in order to compensate in a conservative direction for uncertainty. (*See also* lowest-observed-effect level, no-observed-effect level.) (Chapter 13)

uptake. The movement of a contaminant into or onto an organism. (Chapter 3)

uridinediphospho glucuronosyltransferase (UDP-glucuronosyltransferase, UDP-GT). A Phase II enzyme that transfers glucuronic acid from uridine diphosphate glucuronic acid to electrophilic xenobiotics or their metabolites. It also binds covalently with electrophilic compounds such as PAHs. (Chapter 6)

viability selection. A component of the life cycle of an individual in which natural selection can occur through the differential survival of individuals. It begins at the formation of the zygote and continues throughout the life of the individual. (Chapter 10)

vital rates. Rates at which important life-cycle processes such as birth, migration, and death occur for individuals in populations. (Chapter 10)

vulnerability (community). Susceptibility to irreversible damage by toxicants. (Chapter 11)

Wahlund effect. There will be a net deficit of heterozygotes when two populations, each in Hardy-Weinberg equilibrium but with different allele frequencies, are mixed and the genotype frequencies quantified in a combined population sample. (Chapter 10)

waldsterben. "The widespread and substantial decline in growth and the change in behavior of many softwood and hardwood forest ecosystems in central Europe" (Schütt and Cowling,1985). (Chapter 12)

weakest link incongruity. An incongruous extension of the critical life stage concept that protection of the most sensitive stage will ensure protection of all life stages. The dubious extension is made in which one assumes that exposure of field populations to concentrations identified in testing as causing significant mortality at a critical life stage will result in significant impact on the field population. This may or may not be true. (Chapter 9)

weathering. The long term process where the relative concentrations of contaminants in a complex mixture change over time due to the different physical and chemical properties of the individual compounds. (Chapter 2)

Weibull metameter. A metameter used occasionally in dose- or concentration-effect data analysis. It has the form, $U = \ln[-\ln(1 - P)]$, where P is the proportion dead. (Chapter 9)

weight of evidence. In risk assessment, this phrase appears to refer to whether a *reasonable person* reviewing the available information *could* agree that the conclusion was plausible. The more the evidence supports the conclusion, the stronger is the "weight of evidence." It could mean a quantitative, semiquantitative, or qualitative estimate of the degree to which the evidence supports or undermines the conclusion. (Chapter 13)

wet deposition. Deposition in precipitation of pollutants that were formed in the liquid media of the precipitation and that were incorporated into the precipitation during rain-out. (Chapter 12)

Wilcoxon rank sum test with Bonferroni's adjustment. A nonparametric, post-ANOVA test often employed in the analysis of sublethal and chronic lethal effects data. (Chapter 8)

Williams's test. A parametric test that is more powerful than other post-ANOVA tests used to analyze sublethal and chronic lethal data. It assumes that a monotonic trend (increase or decrease) can occur with increasing concentration. (Chapter 8)

working hypothesis. A hypothesis that is used to determine fact during scientific inquiry. It is not assumed to be true and only serves to test facts. (Chapter 1)

xenobiotic. A "foreign chemical or material not produced in nature and not normally considered a constitutive component of a specified biological system. [It is] usually applied to manufactured chemicals" (Rand and Petrocelli, 1985). (Chapters 1, 2)

x-rays. Electromagnetic photons emitted from the shells of electrons that surround the nucleus. (Chapter 14)

Zenker's necrosis. Necrosis (cell death) that occurs in skeletal muscle and is similar to coagulation necrosis. (Chapter 7)

References

Adams, S.M., Biological indicators of stress in fish, *Am. Fish. Soc. Symp.*, 8, 1–8, 1990.

Adams, S.M., W.D. Cumby, M.S. Greeley, Jr., M.G. Ryon, and E.M. Schilling, Relationships between physiological and fish population responses in a contaminated stream, *Environ. Toxicol. Chem.*, 11, 1549–1557, 1992.

Adamus, C.L. and M.J. Bergman, Estimating nonpoint source pollution loads with a GIS screening model, *Water Resour. Bull.*, 31, 647–655, 1995.

Adolph, E.F., Quantitative relations in the physiological constitutions of mammals, *Science* (Washington, D.C.), 109, 579–585, 1949.

Agard, D.A., To fold or not to fold..., *Science* (Washington, D.C.), 260, 1903–1904, 1993.

Aguilar, A. and A. Borrell, Reproductive transfer and variation of body load of organochlorine pollutants with age in fin whales (*Balaenoptera physalus*), *Arch. Environ. Contam. Toxicol.*, 27, 546–554, 1994.

Ahlbom, A., *Biostatistics for Epidemiologists,* Lewis Publishers, Boca Raton, FL, 1993.

Alberts, B., D. Bray, J. Lewis, M. Raff, K. Roberts, and J.D. Watson, *Molecular Biology of the Cell,* Garland Publishing, New York, 1983.

Allen, H.E., R.H. Hall, and T.D. Brisbin, Metal speciation: effects on aquatic toxicity, *Environ. Sci. Technol.*, 14, 441–442, 1980.

Aloj Totaro, E., F.A. Pisanti, P. Glees, and A. Continillo, The effect of copper pollution on mitochondrial degeneration, *Mar. Environ. Res.*, 18, 245–253, 1986.

Aloj Totaro, E., F.A. Pisani, and P. Glees, The role of copper level in the formation of neuronal lipofuscin in the spinal ganglia of *Torpedo m., Mar. Environ. Res.*, 15, 153–163, 1985.

Al-Sabti, K., Frequency of chromosomal aberrations in the rainbow trout, *Salmo gairdneri* Rich., exposed to five pollutants, *J. Fish Biol.*, 26, 13–19, 1985.

Al-Sabti, K. and J. Hardig, Micronucleus test in fish for monitoring the genotoxic effects of industrial waste products in the Baltic Sea, Sweden, *Comp. Biochem. Physiol.*, 97C, 179–182, 1990.

Al-Sabti, K. and B. Kurelec, Chromosomal aberrations in onion (*Allium cepa*) induced by water chlorination by-products, *Bull. Environ. Contam. Toxicol.*, 34, 80–88, 1985a.

Al-Sabti, K. and B. Kurelec, Induction of chromosomal aberrations in the mussel *Mytilus galloprovincialis* Watch, *Bull. Environ. Contam. Toxicol.*, 35, 660–665, 1985b.

Ambrose, P., Osprey revival from DDT complete in Chesapeake Bay, *Mar. Pollut. Bull.*, 42, 338, 2001.

Amiard-Triquet, C., D. Pain, G. Mauvais, and L. Pinault, Lead poisoning in waterfowl: field and experimental data, in *Impact of Heavy Metals on the Environment,* Vernet, J.-P., Ed., Elsevier Science Publishers B.V., Amsterdam, 1992.

Amiro, B.D., Radiological dose conversion factors for generic non-human biota used for screening potential ecological impacts, *J. Environ. Radioact.*, 35, 37–51, 1997.

Anderson, D.P., Immunological indicators: effects of environmental stress on immune protection and disease outbreaks, *Am. Fish. Soc. Symp.*, 8, 38–50, 1990.

Anderson, E.V., Phasing Lead Out of Gasoline: Hard Knocks for Lead Alkyls Producers, *Chem. Eng. News,* Feb. 6, 1978, pp. 12–16.

Anderson, P.D. and L.J. Weber, Toxic response as a quantitative function of body size, *Toxicol. Appl. Pharmacol.*, 33, 471–483, 1975.

Anderson, S.L. and F.L. Harrison, Effects of Radiation on Aquatic Organisms and Radiological Methodologies for Effects Assessment, EPA Report 5201-85-016, U.S. Environmental Protection Agency, Washington, D.C., 1986, p. 128.

Andreae, M.O., Biomass burning: its history, use, and distribution and its impact on environmental quality and global climate, in *Global Biomass Burning: Atmospheric, Climatic, and Biospheric Implications,* Levine, J.S., Ed., MIT Press, Cambridge, MA, 1991.

Andrew, R.W., K.E. Biesinger, and G.E. Glass, Effects of inorganic complexing on the toxicity of copper in *Daphnia magna, Water Res.,* 11, 309–315, 1977.

Andrews, G.K., Regulation of metallothionein gene expression by oxidative stress and metal ions, *Biochem. Pharmacol.,* 59, 95–104, 2000.

Andrews, G.K., Cellular zinc sensors: MTF-1 regulation of gene expression, *Biometals,* 14, 223–237, 2001.

Angelone, M. and C. Bini, Trace elements concentrations in soils and plants of Western Europe, in *Biogeochemistry of Trace Metals,* Adriano, D.C., Ed., Lewis Publishers, Boca Raton, FL, 1992.

Ankley, G.T., G.L. Phipps, E.N. Leonard, D.A. Benoit, V.R. Mattson, P.A. Kosian, A.M. Cotter, J.R. Dierkes, D.J. Hansen, and J.D. Mahony, Acid-volatile sulfide as a factor mediating cadmium and nickel bioavailability in contaminated sediments, *Environ. Toxicol. Chem.,* 10, 1299–1307, 1991.

Anon., Heart Disease, Cancer Linked to Trace Metals, *Chem. Eng. News,* May 3, 1976, pp. 24–27.

Anon., Hooker Settles on Hyde Park Dump Cleanup, *Chem. Eng. News,* Jan. 26, 1981, p. 10.

Anon., Most of Love Canal Habitable, EPA Says, *Chem. Eng. News,* July 19, 1982, p. 6.

Anon., Lead Use in Gasoline: EPA Proposes 91% Cut by 1986, *Chem. Eng. News,* Aug. 6, 1984a, p. 4.

Anon., India's Chemical Tragedy: Death Toll at Bhopal Still Rising, *Chem. Eng. News,* Dec. 10, 1984b, pp. 6–7.

Anon., Global Climate Warming: Trace Gases Other than CO_2 Play Role, *Chem. Eng. News,* May 6, 1985, pp. 6–7.

Anon., Appendix II: Method 1311 toxicity characteristic leaching procedure, *Federal Register* 55, 11863–11875, 1990.

Anspaugh, L.R., R.J. Catlin, and M. Goldman, The global impact of the Chernobyl reactor accident, *Science* (Washington, D.C.), 242, 1513–1519, 1988.

APHA, *Standard Methods for the Examination of Water and Wastewater,* 15th ed., American Public Health Association, Washington, D.C., 1981.

Apple, G.J., W.G. Hunter, and S. Bisgaard, Scientific data and environmental regulation, in *Statistics and the Law,* DeGroot, M.H., Fienberg, S.E., and Kadane, J.B., Eds., John Wiley & Sons, New York, 1986.

Armitage, P. and I. Allen, Methods of estimating the LD 50 in quantal response data, *J. Hyg.,* 48, 298–322, 1950.

ASTM, Standard guide for conducting the frog embryo teratogenesis assay: *Xenopus* (FETAX), in *Annual Book of ASTM Standards,* American Society for Testing and Materials, Philadelphia, 1993.

Atlas, R.M. and R. Barta, *Microbial Ecology: Fundamentals and Applications,* Addison-Wesley Publishing Co., Reading, MA, 1981.

Atchison, G.J., M.G. Henry, and M.B. Sandheinrich, Effects of metals on fish: a review, *Environ. Biol. Fishes,* 18, 11–25, 1987.

Atchison, G.J., M.B. Sandheinrich, and M.D. Bryan, Effects of environmental stressors on interspecific interactions of aquatic animals, in *Ecotoxicology: A Hierarchical Treatment,* Newman, M.C. and Jagoe, C.H., Eds., CRC Press, Boca Raton, FL, 1996.

Aust, A.E., Mutations and cancer, in *Genetic Toxicology,* Li, A.P. and Heflich, R.H., Eds., CRC Press, Boca Raton, FL, 1991.

Austin, A. and J. Deniseger, Periphyton community changes along a heavy metals gradient in a long narrow lake, *Environ. Exp. Bot.,* 25, 41–52, 1985.

Austin, A., J. Deniseger, and M.J.R. Clark, Lake algal populations and physico-chemical changes after 14 years input of metallic wastes, *Water Res.,* 19, 299–308, 1985.

Avery, T.E. and G.L. Berlin, *Interpretation of Aerial Photographs,* MacMillan Publishing Co., New York, 1985.

Azenha, M., M.T. Vasconcelos, and J.P.S. Cabral, Organic ligands reduce copper toxicity in *Pseudomonas syringae, Environ. Toxicol. Chem.,* 14, 369–373, 1995.

Babukutty, Y. and J. Chacko, Chemical partitioning and bioavailability of lead and nickel in an estuarine system, *Environ. Toxicol. Chem.,* 14, 427–434, 1995.

Baes, C.F., Jr. and S.B. McLaughlin, Trace elements in tree rings: evidence of recent and historical air pollution, *Science* (Washington, D.C.), 224, 494–497, 1984.

Baeyens, W., C. Meuleman, B. Muhaya, and M. Leermakers, Behaviour and speciation of mercury in the Scheldt estuary (water, sediments, and benthic organisms), *Hydrobiologia,* 366, 63–79, 1998.

Baird, C., *Environmental Chemistry,* W.H. Freeman and Co., New York, 1995.

Baker, A.J.M. and P.L. Walker, Physiological responses of plants to heavy metals and the quantification of tolerance and toxicity, *Chem. Speciation Bioavail.,* 1, 7–18, 1989.

Baker, C.E. and P.B. Dunaway, Retention of ^{134}Cs as an index to metabolism in the cotton rat (*Sigmodon hispidus*), *Health Phys.,* 16, 227–230, 1969.

Baker, L.A., A.T. Herlihy, P.R. Kaufmann, and J.E. Eilers, Acidic lakes and streams in the United States: the role of acidic deposition, *Science* (Washington, D.C.), 252, 1151–1154, 1991.

Baker, R., B. Lavie, and E. Nevo, Natural selection for resistance to mercury pollution, *Experientia,* 41, 697–699, 1985.

Baker, R.J., R.A. Van Den Bussche, A.J. Wright, L.E. Wiggins, M.J. Hamilton, E.P. Reat, M.H. Smith, M.D. Lomakin, and R.K. Chesser, High levels of genetic change in rodents of Chernobyl, *Nature* (London), 380, 707–708, 1996.

Bantle, J.A., FETAX: a developmental toxicity assay using frog embryos, in *Fundamentals of Aquatic Toxicology: Effects, Environmental Fate, and Risk Assessment,* 2nd ed., Rand, G.M., Ed., Taylor & Francis, Washington, D.C., 1995.

Bantle, J.A. and T.D. Sabourin, Standard guide for conducting the frog embryo teratogenesis assay: *Xenopus* (FETAX), ASTM Spec. Pub. E1439-91, American Society for Testing and Materials, Philadelphia, 1991, pp. 1–11.

Barber, M.C., L.A. Suarez, and R.R. Lassiter, Modeling bioconcentration of nonpolar organic pollutants by fish, *Environ. Toxicol. Chem.,* 7, 545–558, 1988.

Barnaby, F., Chernobyl: the consequences to Europe, *Ambio,* 15, 332–334, 1986.

Barnthouse, L.W., Population-level effects, in *Ecological Risk Assessment,* Suter, G.W., II, Ed., Lewis Publishers, Chelsea, MI, 1993.

Barnthouse, L.W., Issues in ecological risk assessment: the CRAM perspective, *Risk Analysis,* 14, 251–256, 1994.

Barnthouse, L.W., G.W. Suter II, A.E. Rosen, and J.J. Beauchamp, Estimating responses of fish populations to toxic contaminants, *Environ. Toxicol. Chem.,* 6, 811–824, 1987.

Barrie, L.A. and J.M. Hales, The spatial distributions of precipitation acidity and major ion wet deposition in North America during 1980, *Tellus,* 36B, 333–335, 1984.

Barron, M.G., Bioaccumulation and bioconcentration in aquatic organisms, in *Handbook of Ecotoxicology,* Hoffman, D.J., Rattner, B.A., Burton, G.A., Jr., and Cairns, J., Jr., Eds., CRC Press, Boca Raton, FL, 1995.

Barron, M.G., G.R. Stehly, and W.L. Hayton, Pharmacokinetic modeling in aquatic animals, I: models and concepts, *Aquatic Toxicol.* (Amsterdam), 18, 61–86, 1990.

Bartholomew, G.A., The roles of physiology and behaviour in the maintenance of homeostasis in the desert environment, in *Symposia of the Society for Experimental Biology,* No. 18, Academic Press, New York, 1964.

Baskin, Y., Ecologists dare to ask: how much does diversity matter? *Science* (Washington, D.C.), 264, 202–203, 1994.

Battaglia, B., P.M. Bisol, V.U. Fossato, and E. Rodino, Studies on the genetic effects of pollution in the sea, *Rapp. Reun. Cons. Int. Explor. Mer.,* 179, 267–274, 1980.

Baumann, P.C., PAH, metabolites, and neoplasia in feral fish populations, in *Metabolism of Polycyclic Aromatic Hydrocarbons in the Aquatic Environment,* Varanasi, U., Ed., CRC Press, Boca Raton, FL, 1989.

Baumann, P.C., M.J. Mac, S.B. Smith, and J.C. Harshbarger, Tumor frequencies in walleye (*Stizostedion vitreum*) and brown bullhead (*Ictalurus nebulosus*) and sediment contaminants in tributaries of the Laurentian Great Lakes, *Can. J. Fish. Aquatic Sci.,* 48, 1804–1810, 1991.

Beach, L.R. and R.D. Palmiter, Amplification of the metallothionein-I gene in cadmium-resistant mouse cells, *Proc. Natl. Acad. Sci. U.S.A.,* 78, 2110–2114, 1981.

Bearhop, S., R.A. Phillips, D.R. Thompson, S. Waldron, and R.W. Furness, Variability in mercury concentrations of great skuas *Catharacta skua*: the influence of colony, diet and trophic status inferred from stable isotope signatures, *Mar. Ecol. Prog. Ser.,* 193, 261–268, 2000.

Bechmann, R.K., Use of life tables and LC50 tests to evaluate chronic and acute toxicity effects of copper on the marine copepod *Tisbe furcata* (Baird), *Environ. Toxicol. Chem.,* 13, 1509–1517, 1994.

Becker, P.H., Seabirds as monitor organisms of contaminants along the German North Sea coast, *Helgoländer Meeresuntersuchungen,* 43, 394–403, 1989.

Becker, P.H., D. Henning, and R.W. Furness, Differences in mercury contamination and elimination during feather development in gull and tern broods, *Arch. Environ. Contam. Toxicol.,* 27, 162–167, 1994.

Beeby, A., Toxic metal uptake and essential metal regulation in terrestrial invertebrates: a review, in *Metal Ecotoxicology: Concepts and Applications,* Newman, M.C. and McIntosh, A.W., Eds., Lewis Publishers, Chelsea, MI, 1991.

Beitinger, T.L., Behavioral reactions for the assessment of stress in fishes, *J. Great Lakes Res.,* 16, 495–528, 1990.

Belfroid, A.C., A. Van der Horst, A.D.Vethaak, A.J. Schafer, G.B.J. Rijs, J. Wegener, and W.P. Cofino, Analysis and occurrence of estrogenic hormones and their glucuronides in surface water and waste water in The Netherlands, *Sci. Total Environ.,* 225, 101–108, 1999.

Benson, A.A. and R.E. Summons, Arsenic accumulation in Great Barrier Reef invertebrates, *Science* (Washington, D.C.), 211, 482–483, 1981.

Bercovitz, K. and D. Laufer, Lead release from human trabecular bone, in *Impact of Heavy Metals on the Environment,* Vernet, J.-P., Ed., Elsevier Science Publishers B.V., Amsterdam, 1992.

Berenbaum, M.C., The expected effect of a combination of agents: the general solution. *J. Theor. Biol.,* 114, 413–431, 1985.

Bergeron, J.M., D. Crews, and J.A. McLachlan, PCBs as environmental estrogens: turtle sex determination as a biomarker of environmental contamination, *Environ. Health Perspect.,* 102, 780–781, 1994.

Berglind, R., Combined and separate effects of cadmium, lead and zinc in ALA: activity, growth and hemoglobin content in *Daphnia magna, Environ. Toxicol. Chem.,* 5, 989–995, 1986.

Berkeley, G., *Principles of Human Knowledge and Three Dialogues between Hylas and Philonous,* 1734, repr., Penguin Books, London, 1988.

Berkson, J., Why I prefer logits to probits, *Biometrics,* 7, 327–339, 1951.

Berkson, J., Maximum likelihood and minimum X_2 estimates of the logistic function, *J. Am. Stat. Assoc.,* 50, 130–162, 1955.

Bernstein, P.L., *Against the Gods: The Remarkable Story of Risk,* John Wiley & Sons, New York, 1996.

Betts, K.S., Rapidly rising PBDE levels in North America, *Environ. Sci. Technol.,* 36, 50A–52A, 2001.

Beyer, W.N., A reexamination of biomagnification of metals in terrestrial food chains, *Environ. Toxicol. Chem.,* 5, 863–864, 1986.

Beyer, W.N., M. Spalding, and D. Morrison, Mercury concentrations in feathers of wading birds from Florida, *Ambio,* 26, 97–100, 1997.

Bezel, V.S. and V.N. Bolshakov, Population ecotoxicology of mammals, in *Bioindications of Chemical and Radioactive Pollution,* Krivolutsky, D.A., Ed., CRC Press, Boca Raton, FL, 1990.

Biesinger, K.E. and G.M. Christensen, Effects of various metals on survival, growth, reproduction, and metabolism of *Daphnia magna, J. Fish. Res. Board Can.,* 29, 1691–1700, 1972.

Biggins, P.D.E. and R.M. Harrison, Chemical speciation of lead compounds in street dusts, *Environ. Sci. Technol.,* 14, 336–339, 1980.

Birge, W.J., R.D. Hoyt, J.A. Black, M.D. Kercher, and W.A. Robison, Effects of chemical stresses on behavior of larval and juvenile fishes and amphibians, in *Water Quality and the Early Life Stages of Fishes,* Fuiman, L.A., Ed., American Fisheries Society, Bethesda, MD, 1993.

Bishop, J.A., L.M. Cook, and J. Muggleton, The response of two species of moths to industrialization in northwest England, I: polymorphisms for melanism, *Philos. Trans. R. Soc. B,* 281, 491–515, 1978.

Bishop, W.E. and A.W. McIntosh, Acute lethality and effects of sublethal cadmium exposure on ventilation frequency and cough rate of bluegill (*Lepomis macrochirus*), *Arch. Environ. Contam. Toxicol.,* 10, 519–530, 1981.

Bjorn T., K. Gunther, and M.J. Schuger, Alkylphenol ethoxylates: trace analysis and environmental behavior, *Chem. Rev.,* 8, 3247–3272, 1997.

Blanck, H., A simple, community level, ecotoxicological test system using samples of periphyton, *Hydrobiologia,* 124, 251–261, 1985.

Blaylock, B.G., Radionuclide data bases available for bioaccumulation factors for freshwater biota, *Nucl. Saf.,* 23, 427–438, 1982.

Blaylock, B.G. and J.R. Trabalka, Evaluating the effects of ionizing radiation on aquatic organisms, *Adv. Radiat. Biol.,* 7, 103, 1978.

Bliss, C.I., The calculation of the dosage-mortality curve, *Ann. Appl. Biol.,* 22, 134–307, 1935.

Bliss, C.I., The size factor in the action of arsenic upon silkworm larvae, *J. Exp. Biol.,* 13, 95–110, 1936.

Blok, J. and F. Balk, Environmental regulation in the European Community, in *Fundamentals of Aquatic Toxicology: Effects, Environmental Fate, and Risk Assessment,* 2nd ed., Rand, G.M., Ed., Taylor & Francis, Washington, D.C., 1995.

Bloom, N.S., On the chemical form of mercury in edible fish and marine invertebrate tissue, *Can. J. Fish. Aquatic Sci.,* 49, 1010–1017, 1992.

Blum, D.J.W. and R.E. Speece, Determining chemical toxicity to aquatic species, *Environ. Sci. Technol.,* 24, 284–293, 1990.

Boesch, D.F., R.B. Brinsfield, and R.E. Magnien, Chesapeake Bay eutrophication: scientific understanding, ecosystem restoration, and challenges for agriculture, *J. Environ. Qual.,* 30, 303–320, 2001.

Bongers, T., The maturity index: an ecological measure of environmental disturbance based on nematode species composition, *Oecologia* (Berlin)*,* 83, 14–19, 1990.

Booth, W., Postmortem on Three Mile Island, *Science* (Washington, D.C.), 238, 1342–1345, 1987.

Borgmann, U., Metal speciation and toxicity of free ions to aquatic biota, in *Aquatic Toxicology,* Nriagu, J.O., Ed., John Wiley & Sons, New York, 1983.

Bornschein, R.L. and S.-R. Kuang, Behavioral effects of heavy metal exposure, in *Biological Effects of Heavy Metals,* Foulkes, E.C., Ed., CRC Press, Boca Raton, FL, 1990.

Borovec, J., Changes in incidence of carcinoma *in situ* after the Chernobyl disaster in Central Europe, *Arch. Environ. Contam. Toxicol.,* 29, 266–269, 1995.

Bortone, S.A., W.P. Davis, and C.M. Bundrick, Morphological and behavioral characters in mosquitofish as potential bioindication of exposure to kraft mill effluent, *Bull. Environ. Contam. Toxicol.,* 43, 370–377, 1989.

Bouquegneau, J.M., Evidence for the protective effect of metallothioneins against inorganic mercury injuries to fish, *Bull. Environ. Contam. Toxicol.,* 23, 218–219, 1979.

Bouton, S.N., P.C. Frederick, M.G. Spalding, and H. McGill, Effects of chronic, low concentrations of dietary methylmercury on the behavior of juvenile great egrets, *Environ. Toxicol. Chem.,* 18, 1934–1939, 1999.

Boutron, C.F. and C.C. Patterson, The occurrence of lead in Antarctic recent snow, firn deposited over the last two centuries and prehistoric ice, *Geochim. Cosmochim. Acta,* 47, 1355–1368, 1983.

Bowerman, W.W., IV, E.D. Evans, J.P. Geisy, and S. Postupalsky, Using feathers to assess risk of mercury and selenium to bald eagle reproduction in the Great Lakes region, *Arch. Environ. Contam. Toxicol.,* 27, 294–298, 1994.

Bowling, J.W., G.J. Leversee, P.F. Landrum, and J.P. Giesy, Acute mortality of anthracene-contaminated fish exposed to sunlight, *Aquatic Toxicol.* (Amsterdam), 3, 79–90, 1983.

Boyden, C.R., Trace element content and body size in molluscs, *Nature* (London)*,* 251, 311–314, 1974.

Boyden, C.R., Effect of size upon metal content of shellfish, *J. Mar. Biol. Assoc. U.K.,* 57, 675–714, 1977.

Boynton, W.R., J.H. Garber, R. Summers, and W.M. Kemp, Inputs, transformations and transport of nitrogen and phosphorus in Chesapeake Bay and selected tributaries, *Estuaries,* 18, 1B, 285–314, 1995.

Bradley, R.W. and J.B. Sprague, The influence of pH, water hardness, and alkalinity on the acute lethality of zinc to rainbow trout (*Salmo gairdneri*), *Can. J. Fish. Aquatic Sci.,* 42, 731–736, 1985.

Branches, F.J.P., T.B. Erickson, S.E. Aks, and D.O. Hryhorczuk, The price of gold: mercury exposure in the Amazonian rain forest, *J. Toxicol. Clin. Toxicol.,* 31, 295–306, 1993.

Brant, H.A., C.H. Jagoe, J.W. Snodgrass, A.L. Bryan, Jr., and J.C. Gariboldi, Potential risk to wood storks (*Mycteria Americana*) from mercury in Carolina bay fish, *Environ. Pollut.,* 120, 405–413, 2002.

Braun, W., M. Vasak, A.H. Robbins, C.D. Stout, G. Wagner, J.H.R. Kägi, and K. Wuthrich, Comparison of the NMR solution structure and the x-ray crystal structure of rat metallothionein-2, *Proc. Natl. Acad. Sci. U.S.A.,* 89, 10124–10128, 1992.

Braune, B.M. and D.E. Gaskin, Mercury levels in Bonaparte's gulls (*Larus Philadelphia*) during the autumn molt in the Quoddy Region, New Brunswick, Canada, *Arch. Environ. Contam. Toxicol.,* 16, 539–549, 1987.

Brezonik, P.L., S.O. King, and C.E. Mach, The influence of water chemistry on trace metal bioavailability and toxicity to aquatic organisms, in *Metal Ecotoxicology: Concepts and Applications,* Newman, M.C. and McIntosh, A.W., Eds., Lewis Publishers, Chelsea, MI, 1991.

Bricelj, V.M., A.E. Bass, and G.R. Lopez, Absorption and gut passage time of microalgae in a suspension feeder: an evaluation of the ^{51}Cr:^{14}C twin tracer technique, *Mar. Ecol. Prog. Ser.,* 17, 57–63, 1984.

Bridgman, H., *Global Air Pollution: Problems for the 1990s,* John Wiley & Sons, New York, 1994.

Broad, W.J., Sir Isaac Newton: mad as a hatter, *Science* (Washington, D.C.), 213, 1341–1344, 1981.

Broderius, S.J., L.L. Smith Jr., and D.T. Lind, Relative toxicity of free cyanide and dissolved sulfide forms to the fathead minnow (*Pimephales promelas*), *J. Fish. Res. Board Can.,* 34, 2323–2332, 1977.

Broman, D., C. Näf, C. Rolff, Y. Zebühr, B. Fry, and J. Hobbie, Using ratios of stable nitrogen to estimate bioaccumulation and flux of polychlorinated dibenzo-p-dioxins (PCDDs) and dibenzofurans (PCDFs) in two food chains from the northern Baltic, *Environ. Toxicol. Chem.*, 11, 331–345, 1992.

Brookins, D.G., *Eh-pH Diagrams for Geochemistry,* Springer-Verlag, New York, 1988.

Brouwer, A., A.J. Murk, and J.H. Koeman, Biochemical and physiological approaches in ecotoxicology, *Functional Ecol.*, 4, 75–281, 1990.

Brown, B.E., Lead detoxification by a copper-tolerant isopod, *Nature* (London), 276, 388–390, 1978.

Brown, T.A. and A. Shrift, Selenium: toxicity and tolerance in higher plants, *Biol. Rev.*, 57, 59–84, 1982.

Bruggeman, W.A., B.J.M. Martron, D. Kooiman, and O. Hutzinger, Accumulation and elimination kinetics of di-, tri- and tetra chlorobiphenyls by goldfish after dietary and aqueous exposure, *Chemosphere*, 10, 811–832, 1981.

Brumley, C.M., V.S. Haritos, J.T. Ahokas, and D.A. Holdway, Validation of biomarkers of marine pollution exposure in sand flathead using Aroclor 1254, *Aquatic Toxicol.* (Amsterdam), 31, 249–262, 1995.

Brunst, V.V., Effects of ionizing radiation on the development of amphibians, *Q. Rev. Biol.*, 40, 1, 1965.

Bryan, A.L., Jr., C.H. Jagoe, H.A. Brant, J.C. Gariboldi, and G.R. Masson, Mercury concentrations in post-fledging wood storks, *Waterbirds*, 24, 277–281, 2001.

Bryan, G.W. and P.E. Gibbs, Impact of low concentrations of tributyltin (TBT) on marine organisms: a review, in *Metal Ecotoxicology. Concepts and Applications,* Newman, M.C. and McIntosh, A.W., Eds., Lewis Publishers, Chelsea, MI, 1991.

Buckland, S.J., H.K. Ellis, and R.T. Salter, Baseline levels of PCDDs, PCDFs, PCBs and organochlorine pesticides in New Zealand rivers and estuaries, *Organohalogen Compounds*, 32, 12–17, 1997.

Buckland, S.J., H.K. Ellis, and R.T. Salter, Ambient concentrations of PCDDs, PCDFs and PCBs in New Zealand soils, *Organohalogen Compounds*, 39, 101–104, 1998.

Buckland, S.J., H.K. Ellis, and R.T. Salter, PCDDs, PCDFs and PCBs in ambient air in New Zealand, *Organohalogen Compounds*, 43, 117–121, 1999.

Bueno, A.M.S., J.M.S. Agostini, K. Gaidzinski, J. Moreira, and I. Brognoli, Frequencies of chromosomal aberrations in rodents collected in the coal-field and tobacco culture region of Cricima, South Brazil, *J. Toxicol. Environ. Health*, 36, 91–102, 1992.

Buikema, A.L., B.R. Niederlehner, and J. Cairns Jr., Biological monitoring, part IV: toxicity testing, *Water Res.*, 16, 239–262, 1982.

Bunce, N., *Environmental Chemistry,* Wuerz Publishing Ltd., Winnipeg, Canada, 1991, p. 339.

Bunzl, K., W. Schimmack, S.V. Krouglov, and R.M. Alexakhin, Changes with time in the migration of radiocesium in the soil, as observed near Chernobyl and in Germany, 1986–1994, *Sci. Total Environ.*, 175, 49–56, 1995.

Burger, J., M.H. Lavery, and M. Gochfeld, Temporal changes in lead levels in common tern feathers in New York and relationship of field levels to adverse effects in the laboratory, *Environ. Toxicol. Chem.*, 13, 581–586, 1994.

Burger, J., J.A. Rodgers, and M. Gochfeld, Heavy metal and selenium levels in endangered wood storks (*Mycteria americana*) from nesting colonies in Florida and Costa Rica, *Arch. Environ. Contam. Toxicol.*, 24, 417–420, 1993.

Burns, L.A., B.J. Meade, and A.E. Munson, Toxic responses of the immune system, in *Casarett and Doull's Toxicology: The Basic Science of Poisons,* 5th ed., Klaassen, C.D., Ed., McGraw-Hill, New York, 1996.

Butler, R.A. and G. Roesijadi, Disruption of metallothionein expression with antisense oligonucleotides abolishes protection against cadmium cytotoxicity in molluscan hemocytes, *Toxicol. Sci.*, 59, 101–107, 2001.

Cabana, G. and J.B. Rasmussen, Modelling food chain structure and contaminant bioaccumulation using stable nitrogen isotopes, *Nature* (London), 372, 255–257, 1994.

Cabana, G., A. Tremblay, J. Kalff, and J.B. Ramussen, Pelagic food chain structure in Ontario lakes: a determinant of mercury in lake trout (*Salvelinus namaycush*), *Can. J. Fish. Aquatic Sci.*, 51, 381–389, 1994.

Cade, T.J., J.L. Lincer, C.M. White, D.G. Roseneau, and L.G. Swartz, DDE residues and eggshell changes in Alaskan falcons and hawks, *Science* (Washington, D.C.), 172, 955–957, 1971.

Cairns, J., Jr., Heated waste-water effects on aquatic ecosystems, in *Thermal Ecology II,* Esch, G.W. and McFarlane, R.W., Eds., National Technical Information Center, Springfield, VA, 1976.

Cairns, J., Jr., The myth of the most sensitive species, *Bioscience, 36,* 670–672, 1986.

Cairns, J., Jr., Will there ever be a field of landscape toxicology? *Environ. Toxicol. Chem.,* 12, 609–610, 1993.

Cairns, J., Jr. and D.I. Mount, Aquatic toxicology, *Environ. Sci. Technol.,* 24, 154–161, 1990.

Cairns, J., Jr. and J.R. Pratt, Biotic impoverishment: effects of anthropogenic stress, in *The Earth in Transition: Patterns and Processes of Biotic Impoverishment,* Woodwell, G.M., Ed., Cambridge University Press, Cambridge, U.K., 1990.

Cairns, J., Jr., J.R. Pratt, B.R. Niederlehner, and P.V. McCormick, A simple cost-effective multispecies toxicity test using organisms with a cosmopolitan distribution, *Environ. Monit. Assess.,* 6, 207–220, 1986.

Calabrese, E.J., M.E. McCarthy, and E. Kenyon, The occurrence of chemically induced hormesis, *Health Phys.,* 52, 531–541, 1987.

Calow, P., R.M. Sibly, and V.E. Forbes, Risk assessment on the basis of simplified population dynamics scenarios, *Environ. Toxicol. Chem.,* 16, 1983, 1997.

Camner, P., T.W. Clarkson, and G.F. Nordberg, Route of exposure, dose and metabolism of metals, in *Handbook on the Toxicology of Metals,* Friberg, L., Nordberg, G.F., and Vouk, V.B., Eds., Elsevier/North-Holland Biomedical Press, Amsterdam, 1979.

Campbell, F.L., Relative susceptibility to arsenic in successive instars of the silkworm, *J. General Physiol.,* 9, 727–733, 1926.

Campbell, P.G.C., A.G. Lewis, P.M. Chapman, A.A. Crowder, W.K. Fletcher, B. Imber, S.N. Luoma, P.M. Stokes, and M. Winfrey, *Biologically Available Metals in Sediments,* NRCC No. 27694, NRCC/CNRC Publications, Ottawa, Canada, 1988.

Campbell, P.G.C. and A. Tessier, Ecotoxicology of metals in the aquatic environment: geochemical aspects, in *Ecotoxicology: A Hierarchical Treatment,* Newman, M.C. and Jagoe, C.H., Eds., CRC Press, Boca Raton, FL, 1996.

Carlson, A.R., G.L. Phipps, V.R. Mattson, P.A. Kosian, and A.M. Cotter, The role of acid-volatile sulfide in determining cadmium bioavailability and toxicity in freshwater sediments, *Environ. Toxicol. Chem.,* 10, 1309–1319, 1991.

Carlson, W.D. and F.X. Gassner, Effects of ionizing radiation on the reproductive system, in *Proc. Int. Symp., Fort Collins, CO,* Pergamon Press, New York, 1964.

Carroll, L., *Through the Looking-Glass,* in *The Best of Lewis Carroll,* 1872, repr., Castle, Secaucus, NJ, 1928.

Carrol, J.J., S.J. Ellis, and W.S. Oliver, Influences of hardness constituents on the acute toxicity of cadmium to brook trout (*Salvelinus fontinalis*), *Bull. Environ. Contam. Toxicol.,* 22, 575–581, 1979.

Carson, R., *Silent Spring,* Houghton-Mifflin Co., Boston, 1962.

Carter, L.J., Michigan's PBB incident: chemical mix-up leads to disaster, *Science,* 192, 240–243, 1976.

Carter, W.H., C. Gennings, J.G. Staniwallis, E.D. Campbell, and K.L. White, A statistical approach to the construction and analysis of isobolograms, *J. Am. Coll. Toxicol.,* 7, 963–973, 1988.

Casarett, A.P., *Radiation Biology,* Prentice-Hall, Englewood Cliffs, NJ, 1968.

Casti, J.L., *Paradigms Lost: Tackling the Unanswered Mysteries of Modern Science,* Avon Books, New York, 1989.

Castro, M.S., C.T. Driscoll, T.E. Jordan, W.R. Reay, W.R. Boyton, S.P. Seitzinger, R.V. Styles, and J.E. Cable, Contribution of atmospheric deposition to the total nitrogen loads of thirty-four estuaries on the Atlantic and Gulf Coast of the United States, in *Atmospheric Nitrogen Deposition in Coastal Waters,* Valigura, R., Ed., Coastal Estuarine Science Series no. 57, 77–106, American Geophysical Union Press, Washington, D.C., 2000.

Caswell, H., Demography meets ecotoxicology: untangling the population level effects of toxic substances, in *Ecotoxicology: A Hierarchical Treatment,* Newman, M.C. and Jagoe, C.H., Eds., Lewis Publishers, Boca Raton, FL, 1996.

Caswell, H., *Matrix Population Models: Construction, Analysers, and Interpretation,* 2nd ed., Sinaver Associates, Sunderland, MA, 2001.

Catallo, W.J., Ecotoxicology and wetland ecosystems: current understanding and future needs, *Environ. Toxicol. Chem.,* 12, 2209–2224, 1993.

Chadwick, J.W. and S.P. Canton, Coal mine drainage on a lotic ecosystem in Northwest Colorado, U.S.A., *Hydrobiologia,* 107, 25–33, 1983.

Chamberlin, T.C., The method of multiple working hypotheses, *J. Geol.,* 5, 837–848, 1897.

Champ, M.A. and P.F. Seligman, Eds., *Organotin: Environmental Fate and Effects,* Chapman & Hall, London, 1996.

Chapman, G.A., Sea urchin sperm cell test, in *Fundamentals of Aquatic Toxicology: Effects, Environmental Fate, and Risk Assessment,* 2nd ed., Rand, G.M., Ed., Taylor & Francis, Washington, D.C., 1995.

Chapman, P.M., R.S. Caldwell, and P.F. Chapman, A warning: NOECs are inappropriate for regulatory use, *Environ. Toxicol. Chem.,* 15, 77–79, 1996.

Chapman, P.M., R.N. Dexter, and E.R. Long, Synoptic measurements of sediment contamination, toxicity, and infaunal community composition (the Sediment Quality Triad) in San Francisco Bay, *Mar. Ecol. Prog. Ser.,* 37, 75–96, 1987.

Chattopadhyay, J., Effect of toxic substances on a two-species competitive system, *Ecol. Model,* 84, 287–289, 1996.

Chen, D.G. and J.G. Pounds, A nonlinear isobologram model with Box-Cox transformation to both sides of chemical mixtures, *Environ. Health Perspect.,* 106, 1367–1371, 1998.

Cherian, M.G. and H.M. Chan, Biological functions of metallothionein: a review, in *Metallothionein III: Biological Roles and Medical Implications,* Suzuki, K.T., Imura, N., and Kimura, M., Eds., Birkhäuser Verlag, Basel, 1993.

Chernyak, S.M., L.L. McConnell, and C.P. Rice, Fate of some chlorinated hydrocarbons in Arctic and Far Eastern ecosystems in the Russian Federation, *Sci. Total Environ.,* 160/161, 75–85, 1995.

Cherry, D.S. and J. Cairns Jr., Biological monitoring. Part V: preference and avoidance studies, *Water Res.,* 16, 263–301, 1982.

Chew, R.D. and M.A. Hamilton, Toxicity curve estimation: fitting a compartment model to median survival times, *Trans. Am. Fish. Soc.,* 114, 403–412, 1985.

Chiou, C.T., Partition coefficients of organic compounds in lipid-water systems and correlations with fish bioconcentration factors, *Environ. Sci. Technol.,* 19, 57–62, 1985.

Choppin, G.R. and J. Rydberg, *Nuclear Chemistry: Theory and Applications,* Pergamon Press, Oxford, U.K., 1980.

Christensen, E.R., Dose-response functions in aquatic toxicity testing and the Weibull model, *Water Res.,* 18, 213–221, 1984.

Christensen, E.R. and N. Nyholm, Ecotoxicological assays with algae: Weibull dose-response curves, *Environ. Sci. Technol.,* 18, 713–718, 1984.

Clark, J.B. and J.C. Harshbarger, Epizootiology of neoplasms in bony fish from North America, *Sci. Total Environ.,* 94, 1–32, 1990.

Clark, K.E. and D. Mackay, Dietary uptake and biomagnification of four chlorinated hydrocarbons by guppies, *Environ. Toxicol. Chem.,* 10, 1205–1217, 1991.

Clark, R.H., Managing radiation risks, presented at Pathway Analysis and Risk Assessment for Environmental Compliance and Dose Reconstruction Workshop, Kiawah Island, SC, Nov. 6–10, 1995.

Clarke, C.A., G.S. Mani, and G. Wynne, Evolution in reverse: clean air and the peppered moth, *Biol. J. Linn. Soc.,* 26, 189–199, 1985.

Clayton, J.R., Jr., S.P. Pavlou, and N.F. Breitner, Polychlorinated biphenyls in coastal marine zooplankton: bioaccumulation by equilibrium partitioning, *Environ. Sci. Technol.,* 11, 676–682, 1977.

Clements, W.H. and P.M. Kiffney, Assessing contaminant effects at higher levels of biological organization, *Environ. Toxicol. Chem.,* 13, 357–359, 1994.

Cockerham, L.G. and B.S. Shane, *Basic Environmental Toxicology,* CRC Press, Boca Raton, FL, 1994.

Cohen, B.L., Perspective on occupational mortality risks, *Health Phys.,* 40, 703–724, 1981.

Cohen, B.L., A test of the linear–no-threshold theory of radiation carcinogenesis, *Environ. Res.,* 53, 193–220, 1990.

Cohen, B.L. and I.-S. Lee, A catalog of risks, *Health Phys.,* 36, 707–722, 1979.

Comins, H.N., The development of insecticide resistance in the presence of migration, *J. Theor. Biol.,* 64, 177–197, 1977.

Connell, D.W., *Bioaccumulation of Xenobiotic Compounds,* CRC Press, Boca Raton, FL, 1990.

Connell, D.W. and D.W. Hawker, Use of polynomial expressions to describe the bioconcentration of hydrophobic chemicals by fish, *Ecotoxicol. Environ. Saf.,* 16, 242–257, 1988.

Connell, D., P. Lam, B. Richardson, and R. Wu, *Introduction to Ecotoxicology,* Blackwell Science, Oxford, U.K., 1999.

Connolly, J.P. and C.J. Pedersen, A thermodynamic-based evaluation of organic chemical accumulation in aquatic organisms, *Environ. Sci. Technol.,* 22, 99–103, 1988.

Cooke, A.S., Shell thinning in avian eggs by environmental pollutants, *Environ. Pollut.,* 4, 85–152, 1973.

Cooke, A.S., Egg shell characteristics of gannets *Sula bassana,* shags *Phalacrocorax aristotelis* and great black-backed gulls *Larus marinus* exposed to DDE and other environmental pollutants, *Environ. Pollut.,* 19, 47–65, 1979.

Cooney, R.V. and A.A. Benson, Arsenic metabolism in *Homarus americanus, Chemosphere,* 9, 335–341, 1980.

Cooper, E.L., *Comparative Immunology,* Prentice-Hall, Englewood Cliffs, NJ, 1976.

Copeland, B.J. and J. Gray, Status and Trends Report of the Albemarle-Pamlico Estuary, Steel, J., Ed., Albemarle-Pamlico Estuarine Study Report 90-01, NC Dept. of Environ. Health and Natl. Resources, Raleigh, NC 1991.

Cordasco, E.M., S.L. Demeter, and C. Zenz, *Environmental Respiratory Diseases,* Van Nostrand Reinhold, New York, 1995.

Corn, M., Corporations Viewed as Environmental Bad Guys, *Chem. Eng. News,* May 3, 1982, pp. 47–48.

Correa, M., Physiological effects of metal toxicity on the tropical freshwater shrimp *Macrobrachium carcinus* (Linneo, 1758), *Environ. Pollut.,* 45, 149–155, 1987.

Correa, M. and H.I. Garcia, Physiological responses of juvenile white mullet, *Mugil curema,* exposed to benzene, *Bull. Environ. Contam. Toxicol.,* 44, 428–434, 1990.

Cosgrove, G.E., Reptilian radiobiology, *J. Am. Vet. Med. Assoc.,* 159, 1678, 1971.

Cossa, D., E. Bourget, D. Pouliot, J. Piuze, and J.P. Chanut, Geographical and seasonal variations in the relationship between trace metal content and body weight in *Mytilus edulis, J. Mar. Biol. Assoc. U.K.,* 58, 7–14, 1980.

Couillard, Y., P.G.C. Campbell, and A. Tessier, Response of metallothionein concentrations in a freshwater bivalve (*Anodonta grandis*) along an environmental cadmium gradient, *Limnol. Oceanogr.,* 38, 299–313, 1993.

Cowling, E.B., Acid precipitation in historical perspective, *Environ. Sci. Technol.,* 16, 110A–123A, 1982.

Cowling, E.B. and R.A. Linthurst, The acid precipitation phenomenon and its ecological consequences, *Bioscience,* 31, 649–654, 1981.

Craig, E.A., The heat shock response, *CRC Critical Reviews in Biochemistry,* 18, 239–280, 1985.

Craig, E.A., Chaperones: helpers along the pathways to protein folding, *Science* (Washington, D.C.), 260, 1902–1903, 1993.

Crane, M., M.C. Newman, P.F. Chapman, and J. Fenlon, *Risk Assessment with Time-to-Event Models,* Lewis Publishers, Boca Raton, FL, 2002a.

Crane, M., N. Sorokin, J.R. Wheeler, A. Grosso, P. Whitehouse, and D. Morritt, European approaches to coastal and estuarine risk assessment, in *Coastal and Estuarine Risk Assessment,* Newman, M.C., Roberts, M.H., Jr., and Hale, R.C., Eds., Lewis Publishers, Boca Raton, FL, 2002b.

Crawford, M., Landmark ozone treaty negotiated, *Science* (Washington, D.C.), 237, 1557–1558, 1987.

Crecelus, E.A., J.T. Hardy, C.I. Bobson, R.L. Schmidt, C.W. Apts, J.M. Gurtisen, and S.P. Joyce, Copper bioavailability to marine bivalves and shrimp: relationship to cupric ion activity, *Mar. Environ. Res.,* 6, 13–26, 1982.

Creed, E.R., D.R. Lees, and J.G. Duckett, Biological method of estimating smoke and sulphur dioxide pollution, *Nature,* 244, 278–280, 1973.

Cremers, A., A. Elsen, P. DePreter, and A. Maes, Quantitative analysis of radiocaesium retention in soils, *Nature* (London), 335, 247–249, 1988.

Crist, R.H., K. Oberhoiser, D. Schwartz, J. Marzoff, D. Ryder, and D.R. Crist, Interactions of metals and protons with algae, *Environ. Sci. Technol.,* 22, 755–760, 1988.

Cristaldi, M.E., L. D'Arcangelo, L.A. Ieradi, D. Mascanzoni, T. Mattei, and I. Castelli van Axel, ^{137}Cs determination and mutagenicity tests in wild *Musculus domesticus* before and after the Chernobyl accident, *Environ. Pollut.,* 64, 1–9, 1990.

Cronan, C.S. and C.L. Schofield, Aluminum leaching response to acid precipitation: effects on high-elevation watersheds in the Northeast, *Science* (Washington, D.C.), 204, 304–306, 1979.

Crop Protection Reference, Chemical and Pharmaceutical Press, New York (published annually).

Culliton, B.J., Continuing confusion over Love Canal, *Science* (Washington, D.C.), 209, 1002–1003, 1980.

Cunningham, P.A., S.L. Smith, J.P. Tippett, and A. Greene, A national fish consumption advisory data base: a step toward consistency, *Fisheries* (Bethesda), 19, 14–23, 1994.

Curtsinger, J.W., H.H. Fukui, D.R. Townsend, and J.W. Vaupel, Demography of genotypes: failure of the limited life-span paradigm in *Drosophila melanogaster, Science* (Washington, D.C.), 258, 461–463, 1992.

Cushing, C.E. and D.G. Watson, Cycling of zinc-65 in a simple food web, in *Proceedings of the Third National Symposium on Radioecology,* Atomic Energy Commission, Oak Ridge National Laboratory, and the Ecological Society of America, Oak Ridge, TN, 1971, pp. 318–322.

Dallinger, R., Y.J. Wang, B. Berger, E.A. Mackay, and J.H.R. Kägi, Spectroscopic characterization of metallothionein from the terrestrial snail, *Helix pomatia, Eur. J. Biochem.,* 268, 4126–4133, 2001.

Daniels, R.E. and J.D. Allan, Life table evaluation of chronic exposure to a pesticide, *Can. J. Fish. Aquatic Sci.,* 38, 485–494, 1981.

Danzmann, R.G., M.M. Feruson, F.W. Allendorf, and K.L. Knudsen, Heterozygosity and developmental rate in a strain of rainbow trout (*Salmo gairdneri*), *Evolution,* 40, 86–93, 1986.

Davies, B.E., Trace metals in the environment: retrospect and prospect, in *Biogeochemistry of Trace Metals,* Adriano, D.C., Ed., Lewis Publishers, Boca Raton, FL, 1992.

Davies, P.H., J.P. Goettl, Jr., J.R. Sinley, and N.F. Smith, Acute and chronic toxicity of lead to rainbow trout *Salmo gairdneri,* in hard and soft water, *Water Res.,* 10, 199–206, 1976.

Davis, J.J. and R.F. Foster, Bioaccumulation of radioisotopes through aquatic food chains, *Ecology,* 39, 530–535, 1958.

Day, K. and N.K. Kaushik, An assessment of the chronic toxicity of the synthetic pyrethroid, Fenvalerate, to *Daphnia galeata mendotae,* using life tables, *Environ. Pollut.,* 44, 13–26, 1987.

Deevey, E.S., Jr., Life tables for natural populations of animals, *Q. Rev. Biol.,* 22, 283–314, 1947.

de Lacerda, L.D., W.C. Pfeiffer, A.T. Ott, and E.G. da Silveira, Mercury contamination in the Madeira River, Amazon: Hg inputs to the environment, *Biotropica,* 21, 91–93, 1989.

D'Elia, C.F., J.G. Sanders, and W.R. Boynton, Nutrient enrichment studies in a coastal plain estuary: phytoplankton growth in large-scale, continuous cultures, *Can. J. Fish. Aquatic Sci.,* 43, 397–406, 1986.

Desbrow, C., E.J. Routledge, G.C. Brighty, J.P. Sumpter, and M. Waldock, Identification of estrogenic chemicals in STW effluent, 1: chemical fractionation and in vitro biological screening, *Environ. Sci. Technol.,* 32, 1549–1558, 1998.

Diamond, J.M., M.J. Parson, and D. Gruber, Rapid detection of sublethal toxicity using fish ventilatory behavior, *Environ. Toxicol. Chem.,* 9, 3–11, 1990.

Diamond, S.A., M.C. Newman, M. Mulvey, P.M. Dixon, and D. Martinson, Allozyme genotype and time to death of mosquitofish, *Gambusia affinis* (Baird and Girard), during acute exposure to inorganic mercury, *Environ. Toxicol. Chem.,* 8, 613–622, 1989.

Diaz, R.J. and R. Rosenberg, Marine benthic hypoxia: a review of its ecological effects and the behavioral responses in benthic macrofauna, *Oceanogr. Mar. Biol. Assoc. Annu. Rev.,* 33, 245–303, 1995.

Dickhut, R.M. and K.E. Gustafson, Atmospheric inputs of selected polycyclic aromatic hydrocarbons and polychlorinated biphenyls to southern Chesapeake Bay, *Mar. Pollut. Bull.,* 30(6), 385–396, 1995.

Dickson, D., Details of 1957 British nuclear accident withheld to avoid endangering U.S. ties, *Science* (Washington, D.C.), 239, 137, 1988.

Di Giulio, R.T., W.H. Benson, B.M. Sanders, and P.A. Van Veld, Biochemical mechanisms: metabolism, adaptation, and toxicity, in *Fundamentals of Aquatic Toxicology: Effects, Environmental Fate, and Risk Assessment,* 2nd ed., Rand, G.M., Ed., Taylor & Francis, Washington, D.C., 1995.

Di Giulio, R.T., P.C. Washburn, R.J. Wenning, G.W. Winston, and C.S. Jewell, Biochemical responses in aquatic animals: a review of determinants of oxidative stress, *Environ. Toxicol. Chem.,* 8, 1103–1123, 1989.

Dillon, T.M. and M.P. Lynch, Physiological responses as determinants of stress in marine and estuarine organisms, in *Stress Effects on Natural Ecosystems,* Barrett, G.W. and Rosenberg, R., Eds., John Wiley & Sons, New York, 1981.

Di Toro, D.M., J.D. Mahony, D.J. Hansen, K.J. Scott, M.B. Hicks, S.M. Mayr, and M.S. Redmond, Toxicity of cadmium in sediments: the role of acid volatile sulfide, *Environ. Toxicol. Chem.,* 9, 1487–1502, 1990.

Di Toro, D.M., C.S. Zarba, D.J. Hansen, W.J. Berry, R.C. Swartz, C.E. Cowan, S.O. Pavlou, H.E. Allen, N.A. Thomas, and P.R. Paquin, Technical basis for establishing sediment quality criteria for nonionic organic chemicals using equilibrium partitioning, *Environ. Toxicol. Chem.,* 10, 1541–1583, 1991.

Dixon, D.G. and J.B. Sprague, Acclimation to copper by rainbow trout (*Salmo gairdneri*): a modifying factor in toxicity, *Can. J. Fish. Aquatic Sci.,* 38, 880–888, 1981a.

Dixon, D.G. and J.B. Sprague, Acclimation-induced changes in toxicity of arsenic and cyanide in rainbow trout, *Salmo gairdneri* Richardson, *J. Fish Biol.,* 18, 579–589, 1981b.

Dixon, D.R. and K.R. Clarke, Sister chromatid exchange: a sensitive method for detecting damage caused by exposure to environmental mutagens in the chromosomes of adult *Mytilus edulis, Mar. Biol. Lett.,* 3, 163–172, 1982.

Dixon, P.M. and M.C. Newman, Analyzing toxicity data using statistical models of time-to-death: an introduction, in *Metal Ecotoxicology: Concepts and Applications,* Newman, M.C. and McIntosh, A.W., Eds., Lewis Publishers, Chelsea, MI, 1991.

Dodd, R.C., P.A. Cunningham, R.J. Curry, and S.J. Stichter, Watershed Planning in the Albemarle-Pamlico Estuarine System, Report No. 93-01, NC Dept. of Environment, Health and Natural Resources, Research Triangle Institute, Research Triangle Park, NC, 1993.

Dodge, E.A. and T.L. Theis, Effect of chemical speciation on the uptake of copper by *Chironomous tentans, Environ. Sci. Technol.,* 13, 1287–1288, 1979.

Dolphin, R., Lake County Mosquito Abatement District Gnat Research Program: Clear Lake Gnat (*Chaoborus astictopus*), in *Proceedings of 27th Annual Conference of the California Mosquito Control Association,* Davis, CA, 1959, pp. 47–48.

Donkin, S.G. and D.B. Dusenbery, Using the *Caenorhabditis elegans* soil toxicity test to identify factors affecting toxicity of four metal ions in intact soil, *Water Air Soil Pollut.,* 86, 359–373, 1994.

Donnelly, K.C., C.S. Anderson, G.C. Barbee, and D.J. Manek, Soil toxicology, in *Basic Environmental Toxicology,* Cockerham, L.G. and Shane, B.S., Eds., CRC Press, Boca Raton, FL, 1994.

Dopp, E., C.M. Barker, D. Schiffmann, and C.L. Reinisch, Detection of micronuclei in hemocytes of *Mya arenaria*: association with leukemia and induction with an alkylating agent, *Aquatic Toxicol.* (Amsterdam), 34, 31–45, 1996.

Dosch, J.J., Salt tolerance of nestling laughing gulls: an experimental field investigation, *Colonial Waterbirds,* 20, 449–457, 1997.

Doust, J.L., M. Schmidt, and L.L. Doust, Biological assessment of aquatic pollution: a review, with emphasis on plants as biomonitors, *Biol. Rev.,* 69, 147–186, 1994.

Douwes, P., K. Mikkola, B. Petersen, and A. Vestergren, Melanism in *Biston betularius* from north-west Europe (Lepidoptera: Geometridae), *Ent. Scand.,* 7, 261–266, 1976.

Downs, T.D. and R.F. Frankowski, Influence of repair processes on dose-response models, *Drug Metab. Rev.,* 13, 839–852, 1982.

Driscoll, C.T., V. Blette, C. Yan, C.L. Schofield, R. Munson, and J. Holsapple, The role of dissolved organic carbon in the chemistry and bioavailability of mercury in remote Adirondack lakes, *Water Air Soil Pollut.,* 80, 499–508, 1995.

Drummond, R.A., G.F. Olson, and A.R. Batterman, Cough response and uptake of mercury by brook trout, *Salvelinus fontinalis,* exposed to mercuric compounds at different hydrogen-ion concentrations, *Trans. Amer. Fish. Soc.,* 2, 244–249, 1974.

Drummond, R.A., C.L. Russom, D.L Geiger, and D.L. DeFoe, Behavioral and morphological changes in fathead minnows, *Pimephales promelas,* as diagnostic endpoints for screening chemicals according to modes of action, in *Aquatic Toxicology: 9th Aquatic Toxicity Symposium,* STP 921, American Society for Testing and Materials, Philadelphia, 1986.

Drummond, R.A. and C.L. Russom, Behavioral toxicity syndromes: a promising tool for assessing toxicity mechanisms in juvenile fathead minnows, *Environ. Toxicol. Chem.,* 9, 37–46, 1990.

Duce, R.A. and W.H. Zoller, Atmospheric trace metals at remote northern and southern hemisphere sites: pollution or natural? *Science* (Washington, D.C.), 187, 59–61, 1975.

Duffus, J.H., *Environmental Toxicology,* John Wiley & Sons, New York, 1980.

Dunson, W.A. and J. Travis, The role of abiotic factors in community organization, *Am. Nat.,* 138, 1067–1091, 1991.

Dwyer, F.J., C.J. Schnitt, S.E. Finger, and P.M. Mehrle, Biochemical changes in longear sunfish, *Lepomis megalotis,* associated with lead, cadmium and zinc from mine tailings, *J. Fish Biol.,* 33, 307–317, 1988.

Eaton, P. and M. Radojevic, Eds., *Forest Fires and Regional Haze in Southeast Asia,* Nova Science Publishers, New York, 2001.

Eberhardt, L.L., Relationship of cesium-137 half-life in humans to body weight, *Health Phys.,* 13, 88–90, 1967.

Eckerman, K.F., A.B. Wolbrast, and A.C.B. Richardson, Limiting Values of Radionuclide Intake and Air Concentration and Dose Conversion Factors for Inhalation, Submersion, and Ingestion, Federal Guidance Report No. 11, EPA-520/1-88-020, U.S. Environmental Protection Agency, Washington, D.C., 1988.

Eckerman, K.F. and J.C. Ryman, External Exposure to Radionuclides in Air, Water and Soil, Federal Guidance Report No. 12, EPA 402-R-93-081, U.S. Environmental Protection Agency, Washington, D.C., 1993.

Eckl, P.M. and D. Riegler, Levels of chromosomal damage in hepatocytes of wild rats living within the area of a waste disposal plant, *Sci. Total. Environ.,* 196, 41–149, 1997.

Edmonds, J.S. and K.A. Francesconi, Isolation and identification of arsenobetaine from the American lobster, *Homarus americanus, Chemosphere,* 10, 1041–1044, 1981.

Edwards, M., Pollution in the former U.S.S.R.: lethal legacy, *Natl. Geogr.,* 186, 70–99, 1994.

Ehrlich, P.R. and A.H. Ehrlich, *Extinction, the Causes and Consequences of the Disappearance of Species,* Random House, New York, 1981.

Eichhorn, G.L., Active sites of biological macromolecules and their interaction with heavy metals, in *Ecological Toxicology: Effects of Heavy Metal and Organohalogen Compounds,* McIntyre, A.D. and Mills, C.F., Eds., Plenum Press, New York, 1975.

Eichhorn, G.L., J.J. Butzow, P. Clark, and Y.A. Shin, Studies on metal ions and nucleic acids, in *Effects of Metals on Cells, Subcellular Elements, and Macromolecules,* Maniloff, J., Coleman, J.R., and Miller, M.W., Eds., Charles C Thomas Publisher, Springfield, IL, 1970.

Eide, I. and H.G. Johnsen, Mixture design and multivariate analysis in mixture research, *Environ. Health Perspect.,* 106, 1373–1376, 1998.

Eisenbud, M., *Environmental Radioactivity,* Academic Press, New York, 1973.

Ejnik, J., A. Munoz, T. Gan, C.F. Shaw, and D.H. Petering, Interprotein metal ion exchange between cadmium-carbonic anhydrase and apo- or zinc-metallothionein, *J. Biol. Inorg. Chem.,* 4, 784–790, 1999.

Ellenton, J.A. and M.F. McPherson, Mutagenicity studies of herring gulls from different locations on the Great Lakes, I: Sister-chromatid exchange rates in herring gull embryos, *J. Toxicol. Environ. Health,* 12, 317–324, 1983.

Ellgehausen, H., J.A. Guth, and H.O. Essner, Factors determining the bioaccumulation potential of pesticides in the individual compartment of aquatic food chains, *Ecotoxicol. Environ. Saf.,* 4, 134–157, 1980.

Elliott, J.M., Tolerance and resistance to thermal stress in juvenile Atlantic salmon, *Salmo salar, Freshwater Biol.,* 25, 61–70, 1991.

Ellis, D., *Environments at Risk: Case Histories of Impact Assessment,* Springer-Verlag, Berlin, 1989.

Elmgren, R., Man's impact on the ecosystem of the Baltic Sea: energy flows today and at the turn of the century, *Environ. Sci. Technol.,* 9, 635–638, 1989.

Elsom, D., *Atmospheric Pollution: Causes, Effects and Control Policies,* Basol Blackwell, New York, 1987.

Ember, L.R., Environmental Lead: Insidious Health Problem, *Chem. Eng. News,* June 23, 1980, pp. 28–35.

Ember, L.R., EPA Study Backs Cut in Lead Use in Gas, *Chem. Eng. News,* Apr. 9, 1984, p. 18.

Endler, J.A., *Natural Selection in the Wild,* Princeton University Press, Princeton, NJ, 1986.

EPA, *Water Quality Standards Handbook,* National Technical Information Service, Springfield, VA, 1983, p. 66.

EPA, Methods for Measuring the Acute Toxicity of Effluents to Freshwater and Marine Organisms, PB85-205383, National Technical Information Service, Springfield, VA, 1985a, p. 216.

EPA, Ambient Water Quality Criteria for Cadmium: 1984, PB85-227031, National Technical Information Service, Springfield, VA, 1985b, p. 127.

EPA, Ambient Water Quality Criteria for Copper: 1984, PB85-227023, National Technical Information Service, Springfield, VA, 1985c, p. 142.

EPA, Ambient Water Quality Criteria for Lead: 1984, PB85-227437, National Technical Information Service, Springfield, VA, 1985d, p. 81.

EPA, Guidelines for Deriving Numerical National Water Quality Criteria for the Protection of Aquatic Organisms and Their Uses, PB85-227049, National Technical Information Service, Springfield, VA, 1985e, p. 98.

EPA, The Enhanced Stream Water Quality Models QUAL2E and QUAL2E-UNCAS: Documentation and User Manual, EPA/600/3-87/007, National Technical Information Service, Springfield, VA, 1987a, p. 189.

EPA, Ambient Water Quality Criteria for Zinc: 1987, PB87-153581, National Technical Information Service, Springfield, VA, 1987b, p. 214.

EPA, Radiation protection guidance to federal agencies for occupational exposure, *Federal Register,* 52(17), 2822, 1987c.

EPA, Short-term Methods for Estimating the Chronic Toxicity of Effluents and Receiving Waters to Marine and Estuarine Organisms, PB89-220503, National Technical Information Service, Springfield, VA, 1988a, p. 415.

EPA, Ambient Water Quality Criteria for Aluminum: 1988, PB88-245998, National Technical Information Service, Springfield, VA, 1988b, p. 47.

EPA, Short-term Methods for Estimating the Chronic Toxicity of Effluents and Receiving Waters to Freshwater Organisms, EPA/600/4-89/001, National Technical Information Service, Springfield, VA, 1989a, p. 249.

EPA, Short-term Methods for Estimating the Chronic Toxicity of Effluents and Surface Waters to Freshwater Organisms, Supplement, PB90-145764, National Technical Information Service, Springfield, VA, 1989b, p. 262.

EPA, Ecological Assessment of Hazardous Waste Sites, EPA 600/3-89/013, National Technical Information Service, Springfield, VA, 1989c, p. 260.

EPA, Risk Assessment Guidance for Superfund, Volume I: Human Health Evaluation Manual (Part A), Interim Final, EPA 540/1-89/002, National Technical Information Service, Springfield, VA, 1989d, p. 290.

EPA, Risk Assessment Guidance for Superfund, Volume II: Environmental Evaluation Manual, EPA 540/1-89/001, National Technical Information Service, Springfield, VA, 1989e, p. 57.

EPA, Summary Report on Issues in Ecological Risk Assessment, EPA 625/3-91/018, National Technical Information Service, Springfield, VA, 1991a, p. 46.

EPA, MINTEQA2/PRODEFA2, a Geochemical Assessment Model for Environmental Systems: Version 3.0 User's Manual, EPA 600/3-91/021, National Technical Information Service, Springfield, VA, 1991b, p. 106.

EPA, Risk Assessment Guidance for Superfund, Volume I: Human Health Evaluation Manual/Supplemental Guidance: Standard Default Exposure Factors, OSWER Directive 9285.6-03, NTIS, Arlington, VA, 1991c.

EPA, Ecological Risk Assessment Guidance for Superfund: Process for Designing and Conducting Ecological Risk Assessment, review draft, U.S. EPA, Washington, D.C., Sep. 26, 1994.

EPA, Proposed Guidelines for Ecological Risk Assessment, EPA/630/R-95/002B, National Technical Information Service, Springfield, VA, 1996, p. 247.

EPA, Guidelines for Ecological Risk Assessment, EPA/630/R-95/002F, Risk Assessment Forum, U.S. Environmental Protection Agency, Washington, D.C., 1998.

EPA, Biopesticides Registration Action Document: *Bacillus thuringiensis* Plant-Incorporated Protectants, http://www.epa.gov/pesticides/biopesticides/reds/brad_bt_pip2.htm, 2001.

EPA/U.S. Army Corps of Engineers, Evaluation of Dredged Material Proposed for Discharge in Waters of the U.S.: Testing Manual, EPA-823-F-98-005, U.S. Environmental Protection Agency, Office of Water, Washington, D.C., 1998.

Erickson, R.J. and J.M. McKim, A simple flow-limited model for exchange of organic chemicals at fish gills, *Environ. Toxicol. Chem.*, 9, 159–165, 1990.

Erickson, J.M., M. Rahire, and J.-D. Rochaix, Herbicide resistance and cross-resistance: changes at three distinct sites in the herbicide-binding protein, *Science* (Washington, D.C.), 228, 204–207, 1985.

Erickson, R.J. and C.E. Stephan, Calculation of the final acute value for water quality criteria for aquatic organisms, PB88-214994, National Technical Information Service, Springfield, VA, 1988.

European Union, *Technical Guidance Document on Risk Assessment for New and Existing Substances, Part 2, Environmental Risk Assessment,* Office for Official Publications of the European Community, Luxembourg, 1996.

Evans, D.H., The fish gill: site of action and model for toxic effects of environmental pollutants, *Environ. Health Perspect.*, 71, 47–58, 1987.

Evans, H.J., Leukaemia and radiation, *Nature* (London), 345, 16–17, 1990.

Evans, M.S., G.E. Noguchi, and C.P. Rice, The biomagnification of polychlorinated biphenyls, toxaphene, and DDT compounds in a Lake Michigan offshore food web, *Arch. Environ. Contam. Toxicol.*, 20, 87–93, 1991.

Exeley, C., J.S. Chappell, and J.D. Birchall, A mechanism of acute aluminum toxicity in fish, *J. Theor. Biol.*, 151, 417–428, 1991.

Fabacher, D.L. and H. Chambers, Rotenone tolerance in mosquitofish, *Environ. Pollut.*, 3, 139–141, 1972.

Fagerström, T., Body weight, metabolic rate, and trace substance turnover in animals, *Oecologia* (Berlin), 29, 99–104, 1977.

Fairchild, J.F., T. Boyle, W.R. English, and C. Rabeni, Effects of sediment and contaminated sediment on structural and functional components of experimental stream ecosystems, *Water Air Soil Pollut.*, 36, 271–293, 1987.

Fent, K., Organotins in municipal wastewater and sewage sludge, in *Organotin: Environmental Fate and Effects,* Champ, M.A. and Seligman, P.F., Eds., Chapman & Hall, London, 1996.

Ferson, S. and H.R. Akçakaya, *Modeling Fluctuations in Age-Structured Populations,* Exeter Software, Seatauket, NY, 1991.

Finney, D.J., *Probit Analysis: A Statistical Treatment of the Sigmoid Response Curve,* Cambridge University Press, Cambridge, U.K., 1947.

Finney, D.J., *Statistical Method in Biological Assay,* Charles Griffin and Co., London, 1971.

Fisher, D.J., M.H. Knott, S.D. Turley, B.S. Turley, L.T. Yonkos, and G.P. Ziegler, The acute whole effluent toxicity of storm water from an international airport, *Environ. Toxicol. Chem.,* 14, 1103–1111, 1995.

Fisher, D.J., M.H. Knott, B.S. Turley, L.T. Yonkos, and G. Ziegler, Acute and chronic toxicity of industrial and municipal effluents in Maryland, USA: results from eight years of toxicity testing, *Water Environ. Res.,* 70, 101–107, 1998.

Fisher, N.S., J.-L. Teyssié, S. Krishnaswami, and M. Baskaran, Accumulation of Th, Pb, U, and Ra in marine phytoplankton and its geochemical significance, *Limnol. Oceanogr.,* 32, 131–142, 1987.

Fisher, R.A., A.S. Corbet, and C.B. Williams, The relation between the number of species and the number of individuals in a random sample of an animal population, *J. An. Ecol.,* 12, 42–58, 1943.

Fitzgerald, W.F., D.R. Engstrom, R.P. Mason, and E.A. Nater, The case for atmospheric mercury contamination in remote areas, *Environ. Sci. Technol.,* 32, 1–7, 1998.

Forbes, V.E. and P. Calow, Is the per capita rate of increase a good measure of population-level effects in ecotoxicology? *Environ. Toxicol. Chem.,* 18, 1544, 1999.

Forbes, V.E. and P. Calow, Species sensitivity distributions revisited: a critical appraisal, *Human and Ecological Risk Assessment,* 8, 473–492, 2002.

Forbes, V.E., P. Calow, and R.M. Sibly, Are current species extrapolation models a good basis for ecological risk assessment? *Environ. Toxicol. Chem.,* 20, 442, 2001a.

Forbes, V.E., R.M. Sibly, and P. Calow, Determining toxicant impacts on density-limited populations: a critical review of theory, practice and results, *Ecol. Appl.,* 11, 1249, 2001b.

Forbes, V.E. and T.L. Forbes, *Ecotoxicology in Theory and Practice,* Chapman & Hall, London, 1994.

Ford, J., The effects of chemical stress on aquatic species composition and community structure, in *Ecotoxicology: Problems and Approaches,* Levin, S.A. et al., Eds., Springer-Verlag, New York, 1989.

Forni, A., Chromosomal effects of lead: a critical review, in *Reviews on Environmental Health,* Vol. III, James, G.V., Ed., Freund Publishing House, Tel Aviv, Israel, 1980.

Foster, R.B., Environmental legislation, in *Fundamentals of Aquatic Toxicology,* Rand, G.M. and Petrocelli, S.R., Eds., Hemisphere Publishing Corp., Washington, D.C., 1985.

Fournie, J.W. and W.K. Vogelbein, Exocrine pancreatic neoplasms in the mummichog (*Fundulus heteroclitus*) from a creosote-contaminated site, *Toxicol. Pathol.,* 22(3), 237–247, 1994.

Fox, G.A., Practical causal inference for ecoepidemiologists, *J. Toxicol. Environ. Health,* 33, 359–373, 1991.

Fox, G.A., S.W. Kennedy, R.J. Norstrom, and D.C. Wigfield, Porphyria in herring gulls: a biochemical response to chemical contamination of Great Lakes food chains, *Environ. Toxicol. Chem.,* 7, 831–839, 1988.

Fox, H.E., S.A. White, M.H.F. Kao, and R.D. Fernald, Stress and dominance in a social fish, *J. Neurosci.,* 17, 6463–6469, 1997.

Frankel, E.G., *Ocean Environmental Management: A Primer on the Role of the Oceans and How to Maintain Their Contributions to Life on Earth,* Prentice Hall PTR, Englewood Cliffs, NJ, 1995.

Frankenburger, W.T. and U. Karlson, Dissipation of soil selenium by microbial volatilization, in *Biogeochemistry of Trace Metals,* Adriano, D.C., Ed., Lewis Publishers, Boca Raton, FL, 1992.

Franson, J.C., L. Sileo, and N.J. Thomas, Causes of eagle deaths, in *Our Living Resources,* LaRoe, E.T. et al., Eds., U.S. Dept. of the Interior, National Biological Service, Washington, D.C., 1995.

Fraústo da Silva, J.J.R. and R.J.P. Williams, *The Biological Chemistry of the Elements: The Inorganic Chemistry of Life,* Oxford University Press, Oxford, U.K., 1991.

Frederick, P.C., M.G. Spalding, M.S. Sepulvada, G.E. Williams, L. Nico, and R. Robins, Exposure of great egret (*Ardea albus*) nestlings to mercury through diet in the Everglades ecosystem, *Environ. Toxicol. Chem.,* 18, 1940–1947, 1999.

Freedman, W., *Federal Statutes on Environmental Protection: Regulation in the Public Interest,* Quorum Books, New York, 1987.

French, N.R., Comparison of radioisotope assimilation by granivorous and herbivorous mammals, in *Radioecological Concentration Processes, Proceedings of an International Symposium, Stockholm, 1966,* Åberg, B., and Hungate, F.P., Eds., Pergamon Press, New York, 1967.

Friedli, H.R., L.F. Radke, and J.Y. Lu, Mercury in smoke from biomass burning, *Geophys. Res. Lett.,* 28, 3223–3226, 2001.

Fromm, P.O., A review of some physiological and toxicological responses of freshwater fish to acid stress, *Environ. Biol. Fishes,* 5, 79–93, 1980.

Fromm, P.O. and J.R. Gillette, Effect of ambient ammonia on blood ammonia and nitrogen excretion of rainbow trout (*Salmo gairdneri*), *Comp. Biochem. Physiol.,* 26, 887–896, 1968.

Frueh, F.W., K.C. Hayashibara, P.O. Brown, and J.P. Whitlock, Use of cDNA microarrays to analyze dioxin-induced changes in human liver gene expression, *Toxicol. Lett.,* 122, 189–203, 2001.

Fry, B., Food web structure on Georges Bank from stable C, N, and S isotopic compositions, *Limnol. Oceanogr.,* 33, 1182–1190, 1988.

Fry, B., Stable isotope diagrams of freshwater food webs, *Ecology,* 72, 2293–2297, 1991.

Fry, D.M. and C.K. Toone, DDT-induced feminization of gull embryos, *Science,* 213, 922–924, 1981.

Furness, R.W., S.A. Lewis, and J.A. Mills, Mercury levels in the plumage of red-billed gulls *Larus novae-hollandiae scopulinus* of known sex and age, *Environ. Pollut.,* 63, 33–39, 1990.

Fürst, P., S. Hu, R. Hackett, and D. Hamer, Copper activates metallothionein gene transcription by altering the conformation of a specific DNA binding protein, *Cell,* 55, 705–717, 1988.

Gächter, R. and W. Geiger, MELIMEX, an experimental heavy metal pollution study: behavior of heavy metals in an aquatic food chain, *Schweiz. Z. Hydrol.,* 41, 277–290, 1979.

Gad, S.C., Statistical analysis of behavioral toxicology data and studies, *Arch. Toxicol. Suppl.,* 5, 256–266, 1982.

Gaddum, J.H., Bioassays and mathematics, *Pharmacol. Rev.,* 5, 87–134, 1953.

Gallagher, K., R.C. Hale, J. Greaves, E.O. Bush, and D.A. Stilwell, Accumulation of polychlorinated terphenyls in aquatic biota of an estuarine creek, *Ecotoxicol. Environ. Safety,* 26, 302–312, 1993.

Gallegos, A.F. and F.W. Whicker, Radiocesium retention by rainbow trout as affected by temperature and weight, in *Proceedings of the Third National Symposium on Radioecology,* Oak Ridge National Laboratories, Oak Ridge, TN, 361–371, 1971.

Gallopin, G. and S. Öberg, Quality of life, in *An Agenda of Science for Environment and Development into the 21st Century,* Dooge, J.C.I. et al., Eds., Cambridge University Press, Cambridge, U.K., 1992.

Galtsoff, P.S., *The American Oyster* Crassostrea virginica *Gmelin,* Fishery Bull. of Fish and Wildlife Service, Vol. 64, U.S. Government Printing Office, Washington, D.C., 1964, p. 480.

Gardner, M.J., M.P. Snee, A.J. Hall, C.A. Powell, S. Downes, and J.D. Terrell, Results of case-control study of leukaemia and lymphoma among young people near Sellafield nuclear plant in West Cumbria, *Br. Med. J.,* 300, 423–434, 1990.

Gardner, W.S., D.R. Kendall, R.R. Odum, H.L. Windom, and J.A. Stephens, The distribution of methyl mercury in a contaminated salt marsh system, *Environ. Pollut.,* 15, 243–251, 1978.

Gariboldi, J.G., A.L. Bryan, Jr., and C.H. Jagoe, Annual and regional variation in mercury concentrations in nestling wood storks, *Environ. Toxicol. Chem.,* 20, 1551–1556, 2001.

Gariboldi, J.G., C.H. Jagoe, and A.L. Bryan, Jr., Dietary exposure to mercury in nestling wood storks (*Mycteria americana*) in Georgia, *Arch. Environ. Contam. Toxicol.,* 34, 398–405, 1998.

Garvey, J.S., Metallothionein: a potential biomonitor of exposure to environmental toxins, in *Biomarkers of Environmental Contamination,* McCarthy, J.F. and Shugart, L.R., Eds., Lewis Publishers, Boca Raton, FL, 1990.

Gaylor, D.W., F.F. Kadlubar, and F.A. Beland, Application of biomarkers to risk assessment, *Environ. Health Perspect.,* 98, 139–141, 1992.

Geckler, J.R., W.B. Hornung, T.M. Neiheisel, Q.H. Pickering, E.L. Robinson, and C.E. Stephan, *Validity of Laboratory Tests for Predicting Copper Toxicity in Streams,* U.S. Environmental Protection Agency, Ecological Research Series EPA-600/3-76, Duluth, MN, 1976.

Geisel, T.S. and A.S. Geisel, *The Lorax,* Random House, New York, 1971.

George, S.G., Enzymology and molecular biology of Phase II xenobiotic-conjugating enzymes in fish, in *Aquatic Toxicology: Molecular, Biochemical and Cellular Perspectives,* Malins, D.C. and Ostrander, G.K., Eds., CRC Press, Boca Raton, FL, 1994.

Geret, F. and R.P. Cosson, Induction of specific isoforms of metallothionein in mussel tissues after exposure to cadmium or mercury, *Arch. Environ. Contam. Toxicol.,* 42, 36–42, 2002.

Gerhart, D.Z., S.M. Anderson, and J. Richter, Toxicity bioassays with periphyton communities: design of experimental streams, *Water Res.,* 11, 567–570, 1977.

Geyer, H., D. Sheehan, D. Kotzias, D. Freitag, and F. Korte, Prediction of ecotoxicological behavior of chemicals: relationship between physiochemical properties and bioaccumulation of organic compounds in the mussel, *Chemosphere,* 11, 1121–1134, 1982.

Giattina, J.D. and R.R. Garton, A review of the preference-avoidance responses of fishes to aquatic contaminants, *Residue Rev.,* 87, 43–90, 1983.

Gibaldi, M., *Biopharmaceutics and Clinical Pharmacokinetics,* Lea and Febiger, Philadelphia, 1991.

Gibaldi, M. and D. Perrier, *Pharmacokinetics,* 2nd ed., Marcel Dekker, New York, 1982.

Gibbs, M.H., L.F. Wicker, and A.J. Stewart, A method for assessing sublethal effects of contaminants in soils to the earthworm, *Eisenia foetida, Environ. Toxicol. Chem.,* 15, 360–368, 1996.

Giesy, J.P., Cadmium inhibition of leaf decomposition in an aquatic microcosm, *Chemosphere,* 6, 467–475, 1978.

Giesy, J.P., S.R. Denzer, C.S. Duke, and G.W. Dickson, Phosphoadenylate concentrations and energy charge in two freshwater crustaceans: responses to physical and chemical stressors, *Verh. Int. Ver. Limnol.,* 21, 205–220, 1981.

Giesy, J.P. and R.A. Hoke, Freshwater sediment quality criteria: toxicity bioassessment, in *Sediments: Chemistry and Toxicity of In-Place Pollutants,* Baudo, R., Giesy, J.P., and Muntau, H., Eds., Lewis Publishers, Chelsea, MI, 1990.

Giesy, J.P. and K. Kannan, Global distribution of perfluorooctane sulfonate in wildlife, *Environ. Sci. Technol.,* 35, 1339–1342, 2001.

Gillespie, R.B., Allozyme frequency variation as an indicator of contaminant-induced impacts in aquatic populations, in *Techniques in Aquatic Toxicology,* Ostrander, G.K., Ed., CRC Press, Boca Raton, FL, 1996.

Gillespie, R.B. and S.I. Guttman, Effects of contaminants on the frequencies of allozymes in populations of the central stoneroller, *Environ. Toxicol. Chem.,* 8, 309–317, 1989.

Gillett, J.W., The role of terrestrial microcosms and mesocosms in ecotoxicological research, in *Ecotoxicology: Problems and Approaches,* Levin, S.A., Harwell, M.A., Kelly, J.R., and Kimball, K.D., Eds., Springer-Verlag, New York, 1989.

Gilliam, J.W., D.L. Osmond, and R.O. Evans, Selected agricultural best management practices to control nitrogen in the Neuse River Basin, NC Agric. Res. Serv. Tech. Bull. 311, NC State University, Raleigh, 1997.

Gilmour, C.C., E.A. Henry, and R. Mitchell, Sulfate stimulation of mercury methylation in freshwater sediments, *Environ. Sci. Technol.,* 26, 2281–2287, 1992.

Glass, N.R., D.E. Arnold, J.N. Galloway, G.R. Hendrey, J.J. Lee, W.W. McFee, S.A. Norton, C.F. Powers, D.L. Rambo, and C.L. Schofield, Effects of acidic precipitation, *Environ. Sci. Technol.,* 16, 163A–169A, 1982.

Gobas, F.A.P.C. and D. Mackay, Dynamics of hydrophobic organic chemical bioconcentration in fish, *Environ. Toxicol. Chem.,* 6, 495–504, 1987.

Gobas, F.A.P.C., J.R. McCorquodale, and G.D. Haffner, Intestinal absorption and biomagnification of organochlorines, *Environ. Toxicol. Chem.,* 12, 567–576, 1993.

Goksøyr, A. and L. Förlin, The cytochrome P-450 system in fish, aquatic toxicology and environmental monitoring, *Aquatic Toxicol.* (Amsterdam), 22, 287–312, 1992.

Goldberg, E.D., The mussel watch concept, *Environ. Monit. Assess.,* 7, 91–103, 1986.

Goldsborough, L.G. and G.G.C. Robinson, Changes in periphytic algal community structure as a consequence of short herbicide exposures, *Hydrobiologia,* 139, 177–192, 1986.

Goldstein, R.S. and R.G. Schnellmann, Toxic responses of the kidney, in *Casarett and Doull's Toxicology: The Basic Science of Poisons,* 5th ed., Klaassen, C.D., Ed., McGraw-Hill, New York, 1996.

Golley, F.B., *A History of the Ecosystem Concept in Ecology: More Than the Sum of the Parts,* Yale University, New Haven, CT, 1993.

Goodyear, C.P., A simple technique for detecting effects of toxicants or other stresses on a predator-prey interaction, *Trans. Am. Fish. Soc.,* 101, 367–370, 1972.

Gore, A., *Earth in the Balance: Ecology and the Human Spirit,* Penguin Books USA, New York, 1992.

Gorman, M., *Environmental Hazards: Marine Pollution,* ABC-CLIO, Santa Barbara, CA, 1993.

Gowen, J.W. and J. Stadler, Acute irradiation effects on reproductivity of different strains of mice, in *Effects of Ionizing Radiation on the Reproductive System,* Carlson, W.D. and Gassner, F.X., Eds., Pergamon Press, New York, 1964.

Graedel, T.E. and B.R. Allenby, *Industrial Ecology,* Prentice-Hall, Upper Saddle River, NJ, 1995.

Graham, J.H., D.C. Freeman, and J.M. Emlen, Developmental stability: a sensitive indicator of populations under stress, in *Environmental Toxicology and Risk Assessment,* ASTM STP 1179, Landis, W.G., Hughes, J.S., and Lewis, M.A., Eds., American Society for Testing and Materials, Philadelphia, 1993a.

Graham, J.H., J.M. Emlen, and D.C. Freeman, Developmental stability and its applications in ecotoxicology, *Ecotoxicology,* 2, 175–184, 1993b.

Graham, J.H., K.E. Roe, and T.B. West, Effects of lead and benzene on the developmental stability of *Drosophila melanogaster, Ecotoxicology,* 2, 185–195, 1993c.

Grandy, N.J., Role of the OECD in chemicals control and international harmonization of testing methods, in *Fundamentals of Aquatic Toxicology: Effects, Environmental Fate, and Risk Assessment,* 2nd ed., Rand, G.M., Ed., Taylor & Francis, Washington, D.C., 1995.

Graney, R.L., Jr., D.S. Cherry, and J. Cairns Jr., The influence of substrate, pH, diet and temperature upon cadmium accumulation in the Asiatic clam (*Corbicula fluminea*) in laboratory artificial streams, *Water Res.,* 18, 833–842, 1984.

Grant, A., Population consequences of chronic toxicity: incorporating density dependence into the analysis of life table response experiments, *Ecol. Model,* 105, 325, 1998.

Grant, B.S., A.D. Cook, D.F. Owen, and C.A. Clarke, Geographic and temporal variation in the incidence of melanism in peppered moth populations in America and Britain, *J. Hered.,* 89, 465–471, 1998.

Grant, B.S., D.F. Owen, and C.A. Clarke, Parallel rise and fall of melanic peppered moths in America and Britain, *J. Hered.,* 87, 351–357, 1996.

Grant, B.S. and L.L. Wiseman, Recent history of melanism in American peppered moths, *J. Hered.,* 93, 86–99, 2002.

Gray, J.S., Pollution-induced changes in populations, *Philos. Trans. R. Soc. London,* 286, 545–561, 1979.

Gray, R.H., Fish behavior and environmental assessment, *Environ. Toxicol. Chem.,* 9, 53–67, 1990.

Green, A.S. and G.T. Chandler, Life-table evaluation of sediment-associated chlorpyrifos chronic toxicity to the benthic copepod, *Amphiascus tenuiremis, Arch. Environ. Contam. Toxicol.,* 31, 77–83, 1996.

Grill, E., E.-L. Winnacker, and M.H. Zenk, Phytochelatins: the principal heavy-metal complexing peptides in higher plants, *Science* (Washington, D.C.), 230, 674–676, 1985.

Grosch, D.S., *Biological Effects of Radiations,* Blaisdell Publishing Co., Waltham, MA, 1965.

Groen, J.P., O. Tajima, V.J. Feron, and E.D. Schoen, Statistically designed experiments to screen chemical mixtures for possible interactions, *Environ. Health Perspect.,* 106, 1361–1365, 1998.

Gunderson, J.L. and W.G. MacIntyre, Dissociation constants of chloroquaiacols in water: a comparison of measured and predicted values, *Environ. Toxicol. Chem.,* 15, 809–813, 1996.

Haasch, M.L., R. Prince, P.J. Wejksnora, K.R. Cooper, and J.J. Lech, Caged and wild fish: induction of hepatic cytochrome P-450 (CYP1A1) as an environmental monitor, *Environ. Toxicol. Chem.,* 12, 885–895, 1993.

Haines, T.A., V.T. Komov, and C.H. Jagoe, Perch mercury content is related to acidity and color in 26 Russian lakes, *Water Air Soil Pollut.,* 80, 823–828, 1995.

Hale, R.C., J. Greaves, K. Gallagher, and G.G. Vadas, Novel chlorinated terphenyls in sediments and shellfish of an estuarine environment, *Environ. Sci. Technol.,* 24, 1727–1731, 1990.

Hale, R.C. et al., Flame retardants: persistent pollutants in land-applied sludges, *Nature,* 412, 140–141, 2001.

Hale, R.C. and M.J. La Guardia, Emerging contaminants of concern in coastal and estuarine environments, in *Coastal and Estuarine Risk Assessment,* Newman, M.C., Roberts, M.H., and Hale, R.C., Eds., CRC Press, Boca Raton, FL, 2002.

Hall, E.J., *Radiobiology for the Radiologist,* Harper & Row Publishers, San Francisco, CA, 1978.

Hall, L.W., Jr., R.D. Anderson, W.D. Killen, M.C. Scott, J.V. Kilian, R. Alden, P. Adolphson, and R. Eskin, A Pilot Study for Ambient Toxicity Testing in Chesapeake Bay: Year 4 Report, EPA 903-R-97-011 CBP/TRS 172/97, U.S. EPA Chesapeake Bay Program Office, Annapolis, MD, 1997.

Hall, R.J., Impact of pesticides on bird populations, in *Silent Spring Revisited,* Marco, G.J., Hollingworth, R.M., and Durham, W., Eds., American Chemical Society, Washington, D.C., 1987.

Ham, L., R. Quinn, and D. Pascoe, Effects of cadmium on the predator-prey interaction between the turbellarian *Dendrocoelum lacteum* (Müller, 1774) and the isopod crustacean *Asellus aquaticus* (L.), *Arch. Environ. Contam. Toxicol.,* 29, 358–365, 1995.

Hamelink, J.L., P.F. Landrum, H.L. Bergman, and W.H. Benson, Eds., *Bioavailability: Physical, Chemical and Biological Interactions,* CRC Press, Boca Raton, FL, 1994.

Hamer, D.H., Metallothioneins, *Ann. Rev. Biochem.,* 55, 913–951, 1986.

Hamer, D.H., D.J. Thiele, and J.E. Lemontt, Function and autoregulation of yeast copper-thionein, *Science,* 228, 685–690, 1985.

Hamilton, M.A., R.C. Russo, and R.V. Thurston, Trimmed Spearman-Karber method for estimating median lethal concentrations in toxicity bioassays, *Environ. Sci. Technol.,* 11, 714–719, 1977.

Hamilton, M.S., R.O. Miller, and A. Whitehouse, Continuing fire threat in Southeast Asia, *Environ. Sci. Technol.,* 19, 82A–85A, 2000.

Hamilton, S.J. and P.M. Mehrle, Metallothionein in fish: review of its importance in assessing stress from metal contaminants, *Trans. Amer. Fish. Soc.,* 115, 596–609, 1986.

Hansen, F., V.E. Forbes, and T.L. Forbes, Using elasticity analysis of demographic models to link toxicant effects on individuals to the population level: an example, *Funct. Ecol.,* 13, 157, 1999.

Hansen, L.G. and B.S. Shane, Xenobiotic metabolism, in *Basic Environmental Toxicology,* Cockerham, L.G. and Shane, B.S., Eds., CRC Press, Boca Raton, FL, 1994.

Hanski, I., Metapopulation ecology, in *Population Dynamics in Ecological Space and Time,* Rhodes, O.E., Jr., Chesser, R.K., and Smith, M.H., Eds., University of Chicago Press, Chicago, IL, 1996.

Harrison, F.L. and I.M. Jones, An in vivo sister-chromatid exchange assay in the larvae of the mussel *Mytilus edulis*: response to 3 mutagens, *Mutation Res.,* 105, 235–242, 1982.

Hart, A., D. Balluff, R. Barfknecht, P. Chapman, T. Hawkes, G. Joermann, A. Leopold, and R. Luttik, *Avian Effects Assessment: A Framework for Contaminants Studies,* SETAC Press, Pensacola, FL, 2001.

Hatakeyama, S. and M. Yasuno, Effects of cadmium on the periodicity of parturition and brood size of *Moina macrocopa* (Cladocera), *Environ. Pollut.,* 26, 111–120, 1981.

Hausbeck, J.S., Analysis of Trace Metal Contamination of Coal Strip Mines and Bioaccumulation of Trace Metals by *Peromyscus leucopus,* M.S. thesis, Oklahoma State University, Stillwater, 1995.

Haux, C. and L. Förlin, Biochemical methods for detecting effects of contaminants on fish, *Ambio,* 17, 376–380, 1988.

Hawken, P., A. Lovins, and L.H. Lovins, *Natural Capitalism: Creating the Next Industrial Revolution,* Little, Brown and Co., New York, 1999.

Hawkes, N., Waterfowl hunters must give up lead shot, *Science,* 198, 1232–1233, 1977.

Hawkins, W.E., W.W. Walker, R.M. Overstreet, J.S. Lytle, and T.F. Lytle, Carcinogenic effects of some polynuclear aromatic hydrocarbons on the Japanese medaka and guppy in waterborne exposures, *Sci. Total Environ.,* 94, 155–167, 1990.

Hawkins, W.E., W.W. Walker, and R.M. Overstreet, Carcinogenicity tests using aquarium fish, *Toxicol. Methods,* 5, 225–263, 1995.

Haygarth, P.M. and K.C. Jones, Atmospheric deposition of metals to agricultural surfaces, in *Biogeochemistry of Trace Metals,* Adriano, D.C., Ed., Lewis Publishers, Boca Raton, FL, 1992.

Hayton, W.L., Pharmacokinetic parameters for interspecies scaling using allometric techniques, *Health Phys.,* 57 (Suppl. 1), 159–164, 1989.

Heading, R.C., J. Nimmo, L.F. Prescott, and P. Tothill, The dependence of paracetamol absorption on the rate of gastric emptying, *Br. J. Pharmacol.,* 47, 415–421, 1973.

Heagler, M.G., M.C. Newman, M. Mulvey, and P.M. Dixon, Allozyme genotype in mosquitofish, *Gambusia holbrooki*: temporal stability, concentration effects and field verification, *Environ. Toxicol. Chem.,* 12, 385–395, 1993.

Hedtke, S.F., Structure and function of copper-stressed aquatic microcosms, *Aquatic Toxicol.* (Amsterdam), 5, 227–244, 1984.

Heinz, G.H., Effects of low dietary levels of methylmercury on mallard reproduction, *Bull. Environ. Contam. Toxicol.,* 11, 386–392, 1974.

Heinz, G.H., Methylmercury: reproductive and behavioral effects on three generations of mallard ducks, *J. Wildl. Manage.,* 43, 394–401, 1979.

Heinz, G.H. and D.J. Hoffman, Methylmercury chloride and selenomethionine interactions on health and reproduction of mallards, *Environ. Toxicol. Chem.,* 17, 139–145, 1998.

Hendricks, J.D., T.R. Meyers, D.W. Shelton, J.L. Casteel, and G.S. Bailey, Hepatocarcinogenicity of benzo[a]pyrene to rainbow trout by dietary exposure and intraperitoneal injection, *J. Natl. Cancer Inst.,* 74, 839–851, 1985.

Hennig, H.F.-K.O., Metal-binding proteins as metal pollution indicators, *Environ. Health Perspect.,* 65, 175–187, 1986.

Henny, C.J. and G.B. Herron, DDE, selenium, mercury, and white-faced ibis reproduction at Carson Lake, Nevada, *J. Wildl. Manage.,* 53, 1032–1045, 1989.

Henry, M.G. and G.J. Atchison, Influence of social rank on the behavior of bluegill, *Lepomis macrochirus* Rafinesque, exposed to sublethal concentrations of cadmium and zinc, *J. Fish. Biol.,* 15, 309–315, 1979.

Henry, M.G. and G.J. Atchison, Metal effects on fish behavior: advances in determining the ecological significance of responses, in *Metal Ecotoxicology: Concepts and Applications,* Lewis Publishers, Chelsea, MI, 1991.

Herricks, E.E. and J. Cairns Jr., Biological monitoring Part III: receiving system methodology based on community structure, *Water Res.,* 16, 141–153, 1982.

Hesslein, R.H., M.J. Capel, D.E. Fox, and K.A. Hallard, Stable isotopes of sulfur, carbon, and nitrogen as indicators of trophic level and fish migration in the lower Mackenzie River basin, Canada, *Can. J. Fish. Aquatic Sci.,* 48, 2258–2265, 1991.

Heusner, A.A., What does the power function reveal about structure and function in animals of different size? *Annu. Rev. Physiol.,* 49, 121–133, 1987.

Heylin, M., Bhopal, *Chem. Eng. News,* Feb. 11, 1985, pp. 14–15.

Hickey, C.W., D.S. Roper, P.T. Holland, and T.M. Trower, Accumulation of organic contaminants in two sediment-dwelling shellfish with contrasting feeding modes: deposit- (*Macomona liliana*) and filter-feeding (*Austrovenus stutchburyi*), *Arch. Environ. Contam. Toxicol.,* 29, 221–231, 1995.

Hickey, D.A. and T. McNeilly, Competition between metal tolerant and normal plant populations: a field experiment on normal soil, *Evolution,* 29, 458–464, 1975.

Hickey, J.J. and D.W. Anderson, Chlorinated hydrocarbons and eggshell changes in raptorial and fish-eating birds, *Science* (Washington, D.C.), 162, 271–273, 1968.

Hightower, L.E., Heat shock, stress proteins, chaperons, and proteotoxicity, *Cell,* 66, 191–197, 1991.

Hileman, B., Acid fog, *Environ. Sci. Technol.,* 17, 117A–120A, 1983.

Hileman, B., Recent reports on the greenhouse effect, *Environ. Sci. Technol.,* 18, 454–455, 1984.

Hileman, B., Global Warming, *Chem. Eng. News,* Mar. 13, 1989, pp. 25–44.

Hill, A.B., The environment and disease: association or causation? *Proc. R. Soc. Med.,* 58, 295–300, 1965.

Hill, W.R., A.J. Stewart, and G.E. Napolitano, Mercury speciation and bioaccumulation in lotic primary producers and primary consumers, *Can. J. Fish. Aquatic Sci.,* 53, 812–819, 1996.

Hinton, D.E., Cells, cellular responses, and their markers in chronic toxicity of fishes, in *Aquatic Toxicology: Molecular, Biochemical and Cellular Perspectives,* Malins, D.C. and Ostrander, G.K., Eds., CRC Press, Boca Raton, FL, 1994.

Hinton, D.E. and D.J. Laurén, Integrative histopathological approaches to detecting effects of environmental stressors on fishes, *Am. Fish. Soc. Symp.,* 8, 51–66, 1990a.

Hinton, D.E. and D.J. Laurén, Liver structural alterations accompanying chronic toxicity in fishes: potential biomarkers of exposure, in *Biomarkers of Environmental Contamination,* McCarthy, J.F. and Shugart, L.R., Eds., Lewis Publishers, Boca Raton, FL, 1990b.

Hirao, Y. and C.C. Patterson, Lead aerosol pollution in the High Sierra overrides natural mechanisms which exclude lead from a food chain, *Science* (Washington, D.C.), 184, 989–992, 1974.

Hobson, J.F. and W.J. Birge, Acclimation-induced changes in toxicity and induction of metallothionein-like proteins in the fathead minnow following sublethal exposure to zinc, *Environ. Toxicol. Chem.,* 8, 157–169, 1989.

Hoekstra, J.A. and P.H. Van Ewijk, Alternatives for the no-observed-effect level, *Environ. Toxicol. Chem.,* 12, 187–194, 1993.

Hoffman, D.J., B.A. Rattner, G.A. Burton, Jr., and J. Cairns, Jr., *Handbook of Ecotoxicology,* CRC Press, Boca Raton, FL, 1995.

Hoffman, F.O., paper presented at Pathway Analysis and Risk Assessment for Environmental Compliance and Reconstruction Workshop, Kiawah Island, SC, 1991.

Holl, K.D. and J. Cairns Jr., Landscape indicators in ecotoxicology, in *Handbook of Ecotoxicology,* Hoffman, D.J. et al., Eds., Lewis Publishers, Boca Raton, FL, 1995.

Holloway, G.J., R.M. Sibly, and S.R. Povey, Evolution in toxin-stressed environments, *Functional Ecol.,* 4, 289–294, 1990.

Honda, K., J.E. Marcovecchio, S. Kan, R. Tatsukama, and H. Ogi, Metal concentrations in pelagic seabirds from the north Pacific Ocean, *Arch. Environ. Contam. Toxicol.* 19, 704–711, 1990.

Hopkin, S.P., Ecophysiological strategies of terrestrial arthropods for surviving heavy metal pollution, in *Proceedings of the 3rd European Congress of Entomology, Amsterdam, 1986,* Nederlandse Entomologische Vereniging, Amsterdam, 1986, pp. 263–266.

Hopkin, S.P., *Ecophysiology of Metals in Terrestrial Invertebrates,* Elsevier Applied Science, London, 1989.

Hopkin, S.P., Ecological implications of "95% protection levels" for metals in soil, *Oikos,* 66, 137–141, 1993a.

Hopkin, S.P., *In situ* biological monitoring of pollution in terrestrial and aquatic ecosystems, in *Handbook of Ecotoxicology,* Calow, P., Ed., Blackwell Scientific Publications, London, 1993b.

Hopkin, S.P., C.A.C. Hames, and A. Dray, X-ray microanalytical mapping of the intracellular distribution of pollutant metals, *Microsc. Anal.,* November, 1989.

Hopkin, S.P. and J.A. Nott, Some observations on concentrically structured, intracellular granules in the hepatopancreas of the shore crab, *Carcinis maenas* (L.), *J. Mar. Biol. Assoc. U.K.,* 59, 867–877, 1979.

Howard, B., P.C.H. Mitchell, A. Ritchie, K. Simkiss, and M. Taylor, The composition of invertebrate granules from metal-accumulating cells of the common garden snail (*Helix aspersa*), *Biochem. J.,* 194, 507–511, 1981.

Howarth, R.S. and J.B. Sprague, Copper lethality to rainbow trout in waters of various hardness and pH, *Water Res.,* 12, 455–462, 1978.

Howarth, R.W., G. Billen, D. Swaney, A. Townsend, N. Jaworski, K. Lajtha, J.A. Downing, R. Elmgren, N. Caraco, T. Jordan, F. Berendse, J. Freney, V. Kudeyarov, P. Murdoch, and Z. Zhao-Liang, Regional nitrogen budgets and riverine N and P fluxes for the drainages to the North Atlantic Ocean: natural and human influences, *Biogeochemistry,* 35, 7–79, 1996.

Howell, W.M., D.A. Black, and S.A. Bortone, Abnormal expression of secondary sex characteristics in a population of mosquitofish, *Gambusia affinis holbrooki*: evidence for environmentally induced masculinization, *Copeia,* 4, 676–681, 1980.

Howson, C. and P. Urbach, *Scientific Reasoning: The Bayesian Approach,* Open Court Publishing Co., La Salle, IL, 1989.

Huang, P.C., Metallothionein structure/function interface, in *Metallothionein III: Biological Roles and Medical Implications,* Suzuki, K.T., Imura, N., and Kimura, M., Eds., Birkhäuser Verlag, Basel, 1993, pp. 407–426.

Huckabee, J.W., J.W. Elwood, and S.G. Hildebrand, Accumulation of mercury in freshwater biota, in *The Biogeochemistry of Mercury in the Environment,* Nriagu, J.O., Ed., Elsevier/North-Holland Biomedical Press, Amsterdam, 1979.

Huggett, R.J. and M.E. Bender, Kepone in the James River, *Environ. Sci. Technol.,* 14, 918–923, 1980.

Huggett, R.J., M.E. Bender, and M.A. Unger, Polynuclear aromatic hydrocarbons in the Elizabeth River, Virginia, in *Fate and Effects of Sediment Bound Chemicals in Aquatic Systems,* Dickson, K.L., Maki, A.W., and Brungs, W.A., Eds., Pergamon Press, New York, 1997.

Hughes, R.M. and D.P. Larsen, Ecoregions: an approach to surface water protection, *J. Water Pollut. Control Fed.,* 60, 486–493, 1988.

Hulett, L.D., Jr., A.J. Weinberger, K.J. Northcutt, and M. Ferguson, Chemical species in fly ash from coal-burning power plants, *Science* (Washington, D.C.), 210, 1356–1358, 1980.

Hunt, E.G. and A.I. Bischoff, Inimical effects on wildlife of periodic DDD applications to Clear Lake, *Calif. Fish Game,* 46, 91–106, 1960.

Hunt, G.L. and M.W. Hunt Jr., Female-female pairing in western gulls (*Larus occidentalis*) in southern California, *Science* (Washington, D.C.), 196, 1466–1467, 1977.

Hunter, G.A. and E.M. Donaldson, Hormonal sex control and its application to fish culture, in *Fish Physiology,* Hoar, W.S., Randall, D.J., and Donaldson, E.M., Eds., Academic Press, San Diego, CA, 1983.

Hursh, J.B.,M.R. Greenwood, T.W. Clarkson, J. Allen, and S. Demuth, The effect of ethanol on the fate of mercury vapor inhaled by man, *J. Pharmacol. Exp. Ther.,* 214, 520–527, 1980.

Husby, M.P., J.S. Hausbeck, and K. McBee, Chromosomal aberrancy in white-footed mice (*Peromyscus leucopus*) collected on abandoned coal strip mines, Oklahoma, USA, *Environ. Toxicol. Chem.,* 18, 919–925, 1999.

Husby, M.P. and K. McBee, Nuclear DNA content variation and double-strand DNA breakage in white-footed mice (*Peromyscus leucopus*) collected on abandoned coal strip mines, Oklahoma, USA, *Environ. Toxicol. Chem.,* 18, 926–931, 1999.

Huxley, J.S., Relative growth and form transformation, *Proc. R. Soc. London B Biol.,* 137, 465–470, 1950.

Hylland, K., T. Nissen-Lie, P.G. Christensen, and M. Sandvik, Natural modulation of hepatic metallothionein and cytochrome P4501A in flounder, *Platichthys flesus* L., *Mar. Environ. Res.,* 46, 51–55, 1998.

IAEA, *Effects of Ionizing Radiation on Plants and Animals at Levels Implied by Current Radiation Protection Standards,* International Atomic Energy Agency, Vienna, Austria, 1992, p. 73.

Ibsen, I., *Peer Gynt,* 1875, repr., New American Library of World Literature, New York, 1964.

ICRP (International Commission on Radiological Protection), *Recommendations of the International Commission on Radiological Protection,* ICRP Publication 26, Annals of the ICRP 1, Pergamon Press, New York, 1977, p. 53.

ICRP (International Commission on Radiological Protection), *Radionuclide Transformations: Energy and Intensity of Emissions,* ICRP Publication 38, Annals of the ICRP, Vols. 11–13, Pergamon Press, New York, 1983, p. 1250.

ICRP (International Commission on Radiological Protection), *1990 Recommendations of the International Commission on Radiological Protection,* ICRP Publication 60, Annals of the ICRP 21, Pergamon Press, New York, 1991, p. 211.

Jackson, A.P. and B.J. Alloway, The transfer of cadmium from agricultural soils to the human food chain, in *Biogeochemistry of Trace Metals,* Adriano, D.C., Ed., Lewis Publishers, Boca Raton, FL, 1992.

Jagoe, C.H., Responses at the tissue level: quantitative methods in histopathology applied to ecotoxicology, in *Ecotoxicology: A Hierarchical Treatment,* Newman, M.C. and Jagoe, C.H., Eds., CRC Press, Boca Raton, FL, 1996.

Jagoe, C.H., A. Faivre, and M.C. Newman, Morphological and morphometric changes in the gills of mosquitofish (*Gambusia holbrooki*) after exposure to mercury (II), *Aquatic Toxicol.* (Amsterdam), 34, 163–183, 1996.

Jagoe, R. and M.C. Newman, Bootstrap estimation of community NOEC values, *Ecotoxicology,* 6, 293–306, 1997.

Janes, N. and R.C. Playle, Modeling silver binding to gills of rainbow trout (*Oncorhynchus mykiss*), *Environ. Toxicol. Chem.,* 14, 1847–1858, 1995.

Janicki, R.H. and W.B. Kinter, DDT: disrupted osmoregulatory events in the intestine of the eel *Anguilla rostrata* adapted to seawater, *Science* (Washington, D.C.), 173, 1146–1148, 1971.

Jenny, M.J., A.H. Ringwood, E.R. Lacy, A.J. Lewitus, J.W. Kempton, P.S. Gross, G.W. Warr, and R.W. Chapman, Potential indicators of stress response identified by expressed sequence tag analysis of hemocytes and embryos from the American oyster, *Crassostrea Virginica, Mar. Biotechnol.,* 4, 81–93, 2002.

Jensen, K.M., J.J. Korte, M.D. Kahl, M.S. Pasha, and G.T. Ankley, Aspects of basic reproductive biology and endocrinology in the fathead minnow (*Pimephales promelas*), *Comp. Biochem. Physiol. Part C Toxicol. Pharmacol.,* 128C, 127–141, 2001.

Jobling, M., *Environmental Biology of Fishes,* Chapman & Hall, London, 1995.

Jobling, S., M. Nolan, C.R. Tyler, G. Brighty, and J.P. Sumpter, Widespread sexual disruption in wild fish, *Environ. Sci. Technol.,* 32, 2498–2506, 1998.

Jobling, S., D. Sheahan, J.A. Osborne, P. Matthiessen, and J.P. Sumpter, Inhibition of testicular growth in rainbow trout (*Oncorhynchus mykiss*) exposed to estrogenic alkylphenolic chemicals, *Environ. Toxicol. Chem.,* 15, 194–202, 1996.

Johansson-Sjöbeck, M.-L. and Å. Larsson, The effect of cadmium on the hematology and on the activity of δ-aminolevulinic acid dehydratase (ALA-D) in blood and hematopoietic tissues of the flounder, *Pleuronectes flesus* L., *Environ. Res.,* 17, 191–204, 1978.

Johansson-Sjöbeck, M.-L. and Å. Larsson, Effects of inorganic lead on delta-aminolevulinic acid dehydratase activity and hemotological variables in the rainbow trout, *Salmo gairdneri, Arch. Environ. Contam. Toxicol.,* 8, 419–431, 1979.

Johns, J.E. and J.R. Cunningham, *The Physics of Radiology,* Charles C Thomas, Springfield, IL, 1980.

Johnson, A.R., Landscape ecotoxicology and assessment of risk at multiple scales, *Hum. Ecological Risk Assessment,* 8, 127–146, 2002.

Johnston, J.W. and K.L. Bildstein, Dietary salt as a physiological constraint in white ibis breeding in an estuary, *Physiol. Zool.,* 63, 190–207, 1990.

Jones, K.A. and T.J. Hara, Behavioral alterations in Arctic char (*Salvelinus alpinus*) briefly exposed to sublethal chlorine levels, *Can. J. Fish. Aquatic Sci.,* 45, 749–752, 1988.

Jones, N.J. and J.M. Parry, The detection of DNA adducts, DNA base changes and chromosome damage for the assessment of exposure to genotoxic pollutants, *Aquatic Toxicol.,* 22, 323–344, 1992.

Jones, P.D., D.J. Hannah, S.J. Buckland, T. van Maanen, S.V. Leathem, S. Dawson, E. Slooten, A. van Helden, and M. Donoghue, Polychlorinated dibenzo-p-dioxins, dibenzofurans and polychlorinated biphenyls in New Zealand cetaceans, *J. Cetacean Res. Manage.,* Special Issue 1, 157–167, 1999.

Jørgensen, S.E., *Modelling in Ecotoxicology,* Elsevier, New York, 1990.

Jørgensen, S.E. and B. Sorensen, Eds., Drugs in the environment, *Chemosphere,* 40 (special issue), 691–699, 2000.

Josephson, J.R. and S.G. Josephson, *Abductive Inference: Computation, Philosophy, Technology,* Cambridge University Press, Cambridge, U.K., 1996, p. 306.

Jung, R.E. and C.H. Jagoe, Effects of low pH and aluminum on body size, swimming performance, and susceptibility to predation of green tree frog (*Hyla cinerea*) tadpoles, *Can. J. Zool.,* 73, 2171–2183, 1995.

Kac, M., Some mathematical models in science, *Science* (Washington, D.C.), 166, 695–699, 1969.

Kägi, J.H.R., Overview of metallothionein, in *Methods in Enzymology,* Vol. 205, *Metallobiochemistry, Part B: Metallothionein and Related Molecules,* Riordan, J.F. and Vallee, B.L., Eds., Academic Press, San Diego, CA, 1991.

Kahn, B. and K.S. Turgeon, The bioaccumulation factor for phosphorus-32 in edible fish tissue, *Health Phys.,* 46, 321–333, 1984.

Kaiser, K.L., M.B. McKinnon, D.H. Stendahl, and W.B. Pett, Response threshold levels of selected organic compounds for rainbow trout (*Oncorhynchus mykiss*), *Environ. Toxicol. Chem.,* 14, 2107–2113, 1995.

Kammenga, J.E., M. Busschers, N.M. Van Straalen, P.C. Jepson, and J. Bakker, Stress induced fitness is not determined by the most sensitive life-cycle trait, *Functional Ecol.,* 10, 106–111, 1996.

Kammenga, J. and R. Laskowski, Eds., *Demography in Ecotoxicology,* John Wiley & Sons, Chichester, U.K., 2000.

Kania, H.K. and J. O'Hara, Behavioral alterations in a simple predator-prey system due to sublethal exposure to mercury, *Trans. Am. Fish. Soc.,* 103, 134–136, 1974.

Karin, M. and H.R. Herschman, Induction of metallothionein in HeLa cells by dexamethasone and zinc, *Euro. J. Biochem.,* 113, 267–272, 1981.

Karr, J.R. and E.W. Chu, *Restoring Life in Running Waters: Better Biological Monitoring,* Island Press, Washington, D.C., 1999.

Karr, J.R. and E.M. Rossano, Applying public health lessons to protect river health, *Ecol. Civ. Eng.,* 4, 3–18, 2001.

Kaufman, L.H., Stream *aufwuchs* accumulation: disturbance frequency and stress resistance and resilience, *Oecologia* (Berlin), 52, 57–63, 1982.

Keith, L.H. and W.A. Telliard, Priority pollutants, I: a perspective view, *Environ. Sci. Technol.,* 13, 416–423, 1979.

Keklak, M.M., M.C. Newman, and M. Mulvey, Enhanced uranium tolerance of an exposed population of the Eastern mosquitofish (*Gambusia holbrooki* Girard 1859), *Arch. Environ. Contam. Toxicol.,* 27, 20–24, 1994.

Kelce, W.R., C.R. Stone, S.C. Laws, L.E. Gray, J.A. Kemppainen, and E.M. Wilson, Persistent DDT metabolite p,p′-DDE is a potent androgen receptor antagonist, *Nature* (London), 375, 581–585, 1995.

Kelly, E.J., C.J. Quaife, G.J. Froelick, and R.D. Palmiter, Metallothionein I and II protect against zinc deficiency and zinc toxicity in mice, *J. Nutr.,* 126, 1782–1790, 1996.

Kerr, R.A., Ozone hole bodes ill for the globe, *Science* (Washington, D.C.), 241, 785–786, 1988a.

Kerr, R.A., Is there life after climate change? *Science* (Washington, D.C.), 242, 1010–1013, 1988b.

Kerr, R.A., New assaults seen on Earth's ozone shield, *Science* (Washington, D.C.), 255, 797–798, 1992.

Kerr, R.A., Ozone-destroying chlorine tops out, *Science* (Washington, D.C.), 271, 32, 1996.

Kersting, K., Normalizing ecosystem strain: a system parameter for analysis of toxic stress in (micro-) ecosystems, *Ecol. Bull.,* 36, 150–153, 1984.

Kettlewell, H.B.D., Selection experiments on industrial melanism in the *Lepidoptera, Heredity,* 9, 323–342, 1955.

Kettlewell, B., *The Evolution of Melanism,* Clarendon Press, Oxford, U.K., 1973.

Khalil, A.M., Chromosome aberrations in blood lymphocytes from petroleum refinery workers, *Arch. Environ. Contam. Toxicol.,* 28, 236–239, 1995.

Khalil, M.A.K. and R.A. Rasmussen, Carbon monoxide in the earth's atmosphere: increasing trend, *Science* (Washington, D.C.), 224, 54–56, 1984.

Khayat, A.I. and Z.A. Shaikh, Dose-effect relationship between ethyl alcohol pretreatment and retention and tissue distribution of mercury vapor in rats, *J. Pharmacol. Exp. Ther.,* 223, 649–653, 1982.

Khera, K.S., Teratogenic and genetic effects of mercury, in *The Biogeochemistry of Mercury in the Environment,* Nriagu, J.O., Ed., Elsevier/North-Holland Biomedical Press, Amsterdam, 1979.

Kidd, K.A., R.H. Hesslein, R.J.P. Fudge, and K.A. Hallard, The influence of trophic level as measured by $\delta^{15}N$ on mercury concentrations in freshwater organisms, *Water Air Soil Pollut.,* 80, 1011–1015, 1995a.

Kidd, K.A., D.W. Schindler, R.H. Hesslein, and D.C.G. Muir, Correlation between stable nitrogen isotope ratios and concentrations of organochlorines in biota from a freshwater food web, *Sci. Total Environ.,* 160/161, 381–390, 1995b.

Kim, E.Y., T. Murakami, K. Saeki, and R. Tatsukama, Mercury levels and its chemical form in tissues and organs of seabirds, *Arch. Environ. Contam. Toxicol.,* 30, 259–266, 1996.

King, J.J., *The Environmental Dictionary,* 2nd ed., Executive Enterprises Publications Co., New York, 1993.

King, J.K., F.M. Saunders, R.F. Lee, and R.A. Jahnke, Coupling mercury methylation rates to sulfide reduction rates in marine sediments, *Environ. Toxicol. Chem.,* 18, 1362–1369, 1999.

Klaassen, C.D., M.O. Amdur, and J. Doull, Eds., *Casarett and Doull's Toxicology: The Basic Science of Poisons,* MacMillan Publishing Co., New York, 1987.

Klaassen, C.D., J. Liu, and S. Choudhuri, Metallothionein: an intracellular protein to protect against cadmium toxicity, *Annu. Rev. Pharmacol. Toxicol.,* 39, 267–294, 1999.

Klaverkamp, J.F., M.D. Dutton, H.S. Majewski, R.V. Hunt, and L.J. Wesson, Evaluating the effectiveness of metal pollution controls in a smelter by using metallothionein and other biochemical responses in fish, in *Metal Ecotoxicology: Concepts and Applications,* Newman, M.C. and McIntosh, A.W., Eds., Lewis Publishers, Chelsea, MI, 1991.

Kleinow, K.M., M.J. Melancon, and J.J. Lech, Biotransformation and induction: implications for toxicity, bioaccumulation and monitoring of environmental xenobiotics in fish, *Environ. Health Perspect.,* 71, 105–119, 1987.

Klerks, P.L. and J.S. Levinton, Rapid evolution of metal resistance in a benthic oligochaete inhabiting a metal-polluted site, *Biol. Bull.,* 176, 135–141, 1989.

Klerks, P.L. and J.S. Weis, Genetic adaptation to heavy metals in aquatic organisms: a review, *Environ. Pollut.,* 45, 173–205, 1987.

Kling, G.W., B. Fry, and W.J. O'Brien, Stable isotopes and planktonic trophic structure in Arctic lakes, *Ecology,* 73, 561–566, 1992.

Kloas, W., Schrag, C. Ehnes, and H. Segner, Binding of xenobiotics to hepatic estrogen receptor and plasma sex steroid binding protein in the teleost fish, the common carp (*Cyprinus carpio*), *Gen. Comp. Endocrinol.,* 119, 287–299, 2000.

Knight, M., A.N. Miller, N.S. Geoghagen, F.A. Lewis, and A.R. Kerlavage, Expressed sequence TAGS (ESTs) of *Biomphalaria glabrata*, an intermediate snail host of *Schistosoma mansoni*: use in the identification of RFLP markers, *Malacologia,* 39, 175–182. 1998.

Knoll, G.F., *Radiation Detection and Measurement,* John Wiley & Sons, New York, 1989.

Knoph, M.B. and Y.A. Olsen, Subacute toxicity of ammonia to Atlantic salmon (*Salmo salar* L.) in seawater: effects on water and salt balance, plasma cortisol and plasma ammonia levels, *Aquatic Toxicol.* (Amsterdam), 30, 295–310, 1994.

Koehn, R.K. and B.L. Bayne, Towards a physiological and genetical understanding of the energetics of the stress response, *Biol. J. Linn. Soc.,* 37, 157–171, 1989.

Koehn, R.K. and P.M. Gaffney, Genetic heterozygosity and growth rate in *Mytilus edulis, Mar. Biol.,* 82, 1–7, 1984.

Koivisto, S. and M. Ketola, Effects of copper on life-history traits of *Daphnia pulex* and *Bosmina longirostris, Aquatic Toxicol.,* 32, 255–269, 1995.

Kojima, Y., P.-A. Binz, and J.H.R. Kägi, Nomenclature of metallothionein: proposal for revision, in *Metallothionein IV,* Klaassen, C., Ed., Birkhäuser Verlag, Basel, 1999.

Kolaja, G.J. and D.E. Hinton, DDT-induced reduction in eggshell thickness, weight, and calcium is accompanied by calcium ATPase inhibition, in *Animals as Monitors of Environmental Pollutants,* National Academy of Sciences, Washington, D.C., 1979.

Kolkwitz, R. and M. Marsson, Ökologie der pflanzlichen Saprobien, *Ber. Dtsch. Bot. Ges.,* 26, 505–519, 1908.

Kondo, S., *Health Effects of Low-level Radiation,* Kinki University Press, Osaka, Japan, 1993.

Kooijman, S.A.L.M., Parametric analyses of mortality rates in bioassays, *Water Res.,* 15, 107–119, 1981.

Kooijman, S.A.L.M., *Dynamic Energy Budgets in Biological Systems,* Cambridge University Press, Cambridge, U.K., 1993.

Kooijman, S.A.L.M. and J.J.M. Bedaux, *The Analysis of Aquatic Toxicity Data,* VU University Press, Amsterdam, 1996.

Kopp, R.L., S.I. Guttman, and T.E. Wissing, Genetic indicators of environmental stress in central mudminnow (*Umbra limi*) populations exposed to acid deposition in the Adirondack Mountains, *Environ. Toxicol. Chem.,* 11, 665–676, 1992.

Koropatnick, J., Amplification of metallothionein-1 genes in mouse liver cells in situ: extra copies are transcriptionally active, *Proc. Soc. Exp. Biol. Med.,* 188, 287–300, 1988.

Koss, G., E. Schuler, B. Arndt, J. Seidel, S. Seubert, and A. Seubert, A comparative toxicological study of pike (*Esox lucius* L.) from the River Rhine and River Lahn, *Aquatic Toxicol.* 8, 1–9, 1986.

Kotz, S. and N.L. Johnson, *Encyclopedia of Statistical Sciences,* Vol. 8, John Wiley & Sons, New York, 1988.

Krahn, M.M., M.S. Myers, D.G. Burrows, and D.C. Malins, Determination of metabolites of xenobiotics in the bile of fish from polluted waterways, *Xenobiotica,* 16, 957–973, 1984.

Kramer, V.J. and M.C. Newman, Inhibition of glucosephosphate isomerase allozymes of the mosquitofish, *Gambusia holbrooki,* by mercury, *Environ. Toxicol. Chem.,* 13, 9–14, 1994.

Kramer, V.J., M.C. Newman, M. Mulvey, and G.R. Ultsch, Glycolysis and Krebs cycle metabolites in mosquitofish, *Gambusia holbrooki,* Girard 1859, exposed to mercuric chloride: allozyme genotype effects, *Environ. Toxicol. Chem.,* 11, 357–364, 1992.

Krantzberg, G., Spatial and temporal variability in metal bioavailability and toxicity of sediment from Hamilton Harbour, Lake Ontario, *Environ. Toxicol. Chem.,* 13, 1685–1698, 1994.

Krause, P.R., Effects of an oil production effluent on gametogenesis and gamete performance in the purple sea urchin (*Strongylocentrotus purpuratus* Stimpson), *Environ. Toxicol. Chem.,* 13, 1153–1161, 1994.

Kuhn, T.S., *The Structure of Scientific Revolutions,* University of Chicago, Chicago, 1970.

Lamb, T., J.W. Bickham, J.W. Gibbons, M.J. Smolen, and S. McDowell, Genetic damage in a population of slider turtles (*Trachemys scripta*) inhabiting a radioactive reservoir, *Arch. Environ. Contam. Toxicol.,* 20, 138–142, 1991.

Lance, B.K., D.B. Irons, S.J. Kendall, and L.L. McDonald, An evaluation of marine bird population trends following the Exxon Valdez oil spill, Prince William Sound, Alaska, *Mar. Pollut. Bull.,* 42, 298–309, 2001.

Landis, W.G., Uncertainty in the extrapolation from individual effects to impacts upon landscapes, *Hum. Ecol. Risk Assessment,* 8, 193–204, 2002.

Landis, W.G. and J.F. McLaughlin, Design criteria and derivation of indicators for ecological position, direction and risk, *Environ. Toxicol. Chem.,* 19, 1059–1065, 2000.

Landis, W.G. and M.-H. Yu, *Introduction to Environmental Toxicology,* Lewis Publishers, Boca Raton, FL, 1995.

Landrum, P.F., G.A. Karkey, and J. Kukkonen, Evaluation of organic contaminant exposure in aquatic organisms: the significance of bioconcentration and bioaccumulation, in *Ecotoxicology: A Hierarchical Treatment,* Newman, M.C. and Jagoe, C.H., Eds., CRC Press, Boca Raton, FL, 1996.

Landrum, P.F., H. Lee II, and M.J. Lydy, Toxicokinetics in aquatic systems: model comparisons and use in hazard assessment, *Environ. Toxicol. Chem.,* 11, 1709–1725, 1992.

Landrum, P.F. and M.J. Lydy, personal communication during toxicokinetics short course, Nov. 3, 1991, Society of Environmental Toxicology and Chemistry 12th Annual Meeting, 1991.

Landrum, P.F. and J.A. Robbins, Bioavailability of sediment-associated contaminants to benthic invertebrates, in *Sediments: Chemistry and Toxicity of In-Place Pollutants,* Baudo, R., Giesy, J., and Muntau, H., Eds., Lewis Publishers, Chelsea, MI, 1990.

Lange, B.W., *Drosophila melanogaster* Metallothionein Genes: Selection for Duplications? Ph.D. dissertation, Duke University, Durham, NC, 1989.

Lange, T.R., H.E. Royals, and L.L. Conner, Mercury accumulation in largemouth bass (*Micropterus salmoides*) in a Florida lake, *Arch. Environ. Contam. Toxicol.,* 27, 466–471, 1994.

Langston, W.J. and G.R. Burt, Bioavailability and effects of sediment-bound TBT in deposit-feeding clams, *Scrobicularia plana Mar. Environ. Res.,* 32, 61–77, 1991.

Lapp, R.E. and H.L. Andrews, *Nuclear Radiation Physics,* Prentice-Hall, Englewood Cliffs, NJ, 1972.

Larson, R.A. and E.J. Weber, *Reaction Mechanisms in Environmental Organic Chemistry,* CRC Press, Boca Raton, FL, 1994.

Larsson, Å., B.-E. Bengtsson, and C. Haux, Disturbed ion balance in flounder, *Platichthys flesus* L. exposed to sublethal levels of cadmium, *Aquatic Toxicol.* (Amsterdam), 1, 19–35, 1981.

Larsson, D.G.J., M. Adolfsson-Erici, J. Parkkonen, M. Pettersson, A.H. Berg, P.E. Olsson, and L. Forlin, Ethinyloestradiol: an undesired fish contraceptive? *Aquatic Toxicol.,* 45, 91–97, 1999.

Laskowski, R., Are the top carnivores endangered by heavy metal biomagnification? *Oikos,* 60, 387–390, 1991.

Laurén, D.J. and D.G. McDonald, Effects of copper on branchial ionoregulation in the rainbow trout, *Salmo gairdneri* Richardson: modulation by water hardness and pH, *J. Comp. Physiol. B,* 155, 635–644, 1985.

La Via, M.F. and R.B. Hill Jr., *Principles of Pathobiology,* Oxford University Press, New York, 1971.

Lavie, B. and E. Nevo, Heavy metal selection of phosphoglucose isomerase allozymes in marine gastropods, *Mar. Biol.,* 71, 17–22, 1982.

Lavie, B. and E. Nevo, Genetic selection of homozygote allozyme genotypes in marine gastropods exposed to cadmium pollution, *Sci. Total Environ.,* 57, 91–98, 1986.

Laws, E.A., *Aquatic Pollution: An Introductory Text,* 2nd ed., John Wiley & Sons, New York, 1993.

Laxen, D.P.H. and R.M. Harrison, The highway as a source of water pollution: an appraisal with the heavy metal lead, *Water Res.,* 11, 1–11, 1977.

Lech, J.J. and M.J. Vodicnik, Biotransformation, in *Fundamentals of Aquatic Toxicology,* Rand, G.M. and Petrocelli, S.R., Eds., Hemisphere Publishing Corp., Washington, D.C., 1985.

Lechelt, M.W. Blohm, B. Kirschneit, M. Pfeiffer, E. Gresens, J. Liley, R. Holz, C. Lüering, and C. Moldaenke, Monitoring of surface water by ultra-sensitive *Daphnia* toximeter, *Environ. Toxicol.,* 15, 390–400, 2000.

Lee, B., S.B. Griscom, J. Lee, H.J. Choi, C. Koh, S.N. Luoma, and N.S. Fisher, Influences of dietary uptake and reactive sulfides on metal bioavailability from aquatic sediments, *Science,* 287, 282–284, 2000.

Lee, C., M.C. Newman, and M. Mulvey, Time to death of mosquitofish (*Gambusia holbrooki*) during acute inorganic mercury exposure: population structure effects, *Arch. Environ. Contam. Toxicol.,* 22, 284–287, 1992.

Lee, G., Lawmakers Move to Check CFC Phaseout, *Washington Post,* Sep. 21, 1995, p. A13.

Lee, L.S., P.S.C. Rao, P. Nkedi-Kizza, and J.J. Delfino. Influence of solvent and sorbent characteristics on the distribution of PCP in octanol-water and soil-water systems, *Environ. Sci. Technol.,* 24, 654–661, 1990.

Lees, D.R. and E.R. Creed, The genetics of the *insularia* forms of the peppered moth, *Biston betularia, Heredity,* 39, 67–73, 1977.

Leggett, R.W., Predicting the retention of Cs in individuals, *Health Phys.,* 50, 747–759, 1986.

Leland, H.V. and J.L. Carter, Effects of copper on species composition of periphyton in a Sierra Nevada, California, stream, *Freshwater Biol.,* 14, 281–296, 1984.

Leland, H.V. and J.S. Kuwabara, Trace elements, in *Fundamentals of Aquatic Toxicology,* Rand, G.M. and Petrocelli, S.R., Eds., Hemisphere Publishing Corp., Washington, D.C., 1985.

Lenihan H. and C. Peterson, How habitat degradation through fishery disturbance enhances impacts of hypoxia on oyster reefs, *Ecological Appl.,* 8, 128–140, 1998.

Lenz, W., The susceptible period for thalidomide malformations in man and monkey, *Ger. Med. Mon.,* 4, 197–198, 1968.

Lenz, W., Malformations caused by drugs in pregnancy, *Am. J. Dis. Child.,* 2, 99–106, 1996.

Leopold, A., *A Sand County Almanac,* Oxford University Press, Oxford, U.K., 1949, repr., Sierra Club/Ballantine Books, New York, 1966.

Lepkowski, W., Bhopal: Indian City Begins to Heal but Conflicts Remain, *Chem. Eng. News,* Dec. 2, 1985, 18–32.

Leslie, P.H., On the use of matrices in certain population mathematics, *Biometrika,* 33, 183–212, 1945.

Leslie, P.H., Some further notes on the use of matrices in population mathematics, *Biometrika,* 35, 213–245, 1948.

Leung, K.M.Y., R. Wheeler, D. Morritt, and M. Crane, Endocrine disruption in fishes and invertebrates: issues for saltwater ecological risk assessment, in *Coastal and Estuarine Risk Assessment,* Newman, M.C., Roberts, M.H., Jr., and Hale, R.C., Eds., CRC Press, Boca Raton, FL, 2002.

Levin, S.A., M.A. Harwell, J.R. Kelly, and K.D. Kimball, *Ecotoxicology: Problems and Approaches,* Springer-Verlag, New York, 1989, p. 547.

Levine, J.S., Ed., *Global Biomass Burning: Atmospheric, Climatic, and Biospheric Implications,* MIT Press, Cambridge, MA, 1991.

Levine, J.S., Ed., *Biomass Burning and Global Change: Remote Sensing, Modeling and Inventory Development, and Biomass Burning in Africa,* MIT Press, Cambridge, MA, 1996a.

Levine, J.S., Ed., *Biomass Burning and Global Change: Biomass Burning in South America, Southeast Asia, and Temperate and Boreal Ecosystems, and the Oil Fires of Kuwait,* MIT Press, Cambridge, MA, 1996b.

Levine, J.S., The 1997 fires in Kalimantan and Sumatra, Indonesia: gaseous and particulate emissions, *Geophys. Res. Lett.,* 26, 815–818, 1999.

Levine, J.S., Global impacts and climate change, in *Forest Fires and Regional Haze in Southeast Asia,* Eaton, P. and Radojevic, M., Eds., Nova Science Publishers, New York, 2001.

Levins, R., Some demographic and genetic consequences of environmental heterogeneity for biological control, *Bull. Entomol. Soc. Am.,* 15, 237–240, 1969.

Lewis, S.A., P.H. Becker, and R.W. Furness, Mercury levels in eggs, tissues and feathers of herring gulls (*Larus argentatus*) from the German Wadden sea coast, *Environ. Pollut.,* 80, 293–299, 1993.

Li, A.P. and R.H. Heflich, *Genetic Toxicology*, CRC Press, Boca Raton, FL, 1991, p. 493.

Liber, K., N.K. Kaushik, K.R. Solomon, and J.H. Carey, Experimental designs for aquatic mesocosm studies: a comparison of the "ANOVA" and "regression" design for assessing the impact of tetrachlorophenol on zooplankton populations in limnocorrals, *Environ. Toxicol. Chem.,* 11, 61–77, 1992.

Lide, D.R., *CRC Handbook of Chemistry and Physics,* 73rd ed., CRC Press, Boca Raton, FL, 1992.

Likens, G.E., Ed., Nutrients and Eutrophication, *Limnol. Oceanogr.,* 17 (special issue), 3–13, 1972.

Likens, G.E., Acid Precipitation, *Chem. Eng. News,* Nov. 22, 1976, pp. 29–44.

Likens, G.E. and F.H. Borman, Acid rain: a serious environmental problem, *Science,* 184, 1176–1179, 1974.

Lindberg, S.E., R.C. Harris, and R.R. Turner, Atmospheric deposition of metals to forest vegetation, *Science* (Washington, D.C.), 215, 1609–1612, 1982.

Lindqvist, L. and M. Block, Excretion of cadmium and zinc during moulting in the grasshopper *Omocestus viridulus* (Orthoptera), *Environ. Toxicol. Chem.,* 13, 1669–1672, 1994.

Lindzen, R.S., On the scientific basis for global warming scenarios, *Environ. Pollut.,* 83, 125–134, 1994.

Linke-Gamenick, I., V.E. Forbes, and R.M. Sibly, Density-dependent effects of a toxicant on life-history traits and population dynamics of a capitellid polychaete, *Mar. Ecol. Prog. Ser.,* 184, 139, 1999.

Lipnick, R.L., A perspective on quantitative structure-activity relationships in ecotoxicology, *Environ. Toxicol. Chem.,* 4, 255–257, 1985.

Lipnick, R.L., Structure-activity relationships, in *Fundamentals of Aquatic Toxicology: Effects, Environmental Fate, and Risk Assessment,* 2nd ed., Rand, G.M., Ed., Taylor & Francis, Washington, D.C., 1995.

Litchfield, J.T., A method for rapid graphic solution of time-percent effect curves, *J. Pharm. Exp. Ther.,* 97, 399–408, 1949.

Litchfield, J.T. and F. Wilcoxon, A simplified method of evaluating dose-effect experiments, *J. Pharm. Exp. Ther.,* 96, 99–113, 1949.

Little, E.E., Behavioral toxicology: stimulating challenges for a growing discipline, *Environ. Toxicol. Chem.,* 9, 1–2, 1990.

Little, E.E. and S.E. Finger, Swimming behavior as an indicator of sublethal toxicity in fish, *Environ. Toxicol. Chem.,* 9, 13–20, 1990.

Little, E.E., J.F. Fairchild, and A.J. DeLonay, Behavioral methods for assessing impacts of contaminants on early life stage fishes, in *Water Quality and the Early Life Stages of Fishes,* Fuiman, L.A., Ed., American Fisheries Society, Bethesda, MD, 1993.

Liu, Y.P., J. Liu, M.B. Iszard, G.K. Andrews, R.D. Palmiter, and C.D. Klaassen, Transgenic mice that overexpress metallothionein-I are protected from cadmium lethality and hepatotoxicity, *Toxicol. Appl. Pharmacol.,* 135, 222–228, 1995.

Lloyd, R. and D.W.M. Herbert, The influence of carbon dioxide on the toxicity of un-ionized ammonia to rainbow trout (*Salmo gairdneri* Richardson), *Ann. Appl. Biol.,* 48, 399–404, 1960.

Lloyd, R. and L.D. Orr, The diuretic response by rainbow trout to sub-lethal concentrations of ammonia, *Water Res.,* 3, 335–344, 1969.

Lobert, J.M., D.H. Scharffe, W.-M. Hao, T.A. Kuhlbusch, R. Seuwen, P. Warneck, and P.J. Crutzen, Experimental evaluation of biomass burning emissions: nitrogen and carbon containing compounds, in *Global Biomass Burning: Atmospheric, Climatic, and Biospheric Implications,* Levine, J.S., Ed., MIT Press, Cambridge, MA, 1991.

Lomborg, B., *The Skeptical Environmentalist,* Cambridge University Press, Cambridge, MA, 2001.

Long, E.R., D.D. MacDonald, J.C. Cubbage, and C.G. Ingersoll, Predicting the toxicity of sediment-associated trace metals with simultaneously extracted trace metal:acid-volatile sulfide concentrations and dry weight-normalized concentrations: a critical comparison, *Environ. Toxicol. Chem.,* 17, 972–974, 1998.

Lowrance, R., R. Todd, J. Fair, Jr., O. Hendrickson, R. Leonard, and L. Asmussen, Riparian forests as nutrient filters in agricultural watersheds, *Bioscience,* 34, 374–377, 1984.

Lovelock, J.E., Gaia as seen through the atmosphere, *Atmos. Environ.,* 6, 579–580, 1972.

Lovelock, J.E., *The Ages of Gaia: A Biography of Our Living Earth,* Oxford University Press, Oxford, U.K., 1988.

Lovelock, J.E., *Healing Gaia: Practical Medicine for the Planet,* Gaia Books Limited, New York, 1991.

Luoma, J.R., New Effect of Pollutants: Hormone Mayhem, *New York Times,* May 24, 1992.

Luoma, S.N., Can we determine the biological availability of sediment-bound trace elements? *Hydrobiolia,* 176/177, 379–396, 1989.

Luoma, S.N. and G.W. Bryan, Factors controlling the availability of sediment-bound lead to the estuary bivalve, *Scrobicularia plana, J. Mar. Biol. Assoc. U.K.,* 58, 793–802, 1978.

Lyon, R., M. Taylor, and K. Simkiss, Ligand activity in the clearance of metals from the blood of the crayfish (*Austropotamobius pallipes*), *J. Exp. Biol.,* 113, 19–27, 1984.

Mackay, D., Correlation of bioconcentration factors, *Environ. Sci. Technol.,* 16, 274–278, 1982.

Mackay, D., *Multimedia Environmental Models: The Fugacity Approach,* Lewis Publishers, Chelsea, MI, 1991.

Mackay, D. and S. Paterson, Fugacity revisited: the fugacity approach to environmental transport, *Environ. Sci. Technol.,* 116, 654A–660A, 1982.

Mackay, D. and F. Wania, Transport of contaminants to the Arctic: partitioning processes and models, *Sci. Total Environ.,* 160/161, 25–38, 1995.

Mackenthun, K.M. and J.I. Bregman, *Environmental Regulations Handbook,* Lewis Publishers, Boca Raton, FL, 1992.

Macklis, R.M., The great radium scandal, *Sci. Am.,* August, 94–99, 1993.

Macovsky, L.M., The Effects of Toxicant-Related Mortality upon Metapopulation Dynamics: A Laboratory Model, M.S. thesis, Western Washington University, Bellingham, WA, 1999.

Magurran, A.E., *Ecological Diversity and Its Measurement,* Princeton University Press, Princeton, NJ, 1988, p. 179.

Majerus, M.E.N., *Melanism: Evolution in Action,* Oxford University Press, Oxford, U.K., 1998.

Malins, D.C., Identification of hydroxyl radical-induced lesions in DNA base structure: biomarkers with a putative link to cancer development, *J. Toxicol. Environ. Health,* 40, 247–261, 1993.

Malins, D.C. and G.K. Ostrander, *Aquatic Toxicology: Molecular, Biochemical, and Cellular Perspectives,* CRC Press, Boca Raton, FL, 1994, p. 539.

Malins, D.C., B.B. McCain, D.W. Brown, S.L. Chan, M.S. Myers, and J.T. Landahl, Chemical pollutants in sediments and diseases of bottom-dwelling fish in Puget Sound, Washington, *Environ. Sci. Technol.,* 18, 705–713, 1984.

Mallatt, J., Fish gill structural changes induced by toxicants and other irritants: a statistical review, *Can. J. Fish. Aquatic Sci.,* 42, 630–648, 1985.

Mallet, J., The evolution of insecticide resistance: have the insects won? *Trends Ecol. Evol.,* 4, 336–340, 1989.

Manahan, S.E., *Fundamentals of Environmental Chemistry,* Lewis Publishers, Chelsea, MI, 1993.

Manahan, S.E., *Environmental Chemistry,* CRC Press, Boca Raton, FL, 2000.

Mance, G., *Pollution Threat of Heavy Metals in Aquatic Environments,* Elsevier Applied Science, London, 1987.

Marco, G.J., R.M. Hollingworth, and W. Durham, *Silent Spring Revisited,* American Chemical Society, Washington, D.C., 1987.

Margoshes, M. and B.L. Vallee, A cadmium protein from equine kidney cortex, *J. Am. Chem. Soc.,* 79, 4813–4814, 1957.

Margulis, L. and J.E. Lovelock, Gaia and geognosy, in *Global Ecology: Towards a Science of the Biosphere,* Rambler, M.B., Margulis, L., and Fester, R., Eds., Academic Press, Boston, MA, 1989.

Marking, L.L., Toxicity of chemical mixtures, in *Fundamentals of Aquatic Toxicology,* Rand, G.M. and Petrocelli, S.R., Eds., Hemisphere Publishing Corp., Washington, D.C., 1985.

Marking, L.L. and V.K. Dawson, Method for assessment of toxicity or efficacy of mixtures of chemicals, *U.S. Fish Wildl. Serv. Invest. Fish Control,* 67, 1–8, 1975.

Maroni, G., J. Wise, J.E. Young, and E. Otto, Metallothionein gene duplications and metal tolerance in natural populations of *Drosophila melanogaster, Genetics,* 117, 739–744, 1987.

Marshall, E., EPA may allow more lead in gasoline, *Science* (Washington, D.C.), 215, 1375–1378, 1982.

Marshall, J.S., The effects of continuous gamma radiation on the intrinsic rate of natural increase of *Daphnia pulex, Ecology,* 43, 598–607, 1962.

Martínez-Jerónimo, F., R. Villaseñor, F. Espinosa, and G. Rios, Use of life-tables and application factors for evaluating chronic toxicity of kraft mill wastes on *Daphnia magna, Bull. Environ. Contam. Toxicol.,* 50, 377–384, 1993.

Marubini, E. and M.G. Valsecchi, *Analysing Survival Data from Clinical Trials and Observational Studies,* John Wiley & Sons, New York, 1995.

Mason, A.Z. and J.A. Nott, The role of intracellular biomineralized granules in the regulation and detoxification of metals in gastropods with special reference to the marine prosobranch *Littorina littorea, Aquatic Toxicol.* (Amsterdam), 1, 239–256, 1981.

Mason, A.Z., K. Simkiss, and K.P. Ryan, The ultrastructural localization of metals in specimens of *Littorina littorea* collected from clean and polluted sites, *J. Mar. Biol. Assoc. U.K.,* 64, 699–720, 1984.

Mason, R.P. and W.F. Fitzgerald, Alkylmercury species in the equatorial Pacific, *Nature,* 347, 457–459, 1990.

Masood, E., Climate panel confirms human role in warming, fights off oil states, *Nature* (London), 378, 524, 1995.

Masters, B.A., E.J. Kelly, C.J. Quaife, R.L. Brinster, and R.D. Palmiter, Targeted disruption of metallothionein I and II genes increases sensitivity to cadmium, *Proc. Natl. Acad. Sci. USA,* 91, 584–588, 1994.

Matthews, R.A., W.G. Landis, and G.B. Matthews, The community conditioning hypothesis and its application to environmental toxicology, *Environ. Toxicol. Chem.,* 15, 597–603, 1996.

Maugh, T.H., It isn't easy being king, *Science* (Washington, D.C.), 203, 637, 1974.

Maur, A.A.D., T. Belser, G. Elgar, O. Georgiev, and W. Schaffner, Characterization of the transcription factor MTF-1 from the Japanese pufferfish (*Fugu rubripes*) reveals evolutionary conservation of heavy metal stress response, *Biol. Chem.,* 380, 175–185, 1999.

Maurer, B.A. and R.D. Holt, Effects of chronic pesticide stress on wildlife populations in complex landscapes: processes at multiple scales, *Environ. Toxicol. Chem.,* 15, 420–426, 1996.

May, R.M., Biological populations with nonoverlapping generations: stable points, stable cycles, and chaos, *Science* (Washington, D.C.), 186, 645–647, 1974.

May, R.M., *Theoretical Ecology: Principles and Applications,* W.B. Saunders, Philadelphia, 1976a.

May, R.M., Simple mathematical models with very complicated dynamics, *Nature* (London), 261, 459–467, 1976b.

McBee, K. and R.L. Lochmiller, Wildlife toxicology in biomonitoring and bioremediation: implications for human health, in *Ecotoxicity and Human Health: A Biological Approach to Environmental Remediation,* de Serres, F.J. and Bloom, A.D., Eds., CRC Lewis Publishers, Boca Raton, FL, 1996, pp. 163–181.

McBee, K., J.W. Bickham, K.W. Brown, and K.C. Donnelly, Chromosomal aberrations in native small mammals (*Peromyscus leucopus* and *Sigmodon hispidus*) at a petrochemical waste disposal site, I: standard karyology, *Arch. Environ. Contam. Toxicol.,* 16, 681–688, 1987.

McBride, W.G., Thalidomide and congenital anomalies, *Lancet,* 2, 1358, 1961.

McCarthy, J.F. and L.R. Shugart, *Biomarkers of Environmental Contamination,* Lewis Publishers, Chelsea, MI, 1990.

McCarthy, J.R., R.S. Halbrook, and L.R. Shugart, *Conceptual Strategy for Design, Implementation, and Validation of a Biomarker-Based Biomonitoring Capability,* Oak Ridge National Laboratory, Oak Ridge, TN, 1991, p. 85.

McCormick, P.V., J. Cairns, Jr., S.E. Belanger, and E.P. Smith, Response of protistan assemblages to a model toxicant, the surfactant C12-TMAC (dodecyl trimethyl ammonium chloride), in laboratory streams, *Aquatic Toxicol.,* 21, 41–70, 1991.

McFarlane, G.A. and W.G. Franzin, Elevated heavy metals: a stress on a population of white suckers, *Catostomus commersoni,* in Hamell Lake, Saskatchewan, *J. Fish. Res. Board Can.,* 35, 963–970, 1978.

McGee, B.L., D.J. Fisher, L.T. Yonkos, G.P. Ziegler, and S.D. Turley, Assessment of sediment contamination, acute toxicity and population viability of the estuarine amphipod *Leptocheirus plumulosus* in Baltimore Harbor, *Environ. Toxicol. Chem.,* 18, 2151–2160, 1999.

McGregor, G.I., *Environmental Law and Enforcement,* Lewis Publishers, Boca Raton, FL, 1994.

McIntosh, A., Trace metals in freshwater sediments: a review of the literature and an assessment of research needs, in *Metal Ecotoxicology: Concepts and Applications,* Newman, M.C. and McIntosh, A.W., Eds., Lewis Publishers, Chelsea, MI, 1991.

McKim, J.M., Early life stage tests, in *Fundamentals of Aquatic Toxicology,* Rand, G.M. and Petrocelli, S.R., Eds., Hemisphere Publishing Corp., Washington, D.C., 1985.

McLachlan, J.A., Functional toxicology: a new approach to detect biologically active xenobiotics, *Environ. Health Perspect.,* 101, 386–387, 1993.

McLaughlin, J.F. and W.G. Landis, Effects of environmental contaminants in spatially structured environments, in *Environmental Contaminants in Terrestrial Vertebrates: Effects on Populations, Communities, and Ecosystems,* Albers, P.H. et al., Eds., Society of Environmental Toxicology and Chemistry, Pensacola, FL, 2000.

McMurry, S.T., Development of an In Situ Mammalian Biomonitor to Assess the Effect of Environmental Contaminants on Population and Community Health, Ph.D. dissertation, Oklahoma State University, Stillwater, 1993.

McNeill, K.G. and G.A.D. Trojan, The cesium-potassium discrimination ratio, *Health Phys.,* 4, 109–112, 1960.

Meador J.P., J.E. Stein, W.L. Reichert, and U. Varanasi, A review of bioaccumulation of polycyclic aromatic hydrocarbons by marine organisms, *Rev. Environ. Contam. Toxicol.,* 143, 79–165, 1995a.

Meador J.P., E. Casillas, C.A. Sloan, and U. Varanasi, Comparative bioaccumulation of polycyclic aromatic hydrocarbons from sediment by two infaunal invertebrates, *Mar. Ecol. Prog. Ser.*, 123, 107–124, 1995b.

Medawar, P.B., *The Art of the Soluble*, Methuen Co., London, 1967.

Medawar, P.B., *Pluto's Republic*, Oxford University Press, Oxford, U.K., 1982.

Medvedev, Z.A., The Ural and Chernobyl Nuclear Accidents, seminar presented at University of Georgia's Savannah River Ecology Laboratory, Sep. 5, 1995.

Mehrle, P.M. and F.L. Mayer, Biochemistry/physiology, in *Fundamentals of Aquatic Toxicology: Methods and Applications*, Rand, G.M. and Petrocelli, S.R., Eds., Hemisphere Publishing Corp., Washington, D.C., 1985.

Melancon, M.J., Bioindicators used in aquatic and terrestrial monitoring, in *Handbook of Ecotoxicology*, Hoffman, D.J., Rattner, B.A., Burton, G.A., Jr., and Cairns, J., Jr., Eds., CRC Press, Boca Raton, FL, 1995.

Mellinger, P.J. and V. Schultz, Ionizing radiation and wild birds: a review, *Crit. Rev. Environ. Control*, 5, 397, 1975.

Mertz, W., The essential trace elements, *Science* (Washington, D.C.), 213, 1332–1338, 1981.

Metalli, P. and E. Ballardin, Radiobiology of *Artemia*: radiation effects and ploidy, *Curr. Top. Radiat. Res. Q.*, 7, 181, 1972.

Metcalfe, C.D., V.W. Cairns, and J.D. Fitzsimons, Experimental induction of liver tumors in rainbow trout (*Salmo gairdneri*) by contaminated sediment from Hamilton Harbor, Ontario. *Can. J. Fish. Aquatic Sci.* 45, pp. 2161–2167, 1988.

Meyers, T.R. and J.D. Hendricks, Histopathology, in *Fundamentals of Aquatic Toxicology: Methods and Applications*, Rand, G.M. and Petrocelli, S.R., Eds., Hemisphere Publishing Corp., Washington, D.C., 1985.

Meyers-Schöne, L.R. Shugart, J.J. Beauchamp, and B.T. Walton, Comparison of two freshwater turtle species as monitors of radionuclide and chemical contamination: DNA damage and residue analysis. *Environ. Toxicol. Chem.* 12, pp. 1487–1496, 1993.

Michalska, A.E. and K.H.A. Choo, Targeting and germ-line transmission of a null mutation at the metallothionein I and II loci in mice, *Proc. Natl. Acad. Sci. U.S.A.*, 90, 8088–8092, 1993.

Michener, W.K., J.W. Brunt, and S.G. Stafford, *Environmental Information Management and Analysis: Ecosystem to Global Scales*, Taylor & Francis, London, 1994.

Millar, I.B. and P.A. Cooney, Urban lead: a study of environmental lead and its significance to school children in the vicinity of a major trunk road, *Atmos. Environ.*, 16, 615–620, 1982.

Miller, R.G., Jr., *Beyond ANOVA, Basics of Applied Statistics*, John Wiley & Sons, New York, 1986.

Milodowski, A.E., J.M. West, J.M. Pearce, E.K. Hyslop, I.R. Basham, and P.J. Hooker, Uranium-mineralized microorganisms associated with uraniferous hydrocarbons in southeast Scotland, *Nature* (London), 347, 465–467, 1990.

Minagawa, M. and E. Wada, Stepwise enrichment of ^{15}N along food chains: further evidence and the relation between δ^{15}N and animal age, *Geochim. Cosmochim. Acta*, 48, 1135–1140, 1984.

Mineau, P., A. Baril, B.T. Collins, J. Duffe, G. Hoermann, and R. Luttik, Reference values for comparing the acute toxicity of pesticides to birds, *Rev. Environ. Contam. Toxicol.*, 24, 24, 2001.

Miranda, C.D. and R. Zemelman, Antibiotic resistant bacteria in fish from the Concepcion Bay, Chile, *Mar. Pollut. Bull.*, 42, 1096–1102, 2001.

Mishima, J. and E.P. Odum, Excretion rate of Zn^{65} by *Littorina irrorata* in relation to temperature and body size, *Limnol. Oceanogr.*, 8, 39–44, 1963.

Molander, S., H. Blanck, and M. Söderström, Toxicity assessment by pollution-induced community tolerance (PICT), and identification of metabolites in periphyton communities after exposure to 4,5,6-trichloroguaiacol, *Aquatic Toxicol.* (Amsterdam), 18, 115–136, 1990.

Møller, V., V.E. Forbes, and M.H. DePledge, Influence of acclimation and exposure temperature on the acute toxicity of cadmium to the freshwater snail *Potamophygus antipodarum* (Hydrobiidae), *Environ. Toxicol. Chem.*, 13, 1519–1524, 1994.

Monk, C.D., Effects of short-term gamma irradiation on an old field, *Radiat. Bot.*, 6, 329–335, 1966.

Monkhouse, F.J., *A Dictionary of Geography*, Aldine Publishing Co., Chicago, 1965, p. 344.

Monteiro, L.R. and R.W. Furness, Seabirds as monitors of mercury in the marine environment, *Water Air Soil Pollut.*, 80, 851–870, 1995.

Monteiro, L.R., J.P. Granadeiro, and R.W. Furness, Relationship between mercury levels and diet in Azores seabirds, *Mar. Ecol. Prog. Ser.*, 166, 259–265, 1998.

Moolgavkar, S.H., Carcinogenesis modeling: from molecular biology to epidemiology, *Ann. Rev. Public Health,* 7, 151–169, 1986.

Moore, M.J. and M.S. Myers, Pathobiology of chemical-associated neoplasia in fish, in *Aquatic Toxicology: Molecular, Biochemical and Cellular Perspectives,* Malins, D.C. and Ostrander, G.K., Eds., CRC Press, Boca Raton, FL, 1994.

Moore, M.J. and J.J. Stegeman, Hepatic neoplasms in winter flounder, *Pleuronectes americanus,* from Boston Harbor, Massachusetts, USA, *Dis. Aquatic Org.,* 20, 33–48, 1994.

Moore, M.J., D. Shea, R.E. Hillman, and J.J. Stegeman, Trends in hepatic tumors and hydropic vacuolation, fin erosion, organic chemical and stable isotope ratios in winter flounder, from Massachusetts, USA. *Mar. Poll. Bull.* 32, pp. 458–470, 1996.

Moriarty, F., *Ecotoxicology: The Study of Pollutants in Ecosystems,* Academic Press, London, 1983.

Morton, B., The tidal rhythm and rhythm of feeding and digestion in *Cardium edule, J. Mar. Biol. Assoc. U.K.,* 50, 499–512, 1970.

Mossman, K.L., M. Goldman, F. Masse, W. Mills, K. Schiager, and R. Vetter, Radiation risk in perspective, *Health Physics Newsletter,* Health Physics Society, McLean, VA, Mar. 1996.

Mouneyrac, C., C. Amiard-Triquet, J.C. Amaird, and P.S. Rainbow, Comparison of metallothionein concentrations and tissue distribution of trace metals in crabs (*Pachygrapsus marmoratus*) from a metal-rich estuary, in and out of the reproductive season, *Comp. Biochem. Physiol. C-Toxicol. Pharmacol.,* 129, 193–209, 2001.

Mount, D.I. and C.E. Stephan, A method for establishing acceptable toxicant limits for fish: malathion and 2,4-D, *Trans. Am. Fish. Soc.,* 96, 185–193, 1967.

Muir, D.C.G., R.J. Norstrom, and M. Simon, Organochlorine contaminants in Arctic marine food chains: accumulation of specific polychlorinated biphenyls and chlordane-related compounds, *Environ. Sci. Technol.,* 22, 1071–1079, 1988.

Muir, D.C.G., N.P. Grift, W.L. Lockhart, P. Wilkinson, B.N. Billeck, and G.J. Brunskill, Spatial trends and historical profiles of organochlorine pesticides in Arctic lake sediments, *Sci. Total Environ.,* 160/161, 447–457, 1995.

Muirhead, S.J. and R.W. Furness, Heavy metal concentrations in the tissues of seabirds from Gough Island, south Atlantic Ocean, *Mar. Pollut. Bull.,* 19, 278–283, 1988.

Mulholland, P.J., A.V. Palumbo, J.W. Elwood, and A.D. Rosemond, *J. North Am. Benthol. Soc.,* 6, 147–158, 1987.

Müller, H. and G. Pröhl, ECOSYS-87: a dynamic model for the assessment of the radiological consequences of nuclear accidents, *Health Phys.,* 64, 232–252, 1993.

Mulvey, M. and S.A. Diamond, Genetic factors and tolerance acquisition in populations exposed to metals and metalloids, in *Metal Ecotoxicology: Concepts and Applications,* Newman, M.C. and McIntosh, A.W., Eds., Lewis Publishers, Chelsea, MI, 1991.

Mulvey, M., M.C. Newman, A. Chazal, M.M. Keklak, M.G. Heagler, and L.S. Hales, Jr., Genetic and demographic responses of mosquitofish (*Gambusia holbrooki* Girard 1859) populations stressed by mercury, *Environ. Toxicol. Chem.,* 14, 1411–1418, 1995.

Mumtaz, M.M., C.T. De Rosa, J. Groten, V.J. Feron, H. Hansen, and P.R. Durkin, Estimation of toxicity of chemical mixtures through modeling of chemical interactions, *Environ. Health Perspect.,* 106, 1353–1360, 1998.

Munkittrick, K.R. and D.G. Dixon, Growth, fecundity, and energy stores of white suckers (*Catostomus commersoni*) from lakes containing elevated levels of copper and zinc, *Can. J. Fish. Aquatic Sci.,* 45, 1355–1365, 1988.

Murdoch, M.H. and P.D.N. Hebert, Mitochondrial DNA diversity of brown bullhead from contaminated and relatively pristine sites in the Great Lakes, *Environ. Toxicol. Chem.,* 8, 1281–1289, 1994.

Murphy, D.L. and J.W. Gooch, Accumulation of *cis* and *trans* chlordane by channel catfish during dietary exposure, *Arch. Environ. Contam. Toxicol.,* 29, 297–301, 1995.

Myers, M.S., L.D. Rhodes, and B.B. McCain, Pathologic anatomy and patterns of occurrence of hepatic neoplasms, putative preneoplastic lesions and other idiopathic hepatic conditions in English sole (*Parophrys vetulus*) from Puget Sound, Washington, *J. Natl. Cancer Inst.,* 78, 333–363, 1987.

Myers, M.S., C.M. Stehr, O.P. Olson, L.L. Johnson, B.B. McCain, S.L. Chan, and U. Varanasi, National Benthic Surveillance Project: Pacific Coast, Fish Histopathology and Relationships between Toxicopathic Lesions and Exposure to Chemical Contaminants for Cycles I to V (1984–88), NOAA Tech. Memo. NMFS/NWFSC-6, U.S. Dept. Commerce, Washington, D.C., 1993, 160 pp.

Myers, M.S., L.L. Johnson, O.P. Olson, C.M. Stehr, B.H. Horness, T.K. Collier, and B.B. McCain, Toxicopathic hepatic lesions as biomarkers of chemical contaminant exposure and effects in marine bottomfish species from the Northeast and Pacific coasts, USA, *Mar. Pollut. Bull.*, 37, 92–113, 1998.

Nagel, E., *The Structure of Science: Problems in the Logic of Scientific Explanation*, Harcourt, Brace and World, New York, 1961.

NAS/NRC (National Academy of Sciences/National Research Council, Committee on the Biological Effects of Ionizing Radiations), *Health Risks of Exposure to Low Levels of Ionizing Radiation*, BEIR V, National Academy Press, Washington, D.C., 1990.

National Research Council, *Risk Assessment in the Federal Government: Managing the Process*, National Academy Press, Washington, D.C., 1983.

National Toxicology Program, The Ninth Report on Carcinogens, U.S. Department of Health and Human Services, Public Health Service National Toxicology Program. Washington, D.C., 1999.

Natural Resource Damage Assessments, Final rule, *Federal Register*, 51, 27674–27753, 1986.

Nayak, B.N. and M.L. Petras, Environmental monitoring for genotoxicity: in vivo sister chromatid exchange in house mouse (*Mus musculus*), *Can. J. Genet. Cytol.*, 27, 351–356, 1985.

NCRP (National Council on Radiation Protection and Measurements), Radiological Assessment: Predicting the Transport, Bioaccumulation, and Uptake by Man of Radionuclides Released to the Environment, NCRP Report No. 76, Bethesda, MD, 1984, p. 304.

NCRP (National Council on Radiation Protection and Measurements), Exposure of the Population in the United States and Canada from Natural Background Radiation, NCRP Report No. 94, Bethesda, MD, 1987, p. 212.

NCRP (National Council on Radiation Protection and Measurements), Screening Techniques for Determining Compliance with Environmental Standards, NCRP Commentary No. 3, Bethesda, MD, 1989, p. 134.

NCRP (National Council on Radiation Protection and Measurements), Effects of Ionizing Radiation on Aquatic Organisms, NCRP Report No. 109, Bethesda, MD, 1991, p. 115.

NCRP (National Council on Radiation Protection and Measurements), Risk Estimates for Radiation Protection, NCRP Report No. 115, Bethesda, MD, 1993, p. 148.

NCRP (National Council on Radiation Protection and Measurements), A Guide for Uncertainty Analysis in Dose and Risk Assessments Related to Environmental Contamination, NCRP Commentary No. 14, Bethesda, MD, 1996, p. 54.

Neathery, M.W. and W.J. Miller, Metabolism and toxicity of cadmium, mercury, and lead in animals: a review, *J. Dairy Sci.*, 58, 1767–1781, 1975.

Neely, W.B., D.R. Bronson, and G.E. Blau, Partition coefficient to measure bioconcentration potential of organic chemicals in fish, *Environ. Sci. Technol.*, 8, 1113–1115, 1974.

Nestler, F., The Characterization of Wood-Preserving Creosote by Physical and Chemical Methods of Analysis, USDA Forest Service Research Paper, FPL 195, U.S. Department of Agriculture Forest Service, Forests Products Laboratory, Madison, WI, 1974.

Neter, J., W. Waserman, and M.H. Kutner, *Applied Linear Statistical Models: Regression, Analysis of Variance, and Experimental Design*, 3rd ed., Richard D. Irwin, Inc., Homewood, IL, 1990.

Neuhold, J.M., The relationship of life history attributes to toxicant tolerance in fishes, *Environ. Toxicol. Chem.*, 6, 709–716, 1987.

Neville, C.M. and P.G.C. Campbell, Possible mechanisms of aluminum toxicity in dilute, acidic environment to fingerlings and older life stages of salmonids, *Water Air Soil Pollut.*, 42, 311–327, 1988.

Nevo, E. et al., Mercury selection of allozyme genotypes in shrimps, *Experientia*, 37, 1152–1154, 1981.

Nevo, E., T. Shimony, and M. Libni, Thermal selection of allozyme polymorphisms in barnacles, *Nature*, 267, 699–701, 1977.

Nevo, E., T. Shimony, and M. Libni, Pollution selection of allozyme polymorphisms in barnacles, *Experientia* 34, 1562–1564, 1978.

Newman, M.C., A statistical bias in the derivation of hardness-dependent metals criteria, *Environ. Toxicol. Chem.*, 10, 1295–1297, 1991.

Newman, M.C., Regression analysis of log-transformed data: statistical bias and its correction, *Environ. Toxicol. Chem.*, 12, 1129–1133, 1993.

Newman, M.C., *Quantitative Methods in Aquatic Ecotoxicology*, Lewis Publishers, Boca Raton, FL, 1995.

Newman, M.C., Ecotoxicology as a science, in *Ecotoxicology: A Hierarchical Treatment*, Newman, M.C. and Jagoe, C.H., Eds., Lewis Publishers, Boca Raton, FL, 1996.

Newman, M.C., *Population Ecotoxicology*, John Wiley & Sons, Chichester, U.K., 2001.

Newman, M.C., J.J. Alberts, and V.A. Greenhut, Geochemical factors complicating the use of *aufwuchs* to monitor bioaccumulation of arsenic, cadmium, chromium, copper and zinc, *Water. Res.,* 19, 111–128, 1985.

Newman, M.C. and M.S. Aplin, Enhancing toxicity data interpretation and prediction of ecological risk with survival time modeling: an illustration using sodium chloride toxicity to mosquitofish (*Gambusia holbrooki*), *Aquatic Toxicol.* (Amsterdam), 23, 85–96, 1992.

Newman, M.C., S.A. Diamond, M. Mulvey, and P. Dixon, Allozyme genotype and time to death of mosquitofish, *Gambusia affinis* (Baird and Girard), during acute toxicant exposure: a comparison of arsenate and inorganic mercury, *Aquatic Toxicol.* (Amsterdam), 15, 141–156, 1989.

Newman, M.C. and P.M. Dixon, Ecologically meaningful estimates of lethal effect in individuals, in *Ecotoxicology: A Hierarchical Treatment,* Newman, M.C. and Jagoe, C.H., Eds., Lewis Publishers, Boca Raton, FL, 1996.

Newman, M.C. and M.G. Heagler, Allometry of metal bioaccumulation and toxicity, in *Metal Ecotoxicology, Concepts and Applications,* Newman, M.C. and McIntosh, A.W., Eds., Lewis Publishers, Chelsea, MI, 1991.

Newman, M.C. and C.H. Jagoe, Ligands and the bioavailability of metals in aquatic environments, in *Bioavailability: Physical, Chemical and Biological Interactions,* Hamelink, J.L. et al., Eds., CRC Press, Boca Raton, FL, 1994.

Newman, M.C. and R.H. Jagoe, Bioaccumulation models with time lags: dynamics and stability criteria, *Ecol. Model,* 1424, 281–286, 1996.

Newman, M.C., M.M. Keklak, and M.S. Doggett, Quantifying animal size effects on toxicity: a general approach, *Aquatic Toxicol.* (Amsterdam), 28, 1–12, 1994.

Newman, M.C. and J.T. McCloskey, Predicting relative toxicity and interactions of divalent metal ions: Microtox[7] bioluminescence assay, *Environ. Toxicol. Chem.,* 15, 275–281, 1966a.

Newman, M.C. and J.T. McCloskey, Time-to-event analyses of ecotoxicology data, *Ecotoxicology,* 5, 187–196, 1996b.

Newman, M.C. and J.T. McCloskey, The individual tolerance concept is not the sole explanation for the probit dose-effect model, *Environ. Toxicol. Chem.,* 19, 520–526, 2000.

Newman, M.C. and A.W. McIntosh, Slow accumulation of lead from contaminated food sources by the freshwater gastropods, *Physa integra* and *Campeloma decisum, Arch. Environ. Contam. Toxicol.,* 12, 685–692, 1983.

Newman, M.C. and A.W. McIntosh, Appropriateness of *aufwuchs* as a monitor of bioaccumulation, *Environ. Pollut.,* 60, 83–100, 1989.

Newman, M.C., A.W. McIntosh, and V.A. Greenhut, Geochemical factors complicating the use of *aufwuchs* as a biomonitor for lead levels in two New Jersey reservoirs, *Water Res.,* 17, 625–630, 1983.

Newman, M.C. and S.V. Mitz, Size dependence of zinc elimination and uptake from water by mosquitofish *Gambusia affinis* (Baird and Girard), *Aquatic Toxicol.* (Amsterdam), 12, 17–32, 1988.

Newman, M.C., M. Mulvey, A. Beeby, R.W. Hurst, and L. Richmond, Snail (*Helix aspersa*) exposure history and possible adaptation to lead as reflected in shell composition, *Arch. Environ. Contam. Toxicol.,* 27, 346–351, 1994.

Newman, M.C., D.R. Ownby, L.C.A. Mézin, D.C. Powell, T.R.L. Christensen, S.B. Lerberg, B.-A. Anderson, and T.V. Padma, Species sensitivity distributions in ecological risk assessment: distributional assumptions, alternate bootstrap techniques, and estimation of adequate number of species, in *Species Sensitivity Distributions in Ecotoxicology,* Posthuma, L., Suter, G.W., II, and Traas, T.P., Eds., Lewis Publishers, Boca Raton, FL, 2002.

Nichols, J.W., J.M. McKim, M.E. Andersen, M.L. Gargas, H.J. Clewell III, and R.J. Erickson, A physiologically based toxicokinetic model for the uptake and disposition of waterborne organic chemicals in fish, *Toxicol. Appl. Pharmacol.,* 106, 433–447, 1990.

Nieboer, E. and D.H.S. Richardson, The replacement of the nondescript term "heavy metals" by a biologically and chemically significant classification of metal ions, *Environ. Pollut.,* 1B, 3–26, 1980.

Niederlehner, B.R., J.R. Pratt, A.L. Buikema, Jr., and J. Cairns, Jr., Laboratory tests evaluating the effects of cadmium on freshwater protozoan communities, *Environ. Toxicol. Chem.,* 4, 155–165, 1985.

Nihlgård, B., The ammonium hypothesis: an additional explanation to the forest dieback in Europe, *Ambio,* 14, 2–8, 1985.

Nikinmaa, K., How does environmental pollution affect red cell function in fish? *Aquatic Toxicol.* (Amsterdam), 22, 227–238, 1992.

Nissen, P. and A.A. Benson, Arsenic metabolism in freshwater and terrestrial plants, *Physiol. Plant.,* 54, 446–450, 1982.

Nixon, S., Coastal marine eutrophication: a definition, social causes, and future concerns, *Ophelia,* 41, 199–220, 1995.

Norheim, G., Levels and interactions of heavy-metals in seabirds from Svalbard and the Antarctic, *Environ. Pollut.,* 47, 83–94, 1987.

Norton, S.B., D.J. Rodier, J.H. Gentile, W.H. Van Der Schalie, W.P. Wood, and M.W. Slimak, A framework for ecological risk assessment at the EPA, *Environ. Toxicol. Chem.,* 11, 1663–1672, 1992.

Nott, J.A. and A. Nicolaidou, Bioreduction of zinc and manganese along a molluscan food chain, *Comp. Biochem. Physiol.,* 104A, 235–238, 1993.

O'Brien, R.D. and L.S. Wolfe, Nongenetic effects of radiation, in *Radiation, Radioactivity, and Insects,* Academic Press, New York, 1964.

O'Connor, R.J., Toward the incorporation of spatiotemporal dynamics into ecotoxicology, in *Population Dynamics in Ecological Space and Time,* Rhodes, O.E., Jr., Chesser, R.K., and Smith, M.H., Eds., University of Chicago Press, Chicago, 1996.

Odin, M., A. Feurtet-Mazel, F. Ribeyre, and A. Boudou, Actions and interactions of temperature, pH and photoperiod on mercury bioaccumulation by nymphs of the burrowing mayfly, *Hexagenia rigida,* from the sediment contamination source, *Environ. Toxicol. Chem.,* 13, 1291–1302, 1994.

Odum, E.P., *Fundamentals of Ecology,* W.B. Saunders Co., Philadelphia, 1971.

Odum, E.P., Trends expected in stressed ecosystems, *Bioscience,* 35, 419–422, 1985.

Odum, E.P., Preface, in *Ecotoxicology: A Hierarchical Treatment,* Newman, M.C. and Jagoe, C.H., Eds., Lewis Publishers, Boca Raton, FL, 1996.

Officer, C.B., R.B. Riggs, J.L. Taft, L.E. Cronin, M.A. Tyler, and W.R. Boynton, Chesapeake Bay anoxia: origin, development and significance, *Science,* 223, 22–27, 1984.

Ogilvie, D.M. and D.L. Miller, Duration of a DDT-induced shift in the selected temperature of Atlantic salmon (*Salmo salar*), *Bull. Environ. Contam. Toxicol.,* 16, 86–89, 1976.

Ohlendorf, H.M., D.J. Hoffman, M.K. Saiki, and T.A. Aldrich, Embryonic mortality and abnormalities of aquatic birds: apparent impacts of selenium from irrigation drainwater, *Sci. Total Environ.,* 52, 49–63, 1986.

Öhman, L.-O. and S. Sjöberg, Thermodynamic calculations with special reference to the aqueous aluminum system, in *Metal Speciation: Theory, Analysis and Application,* Kramer, J.R. and Allen, H.E., Eds., Lewis Publishers, Chelsea, MI, 1988.

Olivieri, G., J. Bodycote, and S. Wolff, Adaptive response of human lymphocytes to low concentrations of radioactive thymidine, *Science* (Washington, D.C.), 223, 594–597, 1984.

Omernik, J.M., Ecoregions of the conterminous United States, *Ann. Assoc. Amer. Geographers,* 77, 118–125, 1987.

Organization for Economic Cooperation and Development, Acute Oral Toxicity: Up-and-Down Procedure, OECD Guideline for Testing of Chemicals, Guideline no. 425, OECD, Paris, 2001.

Osano, O., W. Admiraal, and D. Otieno, Developmental disorders in embryos of the frog, *Xenopus laevis,* induced by chloroacetanilide herbicides and their degradation products, *Environ. Toxicol. Chem.,* 21, 375–379, 2002.

Otto, E., J.E. Young, and G. Maroni, Structure and expression of a tandem duplication of the *Drosophila* metallothionein gene, *Proc. Natl. Acad. Sci. U.S.A.,* 83, 6025–6029, 1986.

Otvos, J.D., R.W. Olafson, and I.M. Armitage, Structure of an invertebrate metallothionein from *Scylla serrata,* *J. Biol. Chem.,* 257, 2427–2431, 1982.

Owen, D.F., Industrial melanism in North American moths, *Am. Nat.,* 95, 227–233, 1961.

Pacyna, J.M., Atmospheric trace elements from natural and anthropogenic sources, in *Toxic Metals in the Atmosphere,* Nriagu, J.O. and Davidson, C.I., Eds., John Wiley & Sons, New York, 1986.

Pacyna, J.M., The origin of Arctic air pollutants: lessons learned and future research, *Sci. Total Environ.,* 160/161, 39–53, 1995.

Paerl, H.W., Nuisance phytoplankton blooms in coastal, estuarine, and inland waters, *Limnol. Oceanogr.,* 33, 823–847, 1988.

Paerl, H.W., Coastal eutrophication and harmful algal blooms: importance of atmospheric deposition and groundwater as "new" nitrogen and other nutrient sources, *Limnol. Oceanogr.,* 42, 1154–1165, 1997.

Paerl, H.W. and M.L. Fogel, Isotopic characterization of atmospheric nitrogen inputs as sources of enhanced primary production in coastal Atlantic Ocean waters, *Mar. Biol.,* 119, 635–645, 1994.

Paerl, H.W., J.L. Pinckney, J.M. Fear, and B.L. Peierls, Ecosystem response to internal and watershed organic matter loading: consequences for hypoxia in the eutrophying Neuse River estuary, NC, USA, *Mar. Ecol. Ser.,* 166, 17–25, 1998.

Paerl, H.W., R.L. Dennis, and D.R. Whitall, Atmospheric deposition of nitrogen: implications for nutrient over-enrichment of coastal waters, *Estuaries,* 25(4B), 677–693, 2002.

Pagenkopf, G.K., Gill surface interaction model for trace-metal toxicity to fishes: role of complexation, pH, and water hardness, *Environ. Sci. Technol.,* 17, 342–347, 1983.

Pain, D.J., Lead in the environment, in *Handbook of Ecotoxicology,* Hoffman, D.J., B.A. Rattner, G.A. Burton, Jr., and J. Cairns, Jr., Eds., CRC Press, Boca Raton, FL, 1995.

Palmer, A.R., Waltzing with asymmetry, *Bioscience,* 46, 518–532, 1996.

Palmiter, R.D., Regulation of metallothionein genes by heavy metals appears to be mediated by a zinc-sensitive inhibitor that interacts with a constitutively active transcription factor, MTF-1, *Proc. Natl. Acad. Sci. U.S.A.,* 91, 1219–1223, 1994.

Pao, E.M., K.H. Fleming, P.M. Gueuther, and S.J. Mickle, Food Commonly Eaten by Individuals: Amount Per Day and Per Eating Occasion, U.S. Department of Agriculture, Springville, VA, 1982.

Papoulias, D.M., D.B. Noltie, and D.E. Tillitt, An in vivo model fish system to test chemical effects on sexual differentiation and development: exposure to ethinyl estradiol, *Aquatic Toxicol.,* 48, 37–50, 2000.

Parke, D., Cytochrome P-450 and the detoxication of environmental chemicals, *Aquatic Toxicol.* (Amsterdam), 1, 367–376, 1981.

Parkinson, A., Biotransformations of xenobiotics, in *Casarett and Doull's Toxicology: The Basic Science of Poisons,* 5th ed., Klaassen, C.D., Ed., McGraw-Hill, New York, 1996.

Parsons, P.A., Inherited stress resistance and longevity: a stress theory of ageing, *Heredity,* 75, 216–221, 1995.

Partridge, G.G., Relative fitness of genotypes in a population of *Rattus norvegicus* polymorphic for warfarin resistance, *Heredity,* 43, 239–246, 1979.

Paterson, S. and D. Mackay, A steady-state fugacity-based pharmacokinetic model with simultaneous multiple exposure routes, *Environ. Toxicol. Chem.,* 6, 395–408, 1987.

Patrick, R., *Use of Algae, Especially Diatoms, in the Assessment of Water Quality,* ASTM Special Technical Publication 528, American Society for Testing and Materials, Philadelphia, 1973, pp. 76–95.

Peakall, D., *Animal Biomarkers as Pollution Indicators,* Chaman & Hall, London, 1992, p. 291.

Peakall, D.B. and K. McBee, Biomarkers for contaminant exposure and effects in mammals, in *Ecotoxicology of Wild Mammals,* Shore, R.F. and Rattner, B.A., Eds., John Wiley & Sons, Chichester, U.K., 2001.

Peltier, W.H. and C.I. Weber, Methods for Measuring the Acute Toxicity of Effluents to Freshwater and Marine Organisms, EPA/600/4-85/013, EPA Environmental Monitoring and Support Laboratory, Cincinnati, OH, 1985.

Pendleton, R.C., R.D. Lloyd, C.W. Mays, and B.W. Church, Trophic level effect on the accumulation of caesium-137 in cougars feeding on mule deer, *Nature* (London), 204, 708–709, 1964.

Peoples, S.A., The metabolism of arsenic in man and animals, in *Arsenic: Industrial, Biomedical, Environmental Perspectives,* Lederer, W.H. and Fensterheim, R.J., Eds., Van Nostrand Reinhold Co., New York, 1983.

Peters, T., The effect of x-rays on fertilized egg cells of *Triton clopostris* with special emphasis on the effects of small doses and less damage, *Strahlentherapie,* 112, 525–542, 1960.

Phillips, D.J.H., The use of biological indicator organisms to monitor trace metal pollution in marine and estuarine environments: a review, *Environ. Pollut.,* 13, 281–317, 1977.

Piatt, J.F., C.J. Lensink, W. Butler, M. Kendziorek, and D.R. Nysewander, Immediate impact of the *Exxon Valdez* oil spill on marine birds, *Auk,* 10, 387–397, 1990.

Pielou, E.C., *Population and Community Ecology: Principles and Methods,* Gordon and Breach Science Publishers, New York, 1974.

Piferrer, F., Endocrine sex control strategies for the feminization of teleost fish, *Aquaculture,* 197, 229–281, 2001.

Pihl, L., S.P. Baden, and R.J. Diaz, Effects of periodic hypoxia on distributions of demersal fish and crustaceans, *Mar. Biol.,* 108, 349–360, 1991.

Pimentel, D. and C.A. Edwards, Pesticides and ecosystems, *Bioscience,* 32, 595–600, 1982.

Pimentel, D., L. Westra, and R.F. Noss, Eds., *Ecological Integrity: Integrating Environment, Conservation, and Health,* Island Press, Washington, D.C., 2000.

Piszkiewicz, D., *Kinetics of Chemical and Enzyme-Catalyzed Reactions,* Oxford University Press, New York, 1977.

Plackett, R.L. and P.S. Hewlett, Quantal responses to mixtures of poisons, *J. R. Stat. Soc.,* 14, 141–163, 1952.

Platt, J.R., Strong inference, *Science* (Washington, D.C.), 146, 347–353, 1964.

Playle, R.C. and C.M. Wood, Water chemistry changes in the gill microenvironment of rainbow trout: experimental observations and theory, *J. Comp. Physiol.,* 159B, 527–537, 1989.

Playle, R.C. and C.M. Wood, Mechanisms of aluminum extraction and accumulation at the gills of rainbow trout, *Oncorhynchus mykiss* (Walbaum), fingerlings, *Aquatic Toxicol.* (Amsterdam), 21, 267–278, 1991.

Polikarpov, G.G., *Radioecology of Aquatic Organisms,* North-Holland Publishing Co., Amsterdam, 1966.

Pontasch, K.W., B.R. Niederlehner, and J. Cairns Jr., Comparisons of single-species, microcosm and field responses to a complex effluent, *Environ. Toxicol. Chem.,* 8, 521–532, 1989.

Porvari, P. and M. Verta, Methylmercury production in flooded soils: a laboratory study, *Water Air Soil Pollut.,* 80, 765–773, 1995.

Posthuma, L., R.F. Hogervorst, E.N.G. Joose, and N.M. Van Straalen, Genetic variation and covariation for characteristics associated with cadmium tolerance in natural populations of the springtail *Orchesella cincta* (L.), *Evolution,* 47, 619–631, 1993.

Posthuma, L., G.W. Suter II, and T.P. Traas, Eds., *Species Sensitivity Distributions in Ecotoxicology,* Lewis Publishers, Boca Raton, FL, 2002.

Postma, J.F., A. van Kleunen, and W. Admiraal, Alterations in life-history traits of *Chironomus riparius* (Diptera) obtained from metal contaminated rivers, *Arch. Environ. Contam. Toxicol.,* 29, 469–475, 1995a.

Postma, J.F., S. Mol, H. Larsen, and W. Admiraal, Life-cycle changes and zinc shortage in cadmium-tolerant midges, *Chironomus riparius* (Diptera), reared in the absence of cadmium, *Environ. Toxicol. Chem.,* 14, 117–122, 1995b.

Pratt, J.R. and J. Cairns, Ecotoxicology and the redundancy problem: understanding effects on community structure and function, in *Ecotoxicology: A Hierarchical Treatment,* Newman, M.C. and Jagoe, C.H., Eds., Lewis Publishers, Boca Raton, FL, 1996.

Preston, F.W., The commonness, and rarity, of species, *Ecology,* 29, 254–283, 1948.

Price, W.J., *Analytical Atomic Absorption Spectrometry,* Heyden & Son, London, 1972.

Prugh, T., *Natural Capital and Human Economic Survival,* 2nd ed., Lewis Publishers, Boca Raton, FL, 1999.

Pulliam, H.R., G.W. Barrett, and E.P. Odum, Bioelimination of tracer ^{65}Zn in relation to metabolic rates in mice, in *Symposium on Radioecology,* Nelson, D.J. and Evans, F.C., Eds., University of Michigan, Ann Arbor, 1967.

Pulliam, H.R. and B.J. Danielson, Source, sinks, and habitat selection: a landscape perspective on population dynamics, *Am. Nat.,* 137 (Suppl.), S50–S66, 1991.

Putka, G., Research on Lead Poisoning Is Questioned, *Wall Street Journal,* Mar. 6, 1992, p. B1.

Pynnönen, K., D.A. Holwerda, and D.I. Zandee, Occurrence of calcium concretions in various tissues of freshwater mussels, and their capacity for cadmium sequestration, *Aquatic Toxicol.* (Amsterdam), 10, 101–114, 1987.

Quine, W.V., *Methods of Logic,* Harvard University Press, Cambridge, MA, 1982.

Rabalais, N.N., W.J. Wiseman Jr., and R.E. Turner, Comparison of continuous records of near-bottom dissolved oxygen from hypoxia zone of Louisiana, *Estuaries,* 17, 850–861, 1994.

Rabalais, N.N., R.E. Turner, and W.J. Wiseman Jr., Hypoxia in the Gulf of Mexico, *J. Environ. Qual.,* 30, 320–329, 2001.

Rago, P.J. and R.M. Dorazio, Statistical inference in life-table experiments: the finite rate of increase, *Can. J. Fish. Aquatic Sci.,* 41, 1361–1374, 1984.

Ramade, F., *Ecotoxicology,* John Wiley & Sons, New York, 1987.

Rand, G.M., Behavior, in *Fundamentals of Aquatic Toxicology,* Rand, G.M. and Petrocelli, S.R., Eds., Hemisphere Publishing Corp., Washington, D.C., 1985.

Rand, G.M. and S.R. Petrocelli, Eds., *Fundamentals of Aquatic Toxicology,* Hemisphere Publishing Corp., Washington, D.C., 1985.

Ranney, R.E., Comparative metabolism of 17-alpha-ethynylsteroids used in oral contraceptives, *J. Toxicol. Environ. Health,* 3, 139–166, 1977.

Rasmussen, J.B., D.J. Rowan, D.R.S. Lean, and J.H. Carey, Food chain structure in Ontario lakes determines PCB levels in lake trout (*Salvelinus namaycush*) and other pelagic fish, *Can. J. Fish. Aquatic Sci.,* 47, 2030–2038, 1990.

Ratcliffe, D.A., Decrease in eggshell weight in certain birds of prey, *Nature,* 215, 208–210, 1967.

Ratcliffe, D.A., Changes attributable to pesticides in egg breakage frequency and eggshell thickness in some British birds, *J. Appl. Ecol.,* 7, 67–107, 1970.

Rau, G.H., Low $^{15}N/^{14}N$ in hydrothermal vent animals: ecological implications, *Nature* (London), 289, 484–485, 1981.

Rauser, W.E. and E.B. Dumbroff, Effects of excess cobalt, nickel and zinc on the water relations of *Phaseolus vulgaris, Environ. Exp. Bot.,* 21, 249–255, 1981.

Ravera, O., The ecological effects of acid deposition, part 1: an introduction, *Experientia,* 42, 329–330, 1986.

Ray, S., D.W. McLeese, and M.R. Peterson, Accumulation of copper, zinc, cadmium and lead from two contaminated sediments by three marine invertebrates: a laboratory study, *Bull. Environ. Contam. Toxicol.,* 26, 315–322, 1981.

Redding, J.M. and Patino, R., Reproductive physiology, in *The Physiology of Fishes,* Evans, D.H., Ed., CRC Press, Boca Raton, FL, 1993.

Reddy A.P., J.M. Spitsbergen, C. Mathews, J.D. Hendricks, and G.S. Bailey, Experimental hepatic tumorigenicity by environmental hydrocarbon dibenzo[a,l]pyrene, *J. Environ. Pathol. Toxicol. Oncol.,* 18(4), 261–269, 1999.

Regoli, F. and G. Principato, Glutathione, glutathione-dependent and antioxidant enzymes in mussel, *Mytilus galloprovincialis,* exposed to metals under field and laboratory conditions: implications for the use of biochemical biomarkers, *Aquatic Toxicol.* (Amsterdam), 31, 143–164, 1995.

Reichle, D.E., Relation of body size to food intake, oxygen consumption, and trace element metabolism in forest floor arthropods, *Ecology,* 49, 538–541, 1968.

Reichle, D.E., P.B. Dunaway, and D.J. Nelson, Turnover and concentration of radionuclides in food chains, *Nucl. Saf.,* 11, 43–55, 1970.

Reichle, D.E. and R.I. Van Hook Jr., Radionuclide dynamics in insect food chains, *Manit. Entomol.,* 4, 22–32, 1970.

Reichhardt, T., Environmental GIS: the world in a computer, *Environ. Sci. Technol.,* 30, 340A–343A, 1996.

Reinert, R.E., L.J. Stone, and W.A. Willford, Effect of temperature on accumulation of methylmercuric chloride and p,p′-DDT by rainbow trout (*Salmo gairdneri*), *J. Fish. Res. Board Can.,* 31, 1649–1652, 1974.

Reinfelder, J.R. and N.S. Fisher, The assimilation of elements ingested by marine copepods, *Science* (Washington, D.C.), 251, 794–796, 1991.

Rench, J.D., Environmental epidemiology, in *Basic Environmental Toxicology,* Cockerham, L.G. and Shane, B.S., Eds., CRC Press, Boca Raton, FL, 1994.

Renzoni, A., S. Focardi, C. Fossi, C. Leonzio, and J. Mayol, Comparison between concentrations of mercury and other contaminants in eggs and tissues of Cory's shearwaters *Calonectris diomedea* collected on Atlantic and Mediterranean islands, *Environ. Pollut. (Ser. A),* 40, 17–35, 1986.

Reuther, R., Mercury accumulation in sediment and fish from rivers affected by alluvial gold mining in the Brazilian Madeira River basin, Amazon, *Environ. Monit. Assess.,* 32, 239–258, 1994.

Rhodes, R., *The Making of the Atomic Bomb,* Simon and Schuster, New York, 1986.

Rice, P.J., C.D. Drewes, T.M. Klubertanz, S.P. Bradbury, and J.R. Coats, Acute toxicity and behavioral effects of chlorpyrifos, permethrin, phenol, strychnine, and 2,4-dinitrophenolto 30-day-old Japanese medaka (*Oryzias latipes*), *Environ. Toxicol. Chem.,* 16, 696–704, 1997.

Richards, C. and G. Host, Examining land use influences on stream habitats and macroinvertebrates: a GIS approach, *Water Resour. Bull.,* 30, 729–738, 1994.

Richards, C., G.E. Host, and J.W. Arthur, Identification of predominant environmental factors structuring stream macroinvertebrate communities within a large agricultural catchment, *Freshwater Biol.,* 29, 285–294, 1993.

Richardson, P., T. Hideshima, and K. Anderson, Thalidomide: emerging role in cancer medicine, *Annu. Rev. Med.,* 53, 629–657, 2002.

Rinsland, C.P., A. Goldman, F.J. Murcray, T.M. Stephen, N.S. Pougatchev, J. Fishman, S.J. David, R.D. Blatherwick, P.C. Novelli, N.B. Jones, and B.J. Connor, Infrared solar spectroscopic measurements of free tropospheric CO, C_2H_6, and HCN above Mauna Loa, Hawaii: seasonal variations and evidence for enhanced emissions from the Southeast Asian tropical fires of 1997–98, *J. Geophys. Res.,* 104, 667–718, 1999.

Ritterhoff, J., G.P. Zauke, and R. Dallinger, Calibration of the estuarine amphipods, *Gammarus zaddachi* Sexton (1912), as biomonitors: toxicokinetics of cadmium and possible role of inducible metal-binding proteins in Cd detoxification, *Aquatic Toxicol.,* 34, 351–369, 1996.

Roberts, D.K., T.C. Hutchinson, J. Paciga, A. Chattopadhyay, R.E. Jervis, and J. Van Loon, Lead contamination around secondary smelters: estimation of dispersal and accumulation by humans, *Science,* 186, 1120–1124, 1974.

Roberts, L., Hard choices ahead on biodiversity, *Science* (Washington, D.C.), 241, 1759–1761, 1988.

Robison, S.H., O. Cantoni, and M. Costa, Analysis of metal-induced DNA lesions and DNA-repair replication in mammalian cells, *Mutat. Res.,* 131, 173–181, 1984.

Robohm, R.A., Paradoxical effects of cadmium exposure on antibacterial antibody responses in two fish species: inhibition in cunners (*Tautogolabrus adspersus*) and enhancement in striped bass (*Morone saxatilis*), *Vet. Immunol. Immunopathol.,* 12, 251–262, 1986.

Roch, M., J.A. McCarter, A.T. Matheson, M.J.R. Clark, and R.W. Olafson, Hepatic metallothionein in rainbow trout (*Salmo gairdneri*) as an indicator of metal pollution in the Campbell River system, *Can. J. Fish. Aquatic Sci.,* 39, 1596–1601, 1982.

Rodgers, B.E. and R.J. Baker, Frequencies of micronuclei in bank voles from zones of high radiation at Chernobyl, Ukraine, *Environ. Toxicol. Chem.,* 19, 1644–1648, 2000.

Roefer, P., S. Snyder, R.E. Zegers, D.J. Rexing, and J.L. Fronk, Endocrine-disrupting chemicals in source water, *J. Am. Water Works Assoc.,* 92, 52–58, 2000.

Roesijadi, G., Metallothioneins in metal regulation and toxicity in aquatic animals, *Aquatic Toxicol.* (Amsterdam), 22, 81–114, 1992.

Roesijadi, G., The basis for increased metallothionein in a natural population of *Crassostrea virginica, Biomarkers,* 4, 467–472, 1999.

Roesijadi, G. and P. Klerks, A kinetic analysis of Cd-binding to metallothionein and other intracellular ligands in oyster gills, *J. Exp. Zool.,* 251, 1–12, 1989.

Roesijadi, G. and W.E. Robinson, Metal regulation in aquatic animals: mechanisms of uptake, accumulation, and release, in *Aquatic Toxicology. Molecular, Biochemical and Cellular Perspectives,* Malins, D.C. and Ostrander, G.K., Eds., CRC Press, Boca Raton, FL, 1994.

Roesijadi, G., R. Bogumil, M. Vasák, and J.H.R. Kägi, Modulation of DNA-binding of a Tramtrack zinc-finger peptide by the metallothionein-thionein conjugate pair, *J. Biol. Chem.,* 273, 17425–17432, 1998.

Rogers, J.M. and R.J. Kavlock, Developmental toxicology, in *Casarett and Doull's Toxicology: The Basic Science of Poisons,* Klaassen, C.D., Ed., McGraw-Hill, New York, 1996.

Rolff, C., D. Broman, C. Näf, and Y. Zebühr, Potential biomagnification of PCDD/Fs: new possibilities for quantitative assessment using stable isotope trophic position, *Chemosphere,* 27, 461–468, 1993.

Romanek, C.S., K.F. Gaines, A.L. Bryan, Jr., and I.L. Brisbin, Jr., Foraging ecology of the endangered wood stork recorded in the stable isotope signature of feathers, *Oecologia,* 125, 584–594, 2000.

Rose, K.S.B., Lower limits of radiosensitivity in organisms, excluding man, *J. Environ. Radioact.,* 15, 113–133, 1992.

Rousch, W., Putting a price tag on Nature's bounty, *Science* (Washington, D.C.), 276, 1029, 1997.

Rowan, D.J. and J.B. Rasmussen, Bioaccumulation of radiocesium by fish: the influence of physicochemical factors and trophic structure, *Can. J. Fish. Aquatic Sci.,* 51, 2388–2410, 1994.

Roy, R. and P.G.C. Campbell, Survival time modeling of exposure of juvenile Atlantic salmon (*Salmo salar*) to mixtures of aluminum and zinc in soft water at low pH, *Aquatic Toxicol.* (Amsterdam), 33, 155–176, 1995.

Rudd, J.W.M., Sources of methyl mercury to freshwater ecosystems: a review, *Water Air Soil Pollut.,* 80, 697–713, 1995.

Rueter, J.G., Jr. and F.M.M. Morel, The interaction between zinc deficiency and copper toxicity as it affects the silicic acid uptake mechanisms in *Thalassiosira pseudonana, Limnol. Oceanogr.,* 26, 67–73, 1981.

Rugh, R. and J. Wolff, Threshold x-irradiation sterilization of the ovary, *Fertil. Steril.,* 8, 428–430, 1957.

Rule, J.H. and R.W. Alden III, Cadmium bioavailability to three estuarine animals in relation to geochemical fractions to sediments, *Arch. Environ. Contam. Toxicol.,* 19, 878–885, 1990.

Ruohtula, M. and J.K. Miettinen, Retention and excretion of [203]Hg-labelled methylmercury in rainbow trout, *Oikos,* 26, 385–390, 1975.

Rupp, E.M., Age dependent values of dietary intake for assessing human exposures to environmental pollutants, *Health Phys.,* 39, 141–146, 1980.

Rupp, E.M., F.L. Miller, and C.F. Baes III, Some results of recent surveys of fish and shellfish consumption by age and region of U.S. residents, *Health Phys.,* 39, 1965–1970, 1980.

Russell, R.W., R. Lazar, and G.D. Haffner, Biomagnification of organochlorines in Lake Erie white bass, *Environ. Toxicol. Chem.,* 14, 719–724, 1995.

Ryan, J.A. and L.E. Hightower, Stress proteins as molecular biomarkers for environmental toxicology, *EXS,* 77, 411–424, 1996.

Ryther, J. and W. Dunstan, Nitrogen, phosphorus, and eutrophication in the coastal marine environment *Science,* 171, 1008–1112, 1971.

Saavedra Alvarez, M.M. and D.V. Ellis, Widespread neogastropod imposex in the Northeast Pacific: implications for TBT contamination surveys, *Mar. Pollut. Bull.,* 21, 244–247, 1990.

Sabins, F.F., Jr., *Remote Sensing: Principles and Interpretation,* W.H. Freeman and Co., New York, 1987.

Sagan, L.A., What is hormesis and why haven't we heard about it before? *Health Phys.,* 52, 521–525, 1987.

Saiki, M.K. and R.S. Ogle, Evidence of impaired reproduction by Western mosquitofish inhabiting seleniferous agricultural drainwater, *Trans. Am. Fish. Soc.,* 124, 578–587, 1995.

Salbu, B., E.H. Oughton, A.V. Ratnikov, T.L. Zhigareva, S.V. Kruglov, K.V. Petrov, N.V. Grebenshakikova, S.K. Firsakova, N.P. Astasheva, N.A. Loshchilov, K. Hove, and P. Strand, The mobility of ^{137}Cs and ^{90}Sr in agricultural soils in the Ukraine, Belarus, and Russia, 1991, *Health Phys.,* 67, 518–528, 1994.

Salsburg, D.S., *Statistics for Toxicologists,* Marcel Dekker, New York, 1986.

Samallow, P.B. and M.E. Soule, A case of stress related heterozygote superiority in nature, *Evolution,* 37, 646–649, 1983.

Sanborn, J.R., R.L. Metcalf, W.N. Bruce, and P-.Y. Lu, The fate of chlordane and toxaphene in a terrestrial-aquatic model ecosystem, *Environ. Entomol.,* 5, 533–538, 1976.

Sanders, B.M., Stress proteins: potential as multitiered biomarkers, in *Biomarkers of Environmental Contamination,* McCarthy, J.F. and Shugart, L.R., Eds., Lewis Publishers, Boca Raton, FL, 1990.

Sanders, B.M. and S.D. Dyer, Cellular stress response, *Environ. Toxicol. Chem.,* 13, 1209–1210, 1994.

Sanders, B.M. and L.S. Martin, Stress proteins as biomarkers of contaminant exposure in archived environmental samples, *Sci. Total Environ.,* 139/140, 459–470, 1993.

Sanders, B.M., K.D. Jenkins, W.G. Sunda, and J.D. Costlow, Free cupric ion activity in seawater: effects on metallothionein and growth in crab larvae, *Science* (Washington, D.C.), 222, 53–55, 1983.

Sandheinrich, M.B. and G.J. Atchison, Sublethal toxicant effects on fish foraging behavior: empirical vs. mechanistic approaches, *Environ. Toxicol. Chem.,* 9, 107–119, 1990.

Sandstead, H.H., Interactions that influence bioavailability of essential metals to humans, in *Metal Speciation: Theory, Analysis and Application,* Kramer, J.R. and Allen, H.E., Eds., Lewis Publishers, Chelsea, MI, 1988.

Sanzharova, N.I., S.V. Fesenko, R.M. Alexakhin, V.S. Anisimov, V.K. Kuznetsov, and L.G. Chernyayeva, Changes in the forms of ^{137}Cs and its availability for plants as dependent on properties of fallout after the Chernobyl nuclear power plant accident, *Sci. Total Environ.,* 154, 9–22, 1994.

Sargent, C. J., J.C. Bowman, and J.L. Zhou, Levels of antifoulant irgarol 1051 in the Conway Marina, North Wales, *Chemosphere,* 41, 1755–1760, 2000.

Sarokin, D. and J. Schulkin, The role of pollution in large-scale population disturbances, part 1: aquatic populations, *Environ. Sci. Technol.,* 26, 1476–1484, 1992.

SAS Institute Inc., SAS/STAT[7] Software: CALIS and LOGISTIC Procedures, Release 6.04, SAS[7] Technical Report P-200, SAS Institute Inc., Cary, NC, 1990.

Satoh, M., N. Nishimura, Y. Kanayama, A. Naganuma, T. Suzuki, and C. Tohyama, Enhanced renal toxicity by inorganic mercury in metallothionein-null mice, *J. Pharmacol. Exp. Ther.,* 283, 1529–1533, 1997.

Saunders, R.L. and J.B. Sprague, Effects of copper-zinc mining pollution on a spawning migration of Atlantic salmon, *Water Res.,* 1, 419–432, 1967.

Scheninger, M.J., M.J. CeNiro, and H. Tauber, Stable nitrogen isotope ratios of bone collagen reflect marine and terrestrial components of prehistoric human diet, *Science* (Washington, D.C.), 220, 1381–1383, 1983.

Schieve, M.S., D.D. Weber, M.S. Myers, F.J. Jaques, W.L. Riechert, C.A. Krone, D.C. Malins, B.B. McCain, S.-L. Chan, and U. Varanasi, Induction of foci of cellular alteration and other hepatic lesions in English sole (*Parophrys vetulus*) exposed to an extract of an urban marine sediment, *Can. J. Fish. Aquatic Sci.,* 48, 1750–1760, 1991.

Schimmack, W., K. Bunzl, and L. Zelles, Initial rates of migration of radionuclides from the Chernobyl fallout in undisturbed soils, *Geoderma,* 44, 211–218, 1989.

Schindler, D.W., Effects of acid rain on freshwater ecosystems, *Science* (Washington, D.C.), 239, 149–157, 1988.

Schindler, D.W., Ecosystems and ecotoxicology: a personal perspective, in *Ecotoxicology: A Hierarchical Treatment*, Newman, M.C. and Jagoe, C.H., Eds., Lewis Publishers, Boca Raton, FL, 1996.

Schlueter, M.A., S.I. Guttman, J.T. Oris, and A.J. Bailer, Survival of copper-exposed juvenile fathead minnows (*Pimephales promelas*) differs among allozyme genotypes, *Environ. Toxicol. Chem.*, 10, 1727–1734, 1995.

Schmitt, C.J., M.L. Wildhaber, J.B. Hunn, T. Nash, M.N. Tieger, and B.L. Steadman, Biomonitoring of lead-contaminated Missouri streams with an assay for erythrocyte δ-aminolevulinic acid dehydratase activity in fish blood, *Arch. Environ. Contam. Toxicol.*, 25, 464–475, 1993.

Schnell, J.H., Some effects of neutron-gamma radiation on late summer bird populations, *Auk*, 81, 528–533, 1964.

Schnute, J.T. and L.J. Richards, A unified approach to the analysis of fish growth, maturity, and survivorship data, *Can. J. Fish. Aquatic Sci.*, 47, 24–40, 1990.

Schober, U. and W. Lampert, Effects of sublethal concentrations of the herbicide Atrazin[7] on growth and reproduction of *Daphnia pulex*, *Bull. Environ. Contam. Toxicol.*, 17, 269–277, 1977.

Scholz, S. and H.O. Gutzeit, 17-Alpha-ethinylestradiol affects reproduction, sexual differentiation and aromatase gene expression of the medaka (*Oryzias latipes*), *Aquatic Toxicol.*, 50, 363–373, 2000.

Schreck, C.B., Physiological, behavioral, and performance indicators of stress, *Am. Fish. Soc. Symp.*, 8, 29–37, 1990.

Schultz, I.R. and W.L. Hayton, Body size and the toxicokinetics of trifluralin in rainbow trout, *Toxicol. Appl. Pharmacol.*, 129, 138–145, 1994.

Schultz, I.R., G. Orner, J.L. Merdink, and A. Skillman, Dose-response relationships and pharmacokinetics of vitellogenin in rainbow trout after intravascular administration of 17-alpha-ethynylestradiol, *Aquatic Toxicol.*, 51, 305–318, 2001.

Schultz, I.R., A. Skillman, D.G. Cyr, and J.J. Nagler, Sub-chronic exposure to 17-alpha-ethynylestradiol decreases the fertility of sexually maturing male rainbow trout (*Oncorhynchus mykiss*), submitted to *Environ. Toxicol. Chem.*, 2002.

Schütt, P. and E.B. Cowling, Waldsterben, a general decline of forests in central Europe: symptoms, development, and possible causes, *Plant Dis.*, 69, 548–558, 1985.

Schwarzenbach, R.P., P.M. Gschwend, and D.M. Imboden, *Environmental Organic Chemistry*, John Wiley & Sons, New York, 1993.

Selye, H., *The Stress of Life*, McGraw-Hill Book Co., New York, 1956.

Selye, H., The evolution of the stress concept, *Amer. Sci.*, 61, 692–699, 1973.

Semagin, V.N., Influence of prolonged low dose irradiation on brain of rat embryos, *Radiobiologiya*, 15, 583–588, 1986.

Senesi, N., Metal-humic substance complexes in the environment, molecular and mechanistic aspects by multiple spectroscopic approach, in *Biogeochemistry of Trace Metals*, Adriano, D.C., Ed., Lewis Publishers, Boca Raton, FL, 1992.

Sepulvada, M.S., P.C. Frederick, M.G. Spalding, and G.E. Williams, Mercury contamination in free-ranging great egret nestlings (*Ardea albus*) from southern Florida, *Environ. Toxicol. Chem.*, 18, 985–992, 1999.

Settle, D.M. and C.C. Patterson, Lead in albacore: guide to lead pollution in Americans, *Science* (Washington, D.C.), 207, 1167–1176, 1980.

Shane, B.S., Introduction to ecotoxicology, in *Basic Environmental Toxicology*, Cockerham, L.G. and Shane, B.S., Eds., CRC Press, Boca Raton, FL, 1994.

Shaw-Allen, P.L. and K. McBee, Chromosome damage in wild rodents inhabiting a site contaminated with Aroclor 1254, *Environ. Contam. Toxicol.*, 12, 677–684, 1993.

Shea, D., Developing national sediment quality criteria, *Environ. Sci. Technol.*, 22, 1256–1261, 1988.

Shore, P.A., B.B. Brodie, and C.A.M. Hogben, The gastric secretion of drugs: a pH partition hypothesis, *J. Pharmacol. Exp. Ther.*, 119, 361–369, 1957.

Shugart, L.R., Quantitation of chemically induced damage to DNA of aquatic organisms by alkaline unwinding assay, *Aquatic Toxicol.* (Amsterdam), 13, 43–52, 1988.

Shugart, L.R., Environmental genotoxicology, in *Fundamentals of Aquatic Toxicology: Effects, Environmental Fate, and Risk Assessment*, 2nd ed., Rand, G.M., Ed., Taylor & Francis, Washington, D.C., 1995.

Shugart, L.R., Molecular markers to toxic agents, in *Ecotoxicology: A Hierarchical Treatment*, Newman, M.C. and Jagoe, C.H., Eds., CRC Press, Boca Raton, FL, 1996.

Shukla, K.K., C.S. Dombroski, and S.H. Cohn, Fallout ^{137}Cs levels in man over a 12-year period, *Health Phys.*, 24, 555–557, 1973.

Sibly, R.M., Effects of pollutants on individual life histories and population growth rates, in *Ecotoxicology: A Hierarchical Treatment,* Newman, M.C. and Jagoe, C.H., Eds., CRC Press, Boca Raton, FL, 1996.

Sibly, R.M. and P. Calow, A life-cycle theory of responses to stress, *Biol. J. Linn. Soc.,* 37, 101–116, 1989.

Sibly, R.M. and J. Hone, Population growth rate and its determinants: an overview, *Philos. Trans. R. Soc. Ser. B,* 357, 1153–1170, 2002.

Simkiss, K., Calcium, pyrophosphate and cellular pollution, *Trends Biochem. Sci.,* April, 1–3, 1981.

Simkiss, K., Lipid solubility of heavy metals in saline solutions, *J. Mar. Biol. Assoc. U.K.,* 63, 1–7, 1983.

Simkiss, K., Ecotoxicants at the cell-membrane barrier, in *Ecotoxicology: A Hierarchical Treatment,* Newman, M.C. and Jagoe, C.H., Eds., Lewis Publishers, Boca Raton, FL, 1996.

Simkiss, K., S. Daniels, and R.H. Smith, Effects of population density and cadmium on growth and survival of blowflies, *Environ. Pollut.,* 81, 41–45, 1993.

Simkiss, K. and M. Taylor, Cellular mechanisms of metal ion detoxification and some new indices of pollution, *Aquatic Toxicol.* (Amsterdam), 1, 279–290, 1981.

Simonich, S.I. and R.A. Hites, Global distribution of persistent organochlorine compounds, *Science* (Washington, D.C.), 269, 1851–1854, 1995.

Sinclair, W.K., Science, Radiation Protection and the NCRP, Lauriston S. Taylor lectures in Radiation Protection and Measurements, Lecture 17, NCRP, Bethesda, MD, 1993, p. 56.

Skidmore, J.F. and P.W.A. Tovell, Toxic effects of zinc sulphate on the gills of rainbow trout, *Water Res.,* 6, 217–230, 1972.

Slater, T.F., Free-radical mechanisms in tissue injury, *Biochem. J.,* 222, 1–15, 1984.

Slobodkin, L.B. and D.E. Dykhuizen, Applied ecology in practice and philosophy, in *Integrated Environmental Management,* Cairns, J., Jr. and Crawford, T.V., Eds., Lewis Publishers, Chelsea, MI, 1991.

Sloman, K.A., D.W. Baker, C.M. Wood, and D.G. McDonald, Social interactions affect physiological consequences of sublethal copper exposure in rainbow trout, *Oncorhynchus mykiss, Environ. Toxicol. Chem.,* 21, in press.

Sluyts, H., F. Van Hoof, A. Cornet, and J. Paulussen, A dynamic new alarm system for use in biological early warning system, *Environ. Toxicol. Chem.,* 15, 1317–1323, 1996.

Smetacek, V., U. Bathmann, E.M. Nothig, and R. Scharek, Coastal eutrophication: causes and consequences, in *Oceans Margin Processes in Global Change,* Mantoura, R.C.F., Martin, J.-M., and Wollast, R., Eds., John Wiley & Sons, Chichester, U.K., 1991.

Smith, A.L. and R.H. Green, Uptake of mercury by freshwater clams (Family Unionidae), *J. Fish. Res. Board Can.,* 32, 1297–1303, 1975.

Smith, R.J., The risks of living near Love Canal, *Science* (Washington, D.C.), 217, 808–811, 1982.

Smith, R.L., *Elements of Ecology,* 2nd ed., Harper & Row, New York, 1986.

Smith, W.H., *Air Pollution and Forests,* Springer-Verlag, New York, 1981.

Smith, W.E. and A.M. Smith, *Minamata,* Holt, Rinehart and Winston, New York, 1975.

Snodgrass, J.W., C.H. Jagoe, A.L. Bryan, Jr., and H.A. Brant, Effects of trophic status and wetland morphology, hydroperiod, and water chemistry on mercury concentrations in fish, *Can. J. Fish. Aquatic Sci.,* 57, 171–180, 2000.

Snyder, S.A., T.L. Keith, D.A. Verbrugge, E.M. Snyder, T.S. Gross, K. Kannan, and J.P. Giesy, Analytical methods for detection of selected estrogenic compounds in aqueous mixtures, *Environ. Sci. Technol.,* 33, 2814–2820, 1999.

Soimasuo, R., I. Jokinen, J. Kukkoen, T. Petänen, T. Ristola, and A. Oikari, Biomarker responses along a pollution gradient: effects of pulp and paper mill effluents on caged whitefish, *Aquatic Toxicol.* (Amsterdam), 31, 329–345, 1995.

Solomon, K.R., D.B. Baker, R.P. Richards, K.R. Dixon, S.J. Klaine, T.W. La Point, R.J. Kendall, C.P. Weisskopf, J.M. Giddings, J.P. Giesy, L.W. Hall, Jr., and W.M. Williams, Ecological risk assessment of atrazine in North American surface waters, *Environ. Toxicol. Chem.,* 15, 31–76, 1996.

Solomon, K.R. et al., Chlorpyrifos: ecotoxicological risk assessment for birds and mammals in corn agroecosystems, *Hum. Ecol. Risk Assessment,* 7, 497–632, 2001.

Sonnenschein, C. and A.M. Soto, An updated review of environmental estrogen and androgen mimics and antagonists, *J. Steroid Biochem. Molec. Biol.,* 65, 143–150, 1998.

Sorkhoh, N., R. Al-Hasan, S. Radwan, and T. Höpner, Self-cleaning of the Gulf, *Nature* (London), 359, 109, 1992.

Spacie, A. and J.L. Hamelink, Bioaccumulation, in *Fundamentals of Aquatic Toxicology,* Rand, G.M. and Petrocelli, S.R., Eds., Hemisphere Publishing Corp., Washington, D.C., 1985.

Spacie, A., L.S. McCarty, and G.M. Rand, Bioaccumulation and bioavailability in multiphase systems, in *Fundamentals of Aquatic Toxicology: Effects, Environmental Fate, and Risk Assessment,* 2nd ed., Rand, G.M., Ed., Taylor & Francis, Washington, D.C., 1995.

Spalding, M.G., P.C. Frederick, H.C. McGill, S.N. Bouton, and L.R. McDowell, Methylmercury accumulation in tissues and its effects on growth and appetite in captive great egrets, *J. Wildl. Dis.,* 36, 411–422, 2000.

Sparks, A.K., *Invertebrate Pathology: Noncommunicable Diseases,* Academic Press, New York, 1972.

Sparks, R.E., W.T. Waller, and J. Cairns Jr., Effect of shelters on the resistance of dominant and submissive bluegills (*Lepomis macrochirus*) to a lethal concentration of zinc, *J. Fish. Res. Board Can.,* 29, 1356–1358, 1972.

Sparrow, A.H., J.P. Binnington, and V. Pond, *Bibliography on the Effects of Ionizing Radiations on Plants, 1896–1955,* U.S. AEC Rep. BNL-504(L-103), Brookhaven National Laboratory, Upton, NY, 1958.

Spies, R.B., D.W. Rice, Jr., P.J. Thomas, J.J. Stegeman, J.N. Cross, and J.E. Hose, A field test of correlates of poor reproductive success and genetic damage in contaminated populations of starry flounder, *Platichthys stellatus, Mar. Environ. Res.,* 28, 542–543, 1989.

Spitzer, P.R., R.W. Risebrough, W. Walker II, R. Hernandez, A. Poole, D. Puleston, and I.C.T. Nisbet, Productivity of ospreys in Connecticut-Long Island increases as DDE residues decline, *Science* (Washington, D.C.), 202, 333–335, 1978.

Sprague, J.B., Avoidance of copper-zinc solutions by young salmon in the laboratory, *J. Water Pollut. Control Fed.,* 36, 990–1004, 1964.

Sprague, J.B., Measurement of pollutant toxicity to fish, I: bioassay methods for acute toxicity, *Water Res.,* 3, 793–821, 1969.

Sprague, J.B., Measurement of pollutant toxicity to fish, II: utilizing and applying bioassay results, *Water Res.,* 4, 3–32, 1970.

Sprague, J.B., Measurement of pollutant toxicity to fish, III: sublethal effects and "safe" concentrations, *Water Res.,* 5, 245–266, 1971.

Sprague, J.B., Current status of sublethal tests of pollutants on aquatic organisms, *J. Fish. Res. Board Can.,* 33, 1988–1992, 1976.

Sprague, J.B., P.F. Elson, and R.L. Saunders, Sublethal copper-zinc pollution in a salmon river: a field and laboratory, *Int. J. Air Water Pollut.,* 9, 531–543, 1965.

Spromberg, J.A., B.M. John, and W.G. Landis, Metapopulation dynamics: indirect effects and multiple distinct outcomes in ecological risk assessment, *Environ. Toxicol. Chem.,* 17, 1640–1649, 1998.

Squibb, S. and B.A. Fowler, Relationship between metal toxicity to subcellular systems and the carcinogenic response, *Environ. Health Perspect.,* 40, 181–188, 1981.

Stacell, M. and D.G. Huffman, Oxytetracycline-induced photosensitivity of channel catfish, *Prog. Fish-Cult.,* 56, 211–213, 1994.

Stanley-Horn, D.E., H.R. Matilla, M.K. Sears, G. Dively, R. Rose, R.L. Hellmich, and L. Lewis, Assessing impact of Cry 1Ab-expressing corn pollen on monarch butterfly larvae in field studies, *Proc. Natl. Acad. Sci. U.S.A.,* 98(2), 11931–11936, 2001.

Stebbing, A.R.D., Hormesis: the stimulation of growth by low levels of inhibitors, *Sci. Total Environ.,* 22, 213–234, 1982.

Steedman, R.J., Modification and assessment of an index of biotic integrity to quantify stream quality in southern Ontario, *Can. J. Fish. Aquatic Sci.,* 45, 492–501, 1988.

Stegeman, J.J. and M.E. Hahn, Biochemistry and molecular biology of monooxygenases: current perspectives on forms, functions, and regulation of cytochrome P450 in aquatic species, in *Aquatic Toxicology: Molecular, Biochemical and Cellular Perspectives,* Malins, D.C. and Ostrander, G.K., Eds., CRC Press, Boca Raton, FL, 1994.

Stein, J.E., W.L. Reichert, M. Nishimoto, and U. Varanasi, Overview of studies on liver carcinogenesis in English sole from Puget Sound; evidence for a xenobiotic chemical etiology, II: biochemical studies, *Sci. Total Environ.,* 94, 51–69, 1990.

Stenehjem, M., Indecent exposure, *Nat. Hist.,* 9, 6–21, 1990.

Stephan, C.E., Methods for calculating an LC50, in *Aquatic Toxicology and Hazard Evaluation,* ASTM STP 634, Mayer, F.L. and Hamelink, J.L., Eds., American Society for Testing and Materials, Philadelphia, 1977.

Stephan, C.E., D.I. Mount, D.J. Hansen, J.H. Gentile, G.A. Chapman, and W.A. Brungs, Guidelines for Deriving Numerical National Water Quality Criteria for the Protection of Aquatic Organisms and Their Uses, PB85-227049, National Technical Information Service, Springfield, VA, 1985.

Stephan, C.E. and J.W. Rogers, Advantages of using regression analysis to calculate results of chronic toxicity tests, in *Aquatic Toxicology and Hazard Assessment: Eighth Symposium,* ASTM STP 891, Bahner, R.C. and Hansen, D.J., Eds., American Society for Testing and Materials, Philadelphia, 1985.

Steward, R.C., Industrial and non-industrial melanism in the peppered moth, *Biston betularia* (L.), *Ecol. Ent.,* 2, 231–243, 1977.

St. Louis, V.L., J.W. Rudd, C.A. Kelly, K.G. Beaty, R.J. Flett, and N.T. Roulet, Production and loss of methylmercury and loss of total mercury from boreal forest catchments containing different types of wetlands, *Environ. Sci. Technol.* 30, 2719–2729, 1996.

Stone, R., Can a father's exposure lead to illness in his children? *Science* (Washington, D.C.), 258, 31, 1992.

Strong, C.R. and S.N. Luoma, Variations in the correlation of body size with concentrations of Cu and Ag in the bivalve *Macoma balthica, Can. J. Fish. Aquatic Sci.,* 38, 1059–1064, 1981.

Stumm, W. and J.J. Morgan, *Aquatic Chemistry: An Introduction Emphasizing Chemical Equilibrium in Natural Waters,* John Wiley & Sons, New York, 1981.

Stumm, W., L. Sigg, and J.L. Schnoor, Aquatic chemistry of acid deposition, *Environ. Sci. Technol.,* 21, 8–13, 1987.

Sturzenbaum, S.R., C. Winters, M. Galay, A.J. Morgan, and P. Kille, Metal ion trafficking in earthworms: identification of a cadmium-specific metallothionein, *J. Biol. Chem.,* 276, 34013–34018, 2001.

Suedel, B.C., J.A. Boraczek, R.K. Peddicord, P.A. Clifford, and T.M. Dillon, Trophic transfer and biomagnification potential of contaminants in aquatic ecosystems, *Rev. Environ. Contam. Toxicol.,* 136, 22–89, 1994.

Sullivan, J.F., G.J. Atchison, D.J. Kolar, and A.W. McIntosh, Changes in predator-prey behavior of fathead minnows (*Pimephales promelas*) and largemouth bass (*Micropterus salmoides*) caused by cadmium, *J. Fish. Res. Board Can.,* 35, 446–451, 1978.

Sundlof, S.F., M.G. Spalding, J.D. Wentworth, and C.K. Steible, Mercury in livers of wading birds (Ciconiiformes) in southern Florida, *Arch. Environ. Contam. Toxicol.,* 27, 299–305, 1994.

Sutcliffe, F.E., Trans., *Discourse on Method and the Meditations by René Descartes,* Penguin Books, London, 1968.

Suter, G.W., II, *Ecological Risk Assessment,* Lewis Publishers, Boca Raton, FL, 1993.

Szczypka, M.S. and D.J. Thiele, A cysteine-rich nuclear protein activates yeast metallothionein gene transcription, *Mol. Cell. Biol.,* 9, 421–429, 1989.

Tachikawa, M. and R. Sawamura, The effects of salinity on pentachlorophenol accumulation and elimination by killifish (*Oryzias latipes*), *Arch. Environ. Contam. Toxicol.,* 26, 304–308, 1994.

Tackett, S.L., Lead in the environment: effects of human exposure, *Am. Lab.* (Fairfield, CT), July, 32–41, 1987.

Tagatz, M.E., Effect of mirex on predator-prey interaction in an experimental estuarine ecosystem, *Trans. Am. Fish. Soc.,* 4, 546–549, 1976.

Tanguy, A. and D. Moraga, Cloning and characterization of a gene coding for a novel metallothionein in the Pacific oyster *Crassostrea gigas* (CgMT2): a case of adaptive response to metal-induced stress? *Gene,* 273, 123–130, 2001.

Taub, F.B., Standardized aquatic microcosms, *Environ. Sci. Technol.,* 23, 1064–1066, 1989.

Taussig, H.B., The thalidomide syndrome, *Sci. Am.,* 207, 29–35, 1962.

Taylor, E.J., J.E. Morrison, S.J. Blockwell, A. Tarr, and D. Pasoe, Effects of lindane on the predator-prey interaction between *Hydra oligactis* Pallas and *Daphnia magna* Strauss, *Arch. Environ. Contam. Toxicol.,* 29, 291–326, 1995.

Teasdale, S., There Will Come Soft Rains (War Time), in *The Collected Poems of Sara Teasdale,* MacMillan Co., New York, 1937.

Terhaar, C.J., W.S. Ewell, S.P. Dziuba, W.W. White, and P.J. Murphy, A laboratory model for evaluating the behavior of heavy metals in an aquatic environment, *Water Res.,* 11, 101–110, 1977.

Ternes, T.A., M. Stumpf, J. Mueller, K. Haberer, R.D. Wilken, and M. Servos, Behavior and occurrence of estrogens in municipal sewage treatment plants, I: investigations in Germany, Canada and Brazil, *Sci. Total Environ.,* 225, 81–90, 1999.

Tessier, A., P.G.C. Campbell, and M. Bisson, Sequential extraction procedure for the speciation of particulate trace metals, *Anal. Chem.,* 51, 844–851, 1979.

Tessier, A., P.G.C. Campbell, J.C. Auclair, and M. Bisson, Relationships between partitioning of trace metals in sediments and their accumulation in the tissues of the freshwater mollusc *Elliptio complanata* in a mining area, *Can. J. Fish. Aquatic Sci.,* 41, 1463–1472, 1984.

Tessier, L., G. Vaillancourt, and L. Pazdernik, Temperature effects on cadmium and mercury kinetics in freshwater molluscs under laboratory conditions, *Arch. Environ. Contam. Toxicol.,* 26, 179–184, 1994.

Theodorakis, C.W., J.W. Bickham, T. Lamb, P.A. Medica, and T.B. Lyne, Integration of genotoxicity and population genetic analyses in kangaroo rats (*Dipodomys merriami*) exposed to radionuclide contamination at the Nevada test site, USA, *Environ. Toxicol. Chem.,* 20, 317–326, 2001.

Thiele, D.J., M.J. Walling, and D.H. Hamer, Mammalian metallothionein is functional in yeast, *Science,* 231, 854–856, 1986.

Thies, M.L., K. Thies, and K. McBee, Organochlorine pesticide accumulation and genotoxicity in Mexican free-tailed bats from Oklahoma and New Mexico, *Arch. Environ. Contam. Toxicol.,* 30, 178–187, 1996.

Thomann, R.V., Bioaccumulation model of organic chemical distribution in aquatic food chains, *Environ. Sci. Technol.,* 23, 699–707, 1989.

Thomas, M.J. and J.A. Thomas, Hormone assays and endocrine function, in *Principles and Methods of Toxicology,* 4th ed., Hayes, A.W., Ed., Taylor & Francis, Philadelphia, 2001.

Thomas, P., Molecular and biochemical responses of fish to stressors and their potential use in environmental monitoring, *Am. Fish. Soc. Symp.,* 8, 9–28, 1990.

Thompson, D.R. and R.W. Furness, Comparison of levels of total and organic mercury in seabird feathers, *Mar. Pollut. Bull.,* 20, 577–579, 1989.

Thompson, H.M., Interactions between pesticides: a review of reported effects and their implications for wildlife risk assessment, *Ecotoxicology,* 5, 59–81, 1996.

Thompson, K.W., A.C. Hendricks, G.L. Nunn, and J. Cairns, Jr., Ventilatory responses of bluegill sunfish to sublethal fluctuating exposures to heavy metals (Zn^{++} and Cu^{++}), *Water Resour. Bull.,* 19, 719–727, 1983.

Thompson, R.A., G.D. Schroder, and T.H. Conner, Chromosomal aberrations in the cotton rat (*Sigmodon hispidus*) exposed to hazardous waste, *Environ. Mol. Mutagen.,* 11, 359–367, 1988.

Thompson, R.D. The changing atmosphere and its impact on planet Earth, in *Environmental Issues in the 1990s,* Mannion, A.M. and Bowlby, S.R., Eds., John Wiley & Sons, West Sussex, U.K., 1992.

Thoreau, H.D., *Essay on the Duty of Disobedience and Walden,* 1854, repr., Lancer Books, New York, 1968.

Thorrold, S.R., C. Latkoczy, P.K. Swart, and C.M. Jones, Natal homing in a marine fish population, *Science,* 291, 297–299, 2001.

Thurston, R.V., R.C. Russo, and G.A. Vonogradov, Ammonia toxicity to fishes: effect of pH on the toxicity of the un-ionized ammonia species, *Environ. Sci. Technol.,* 15, 837–840, 1981.

Tice, R.R., B.G. Ormiston, R. Boucher, C.A. Luke, and D.E. Paquette, Environmental biomonitoring with feral rodent species, in *Short-Term Assays in the Analysis of Complex Mixtures II,* Sandhu, S.S., DeMarini, D.M., Mass, M.J., Moore, M.M., and Mumford, J.L., Eds., Plenum Press, New York, 1987.

Till, J.E. and H.R. Meyer, *Radiological Assessment: A Textbook on Environmental Dose Analysis,* U.S. Nuclear Regulatory Commission, Washington, D.C., and National Technical Information Service, Springfield, VA, 1983.

Tilton, F., W.H. Benson, and D. Schlenk, Elevation of serum 17-beta-estradiol in channel catfish following injection of 17-beta-estradiol, ethynyl estradiol, estrone, estriol and estradiol-17-beta-glucuronide, *Environ. Toxicol. Pharmacol.,* 9, 169–172, 2001.

Timbrell, J., *Principles of Biochemical Toxicology,* 3rd ed., Taylor & Francis, Philadelphia, 2000.

Tolmazin, D., Soviet environmental practices, *Science* (Washington, D.C.), 221, 1136, 1983.

Touart, L.W., The federal insecticide, fungicide, and rodenticide act, in *Fundamentals of Aquatic Toxicology: Effects, Environmental Fate, and Risk Assessment,* 2nd ed., Rand, G.M., Ed., Taylor & Francis, Washington, D.C., 1995.

Trabalka, J.R., L.D. Eyman, and S.I. Aurbach, Analysis of the 1957–1958 Soviet nuclear accident, *Science* (Washington, D.C.), 209, 345–353, 1980.

Truhaut, R., Ecotoxicology: objectives, principles and perspectives, *Ecotoxicol. Environ. Saf.,* 1, 151–173, 1977.

Tucker, J.D., A. Auletta, M.C. Cimino, K.L. Dearfield, D. Jacobson-Kram, R.R. Tice, and A.V. Carrano, Sister-chromatid exchange: second report of the Gene-Tox program, *Mutat. Res.,* 297, 101–180, 1993.

Tull-Singleton, S., S. Kimball, and K. McBee, Correlative analysis of heavy metal bioconcentration and genetic damage in white-footed mice (*Peromyscus leucopus*) from a hazardous waste site, *Bull. Environ. Contam. Toxicol.*, 52, 667–672, 1994.

Turner, F.B., Effects of continuous irradiation on animal populations, *Adv. Radiat. Biol.*, 5, 83, 1975.

Tutt, J.W., *British Moths*, Routledge, London, 1896.

Tuurala, H. and A. Soivio, Structural and circulatory changes in the secondary lamellae of *Salmo gairdneri* gills after sublethal exposures to dehydroabietic acid and zinc, *Aquatic Toxicol.* (Amsterdam), 2, 21–29, 1982.

Ugedal, O., B. Jonsson, O. Njåstad, and R. Neumann, Effects of temperature and body size on radiocaesium retention in brown trout, *Salmo trutta, Freshwater Biol.*, 28, 165–171, 1992.

Ugolini, F.C. and H. Spaltenstein, Pedosphere, in *Global Biogeochemical Cycles*, Butcher, S.S. et al., Eds., Academic Press, London, 1992.

Ullrich, R.L., M.C. Jernigan, L.C. Satterfield, and N.D. Bowles, Radiation carcinogenesis: time-dose relationships, *Radiat. Res.*, 111, 179–184, 1987.

Ulmer, D.D., Effects of metals on protein structure, in *Effects of Metals on Cells, Subcellular Elements, and Macromolecules*, Maniloff, J., Coleman, J.R., and Miller, M.W., Eds., Charles C Thomas Publisher, Springfield, IL, 1970.

Underwood, A.J. and C.H. Peterson, Towards an ecological framework for investigating pollution, *Mar. Ecol. Prog. Ser.*, 46, 227–234, 1988.

UNSCEAR (United Nations Scientific Committee on the Effects of Atomic Radiation), *Genetic and Somatic Effects of Ionizing Radiation*, Publication E.86.IX.9, United Nations, New York, 1986.

Valigura, R.A., R.B. Alexander, M.S. Castro, T.P. Meyers, H.W. Paerl, P.E. Stacey, and R.E. Turner, Eds., *Nitrogen Loading in Coastal Water Bodies: An Atmospheric Perspective*, Coastal and Estuarine Studies No. 57, American Geophysical Union Press, Washington, D.C., 2000.

Van Beneden, R.J. and G.K. Ostrander, Expression of oncogenes and tumor suppressor genes in teleost fishes, in *Aquatic Toxicology: Molecular, Biochemical and Cellular Perspectives*, Malins, D.C. and Ostrander, G.K., Eds., CRC Press, Boca Raton, FL, 1994.

Van Cleef-Toedt, K.A., L.A.E. Kaplan, and J.F. Crivello, Killifish metallothionein messenger RNA expression following temperature perturbation and cadmium exposure, *Cell Stress Chaperones*, 6, 351–359, 2001.

Van den Belt, K., R. Verheyen, and H. Witters, Reproductive effects of ethynylestradiol and 4t-octylphenol on the zebrafish (*Danio rerio*), *Arch. Environ. Contam. Toxicol.*, 41, 458–467, 2001.

Van den Berg, M., L. Birnbaum, A.T.C. Bosveld, B. Brunström, P. Cook, M. Feeley, G.P. Giesy, A. Hanberg, R. Hasagawa, S.W. Kennedy, T. Kubiak, J.C. Larsen, F.X.R. van Leeuwen, A.K.D. Liem, C. Nolt, R.E. Peterson, L. Poellinger, S. Safe, D. Schrenk, D. Tillitt, M. Tysklind, M. Younes, F. Waern, and T. Zacharewski, Toxic equivalency factors (TEFs) for PCBs, PCDDs, PCDFs for human and wildlife, *Environ. Health Perspect.*, 106, 775–792, 1998.

van den Heuvel, M.R., L.S. McCarty, R.P. Lanno, B.E. Hickie, and D.G. Dixon, Effect of total body lipid on the toxicity and toxicokinetics of pentachlorophenol in rainbow trout (*Oncorhynchus mykiss*), *Aquatic Toxicol.* (Amsterdam), 20, 235–252, 1991.

Van Leeuwen, C.J., F. Moberts, and G. Niebeek, Aquatic toxicological aspects of dithiocarbamates and related compounds, II: effects on survival, reproduction and growth of *Daphnia magna, Aquatic Toxicol.* (Amsterdam), 7, 165–175, 1985.

Van Straalen, N.M. and C.A. Denneman, Ecotoxicological evaluation of soil quality criteria, *Ecotoxicol. Environ. Saf.*, 18, 241–251, 1989.

Van Veld, P.A., D.J. Westbrook, B.R. Woodin, R.C. Hale, C.L. Smith, R.J. Huggett, and J.J. Stegeman, Induced cytochrome P-450 in intestine and liver of spot (*Leiostomus xanthurus*) from a polycyclic aromatic hydrocarbon contaminated environment, *Aquatic Toxicol.* (Amsterdam), 17, 119–132, 1990.

Van Veld, P.A., W.K. Vogelbein, R. Smolowitz, B.R. Woodin, and J.J. Stegeman, Cytochrome P4501A1 in hepatic lesions of a teleost fish (*Fundulus heteroclitus*) collected from a polycyclic aromatic hydrocarbon-contaminated site, *Carcinogenesis*, 13(3), 505–507, 1992.

Van Veld, P.A., W.K. Vogelbein, M.K. Cochran, A. Goksoyr, and J.J. Stegeman, Route-specific cellular expression of cytochrome P4501A (CYP1A) in fish (*Fundulus heteroclitus*) following exposure to aqueous and dietary benzo[a]pyrene, *Toxicol. Appl. Pharmacol.*, 142, 348–359, 1997.

Varanasi, U., M. Nishimoto, W.L. Reichert, and B.T. Le Eberhart, Comparative metabolism of benzo[a]pyrene and covalent binding to hepatic DNA in English sole, starry flounder, and rat, *Cancer Res.*, 46, 3817–3824, 1986.

Velasko, R.H., M. Belli, U. Sanasone, and S. Menegon, Vertical transport of radiocesium in surface soils: model implementation and dose-rate computation, *Health Phys.,* 64, 37–44, 1993.

Verschueren, K., *Handbook of Environmental Data on Organic Chemicals,* Van Nostrand Reinhold Co., New York, 1983.

Víg, É. and J. Nemcsók, The effects of hypoxia and paraquat on the superoxide dismutase activity in different organs of carp, *Cyprinus carpio* L., *J. Fish Biol.,* 35, 23–25, 1989.

Vitousek, P.M., H.A. Mooney, J. Lubchenko, and J.M. Mellilo, Human domination of Earth's ecosystems, *Science,* 277, 494–499, 1997.

Vogelbein, W.K. and J.W. Fournie, The ultrastructure of normal and neoplastic exocrine pancreas in the mummichog, *Fundulus heteroclitus, Toxicol. Pathol.,* 22(3), 248–260, 1994.

Vogelbein, W.K., J.W. Fournie, P.A. Van Veld, and R.J. Huggett, Hepatic neoplasms in the mummichog *Fundulus heteroclitus* from a creosote-contaminated site, *Cancer Res.,* 50, 5978–5986, 1990.

Vogelbein, W.K., D.E. Zwerner, M.A. Unger, C.L. Smith, and J.W. Fournie, Hepatic and extra-hepatic neoplasms in a teleost fish from a polycyclic aromatic hydrocarbon contaminated habitat in Chesapeake Bay, USA, in *Spontaneous Animal Tumors: A Survey,* Rossi, L., Richardson, R., and Harshbarger, J., Eds., 1997, pp. 55–64.

Vogelbein, W.K., J.W. Fournie, P.S. Cooper, and P.A. Van Veld, Hepatoblastomas in the mummichog, *Fundulus heteroclitus* (Linnaeus), from a creosote-contaminated environment: a histologic, ultrastructural, and immunohistochemical study, *J. Fish Dis.,* 22, 419–431, 1999.

Vollenweider, R.A., and J.J. Kerekes, *Eutrophication of Waters: Monitoring, Assessment and Control,* OECD, Paris, 1982.

Wagner, C. and H. Løkke, Estimation of ecotoxicological protection levels from NOEC toxicity data, *Water Res.,* 25, 1237–1242, 1991.

Wagner, J.G., *Fundamentals of Clinical Pharmacokinetics,* Drug Intelligence Publications, Hamilton, IL, 1975.

Walker, B., Biodiversity and ecological redundancy, *Conserv. Biol.,* 6, 12–23, 1991.

Walker, C.H., Species differences in microsomal monooxygenase activity and their relationship to biological half-lives, *Drug Metab. Rev.,* 7, 295–323, 1978.

Walker, C.H., I. Newton, S.D. Hallam, and M.J.J. Ronis, Activities and toxicological significance of hepatic microsomal enzymes of the kestrel (*Falco tinnunculus*) and sparrowhawk (*Accipiter nisus*), *Comp. Biochem. Physiol.,* 86C, 379–382, 1987.

Walker, C.H., S.P. Hopkin, R.M. Sibly, and D.B. Peakall, *Principles of Ecotoxicology,* 2nd ed., Taylor & Francis, London, 2001.

Wallace, W.G. and G.R. Lopez, Bioavailability of biologically sequestered cadmium and the implications of metal detoxification, *Mar. Ecol. Prog. Ser.,* 147, 149–157, 1997.

Wallace, W.G., G.R. Lopez, and J.S. Levinton, Cadmium resistance in an oligochaete and its effect on cadmium trophic transfer to an omnivorous shrimp, *Mar. Ecol. Prog. Ser.,* 172, 225–237, 1998.

Wallace, W.G., T.M.H. Brouwer, M. Brouwer, and G.R. Lopez, Alterations in prey capture and induction of metallothioneins in grass shrimp fed cadmium-contaminated prey, *Environ. Toxicol. Chem.,* 19, 962–971, 2000.

Wang, Y.J., E.A. Mackay, O. Zerbe, D. Hess, P.E. Hunziker, M. Vasák, and J.H.R. Kägi, Characterization and sequential localization of the metal clusters in sea urchin metallothionein, *Biochemistry,* 34, 7460–7467, 1995.

Wang, C.H., D.L. Willis, and W.D. Loveland, *Radiotracer Methodology in the Biological, Environmental, and Physical Sciences,* Prentice-Hall, Englewood Cliffs, NJ, 1975.

Wangen, L.E., Elemental composition of size-fractionated aerosols associated with a coal-fired power plant plume and background, *Environ. Sci. Technol.,* 15, 1080–1088, 1981.

Wania, F. and D. Mackay, A global distribution model for persistent organic chemicals, *Sci. Total Environ.,* 160/161, 211–232, 1995.

Wania, F. and D. Mackay, Tracking the distribution of persistent organic pollutants, *Environ. Sci. Technol.,* 30, 390A–396A, 1996.

Ware, D.M., C. Tovey, D. Hay, and B. McCarter, Straying Rates and Stock Structure of British Columbia Herring, Canadian Stock Assessment Secretariat Research Document 2000/006, Fisheries and Ocean Canada, 2000.

Warnau, M., G. Ledent, A. Temara, V. Alva, M. Jangoux, and P. Dubois, Allometry of heavy metal bioconcentration in the echinoid *Paracentrotus lividus, Arch. Environ. Contam. Toxicol.,* 29, 393–399, 1995.

Watkins, B. and K. Simkiss, The effect of oscillating temperatures on the metal ion metabolism of *Mytilus edulis, J. Mar. Biol. Assoc. U.K.,* 68, 93–100, 1988.

Watras, C.J., R.C. Back, S. Halvorsen, R.J.M. Hudsin, K.A. Morrison, and S.P. Wente, Bioaccumulation of mercury in pelagic freshwater food webs, *Sci. Total Environ.,* 219, 183–208, 1998.

Webb, R.E. and F. Horsfall, Jr., Endrin resistance in the pine mouse, *Science* (Washington, D.C.), 156, 1762, 1967.

Weber, C.I., W.H. Peltier, T.J. Norberg-King, W.B. Horning, II, F.A. Kessler, J.R. Menkedick, T.W. Neiheisel, P.A. Lewis, D.J. Klemm, Q.H. Pickering, E.L. Robinson, J.M. Lazorchak, L.J. Wymer, and R.W. Freyberg, Short-term methods for estimating the chronic toxicity of effluents and receiving waters to freshwater organisms, EPA/600/4-89/001, Environmental Monitoring Systems Laboratory, EPA, Cincinnati, OH, 1989.

Weeks, J.M. and P.S. Rainbow, A dual-labelling technique to measure the relative assimilation efficiencies of invertebrates taking up trace metals from food, *Functional Ecol.,* 4, 711–717, 1990.

Weis, J.S. and J. Perlmutter, Effects of tributyltin on activity and burrowing behavior of the fiddler crab, *Uca pugilator, Estuaries,* 10, 342–346, 1987.

Weis, J.S. and P. Weis, Pollutants as developmental toxicants in aquatic organisms, *Environ. Health Perspect.,* 71, 77–85, 1987.

Weis, J.S. and P. Weis, Effects of environmental pollutants on early fish development, *CRC Crit. Rev. Aquatic Sci.,* 1, 45–73, 1989a.

Weis, J.S. and P. Weis, Tolerance and stress in a polluted environment: the case of the mummichog, *Bioscience,* 39, 89–95, 1989b.

Weis, J.S. and P. Weis, Effects of embryonic exposure to methylmercury on larval prey-capture ability in the mummichog, *Fundulus heteroclitus, Environ. Toxicol. Chem.,* 14, 153–156, 1995.

Weis, J.S., G. Smith, T. Zhou, C. Santiago-Bass, and P. Weis, Effects of contaminants on behavior: biochemical mechanisms and ecological consequences, *BioScience,* 51, 209–217, 2001.

Welch, L.J., Contaminant Burdens and Reproductive Rates of Bald Eagles in Maine, M.S. thesis, University of Maine, Orono, 1994.

Welch, W.J., Mammalian stress response: cell physiology and biochemistry of stress proteins, in *Stress Proteins in Biology and Medicine,* Morimoto, R.I., Tissieres, A., and Georgopolis, C.C., Eds., Cold Spring Harbor Press, Cold Spring Harbor, NY, 1990.

Wenning, R.J., R.T. Di Giulio, and E.P. Gallagher, Oxidant-mediated biochemical effects of paraquat in the ribbed mussel, *Geukensia demissa, Aquatic Toxicol.* (Amsterdam), 12, 157–170, 1988.

Westlake, G.F., Behavioral effects of industrial chemicals in aquatic animals, in *Hazard Assessment of Chemicals: Current Developments,* Vol. III, Saxena, J., Ed., Academic Press, Orlando, FL, 1984.

Wetzel, R.G., *Limnology,* Saunders College Publishing, Philadelphia, 1982.

Wheeler, D.L., An Eclectic Biologist Argues that Humans Are Not Evolution's Most Important Result; Bacteria Are, *Chron. Higher Education,* Sep. 6, 1996, pp. A23-A24.

Whicker, F.W. and L. Fraley, Jr., Effects of ionizing radiation on terrestrial plant communities, *Adv. Radiat. Biol.,* 4, 317, 1974.

Whicker, F.W., T.G. Hinton, and D.J. Niquette, Health risk to hypothetical residents of a radioactively contaminated lake bed, in *Proceedings of the Environmental Remediation Conference,* Augusta, GA, US DOE, Oct. 24–28, 1993.

Whicker, F.W. and V. Schultz, *Radioecology: Nuclear Energy and the Environment,* CRC Press, Boca Raton, FL, 1982, Vol. I, p. 212; Vol. II, p. 228.

Whicker, F.W. and T.B. Kirchner, PATHWAY: a dynamic food-chain model to predict radionuclide ingestion after fallout deposition, *Health Phys.,* 52, 717–737, 1987.

Whitall, D.R. and H.W. Paerl, Spatiotemporal variability of wet atmospheric nitrogen deposition to the Neuse River estuary, North Carolina, *J. Environ. Qual.,* 30, 508–1515, 2001.

White, R.M., The great climate debate, *Sci. Am.,* 263, 36–43, 1990.

White, R.M., Preface, in *The Greening of Industrial Ecosystems,* Allenby, B.R. and Richards, D.J., Eds., National Academy Press, Washington, D.C., 1994, pp. v–vi.

Wichterman, R., Biological effects of ionizing radiations on protozoa: some discoveries and unsolved problems, *BioScience,* 22, 281, 1972.

Wilkinson, K.J. and P.G.C. Campbell, Aluminum bioconcentration at the gill surface of juvenile Atlantic salmon in acidic media, *Environ. Toxicol. Chem.,* 12, 2083–2095, 1993.

Williams, L.G. and D.I. Mount, Influence of zinc on periphytic communities, *Am. J. Bot.,* 52, 26–34, 1965.

Williamson, P., Use of ^{65}Zn to determine the field metabolism of the snail *Cepaea nemoralis* L., *Ecology,* 56, 1185–1192, 1975.

Wilson, J.B., The cost of heavy-metal tolerance: an example, *Evolution,* 42, 408–413, 1988.

Winge, D.R. and M. Brouwer, Discussion summary: techniques and problems in metal-binding protein chemistry and implications for proteins in nonmammalian organisms, *Environ. Health Perspect.,* 65, 211–214, 1986.

Winn, R. and D. Knott, An evaluation of the survival of experimental populations exposed to hypoxia in the Savannah River estuary, *Mar. Ecol. Prog. Ser.,* 88, 161–179, 1992.

Winner, R.W., M.W. Boesel, and M.P. Farrell, Insect community structure as an index of heavy-metal pollution in lotic ecosystems, *Can. J. Fish. Aquatic Sci.,* 37, 647–655, 1980.

Winner, R.W., T. Keeling, R. Yeager, and M.P. Farrell, Effect of food type on the acute and chronic toxicity of copper to *Daphnia magna, Freshwater Biol.,* 7, 343–349, 1977.

Winston, G.W. and R.T. Di Giulio, Prooxidant and antioxidant mechanisms in aquatic organisms, *Aquatic Toxicol.* (Amsterdam), 19, 137–161, 1991.

Wirgin, I.I., C. Grunald, S. Courtenay, G.-L. Kreamer, W.L. Reichert, and J.E. Stein, A biomarker approach to assessing xenobiotic exposure in Atlantic tomcod from the North American Atlantic coast, *Environ. Health Perspect.,* 102, 764–770, 1994.

Wise, D., J.D. Yarbrough, and R.T. Roush, Chromosomal analysis of insecticide resistant and susceptible mosquitofish, *J. Hered.,* 77, 345–348, 1986.

Witters, H.E., Acute acid exposure of rainbow trout, *Salmo gairdneri* Richardson: effects of aluminum and calcium on ion balance and haematology, *Aquatic Toxicol.* (Amsterdam), 8, 197–210, 1986.

Wofford, H.W. and P. Thomas, Effect of xenobiotics on peroxidation of hepatic microsomal lipids from striped mullet (*Mugil cephalus*) and Atlantic croaker (*Micropogonus undulatus*), *Mar. Environ. Res.,* 2, 285–289, 1988.

Wolfe, M.F., S. Schwarzbach, and R.A. Sulaiman, Effects of mercury on wildlife: a comprehensive review, *Environ. Toxicol. Chem.,* 17, 146–160, 1998.

Wolff, S., Are radiation-induced effects hormetic? *Science,* 245, 575–621, 1989.

Woltering, D.M., J.L. Hedtke, and L.J. Weber, Predator-prey interactions of fishes under the influence of ammonia, *Trans. Am. Fish. Soc.,* 107, 500–504, 1978.

Wong, C.K. and P.K. Wong, Life table evaluation of the effects of cadmium exposure on the freshwater cladoceran *Moina macrocopa, Bull. Environ. Contam. Toxicol.,* 44, 135–141, 1990.

Wood, J.M. and H.-K. Wang, Microbial resistance to heavy metals, *Environ. Sci. Technol.,* 17, 582A–590A, 1983.

Woods, J.S., M.D. Martin, C.A. Naleway, and D. Echeverria, Urinary porphyrin profiles as a biomarker of mercury exposure: studies on dentists with occupational exposure to mercury vapor, *J. Toxicol. Environ. Health,* 40, 235–246, 1993.

Woodward, L.A., M. Mulvey, and M.C. Newman, Mercury contamination and population-level responses in chironomids: can allozyme polymorphism indicate exposure? *Environ. Toxicol. Chem.,* 15, 1309–1316, 1996.

Woodwell, G.M., Effects of ionizing radiation on terrestrial ecosystems, *Science* (Washington, D.C.), 138, 572–577, 1962.

Woodwell, G.M., The ecological effects of radiation, *Sci. Am.,* 208, 2–11, 1963.

Woodwell, G.M., Toxic substances and ecological cycles, *Sci. Am.,* 216, 24–31, 1967.

Woodwell, G.M., The carbon dioxide question, *Sci. Am.,* 238, 34–43, 1978.

Wotawa, G. and M. Trainer, The influence of Canadian forest fires on pollutant concentrations in the United States, *Science,* 288, 324–326, 2000.

Wren, C.D. and H.R. MacCrimmon, Comparative bioaccumulation of mercury in two adjacent freshwater ecosystems, *Water Res.,* 20, 763–769, 1986.

Wren, C.D., H.R. MacCrimmon, and B.R. Loescher, Examination of bioaccumulation and biomagnification of metals in a Precambrian Shield lake, *Water Air Soil Pollut.,* 19, 277–291, 1983.

Wright, J.F., *Assessing the Biological Quality of Fresh Waters: RIVPACS and Other Techniques,* Freshwater Biological Association, Cumbria, U.K., 2000.

Wright, P.A. and C.D. Zamuda, Copper accumulation by two bivalve molluscs: salinity effect is independent of cupric ion activity, *Mar. Environ. Res.,* 23, 1–14, 1987.

Wu, J., J.L. Vankat, and Y. Barlas, Effects of patch connectivity and arrangement on animal metapopulation dynamics: a simulation study, *Ecol. Model,* 65, 221–254, 1993.

Wu, L., J. Chen, K.K. Tanji, and G.S. Banuelos, Distribution and biomagnification of selenium in a restored upland grassland contaminated by selenium from agricultural drain water, *Environ. Toxicol. Chem.,* 14, 733–742, 1995.

Yamada, H. and S. Koizumi, DNA microarray analysis of human gene expression induced by a non-lethal dose of cadmium, *Ind. Health,* 40, 159–166, 2002.

Yamada, H., M. Tateishi, and K. Takayanagi, Bioaccumulation of organotin compounds in the Red Sea bream (*Pagrus major*) by two uptake pathways: dietary uptake and direct uptake from water, *Environ. Toxicol. Chem.,* 13, 1415–1422, 1994.

Yamaoka, K., T. Nakagawa, and T. Uno, Statistical moments in pharmacokinetics, *J. Pharmacokinet. Biopharm.,* 6, 547–558, 1978.

Yang, Y. and C.B. Nelson, An Estimation of the Daily Average Food Intake by Age and Sex for Use in Assessing the Radionuclide Intake of Individuals in the General Population, EPA Report No. 520/1-84-021, Office of Radiation Programs, U.S. Environmental Protection Agency, Washington, D.C., 1984.

Yap, H.H., D. Desaiah, L.K. Cutkomp, and R.B. Koch, Sensitivity of fish ATPases to polychlorinated biphenyls, *Nature* (London), 233, 61–62, 1971.

Yarbrough, J.D., R.T. Roush, J.C. Bonner, and D.A. Wise, Monogenic inheritance of cyclodiene insecticide resistance in mosquitofish, *Gambusia affinis, Experientia* (Basel), 42, 851–853, 1986.

You, C.H., E.A. Mackay, P.M. Gehrig, P.E. Hunziker, and J.H.R. Kagi, Purification and characterization of recombinant *Caenorhabditis elegans* metallothionein, *Arch. Biochem. Biophys.,* 372, 44–52, 1999.

Young, L.B. and H.H. Harvey, Metal concentrations in chironomids in relation to the geochemical characteristics of surficial sediments, *Arch. Environ. Contam. Toxicol.,* 21, 202–211, 1991.

Zakharov, V.M., Analysis of fluctuating asymmetry as a method of biomonitoring at the population level, in *Bioindications of Chemical and Radioactive Pollution,* Krivolutsky, D.A., Ed., CRC Press, Boca Raton, FL, 1990.

Zhang, B., D. Egli, O. Georgiev, and W. Schaffner, The *Drosophila* homolog of mammalian zinc finger factor MTF-1 activates transcription in response to heavy metals, *Mol. Cell. Biol.,* 21, 4505–4514, 2001.

Zurer, P.S., Chemists Solve Key Puzzle of Antarctic Ozone Hole, *Chem. Eng. News,* Nov. 30, 1987, pp. 25–27.

Zurer, P.S., Studies on Ozone Destruction Expand beyond Antarctic, *Chem. Eng. News,* May 30, 1988, pp. 16–25.

Zurer, P.S., Arctic Ozone Loss: Fact-Finding Mission Concludes Outlook Is Bleak, *Chem. Eng. News,* Mar. 6, 1989, pp. 29–31.

Index

A

abductive inference, **280**
absolute
    bioavailability, **76**
    probabilities, 280
absorbed dose, 307-308, 315
absorbed dose rate, **307**
absorption, 16, 78, 80, 115, 270, 286, 309, 311
absorption rate constant, **78**
abundance, 203, 207, 240, 243-245, 250, 259, 261, 293, 310
abundance classes, 243
accelerated failure time model, **197**
acclimation, 89, 192, **199**-200, 230, 254
acclimatization, **199**
accumulation factor, **71**
acetaldehyde, 183
    dehydrogenase, 183
    syndrome, 183
acetate, 61, 81
acetic acid, 162
acetylcholine, 162-163
acetylcholinesterase, 162-163
acetylcholinesterase inhibitors, 41, **162-**163
acid,
    conditions, 164
    mine drainage, 28
    neutralizing capacity, 264
    precipitation, 7, 24, 30, 261, **263**, 264, 265, 271
    volatile sulfides, **82**
acidic conditions, 82, 150, 265
acidification, 241, 250, 254
acidophilic component, **133**
acrolein, 167
acrylonitrile, 272
action at a distance, 213
    hypothesis, 214-215
activation, 58, **59,** 131-132, 140, 203
active transport, 55, **57,** 72
active tubular secretion, 63
activity, **305**
activity coefficient, 79
activity level, 163-165
acute
    criterion, 189
    lethality, **173**, 174, 188, 321

    toxicity data, 346
acute-to-chronic ratios, 347
additive, **183** 187, 188, 301
additive index, **187**
additivity, 223
adduct, 127
adenine, 125
adenoma, 140
adenosine, 127, 129, 163
adenosine triphosphatases, 129
adenylate energy charge, **163**
adrenal cortex, 150-151
adrenal enlargement, 151
adsorption, **26**, **55**-56, 72, 79, 176, 311,
    coefficient, **26**
AEC, **163**
aerobic respiration, 124, 125
aerosols, 79, 265
AFBI, 116
Afghan, 6
aflatoxin B1 2,3-epoxidase, 116
African clawed frog, 154, 155
Ag, 60, 117
age, Preface, 87, 96, 98, 105, 109, 203, 204, 211, 216-220,
    271, 293, 313, 316
age pigment, **134**
age-specific
    birth rates, **216**
    death rate, **216**
    life expectancies, 282
    number of individuals dying, **216**
Agent Orange, 44
aggressive behavior, 165, 240
aging, 225, **310**, 328
agonistic, 167
agricultural
    lands, 273
    waste, 273
agriculture, 247
agroindustrial farming, 260
AH binding compounds, 186, 187
AHH, **115**
AL, 79
AI, **187**
AIDS, 153, 154
air, 3, 4, 18, 53, 79, 98, 164, 205, 218, 260, 266, 269, 285,
    296, 314, 342

433

D

P